Green Science and Technology

Editors

Enoch Y. Park
Takayuki Saito
Hirokazu Kawagishi
Masakazu Hara

Research Institute of Green Science and Technology
Shizuoka University
Shizuoka, Japan

CRC Press
Taylor & Francis Group
Boca Raton London New York

CRC Press is an imprint of the
Taylor & Francis Group, an **informa** business
A SCIENCE PUBLISHERS BOOK

CRC Press
Taylor & Francis Group
6000 Broken Sound Parkway NW, Suite 300
Boca Raton, FL 33487-2742

Version Date: 20190830

International Standard Book Number-13: 978-0-367-41513-6 (Hardback)

Visit the Taylor & Francis Web site at
http://www.taylorandfrancis.com

and the CRC Press Web site at
http://www.crcpress.com

Preface

The rapid growth of the aging population with declining birth rate and world-wide deterioration of the environment threatens our lifestyle and the survival of all creatures. In order to weather these problems a multidisciplinary approach involving the fusion of various advanced current research fields is required.

Fossil fuel is already limited and we need renewable energy for sustainable energy supply. Many different types of renewable energy are available, but what is still needed is the development of widely acceptable new renewable and non-pollution energy. The Earth's ecosystem has also broken and we have to work on a new system for an environmentally-friendly life style utilizing natural resources in ocean, land and atmosphere. Realization of a recycling-oriented society based on generation and utilization of new renewable products is required. Science and technology relating to this area are referred to as green science and technology in this book.

This volume provides a comprehensive up-to-date review about recent and future trends of green science and technology, aimed at achieving sustainability. This volume consists of three parts. In Part I Green energy, woody biorefinery, biomass energy, save energy and green organic synthesis are dealt in Chapters 1 to 6. In Part II Ecosystem of natural waters, plant stress management technology including heat tolerance mechanism in plant cell and plant hormone signaling, and artificial intelligent (AI)-incorporated food production are described in Chapters 7 to 13. In Part III Health of living, removal of toxic substances, design and synthesis of virus inhibitor, supramolecule and fairy chemicals, nanoparticles for vaccine preparation, protein glycosylation and rapid detection of infectious viruses are introduced in Chapters 14 to 21. This book begins with an overview of the fundamental research on the green science and technology, and extends to applications for global sustainability.

Contents

Preface iii

1. **Novel Energy Production: Conversion of Organic Compounds to Electricity by Microbial Extracellular Electron Transfer Mechanisms** 1
Kei Suzuki, Rubaba Owen, Arashi Yui, Shota Ando, Yuki Kudo, Kazuki Yasuike, Yosuke Tashiro and *Hiroyuki Futamata*

2. **Woody Biorefinery with White-rot Fungi** 14
Toshio Mori and *Hirofumi Hirai*

3. **Energy Conversion of Biomass using Supercritical and Subcritical Fluid** 30
Takeshi Sako and *Idzumi Okajima*

4. **Determination of Diffusion Coefficients in Supercritical Fluid** 46
Chang Yi Kong, Toshitaka Funazukuri and *Koji Miyake*

5. **Catalytic Green Organic Synthesis in Unique Reaction Environments** 65
Nobuyuki Mase and *Kohei Sato*

6. **Sequestration of Carbon Dioxide in Deep Portions of Ocean through a Gas-lift Effect** 83
Takayuki Saito

7. **Nutrient Concentrations in Natural Waters from Mt. Fuji to the Suruga Bay** 100
Rumi Sohrin

8. **The Diversity of Cyanobacterial Photoresponsive Systems** 116
Keita Miyake, Keiji Fushimi and *Rei Narikawa*

9. **Hyperspectral Remote Sensing of Plant Water Status and Plant Water Use under Drought Stress** 127
Quan Wang and *Jia Jin*

10. **Aroma Glycosides Contribute Plant Chemical Defense** 145
Toshiyuki Ohnishi

11. **Heat Tolerance Enhancers of Plants for Agricultural Use** 158
Masakazu Hara

12. **Small Molecules that Regulate Plant Hormone Signaling** 175
Yasushi Todoroki

13. **Wilt Detection Soft Sensor for Plant Water Stress Management** 186
Hiroshi Mineno

14. **Removal of Toxic Anions from Environmental Water by using Cage Compounds** 194
Mitsuru Kondo

15. **Zirconocene Derivatives having α-Keggin-type Mono-Aluminum-coordinated Polyoxometalates: Synthesis, Molecular Structures, and Application to Heterogeneous Catalysts** 215
Chika Nozaki Kato

16. **Design, Synthesis, and Application of Highly Photosensitive Caging Groups for Green Chemistry** 232
Tetsuo Narumi

17. **Development and Applications of Supramolecular Hydrogel** 250
Masamichi Yamanaka

18. **Chemistry of Fairy Rings** 269
Jae-Hoon Choi and *Hirokazu Kawagishi*

19. **Production of Viral Nanoparticles in Silkworms** 284
Tatsuya Kato and *Enoch Y. Park*

20. **Glycobiology and Glycoengineering in Silkworm** 298
Takatsugu Miyazaki and *Enoch Y. Park*

21. **Detection of Infectious Viruses using Advanced Nanobiotechnology for Green Society** 316
Ankan Dutta Chowdhury and *Enoch Y. Park*

Index 333

About the Editors 335

Color Plate Section 337

Chapter 1

Novel Energy Production

Conversion of Organic Compounds to Electricity by Microbial Extracellular Electron Transfer Mechanisms

Kei Suzuki,[1] *Rubaba Owen,*[2] *Arashi Yui,*[3] *Shota Ando,*[3] *Yuki Kudo,*[3] *Kazuki Yasuike,*[3] *Yosuke Tashiro*[1,3] and *Hiroyuki Futamata*[1,3,4,*]

1. Introduction

Microbial interactions are important for microbial community structure, population dynamics, and function in various anaerobic environments, such as wastewater treatment (Oshiki et al. 2013), corrosion (Kato et al. 2015), bioremediation (Futamata et al. 2007), energy production (Kato 2015, Kato and Watanabe 2010, Yamamoto et al. 2014), and other anaerobic environments including subseafloor sediments (Aoki et al. 2015, Aoyagi et al. 2015, Holmes et al. 2004, Tender et al. 2002) and rice paddy fields (Hori et al. 2010). Interspecies metabolite transfer is a key process of material and energy flow in anaerobic microbial ecosystems (Futamata et al. 2009, Hattori 2008, Kato and Watanabe 2010, Kouzuma et al. 2015). H_2, formate and acetate have long been known as those metabolites. Over the last decade, special attention has been paid to Extracellular Electron Transfer (EET) between organisms as alternative fashion of electron transfer and extensive research has been performed over the last decade. Several interesting lifestyles of EET-active microorganisms have been found through studies of Microbial Fuel Cells (MFCs), conductive nanowires (El-naggar et al. 2010, Gorby et al. 2006, Malvankar et al. 2011),

[1] Graduate School of Science and Technology, Shizuoka University, Hamamatsu, Hamamatsu 432-8561, Japan.
[2] College of Health Sciences, University of KwaZulu-Natal, Durban, 4000, South Africa.
[3] Department of Applied Chemistry and Biochemical Engineering, Graduate School of Engineering, Shizuoka University, Hamamatsu, Hamamatsu 432-8561, Japan.
[4] Research Institute of Green Science and Technology, Shizuoka University, Suruga-ku, Shizuoka, Shizuoka 422-8529, Japan.
* Corresponding author: futamata.hiroyuki@shizuoka.ac.jp

respiration systems using solid conductors (Bond and Lovley 2003, Kato et al. 2010, Torres et al. 2009), and biomineralization (Byrne et al. 2011, Chen et al. 2014, Lee et al. 2007).

MFCs are capable of directly converting biochemical energy to electrical energy through microbial metabolism and are therefore expected to become novel energy-producing devices that can simultaneously be used for the treatment of organic wastes (Du et al. 2007, Ishii et al. 2012, Logan et al. 2006). In addition, MFCs are ideal model microbial ecosystems to better understand the physicochemical traits of electricity-producing microorganisms (exoelectrogens) in EET-based anaerobic communities (Ishii et al. 2012, 2013a, 2014, 2015, Okamoto et al. 2014c) (Figs. 1.1 and 1.2). This is significant both for academic research and for practical application (Kouzuma and Watanabe 2014). The interaction between microbes and minerals has been of much concern recently (Suga et al. 2017). Here, we describe

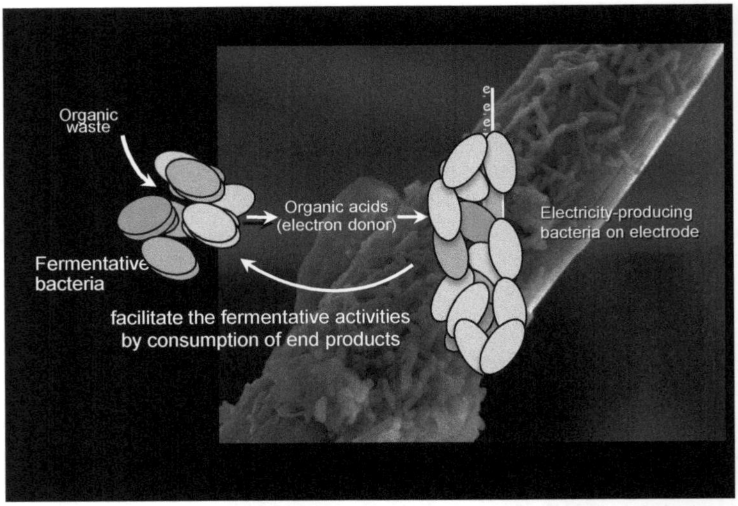

Fig. 1.1: Schematic diagram of electron flow, material conversion, role of microbes in an MFC. Photograph shows the microbial cells grown on the surface of anode in an MFC, suggesting that these cells use electrode as an electron acceptor.

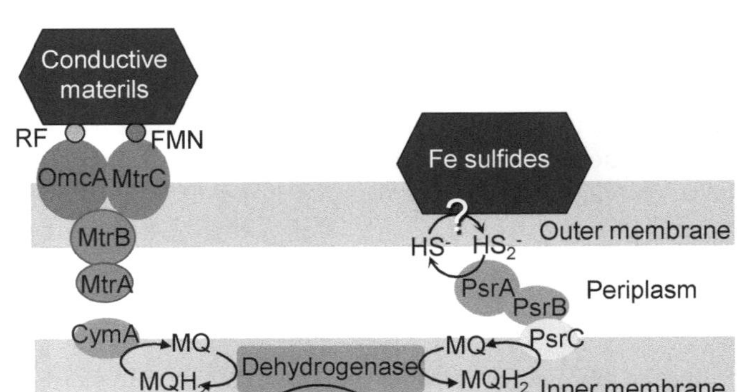

Fig. 1.2: Electron transfer to Fe oxides occurs through CymA, MtrA, MtrB and MtrC or OmcA, whereas the proposed model of electron transfer to Fe sulfides is through PsrC, PsrB, PsrA and HS^- in *S. oneidensis* MR-1. This figure is a slightly modified version of a figure by Kondo et al. (Kondo et al. 2015). RF: riboflavin, FMN: flavin mononucleotide. Numbers in parentheses indicate references.

MFCs for practical application, the EET mechanisms in several model exoelectrogens (e.g., *Shewanella oneidensis* and *Geobacter sulfurreducens*), *in vitro* anaerobic microbial ecosystems based on EET, Direct Interspecies Electron Transfer, and (DIET) mechanisms.

2. MFCs as Engineered Model Microbial Ecosystems

In general, under anaerobic conditions, complex organic compounds are degraded by fermentative microbes, and the by-products produced in the fermentation process, such as volatile fatty acids and hydrogen, are used as electron donors (even acetate) for subsequent microbial respiration through anode reduction via microbial EET systems and syntrophic associations (Kimura and Okabe 2013). We are able to analyze these sequential processes in an MFC by controlling the factors, such as the anode potential (Ishii et al. 2014, Suzuki et al. 2018, Torres et al. 2009), inoculum sources (Holmes et al. 2004, Ishii et al. 2008), electrode materials (Rubaba et al. 2013), and so forth (Ishii et al. 2012, 2013a, b, 2014, 2015, Jung and Regan 2007, Kan et al. 2011, Xing et al. 2009, Yamamoto et al. 2014). Therefore, an MFC is a representative electromicrobiome that can facilitate our understanding of the metabolic relationships between fermenters and the EET-active community in complex systems (Ishii et al. 2014, Jung and Regan 2007, Kan et al. 2011, Rubaba et al. 2013, Yamamoto et al. 2014). Anolytic and biofilm communities, which are mainly fermentative and respiring microbes, are symbiotic systems that use the anode electrode as their final electron acceptor (Ishii et al. 2012, Suzuki et al. 2018, Yamamoto et al. 2014). When an MFC supplied with municipal wastewater was operated in a repeat-batch mode, the electrochemical performance and BOD removal were reproducible throughout the MFC operation regardless of sample variability (Ishii et al. 2012). There is evidence that after the bacterial community structure of a biofilm has converged, the dynamics of the biofilm communities are much more stable than those of anolytic communities, and the final community structures are different from each other (Ishii et al. 2012, Yamamoto et al. 2014). These findings suggest that the potential of the electron acceptor is a critical parameter for constructing electromicrobiome networks. Although the effect of the anode potential on the bacterial community structure was initially controversial (Aelterman et al. 2008), the anode potential was found to provide a strong selective pressure at the species and strain level in anolytic fermenters and biofilm respirators (Aelterman et al. 2008, Rismani-Yazdi et al. 2011, Suzuki et al. 2018, Torres et al. 2009). Since an external resistance affects the anode potential, that is also a selective pressure (Jung and Regan 2011, Rismani-Yazdi et al. 2011, Torres et al. 2009).

3. How are Electrons Transferred from Cells to Soluble or Insoluble Electron Acceptors in Model Microbes?

EET is a microbial respiration system that enables the transfer of electrons between microbial cells and extracellular soluble compounds (Kouzuma et al. 2012, Newman and Kolter 2000), solid conductive materials including artificial electrodes (Canstein et al. 2008, Malvankar et al. 2011, Marsili et al. 2008, Okamoto et al. 2013, Preffer et al. 2012, Reguera et al. 2005), naturally-occurring metal compounds (Kato et al. 2010, 2013), and other bacterial cells (Scheller et al. 2016, Snider et al. 2012) (Fig. 1.2). Microorganisms belonging to the genus *Shewanella* are facultative anaerobes widely used as model exoelectrogens (Bretschger et al. 2007, Gorby et al. 2006). Bacteria within the genus *Geobacter* are metal-reducing obligate anaerobes, which frequently emerge as a major player in the processes of power generation in MFCs with a positive correlation between population density and current density (Ishii et al. 2008). In both genera, several types of nanowire and soluble electron carriers are associated with EET (Burns and DiChristina 2009, Inoue et al. 2010, 2011, Leang et al. 2010, Pirbadiana et al. 2014, Shi et al. 2016).

3.1 Nanowires

Bacterial nanowires offer an EET pathway for linking the respiratory chain of bacteria to an external surface, including oxidized metals in the environment and engineered electrodes in renewable energy devices. Using *in vivo* fluorescence measurements, immunolabeling, quantitative gene expression analysis, and in *in vitro* observation, *S. oneidensis* MR-1 nanowires are shown to be extensions of the outer membrane and periplasm that include the multiheme cytochromes (MtrC and OmcA) responsible for EET (Fig. 1.2). These membrane extensions were associated with outer membrane vesicles (Pirbadiana et al. 2014), which are ubiquitous structures in Gram-negative bacteria. The nanowires of *Shewanella* are involved in long-range EET through a cytochrome-to-cytochrome electron hopping mechanism (Preffer et al. 2012). Therefore, the redox-functionalized membrane and vesicular extensions may be a general microbial strategy for electron transport and energy distribution.

In contrast, the pili of *G. sulfurreducens* (type IV pili), recognized as nanowires, have been demonstrated to have metallic-like conductivity through overlapping π-π orbitals of key aromatic amino acids (tyrosines and phenylalanines) (Lampa-Pastirk et al. 2016). Nanowires do not contain metals, quinone, or flavin (Lampa-Pastirk et al. 2016). As nanowires, the pili of *Geobacter* are involved in long-distance EET to insoluble electron acceptors and interspecies electron transfer (Malvankar and Lovley 2013). The expression of *pilA*, encoding the pilus structural protein, is directly regulated by a two-component regulatory system (PilR as a RpoN-dependent enhancer binding protein), and the *pilR*-deficient mutant diminishes the reducing activity toward soluble Fe(III) and is not capable of reducing insoluble Fe(III) (Juarez et al. 2009).

3.2 Soluble Electron Carriers

Both *G. sulfurreducens* PCA and *S. oneidensis* MR-1 have a similar EET mechanism, employing the electron carrier's riboflavin (RF) and flavin mononucleotide (FMN). RF was initially proposed to be a redox-shuttling molecule when present at high concentrations (Brutinel and Gralnick 2012, Marsili et al. 2008). Unexpectedly, it has recently been shown that flavins also function as cofactors bound to the outer membrane *c*-type cytochromes (*c*-Cyts), leading to more rapid kinetics than can be obtained through electron shuttling (Okamoto et al. 2013, 2014a, 2014c, Xu et al. 2016).

G. sulfurreducens PCA excretes RF during its anaerobic growth with acetate as the electron donor and fumarate as the electron acceptor (Okamoto et al. 2014a). *G. sulfurreducens* PCA utilizes RF as a bound cofactor in outer membrane *c*-Cyts. Differential Pulse Voltammetry (DPV) analysis demonstrated that *G. sulfurreducens* is also capable of utilizing FMN as an efficient electron carrier (Okamoto et al. 2014a, b). Interestingly, the redox potential of RF and FMN solutions measured by DPV in the absence of cells significantly differed from that observed in *Geobacter* biofilms. The peak potentials of 10 μM RF and the wild-type strain of *G. sulfurreducens* were −0.27 V and −0.20 V, respectively, whereas a mutant strain (*ΔomcBEST*) of *G. sulfurreducens* lacking the major multiheme *c*-Cyts OmcB, OmcE, OmcS, and OmcT (Inoue et al. 2011, Leang et al. 2010) exhibited a peak potential of −0.125 V, and a 20-fold decreased redox peak current compared to the wild-type strain. Conversely, in FMN supplemented conditions (2 μM), the peak potential shifts from −0.175 V in the wild-type to −0.225 V in the *ΔomcBEST* mutant strain, with an approximately 200-fold decrease in the redox peak current compared to the wild-type strain. These results demonstrated that both RF and FMN alter their redox properties when bound to outer membrane major multiheme *c*-Cyts and other types of outer membrane *c*-Cyt proteins to enhance the rate of EET in *G. sulfurreducens*. These soluble electron carriers are capable of connecting bacterial cells and electrodes in an extracellular matrix including *c*-Cyts (Reguera et al. 2006, Snider et al. 2012), which enables the long-range transfer of electrons via the redox gradient in *Geobacter* biofilms.

S. oneidensis MR-1 uses self-secreted flavins not only as electron shuttles but also as redox cofactors bound to the outer membrane OmcA-MtrC proteins (Canstein et al. 2008, Marsili et al. 2008, Okamoto et al. 2013, Shi et al. 2009). The bound flavins (RF to OmcA and FMN to MtrC) participate in one-electron redox reactions (−0.145 V redox potential), whereas the free flavins play a role in the shuttling reaction with two-electron transfers (−0.26 V redox potential).

The secreted RF was shown to interact with the outer membrane *c*-Cyts, resulting in stimulation of the EET rate. The bound flavins are a major factor in EET at the interface of cells and electrodes for *S. oneidensis* MR-1 and *G. sulfurreducens*, including nanowires, secreted membrane vesicles, and conductive filaments, referred to as the bound flavin model (Okamoto et al. 2014b). However, it is unknown how direct EET or electron shuttling is regulated (Okamoto et al. 2014c).

Considering the redox potentials of FMN and FAD (−219 mV) and RF (−208 mV) (van der Zee 2002), flavins are more electronegative than the redox couple of poorly crystalline Fe(III) oxide (ferrihydrite)/Fe^{2+} (−100 to +100 mV) (Straub et al. 2001). Therefore, flavins have the potential to act as efficient extracellular redox mediators for the reduction of poorly soluble Fe(III) oxide at neutral pH. Since diverse microbes secrete flavin-like compounds (Rabaey et al. 2005), it is predicted that the flexible EET systems of *G. sulfurreducens* and *S. oneidensis* MR-1 enable the formation of gentle symbiotic relationships between exoelectrogens and other microbes.

3.3 Long-Distance Electron Transfer using Insoluble Iron Compounds

The cell membrane of electricity-producing *S. loihica* PV-4 is an insulator even in biofilms grown on the anode surface. Therefore, the long-distance EET process is much less effective than *c*-Cyt-mediated EET to the electrode surface (Okamoto et al. 2013, 2014b, c). Conversely, *S. loihica* PV-4 is capable of self-assembling into an electric conductive network in the presence of iron (III) oxides (hematite α-Fe_2O_3), resulting in the promotion of long-distance EET in the bacterial network (Nakamura et al. 2009). The addition of α-Fe_2O_3 improved the electrochemical properties of an MFC with a 50-fold increase of current density. Scanning electron microscope images of the electrode surface showed the formation of thick layers composed of cells and α-Fe_2O_3 colloids, in which the outer surfaces of the cells were covered with semi conductive α-Fe_2O_3 colloids serving to interconnect the cells. Cyclic voltammetry analysis indicated that no energy barrier exists to electron-exchange processes between the outer membrane *c*-Cyts and α-Fe_2O_3 colloids, suggesting that a semi conductor-mediated electron hopping process enables long-distance EET through the bacterial network. *S. oneidensis* MR-1 and *S. loihica* PV-4 also develop electrically conductive networks using nanoparticles of crystalline, semi conductive α-Fe_2O_3, serving as electron conduits (Nakamura et al. 2009).

These electrically conductive networks have been characterized only in laboratory pure-culture systems. However the ecological effects of nanosized iron oxides and their electrical conductivity in microbial communities have been investigated (Kato et al. 2015). Iron oxides occur in nature as a spectrum of forms, ranging from amorphous minerals (ferrihydrite) to crystals (hematite [α-Fe_2O_3], goethite [α-FeOOH], and magnetite [Fe_3O_4]) (Cornell and Schwertmann 2003). Crystalline iron oxides are typically more abundant than the amorphous forms (Weber et al. 2006). A number of iron-reducing and -oxidizing microorganisms are known to produce crystalline iron oxides, including magnetite and goethite, through the respiratory reduction and oxidation of soluble and amorphous iron species (Banfield et al. 2000, Chaudhuri et al. 2001, Coker et al. 2008, Vali et al. 2004). It has been suggested that subsurface-clade *Geobacter* species thrive in soil by preferentially using (semi-) conductive iron oxides for respiration (Holmes et al. 2007, Kato et al. 2010). By contrast, other *Geobacter* strains were shown to use several compounds and metals as electron donors and acceptors (Lovley et al. 1993). These results demonstrate that (semi) conductive iron-oxide nanoparticles serve as conduits for electrons between microbes and extracellular electron acceptors. In addition, it is clear that the resulting networks established through microbial interactions with several (semi-) conductive iron-oxide minerals are widely distributed in natural environments (Nielsen et al. 2010).

3.4 EET to Insoluble Biogenic Minerals through Mtr or Psr Pathways in Shewanella

EET can also result in the transfer of electrons to biogenic minerals, such as mackinawite (FeS), greigite (Fe_3S_4), marcasite (FeS_2), and pyrite (FeS_2) (Fig. 1.2). These iron sulfide(s) are ubiquitous and abundant minerals in anoxic marine environments, where they are formed as by-products of microbial metabolism or as a consequence of geothermal activity (Nakagawa and Takai 2008). Furthermore, ferrous sulfide has attracted significant interest for application in environmental remediation for the removal of heavy metal, oxyanions, radionuclides, chlorinated, and other compounds under anaerobic conditions (Gong et al. 2016, Huo et al. 2016). *S. loihica* PV-4 (Jiang et al. 2014) and *S. oneidensis* MR-1 (Kondo et al. 2015) are capable of producing nanoscale mackinawite on the cell surface. Additionally, *S. loihica* PV-4 has the ability to exploit the Fe 3d electrons of biologically synthesized iron monosulfide as extracellular long-distance electron transfer conduits, resulting in an increase of the current density by two orders of magnitude higher than in cell cultures lacking the biogenic minerals. The crystallinity of the produced mackinawite improves gradually with the prolongation of the incubation time (Kondo et al. 2015). The process of mackinawite biomineralization, including crystallization, is not yet completely understood, although the biomineralization of nanosized FeS is a well-investigated phenomenon.

Knowledge of the mechanism of electron transfer from bacterial cells to biogenic FeS is still limited. It has been demonstrated that the Mtr pathway (including MtrABC, CymA, and OmcA) (Coursolle et al. 2010, Okamoto et al. 2013) is not crucial for EET to biogenic FeS in *S. oneidensis* MR-1 (Kondo et al. 2015). A novel model has been proposed in *S. oneidensis* MR-1 (Fig. 1.2). According to the Pourbaix diagram of polysulfides at pH values from 5 to 10 and a potential from −0.13 to 0.5 V vs standard hydrogen electrode (SHE) (Rickard and Luther 2007), protonated disulfide (HS_2^-) would be produced in the periplasmic space. Inorganic compounds including iron or sulfur, change form depending on the pH and the potential, which influences the composition of chemical components in the periplasmic space of cells (Kondo et al. 2015, Rickard et al. 2007). Polysulfide reductase (PsrA) (Burns and DiChristina 2009) has been reported to be anchored on the periplasmic side of the inner membrane, where it mediates the two-electron reduction of the thiosulfate and polysulfide (Jormakka et al. 2008), leading to the conversion of HS_2^- to HS^-. A hypothesis about the mechanism of electron transfer is suggested as follows. Microbially generated HS^- is oxidized on the surface of FeS, yielding elemental sulfur (S^0) and free electrons in FeS. The free electrons are transported to distant electron acceptors via the extracellular FeS, while the S^0 formed on the FeS surfaces reacts with HS to form water-soluble polysulfides (HS_n^-, $n \geq 2$). The generated polysulfides are then reduced again by PsrA to form HS^- in the periplasmic space. It is likely that the Psr-mediated sulfur respiratory system operates as an alternative pathway to the Mtr pathway for EET to the extracellular FeS matrix. *S. oneidensis* MR-1 can generate FeS wires formed through cellular metabolic processes, which enables long-distance EET.

3.5 Direct Interspecies Electron Transfer

Geobacter species are capable of DIET within anaerobic microbial communities (Summers et al. 2010), which may be more effective than interspecies electron transfer via H_2 and formate (Cord-Ruwisch et al. 1998) in aggregates. OmcS of *G. sulfurreducens*, a *c*-Cyt that promotes electron transfer to insoluble Fe(III) oxides (Mehta et al. 2005) and electrodes (Holmes et al. 2006), is predicted to accept electrons from the outer-surface *c*-Cyts of *G. metallireducens*, localized either on the cell surface or along pili in an electron-exchange model (Summers et al. 2010). Interestingly, it has been reported that a mutation enhancing the production of OmcS is selectively introduced into the genome of *G. sulfurreducens* under co-culture conditions.

DIET between *G. sulfurreducens* and *Methanosaeta concilii* could be an important mechanism for electron exchange in methanogenic wastewater aggregates (Morita et al. 2011). Also, *G. metallireducens* and *M. harundinacea* can exchange electrons via DIET and convert ethanol to methane stoichiometrically (Rotaru et al. 2014). The electron-accepting components of *Methanosaeta* species in DIET remain unknown. In a similar system, methanogens were able to accept electrons from abiotic donors, such as metallic iron (Dinh et al. 2004, Uchiyama et al. 2010) or electrodes (Chaudhuri et al. 2001), as well as to enhance electron transfer via conductive minerals (Kato et al. 2010, 2012, 2013) or granular activated carbon (Pirbadiana et al. 2014).

DIET is also observed between anaerobic methanotrophic archaea (ANME) and sulfate-reducing deltaproteobacteria (SRB), which occur in microbial co-aggregates formed under conditions of anaerobic oxidation of methane (McGlynn et al. 2015, McGlynn 2017, Scheller et al. 2016, Wegener et al. 2015). The DIET system has attracted significant research interest regarding control of the emission of the greenhouse gas methane from the ocean floor (Boetius et al. 2000, McGlynn et al. 2015). This system is also an important component of the microbial metabolism in the global carbon cycle. Analyses of co-aggregates using FISH, nanoSIMS, and modeling techniques revealed that the activity of aggregates is independent of both the distance between syntrophic partners and the diffusion of metabolites, including H_2, formate, and acetate, within the consortia (McGlynn et al. 2015). The addition of H_2 induced a significant relative increase of gene expression in SRB over that in ANME, even in the presence of methane (Wegener et al. 2015). This result indicates that when the partner SRB does not play a role as an electron acceptor, the activity of ANME is suppressed. Draft genome information indicates that ANME contain several potential extracellular *c*-Cyt proteins but no genes for pilus formation, whereas SRB contain the gene for the biosynthesis and assembly of type IV pili (Shi et al. 2009) and large *c*-Cyts proteins with high amino acid sequence similarities to the respective proteins in *Geobacter* spp. (Wegener et al. 2015).

4. Perspective

EET and DIET are unique and overlooked mechanisms for production of energy in microbes. Diverse energy-producing mechanisms have evolved in anaerobic microorganisms (Kracke et al. 2015, Lovley 2012). We would be able to understand microbial survival strategies through the analyses of these mechanisms. Furthermore, these findings will guide us in the development of future technologies that are independent from nuclear energy and fossil fuels (Logan 2015), as well as provide us with exciting insights into microbial ecosystems.

Acknowledgements

Some of the important works introduced in this chapter were carried out as grant the ALCA project, Japan Science and Technology Agency, KAKENHI B (18H03400), and Institution for Fermentation, Osaka (G-2018-3-033).

References

Aelterman, P., S. Freguia, J. Keller, W. Verstraete and K. Rabaey. 2008. The anode potential regulates bacterial activity in microbial fuel cells. Appl. Microbiol. Biotechnol. 78: 409–418.

Aoki, M., R. Kakiuchi, T. Yamaguchi, K. Takai, F. Inagaki and H. Imachi. 2015. Phylogenetic diversity of *aprA* genes in subseafloor sediments on the northwestern Pacific margin off Japan. Microbes Environ. 30: 276–280.

Aoyagi, T., M. Kimura, N. Yamada, P.R. Navarro, H. Itoh, A. Ogata et al. 2015. Dynamic transition of chemolithotrophic sulfur–oxidizing bacteria in response to amendment with nitrate in deposited marine sediments. Frontiers Microbiol. 6: 426.

Banfield, J.F., S.A. Welch, H. Zhang, T.T. Ebert and R.L. Penn. 2000. Aggregation-based crystal growth and microstructure development in natural iron oxyhydroxide biomineralization products. Science 289: 751–754.

Boetius, A., K. Ravenschlag, C.J. Schubert, D. Rickert, F. Widdel, A. Gieseke et al. 2000. A marine microbial consortium apparently mediating anaerobic oxidation of methane. Nature 407: 623–6.

Bond, D.R. and D.R. Lovley. 2003. Electricity production by *Geobacter sulfurreducens* attached to electrodes. Appl. Environ. Microbial. 69: 1548–1555.

Bretschger, O., A. Obraztsova, C.A. Sturm, I.S. Chang, Y.A. Gorby, S.B. Reed et al. 2007. Current production and metal oxide reduction by *Shewanella oneidensis* MR-1 wild type and mutants. Appl. Environ. Microbiol. 73: 7003–7012.

Brutinel, E.D. and J.A. Gralnick. 2012. Shuttlng happens: soluble flavin mediators of extracellular electron transfer in *Shewanella*. Appl. Microbiol. Biotechnol. 93: 41–48.

Burns, J.L. and T.J. DiChristina. 2009. Anaerobic respiration of elemental sulfur and thiosulfate by *Shewanella oneidensis* MR-1 requires *psrA*, a homolog of the *phsA* gene of *Salmonella enterica* serovar Typhimurium LT2. Appl. Environ. Microbiol. 75: 5209–5217.

Byrne, J.M., N.D. Telling, V.S. Coker, R.A. Pattrick, G. van der Laan, E. Arenholz et al. 2011. Control of nanoparticle size, reactivity and magnetic properties during the bioproduction of magnetite by *Geobacter sulfurreducens*. Nanobiotech. 22: 455709.

Canstein, H., J. Ogawa, S. Shimizu and J.R. Lloyd. 2008. Secretion of flavins by *Shewanella* species and their role in extracellular electron transfer. Appl. Environ. Microbiol. 74: 615–623.

Chaudhuri, S.K., J.G. Lack and J.D. Coates. 2001. Biogenic magnetite formation through anaerobic biooxidation of Fe(II). Appl. Environ. Microbiol. 67: 2844–2848.

Chen, Y., Y.L. Li, G.T. Zhou, H. Li, Y.T. Lin, X. Xiao et al. 2014. Biomineralization mediated by anaerobic methane-consuming cell consortia. Sci. Reports 4: 5696.

Coker, V.S., A.M.T. Bell, C.I. Pearce, R.A.D. Pattrick, G. van der Laan and J.R. Lloyd. 2008. Time-resolved synchrotron powder X-ray diffraction study of magnetite formation by the Fe(III)-reducing bacterium *Geobacter sulfurreducens*. Am. Mineral 93: 540–547.

Cord-Ruwisch, R., D.R. Lovley and B. Schink. 1998. Growth of *Geobacter sulfurreducens* with acetate in syntrophic cooperation with hydrogen-oxidizing anaerobic partners. Appl. Environ. Microbiol. 64: 992–998.

Cornell, R.M. and U. Schwertmann. 2003. The iron oxides: Structure, Properties, Reactions, Occurrences and Uses, 2nd ed. WILEY-VCH GmbH&Co. KGaA, Weinheim.

Coursolle, D., D.B. Baron, D.R. Bond and J.A. Gralnick. 2010. The Mtr respiratory pathway is essential for reducing flavins and electrodes in *Shewnalla oneidensis*. J. Bacteriol. 192: 467–474.

Dinh, H.T., J. Kuever, M. Muβmann, A.W. Hassei, M. Stratmann and F. Widdel. 2004. Iron corrosion by novel anaerobic microorganisms. Nature 427: 829–832.

Du, Z., H. Li and T. Gu. 2007. A state of the art review on microbial fuel cells: A promising technology for wastewater treatment and bioenergy. Biotechnol. Adv. 25: 464–482.

El-naggar, M.Y., G. Wanger, K.M. Leung, T.D. Yuzvinsky, G. Southam, J. Yang et al. 2010. Electron transport along bacterial nanowires from *Shewanella oneidensis* MR-1. Proc. Natl. Acad. Sci. USA 107: 18127–18131.

Futamata, H., N. Yoshida, T. Kurogi, S. Kaiya and A. Hiraishi. 2007. Reductive dechlroination of chloroethenes by *Dehalococcoides*-containing cultures enriched from a polychlorinated-dioxin-contaminated microcosm. The ISME J. 1: 471–479.

Futamata, H., S. Kaiya, M. Sugawara and A. Hiraishi. 2009. Phylogenetic and transcriptional analyses of a tetrachloroethene-dechlorinating "*Dehalcoccoides*" enrichment culture TUT2264 and its reductive-dehalogenase genes. Microbes Environ. 24: 330–337.

Gong, Y., J. Tang and D. Zhao. 2016. Application of iron sulfide particles for groundwater and soil remediation: A review. Water Res. 89: 309–320.

Gorby, Y.A., S. Yanina, J.S. KcLean, K.M. Rosso, D. Moyles, A. Dohnalkova et al. 2006. Electrically conductive bacterial nanowires produced by *Shewanella oneidensis* strain MR-1 and other microorganisms. Proc. Natl. Acad. Sci. USA 103: 11358–11363.

Hattori, S. 2008. Syntrophic acetate-oxidizing microbes in methanogenic environments. Microbes Environ. 23: 118–127.

Holmes, D.E., D.R. Bond, R.A. O'Neil, C.E. Reimers, L.R. Tender and D.R. Lovley. 2004. Microbial communities associated with electrodes harvesting electricity from a variety of aquatic sediments. Microb. Ecol. 48: 178–190.

Holmes, D.E., S.K. Chaudhuri, K.P. Nevin, T. Mehta, B.A. Methé, A. Liu et al. 2006. Microarray and genetic analysis of electron transfer to electrodes in *Geobacter sulfurreducens*. Environ. Microbiol. 8: 1805–1815.

Holmes, D.E., R.A. O'Neil, H.A. Vrionis, L.A. N'guessan, I. Ortiz-Bemad, M.J. Larrahondo et al. 2007. Subsurface clade of *Geobacteraceae* that predominates in a diversity of Fe(III)-reducing subsurface environments. The ISME J. 1: 663–677.

Hori, T., A. Muller, Y. Igarashi, R. Conrad and M.W. Friedrich. 2010. Identification of iron-reducing microorganisms in anoxic rice paddy soil by ^{13}C-aceate probing. The ISME J. 4: 267–278.

Huo, Y.C., W.W. Li, C.B. Chen, C.X. Li, R. Zeng, T.C. Lau et al. 2016. Biogenic FeS accelerates reductive dechlorination of carbon tetrachloride by *Shewanella putrefaciens* CN32. Enzyme Microb. Tech. 95: 236–241.

Inoue, K., X. Qian, L. Morgado, B.-C. Kim, T. Mester, M. Izallalen et al. 2010. Purification and characterization of OmcZ, an outer-surface, octaheme *c*-type cytochrome essential of optimal current production by *Geobacter sulfurreducens*. Appl. Environ. Microbiol. 76: 3999–4007.

Inoue, K., C. Leang, A.E. Franks, T.L. Woodard, K.P. Nevin and D.R. Lovley. 2011. Specific localization of the c-type cytochrome OmcZ at the anode surface in current-producing biofilms of *Geobacter sulfurreducens*. Envrion. Microbiol. Reports 3: 211–217.

Ishii, S., K. Watanabe, S. Yabuki, B.E. Logan and Y. Sekiguchi. 2008. Comparison of electrode reduction activities of *Geobacter sulfurreducens* and an enriched consortium in an air-cathode microbial fuel cell. Appl. Environ. Microbiol. 74: 7348–7355.

Ishii, S., S. Suzuki, T.M. Norden-Krichmar, K.H. Nealson, Y. Sekiguchi, Y.A. Gorby et al. 2012. Functionally stable and phylogenetically diverse microbial enrichments from microbial fuel cells during wastewater treatment. PLoS One 7: e30495.

Ishii, S., S. Suzuki, T.M. Norden-Krichmar, A. Tenney, P.S.G. Chain, M.B. Scholz et al. 2013a. A novel metatranscriptomic approach to identify gene expression dynamics during extracellular electron transfer. Nature Commun. 4: 1601.

Ishii, S., S. Suzuki, T.M. Norden-Krichmar, A. Wu, Y. Yamanaka, K.H. Nealson et al. 2013b. Identifying the microbial communities and operational conditions for optimized wastewater treatment in microbial fuel cells. Water Res. 47: 7120–7130.

Ishii, S., S. Suzuki, T.M. Norden-Krichmar, T. Phan, G. Wanger, K.H. Nealson et al. 2014. Microbial population and functional dynamics associated with surface potential and carbon metabolism. The ISME J. 8: 963–978.

Ishii, S., S. Suzki, A. Tenney, T.M. Norden-Krichmar, K.H. Nealson and O. Bretschger. 2015. Microbial metabolic networks in a complex electrogenic biofilm recovered from a stimulus-induced metatranscriptomics approach. Sci. Rep. 5: 14840–14853.

Jiang, X., J. Hu, A.M. Lieber, C.S. Jackan, J.C. Biffinger, L.A. Fitzgerald et al. 2014. Nanoparticle facilitated extracellular electron transfer in microbial fuel cells. Nano Lett. 14: 6737–6742.

Jormakka, M., K. Yokoyama, T. Yano, K. Takaoshi, S. Akimoto, T. Shimamura et al. 2008. Molecular mechanism of energy conservation in polysulfide respiration. Nat. Struct. Mol. Biol. 15: 730–737.

Juarez, K., B.-C. Kim, K. Nevin, L. Olvera, G. Reguera, D.R. Lovley et al. 2009. PilR, a transcriptional regulator for pilin and other genes required for Fe(III) reduction in *Geobacter sulfurreducens*. J. Mol. Microbiol. Biotechnol. 16: 146–158.

Jung, S. and J.M. Regan. 2007. Comparison of anode bacterial communities and performance in microbial fuel cells with different electron donors. Appl. Microbiol. Biotechnol. 77: 393–402.

Jung, S. and J.M. Regan. 2011. Influence of external resistance on electrogenesis, methanogenesis, and anode prokaryotic communities in microbial fuel cells. Appl. Environ. Microbiol. 77: 564–571.

Kan, J., L. Hsu, A.C.M. Cheung, M. Pirvazari and K.H. Nealson. 2011. Current production by bacterial communities in microbial fuel cells enriched from wastewater sludge with different electron donors. Environ. Sci. Technol. 45: 1139–1146.

Kato, S. and K. Watanabe. 2010. Ecological and evolutionary interactions in syntrophic methanogenic consortia. Microbes Environ. 25: 145–151.

Kato, S., R. Nakamura, F. Kai, K. Watanabe and K. Hashimoto. 2010. Respiratory interactions of soil bacteria with (semi)conductive iron-oxide minerals. Environ. Microbiol. 12: 3114–3123.

Kato, S., K. Hashimoto and K. Watanabe. 2012. Microbial interspecies electron transfer via electric currents through conductive minerals. Proc. Natl. Acad. Sci. U.S.A. 109: 10042–10046.

Kato, S., K. Hashimoto and K. Watanabe. 2013. Iron-oxide minerals affect extracellular electron-transfer paths of *Geobacter* spp. Microbes Environ. 28: 141–148.

Kato, S. 2015. Biotechnological aspects of microbial extracellular electron transfer. Microbes Environ. 30: 133–139.

Kato, S., Y. Isao and Y. Kamagata. 2015. Isolation of acetogenic bacteria that induce biocorrosion by utilizing metallic iron as the sole electron donor. Appl. Environ. Microbiol. 81: 67–73.

Kimura, Z. and S. Okabe. 2013. Acetate oxidation by syntrophic association between *Geobacter sulfurreducens* and a hydrogen-utilizing exoelectrogen. The ISME J. 7: 1472–1482.

Kondo, K., A. Okamoto, K. Hashimoto and R. Nakamura. 2015. Sulfur-mediated electron shuttling sustains microbial long-distance extracellular electron transfer with the aid of metallic iron sulfides. Langmuir 31: 7427–7433.

Kouzuma, A., K. Hashimoto and K. Watanabe. 2012. Roles of siderophore in manganase-oxide reduction by *Shewanella onidensis* MR-1. FEMS Microbiol. Lett. 326: 91–98.

Kouzuma, A. and K. Watanabe. 2014. Microbial ecology pushes frontiers in biotechnology. Microbes Environ. 29: 1–3.

Kouzuma, A., S. Kato and K. Watanabe. 2015. Microbial interspecies interactions: recent findings in syntrophic consortia. Front. Microbial. 6: 477.

Kracke, F., I. Vassilev and J.O. Krömer. 2015. Microbial electron transport and energy conservation—the foundation for optimizing bioelectrochemical systems. Front. Microbiol. 6: Article 575.

Lampa-Pastirk, S., J.P. Veazey, K.A. Walsh, G.T. Feliciano, R.J. Steidl, S.H. Tessmer et al. 2016. Thermally activated charge transport in microbial protein nanowires. Sci. Rep. 6: 23517.

Leang, C., X. Qian, T. Mester and D.R. Lovley. 2010. Alignment of the *c*-type cytochrome OmcS along pili of *Geobacter sulfurreducens*. Appl. Environ. Microbiol. 76: 4080–4084.

Lee, J.H., M.G. Kim, B. Yoo, N.V. Myung, J. Maeng, T. Lee et al. 2007. Biogeneic formation of photoactive arsenic-sulfide nanotubes by *Shewanalla* sp. strain NH-41. Proc. Natl. Acad. Sci. U.S.A. 104: 20410–20415.

Logan, B. 2015. Japan reaches goal of zero reliance on fossil and nuclear fuels. Environ. Sci. Technol. Lett. 2: 1.

Logan, B.E., B. Hamelers, R. Rozendal, U. Schröder, J. Keller, S. Freguia et al. 2006. Microbial fuel cells: Methodology and technology. Environ. Sci. Technol. 40: 5181–5192.

Lovley, D.R., S.J. Giovannoni, D.C. White, J.E. Champine, E.J.P. Phillips, Y.A. Gorby et al. 1993. *Geobacter metallireducens* gen. nov. sp. nov., a microorganism capable of coupling the complete oxidation of organic compounds to the reduction of iron and other metals. Arch. Microbiol. 159: 336–344.

Lovley, D.R. 2012. Electromicrobiology. Annu. Rev. Microbiol. 66: 391–409.

Malvankar, N.S., M. Vargas, K.P. Nevin, A.E. Franks, C. Leang, B.C. Kim et al. 2011. Tunable metallic-like conductivity in microbial nanowire networks. Nat. Nanotechnol. 6: 573–579.

Malvankar, N.S. and D.R. Lovley. 2013. Microbial nanowires: a new paradigm for biological electron transfer and bioelectronics. Curr. Opin. Biotechnol. 27: 88–95.

Marsili, E., D.B. Baron, I.D. Shikhare, D. Coursolle, J.A. Gralnick and D.R. Bond. 2008. *Shewanella* secretes flavins that mediate extracellular electron transfer. Proc. Natl. Acad. Sci. U.S.A. 105: 3968–3973.

McGlynn, S. 2017. Energy metabolism during anaerobic methane oxidation in ANME archaea. Microbes Environ. 32: 5–13.

McGlynn, S.E., G.L. Chadwick, C.P. Kempes and V.J. Orphan. 2015. Single cell activity reveals direct electron transfer in methanotrophic consortia. Nature 526: 531–535.

Mehta, T., M.V. Coppi, S.E. Childers and D.R. Lovley. 2005. Outer membrane *c*-type cytochromes required for Fe(III) and Mn(IV) oxide reduction in *Geobacter sulfurreducens*. Appl. Environ. Microbiol. 71: 8634–8641.

Morita, M., N.S. Malvankar, A.E. Franks, Z.M. Summers, L. Giloteaux, A.E. Rotaru et al. 2011. Potential for direct interspecies electron transfer in methanogenic wastewater digester aggregates. MBio 2: e00159–11.

Nakagawa, S. and K. Takai. 2008. Deep-sea vent chemoautotrophs: diversity, biochemistry and ecological significance. FEMS Microbiol. Ecol. 65: 1–14.

Nakamura, R., F. Kai, A. Okamoto, G.J. Newton and K. Hashimoto. 2009. Self-constructed electrically conductive bacterial networks. Angew. Chem. Int. Ed. 48: 508–511.

Newman, D.K. and R. Kolter. 2000. A role for excreted quinones in extracellular electron transfer. Nature 405: 94–97.

Nielsen, L.P., N. Risgaard-Petersen, H. Fossing, P.B. Christensen and M. Sayama. 2010. Electric currents couple spatially separated biogeochemical processes in marine sediment. Nature 463: 1071–1074.

Okamoto, A., K. Hashimoto, K.H. Nealson and R. Nakamura. 2013. Rate enhancement of bacterial extracellular electron transport involves bound flavin semiquinones. Proc. Natl. Acad. Sci. U.S.A. 110: 7856–7861.

Okamoto, A., K. Saito, K. Inoue, K.H. Nealson, K. Hashimoto and R. Nakamura. 2014a. Uptake of self-secreted flavins as bound cofactors for extracellular electron transfer in *Geobacter* species. Energy Environ. Sci. 7: 1357–1361.

Okamoto, A., R. Nakamura, K.H. Nealson and K. Hashimoto. 2014b. Bound flavin model suggests similar electron-transfer mechanisms in *Shewanella* and *Geobacter*. ChemElectroChem. 1: 1808–1812.

Okamoto, A., S. Kalathil, X. Deng, K. Hashimoto, R. Nakamura and K.H. Nealson. 2014c. Cell-secreted flavins bound to membrane cytochromes dictate electron transfer reactions to surfaces with diverse charge and pH. Sci. Rep. 4: 5628.

Oshiki, M., T. Awata, T. Kindaichi, H. Satoh and S. Okabe. 2013. Cultivation of planktonic anaerobic ammonium oxidation (Anammox) bacteria using membrane bioreactor. Microbes Environ. 28: 436–443.

Pirbadiana, S., S.E. Barchingerb, K.M. Leunga, H.S. Byun, Y. Jangir, R.A. Bouhenni et al. 2014. *Shewanella oneidensis* MR-1 nanowires are outer membrane and periplasmic extensions of the extracellular electron transport components. Proc. Natl. Acad. Sci. U.S.A. 111: 12883–12888.

Preffer, C., S. Larsen, J. Song, M. Dong, F. Besenbacher, R.L. Meyer et al. 2012. Filamentous bacteria transport electrons over centimeter distances. Nature 491: 218–221.

Rabaey, K., N. Boon, M. Hofte and W. Verstraete. 2005. Microbial phenazine production enhances electron transfer in biofuel cells. Environ. Sci. Technol. 39: 3401–3408.

Reguera, G., K.D. McCarthy, T. Mehta, J.S. Nicoll, M.T. Tuominen and D.R. Lovley. 2005. Extracellular electron transfer via microbial nanowires. Nature 435: 1098–1101.

Reguera, G., K.P. Nevin, J.S. Nicoll, S.F. Covalla, T.L. Woodard and D.R. Lovley. 2006. Biofilm and nanowire production leads to increase current in *Geobacter sulfurreducens* fuel cells. Appl. Environ. Microbiol. 72: 7345–7348.

Rickard, D. and G.W. Luther III. 2007. Chemistry of iron sulfides. Chem. Rev. 107: 514–562.

Rismani-Yazdi, H., A.D. Christy, S.M. Carver, Z. Yu, B.A. Dehority and O.H. Tuovinen. 2011. Effect of external resistance on bacterial diversity and metabolism in cellulose-fed microbial fuel cells. Bioresour. Technol. 102: 278–283.

Rotaru, A.E., P.M. Shrestha, F. Liu, M. Shrestha, D. Shrestha, M. Embree et al. 2014. A new model for electron flow during anaerobic digestion: direct interspecies electron transfer to *Methanosaeta* for the reduction of carbon dioxide to methane. Energy Envion. Sci. 7: 408–415.

Rubaba, O., Y. Araki, S. Yamamoto, K. Suzuki, H. Sakamoto, A. Matsuda et al. 2013. Electricity producing property and bacterial community structure in microbial fuel cell equipped with membrane electrode assembly. J. Biosci. Bioeng. 116: 106–113.

Scheller, S., H. Yu, G.L. Chadwick, S.E. McGlynn and V.J. Orphan. 2016. Artificial electron acceptors decouple archaeal methane oxidation from sulfate reduction. Science 351: 703–707.

Shi, L.A., D.J. Richardson, Z. Wang, S.N. Kerisit, K. Rosso, J.M. Zachara et al. 2009. The roles of outer membrane cytochromes of *Shewanella* and *Geobacter* in extracellular electron transfer. Environ. Microbiol. Rep. 1: 220–227.

Shi, L., H. Dong, G. Reguera, H. Beyenal, A. Lu, J. Liu et al. 2016. Extracellular electron transfer mechanisms between microorganisms and minerals. Nature Review 14: 651–662.

Snider, R.M., S.M. Strycharz-Glaven, S.D. Tsoi, J.S. Erickson and L.M. Tender. 2012. Long-range electron transport in *Geobacter sulfurreducens* biofilms is redox gradient-driven. Proc. Natl. Acad. Sci. USA 109: 15467–15472.

Straub, K.L., M. Benz and B. Schink. 2001. Iron metabolism in anoxic environments at near neutral pH. FEMS Microbiol. Ecol. 34: 181–186.

Suga, H., S. Kikuchi, Y. Takeichi, C. Miyamoto, M. Miyahara, S. Mitsunobu et al. 2017. Spatially resolved distribution of Fe sepcies around microbes at the submicron scale in natural bacteriogenic iron oxides. Microbes Environ. 32: 283–287

Summers, Z.M., H.E. Fogarty, C. Leang, A.E. Franks, N.S. Malvankar and D.R. Lovley. 2010. Direct exchange of electrons within aggregates of an evolved syntrophic coculture of anaerobic bacteria. Science 330: 1413–1415.

Suzuki, K., Y. Kato, A. Yui, S. Yamamoto, S. Ando, O. Rubaba et al. 2018. Bacterial communities adapted to higher external resistance can reduce the onset potential of anode in microbial fuel cells. J. Biosci. Bioeng. 125: 565–571.

Tender, L.M., C.E. Reimers, H.A. Stecher, D.E. Holmes, D.R. Bond, D.A. Lowy et al. 2002. Harnessing microbially generated power on the seafloor. Nat. Biotechnol. 20: 821–825.

Torres, C.I., R. Krajmalnik-Brown, P. Parameswarn, A.K. Marcus, G. Wanger, Y.A. Gorby et al. 2009. Selecting anode-respiring bacteria based on anode potential: phylogenetic, electrochemical, and microscopic characterization. Environ. Sci. Technol. 43: 9519–9524.

Uchiyama, T., K. Ito, K. Mori, H. Tsurumura and S. Harayama. 2010. Iron-corroding methanogen isolated from a crude-oil storage tank. Appl. Environ. Microbiol. 76: 1783–1788.

Vali, H., B. Weiss, Y.L. Li, S.K. Sears, S.S. Kim, J.L. Kirschvink et al. 2004. Formation of tabular single domain magnetite induced by *Geobacter metallireducens* GS-15. Proc. Natl. Acad. Sci. USA 101: 16121–16126.

van der Zee, F.P. 2002. Anaerobic azo dye reduction. Wageningen University, Wageningen, The Neitherlands.

Weber, K.A., L.A. Achenbach and J.D. Coates. 2006. Microorganisms pumping iron: anaerobic microbial iron oxidation and reduction. Nat. Rev. Microbiol. 4: 752–764.

Wegener, G., V. Krukenberg, D. Riedel, H.E. Tegetmeyer and A. Boetius. 2015. Intercellular wiring enables electron transfer between methanotrophic archaea and bacteria. Nature 526: 587–590.

Xing, D., S. Cheng, J.M. Regan and B.E. Logan. 2009. Change in microbial communities in acetate- and glucose-fed microbial fuel cells in the presence of light. Biosens. Bioelectron. 25: 105–111.

Xu, S., Y. Jangir and M.Y. El-Naggar. 2016. Disentangling the roles of free and cytochrome-bound flavins in extracellular electron transport from *Shewanella oneidensis* MR-1. Electrochim. Acta 198: 49–55.

Yamamoto, S., K. Suzuki, Y. Araki, H. Mochihara, T. Hosokawa, H. Kubota et al. 2014. Dynamics of different bacterial communities are capable of generating sustainable electricity from microbial fuel cells with organic waste. Microbes Environ. 29: 145–153.

Chapter 2

Woody Biorefinery with White-rot Fungi

Toshio Mori[1] and *Hirofumi Hirai*[2,*]

1. Introduction

This project aims at the construction of lignocellulosic biorefinery technology in which all processes are carried out biologically. Lignocellulosic biorefinery generally consists of three main sequential processes: two pretreatment processes (delignification, and saccharification of polysaccharide) and a fermentation process. Additional processes, such as purification and neutralization, can also be added to the main processes as necessary, as the pretreatment processes frequently use acid / alkaline and produce unintended chemicals that decrease the efficiency of the fermentation process. On the other hand, it is thought that biological pretreatments produce little or no inhibitory chemicals. White-rot fungi are known to be the only organisms capable of being utilized in all of the lignocellulosic biorefinery processes (delignification, saccharification, and fermentation). However, the activities of white-rot fungi for each process are too weak for commercial use. Therefore, we attempted to identify several enzymes that can contribute to each biorefinery process and to evaluate the effects of recombination of the discovered enzymes for the processes as pre-limbic studies in order to create high-functional white-rot fungi.

Biorefineries represent technologies for producing energy, fuels, and chemicals from renewable biomass and securing energy and resources for establishing a sustainable society. Lignocellulosic materials, including woody biomass, are the most abundant renewable resource on earth. Therefore, development of biorefinery processes utilizing lignocellulosics as raw materials is one of the most important tasks to overcome our dependence on conventional oil. In the lignocellulosic biorefinery process, there are two types of pre-treatment methods: thermo-chemical and biochemical conversion (Naik et al.

[1] Faculty of Agriculture, Shizuoka University, Shizuoka, 422-8529, Japan.
Email: mori.toshio@shizuoka.ac.jp
[2] Research Institute of Green Science and Technology, Shizuoka University, Shizuoka, 422-8529, Japan.
* Corresponding author: hirai.hirofumi@shizuoka.ac.jp

2010). Biochemical conversion processes frequently include thermo-chemical steps, such as delignification pretreatment with alkaline under high temperature and pressure, and acidic hydrolysis of polysaccharides pretreatment (Kamm and Kamm 2004). Although the thermo-chemical pretreatment steps should proceed with high efficiency in a short time, the process requires high temperature and a considerable input of energy and reagents.

It is believed that biological pretreatment has some advantages (e.g., mild condition, low energy input, environmentally friendly and low chemical requirement) when compared with thermo-chemical pretreatment. However, biological pretreatments rely on immature technologies and are the subject of active research to improve cost and reaction efficiency. For example, enzymatic saccharification is carried out under mildly acidic pH, without the addition of toxic chemicals, at relatively low temperature, and produces fewer by-products by inhibiting subsequent fermentation steps. Many researchers continue to develop low cost and effective saccharification enzymes for lignocellulosic biomass. Moreover, fermentative microorganisms capable of producing saccharification enzymes, such as Clostridia and filamentous fungi, have the potential to bring major advantages to the biorefinery process when compared with conventional biological processes, as the processes utilizing these microorganisms do not require prior processes, enzyme production and hydrolysis. Therefore, many researchers are attempting to confer the saccharification ability onto high fermentation microorganisms or to enhance the fermentation capacity of native cellulolytic microorganisms by metabolic engineering to improve the overall efficiency of the process (Hasunuma et al. 2013).

As mentioned above, many studies have aimed for construction of biological processes using biological saccharification pretreatment; however, there have been several studies on biological delignification pretreatment. Lignin, one of main component of lignocellulosics, is a heterogenous aromatic polymer with resistance to biological attacks and is formed by radical polymerization of phenylpropanoid structures. Lignin forms an amorphous matrix with hemicellulose and is deposited on the surface of cellulose fibers. As a result, lignin protects polysaccharides in the plant cell wall and confers mechanical strength and high resistance against chemical/biological decomposition to the lignocellulosic materials, like wood (Bjurhager et al. 2010, Voelker et al. 2011). Therefore, the efficiency of saccharification and fermentation of lignocellulosic biomass is strongly affected by its lignin content and structure, and delignification pretreatment is an important step in the lignocellulosic biorefinery processes (Ruiz-Dueñas and Martínez 2009). Although there are few microorganisms that degrade natural lignin completely in the environment, a group of filamentous fungi called "white-rot fungi" which are wood-decaying basidiomycetes, are able to degrade lignin into water and carbon dioxide. This unique ability of white-rot fungi is expected to be useful in biological delignification pretreatment (Sindhu et al. 2016). In addition, direct ethanol production from lignocellulosic biomass using white-rot fungi has been recently reported (Kamei et al. 2012a, Mattila et al. 2017). It is therefore clear that white-rot fungi are unique microorganisms with vital capabilities for lignocellulosic biorefineries (delignification, saccharification, and fermentation), and are the most suitable candidate microorganisms for low-energy and environmentally friendly biological biorefinery technology.

On the other hand, to practically utilize biorefinery technology by using white-rot fungi, several tasks remain. There are three major problems: (1) poor delignification efficiency for commercial use; (2) low efficiency of fermentation from polysaccharides; and (3) no selectivity of fermentation products other than ethanol and organic acids. We have been trying to overcome these problems using *Phanerochaete sordida* YK-624, which is a white-rot fungus with higher delignification activity and selectivity (Hirai et al. 1994). In this chapter, we summarize our research results since 2014, as related to each of these problems (1)–(3).

2. Exploration of Enzymes to Improve Delignification Efficiency

In the delignification system of white-rot fungi, it is known that delignification reactions are carried out by extracellular ligninolytic oxidoreductases and their accessory enzymes (Dashtban et al. 2010). The main ligninolytic enzymes, lignin peroxidase, manganese peroxidase (MnP), versatile peroxidase, and laccase, oxidize lignin directly or indirectly to form low-molecular weight lignin fragments. These fragments are then decomposed into water and carbon dioxide after being taken up by the fungal cells (Ander et al. 1983).

Each white-rot fungus is considered to produce a specific enzyme set with its own balance. In *P. sordida* YK-624, which shows excellent delignification activity, the fungus mainly secretes MnP under ligninolytic conditions (Hirai et al. 1994). Previously, a *P. sordida* transformant that homologously expresses and produces MnP under ligninolytic conditions was constructed (Sugiura et al. 2012). The transformant showed 22% higher ligninolytic activity than wild type after 4 weeks of cultivation on beech wood meal and maintained higher MnP activity during the whole cultivation period. Based on this result, it is clear that MnP activity has an influence on lignin degradation; however, delignification efficiency of this transformant did not reach sufficient levels for commercial uses. Therefore, in order to greatly improve the delignification activity of white-rot fungi, we speculated that both the main delignification enzymes and the activities of lignin degradation-related enzymes (including accessory and lignin-fragment metabolizing enzymes) are important. Accordingly, we investigated how various lignin degradation-related enzymes affect total delignification activity by using a homologous expression system.

2.1 Glyoxal Oxidase

White-rot fungi secrete some accessory enzymes for production of hydrogen peroxide required for ligninolytic peroxidases. For example, aryl-alcohol oxidase is able to oxidize an aromatic primary alcohol to generate a hydrogen peroxide and an aromatic aldehyde (Guillén et al. 1990). In addition, quinone reductase is involved in hydrogen peroxide production via quinone redox cycling (Gómez-Toribio et al. 2009), and cellobiose dehydrogenase is able to reduce molecular oxygen to hydrogen peroxide by using an electron donor cellobiose as substrate (Nutt et al. 1997). Glyoxal oxidase (GLOX) catalyzes oxidation of glyoxal to form glyoxylate and hydrogen peroxide. In white-rot fungi, this enzyme is one of the sources of extracellular hydrogen peroxide required for activation of ligninolytic peroxidases (Kersten and Kirk 1987).

We constructed the GLOX gene (from *P. chrysosporium*) expression plasmid under control of the *bee-2* promoter (Sugiura et al. 2012), which activates under ligninolytic conditions, and the plasmid was co-transformed with p*PsURA5* plasmid into *P. sordida* YK-624 uracil auxotrophic mutant (Yamada et al. 2014, Yamagishi et al. 2007). After that, 16 GLOX gene-introducing transformants were obtained, and these strains showed higher GLOX activity than wild-type and control transformants harboring marker plasmid. On the other hand, no significant difference was observed in MnP activity between the transformants and control strain. Slightly but significantly higher (108%) average ligninolytic activities was observed in the wood meal cultures of *GLOX* introducing transformants when compared

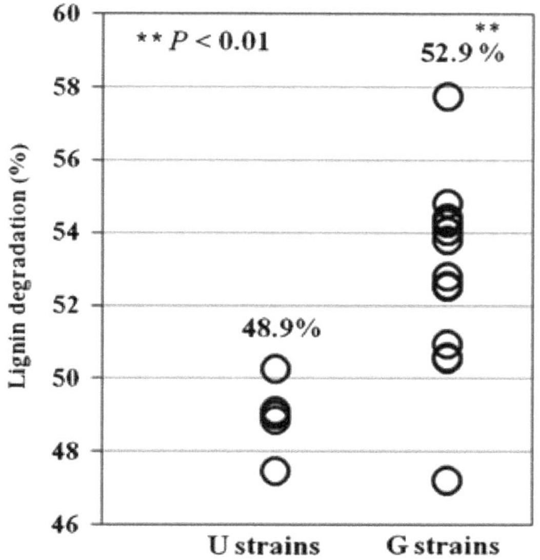

Fig. 2.1: Ligninolytic properties of *GLOX*-introduced (G) and control (U) strains (Yamada et al. 2014). Each open circle indicates lignin degradation of individual transformant. Copyright (2014) Taylor & Francis.

with control strains, and one of the transformants showed 122% of ligninolytic activity (Fig. 2.1). From these results, it was suggested that the amount of extracellular hydrogen peroxide is, in addition to MnP activity, very important to the ligninolytic reaction of *P. sordida* YK-624.

2.2 Benzoquinone Reductase and Homogentisate 1,2-dioxygenase

During delignification, it is known that low molecular weight lignin fragments, such as vanillate, are generated from oxidative cleave of lignin polymer. Because these lignin fragments act as substrates of ligninolytic oxidoreductase and compete with the oxidation of high molecular weight lignin, it is anticipated that elimination of lignin fragments from the working location of ligninolytic enzymes is an important function for accomplishing better delignification efficiency in white-rot fungi.

Thus, the effects of homologous expression of intracellular benzoquinone reductase (BQR) and homogentisate 1,2-dioxygenase (HGD) in *P. sordida* YK-624 were investigated (Mori et al. 2016b). First, we identified putative BQR and HGD genes of *P. sordida* YK-624, and homologously expressed the respective genes under ligninolytic conditions. As a result, the transformants individually introduced to *BQR* and *HGD* showed significantly improved intracellular enzyme activity compared with the control strain, and a significantly higher vanillin metabolism rate was observed in both transformants (Fig. 2.2A). However, homologous recombination of *BQR* and *HGD* limited the effects on the degradation of lignin. In particular, wood weight loss was decreased by *HGD* recombination, while there was almost no effect on delignification. Therefore, *HGD* transformants showed significantly higher lignin degradation selectivity (Fig. 2.2B). Lignin fragments having phenolic structures are easily oxidized to form benzoquinone structures by ligninolytic enzymes. Extracellular benzoquinones are utilized for quinone redox cycling to produce hydroxy radicals (Gómez-Toribio et al. 2009). In brown-rot fungi, it has

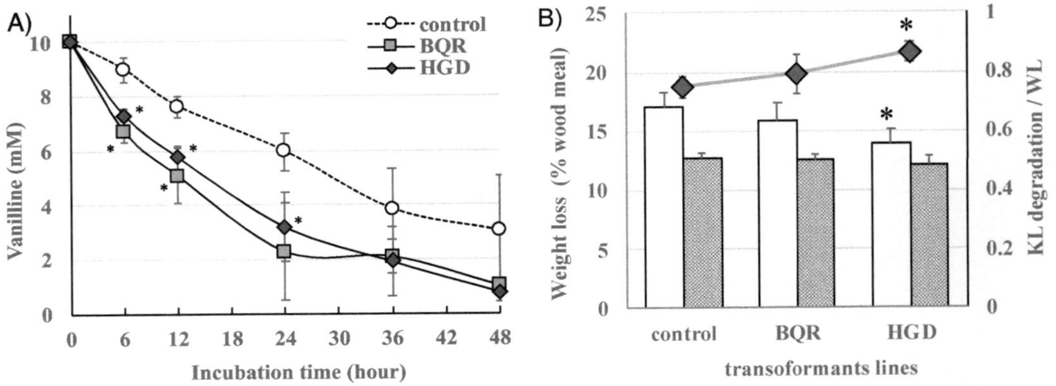

Fig. 2.2: Vanillin (A) and wood (B) degradation properties of BQR- and HGD gene-introduced transformants (Mori et al. 2016b). (A) Time courses of vanillin degradation by individual co-transformants with BQR and HGD gene. (B) Average wood weight loss (white bar), Klason lignin degradation (gray bar) and delignification selectivity (black diamond) of each transformed line. Copyright (2016) Springer Nature.

been reported that a decrease in holocellulose viscosity is correlated with hydroquinone concentration, and hydroxyl radicals derived from quinone redox cycling are responsible for most of the holocellulose depolymerization (Suzuki et al. 2006). Thus, enhancement of intracellular HGD catalyzing aromatic ring-cleavage affects polysaccharide metabolic activity, but not ligninolytic activity, and consequently, lignin degradation selectivity is improved.

2.3 Manganese Transporter

Manganese is known to be an important element for ligninolytic activity of white-rot fungi. Manganese ions not only work as a substrate of MnP, but also induce MnP gene expression (Bonnarme and Jeffries 1990, Brown et al. 1990). We therefore attempted to elucidate the effects of a manganese transporter from *P. sordida* YK-624 on MnP production and ligninolytic activity. A homologue gene of the yeast natural resistance-associated macrophage protein (Nramp) family transporter *smf2* was identified, and was found to be expressed under ligninolytic conditions (Mori et al. 2018). In yeast, SMF2 is considered to be the most important proton-coupled transporter of Mn ions into the cell, as Mn homeostasis is significantly affected in *smf2* deletion yeast, reducing Mn accumulation and the activity of Mn-dependent enzymes (Cyert and Philpott 2013, Reddi et al. 2010). This *smf2* homologue manganese transporter of *P. sordida* YK-624, designated *PsMnt*, showed equivalent relative expression levels in cultures containing different Mn(II) concentrations (0–1 mM). *PsMnt* transformants of *P. sordida* YK-624 produced higher extracellular MnP activity than wild-type and control strains, however, this improvement decreased with increasing Mn concentration (Fig. 2.3A, B). After that, one transformant (MT-20) was selected and characterized. MT-20 expressed higher MnP activity during all incubation periods in nitrogen-limited liquid culture when compared with the control strain. At lower Mn concentrations (10 μM), relative *PsMnt* expression of MT-20 was 10 times higher than controls on day 4. MT-20 showed slightly improved Mn accumulation (136%) when compared with controls, whereas MnP activity was increased greatly to near 300% of that in control strain. Therefore, we concluded that the Nramp manganese transporter *PsMnt* is important for manganese uptake under low Mn conditions, as in wood culture.

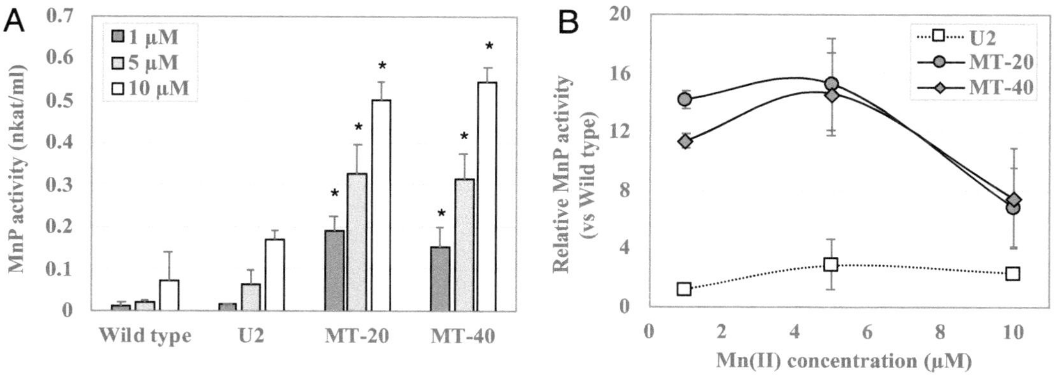

Fig. 2.3: Effect of *PsMnt* recombination for MnP activity at various Mn(II) concentrations (Mori et al. 2018). (A) MnP activity in culture fluids of wild type, control strain (U2), and *PsMnt* recombinant strains (MT-20 and 40). (B) Rate of increase in MnP activity in control and *PsMnt* recombinant strains at each Mn(II) concentration, as compared with wild type. Copyright (2018) Oxford University Press.

Overexpression of *PsMnt* also enhances MnP activity, and is expected to improve delignification activity. However, the mycelial growth of selected *PsMnt* overexpressing strains on wood and liquid cultures was inhibited; thus, the effects of *PsMnt* overexpression for delignification could not be estimated correctly.

2.4 Molecular Breeding of Lignin-degrading Brown-rot Fungus

We then attempted molecular breeding of lignin-degrading brown-rot fungi. Brown-rot fungi usually show little or no ligninolytic activity, because they produce no ligninolytic enzymes. Brown-rot fungi have high depolymerization activity for polysaccharides, including cellulose and hemicellulose in wood, and leave modified lignin in the brown residue (Fackler and Schwanninger 2010). In addition, brown-rot fungi are able to ferment ethanol from corn fiber under anaerobic conditions (Rasmussen et al. 2010). These facts indicate that brown-rot fungi have a set of functions required for ethanol fermentation from lignocellulosic biomass, excluding ligninolytic enzyme systems.

Therefore, the effect of expression of endogenous putative laccase gene in brown-rot fungus *Gloeophyllum trabeum* for ethanol fermentation was investigated (Arimoto et al. 2015). First, we selected one putative laccase gene (*Gtlcc3*, protein ID 127593) from four putative laccase genes of *G. trabeum*, which were identified in the genome database of the Joint Genome Institute, and we constructed a plasmid expressing *Gtlcc3* under control of the *gpd* promoter. Transformants expressing endogenous laccase gene were screened by chromophoric assay using 2,2′-azino-bis(3-ethylbenzothiazoline-6-sulphonic acid) (ABTS). Screened transformants showed slight but clear laccase activity, and endogenous laccase production granted ligninolytic activity to *G. trabeum*. Although forced laccase production increased delignification activity for cedar wood meal from 0.8% in the parent strain to 2.7%, the value was still low when compared with the ligninolytic activity of white-rot fungi. This result was in accordance with the lack of other lignin degradation related enzymes present in white-rot fungi. On the other hand, ethanol production under semi-aerobic conditions following 28-day aerobic cultivation was significantly elevated to almost 1.5 times when compared with the parent strain (Fig. 2.4). From these results, we confirmed that lignin degradation related enzymes are important for accomplishing high

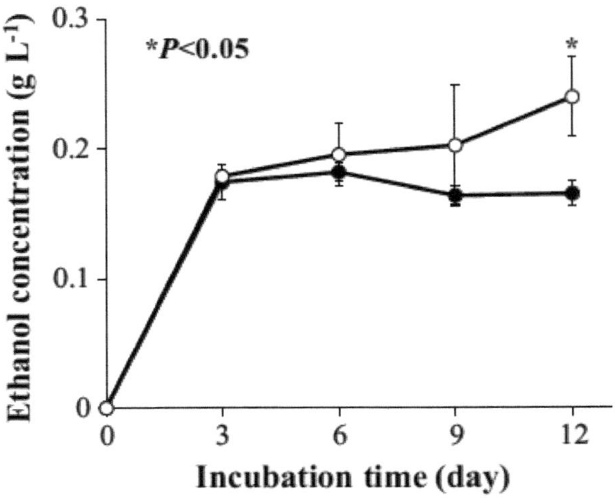

Fig. 2.4: Time courses of ethanol production from cedar wood meal under semi-aerobic condition following 28 days pre-treatment with parent *G. trabeum* (black) and *Gtlcc3* recombinant strain (white) (Arimoto et al. 2015).

delignification activity; hence, white-rot fungi are more suitable than brown-rot fungi for biorefineries using lignocellulosic materials.

3. Functional Analysis of Fermentation Mechanisms of White-rot Fungi

As with the studies intending to utilize white-rot fungi for delignification pretreatment, there have been various studies on the fermentation capacity of white-rot fungi. To our knowledge, the first report of ethanol fermentation of white-rot fungi was published in 2004, and it describes the ethanol production of *Phanerochaete chrysosporium* under oxygen limited culture containing various mono- or disaccharides (Kenealy and Dietrich 2004). In 2010 and 2011, Okamoto et al. reported on ethanol production of white-rot fungi from hexose, starch-based biomass and lignocellulosic biomass (Okamoto et al. 2010, 2011). In 2012, Kamei et al. reported on the excellent white-rot fungus (*Phlebia* sp. MG-60) for ethanol fermentation (Kamei et al. 2012b). This fungus has good fermentability for a broad range of mono- and disaccharides, and showed excellent ethanol productivity from lignocellulosic materials, 44% of theoretical maximum (%TM) from oak wood, and 72% TM from unbleached hardwood Kraft pulp, without addition of any other enzymes or microorganisms (Kamei et al. 2012a, 2012b). Subsequently, several white-rot fungi have been studied for ethanol fermentation activity from lignocellulosic materials (Horisawa et al. 2015, Mattila et al. 2017, Okamoto et al. 2014). However, it remains unelucidated how *Phlebia* sp. MG-60 exhibits such excellent ethanol productivity, and because some white-rot fungi are able to produce ethanol under aerobic conditions (Kamei et al. 2012b, Okamoto et al. 2010, 2014), it is not well understood how white-rot fungi regulate their fermentation mechanisms.

3.1 Analysis of Ethanol Fermentation Mechanism

As mentioned above, the fermentation mechanism of *Phlebia* sp. MG-60 exhibiting high ethanol production activity from lignocellulose is poorly understood. To elucidate the key factors related to the fermentation activity of *Phlebia* sp. MG-60, we performed differential expression analysis between two conditions: ethanol fermentation and non-fermentation (zero glucose) states (Wang et al. 2016b). As an initial attempt, glucose was used as a feedstock for fermentation in this experiment. In RNA-seq analysis, *Phlebia* sp. MG-60 expressed two genes encoding for pyruvate decarboxylase (PDC) and alcohol dehydrogenase (ADH), which are involved in the two-step ethanol fermentation pathway, but did not detect genes involved in the three-step pathway, which is widespread in bacteria (Schworm et al. 1996). *Phlebia* ap. MG-60 greatly increased expression levels of some high-affinity glucose transporters under fermentation conditions, and whole genes involved in the glycolytic pathway from glucose to pyruvate were up-regulated during ethanol production. This up-regulation was not observed in the standard white-rot fungus *P. chrysosporium*; this fungus up-regulated just some genes of the glycolytic pathway during fermentation, and expression of some genes encoding PDCs and ADHs were increased in both fungi under fermentation conditions. On the other hand, aldehyde dehydrogenase was differentially down-regulated in *Phlebia* sp. MG-60 under fermentation conditions. As this enzyme catalyzes the oxidation of acetaldehyde to acetate, it competitively works against the ethanol production pathway. Therefore, low expression levels of aldehyde dehydrogenase genes under fermentation conditions might be one of the causes of the high ethanol productivity of *Phlebia* sp. MG-60. Furthermore, it is also expected that wider up-regulation of the glucose uptake, glycolysis, and ethanol fermentation pathways under fermentation conditions contributed to their high ethanol productivity.

Recently, Mäkinen et al. (2018) reported a transcriptional analysis of another ethanol-producing white-rot fungus, *Phlebia radiata*, on solid-state lignocellulosic culture under semi-aerobic conditions (Mäkinen et al. 2018). As compared with glucose-containing medium, *P. radiata* up-regulated carbohydrate-active enzyme genes on lignocellulosic biomass. However, lignin-degrading enzymes showed differential expression patterns. Taken together, these results suggest that cooperative regulation of the whole ethanol production pathway, including saccharification, glycolysis, and fermentation, is particularly important for achieving better ethanol productivity from lignocellulosic biomass.

3.2 Pyruvate Decarboxylase

PDC is a key enzyme in ethanol fermentation. This enzyme catalyzes the non-oxidative decarboxylation of pyruvate into acetaldehyde, which is converted to ethanol by ADH. *PDC* overexpression in the yeast *Hansenula polymorpha* resulted in improvement of ethanol production (Ishchuk et al. 2008). This suggests the possibility of increased alcoholic fermentation by overexpression of the PDC gene in white-rot fungi.

Accordingly, we identified the PDC gene of *P. sordida* YK-624 and then generated transformants that constantly express the PDC gene (Wang et al. 2016a). Subsequently, we screened the 16 resulting recombinant *PDC* transformants by ethanol production from glucose. All recombinant strains showed higher ethanol productivity than wild-type, and the strain showing the highest ethanol productivity was selected. This transformant exhibited 1.3 times higher PDC activity and 1.22 times higher ethanol production from glucose, as compared with the wild-type under nutrient-rich conditions. Although the cause is unknown, ethanol productivity from fructose and maltose were decreased compared to the wild-type. Increases in PDC activity also improved ethanol productivity from mannose (1.56 times) and cellobiose (1.26 times). Although the amount of ethanol production from beech wood meal showed no significant differences between the transformant and

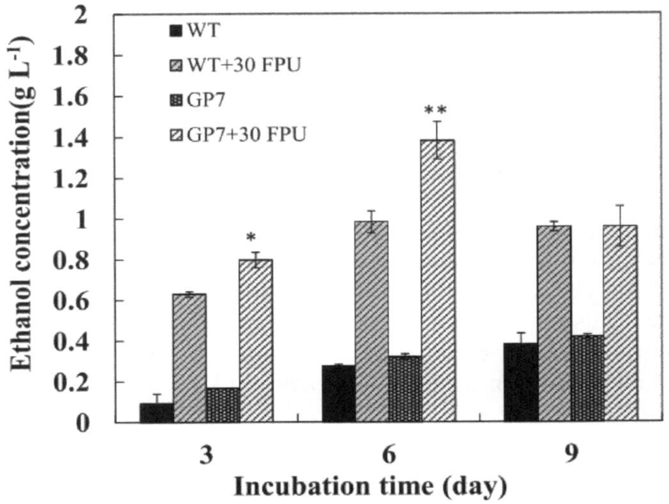

Fig. 2.5: Ethanol production (with or without cellulase) on beech wood after 28 days of pretreatment with WT and PDC gene recombinant transformant (GP7) (Wang et al. 2016a). Copyright (2016) Elsevier.

wild-type during oxygen-limited cultivation after aerobic delignification pre-cultivation (without addition of celluloses), the transformant produced 1.41 times more ethanol under Simultaneous Saccharification and Fermentation (SSF) conditions with addition of commercial cellulase (Fig. 2.5). As shown in Fig. 2.5, enhancement of PDC activity is an effective approach for improvement of fermentation efficiency to lignocellulosic biomass. Furthermore, because the cellulase activity of *P. sordida* YK-624 was insufficient for enzymatic saccharification, it was also suggested that improvement of cellulase activity is another effective approach for lignocellulosic fermentation.

3.3 Ethanol Fermentation of White-rot Fungi under Aerobic Conditions

Studies relating to ethanol fermentation by white-rot fungi have often used oxygen-supply limited conditions, known as "semi-aerobic" conditions (Kamei et al. 2012b, Mattila et al. 2017, Okamoto et al. 2010). Although respiration by wood-rot fungi leads to a low-oxygen concentration state, fungal growth and viability strongly affect the eventual oxygen concentration. It is also conceivable that ethanol fermentation begins upon reaching a specific low oxygen threshold. However, there have been reports describing aerobic ethanol fermentation by white-rot fungi (Kamei et al. 2012b, Okamoto et al. 2010, 2014). Therefore, we investigated the regulation of ethanol fermentation by *P. sordida* YK-624 (Mori et al. 2019).

P. sordida YK-624 produces ethanol under aerobic conditions (containing 20 g/L initial glucose). Higher ethanol production, higher glucose consumption, and better mycelial growth were observed under aerobic culture, as compared to semi-aerobic conditions. However, the conversion efficiency from glucose to ethanol was significantly higher in semi-aerobic culture. This indicates that a large part of the consumed glucose was utilized for aerobic respiration growth. However, *P. sordida* YK-624 did not produce detectable amounts of ethanol at lower initial glucose concentrations (1.3 g/L or lower). In the early cultivation stage, glucose consumption, mycelial growth, and ethanol production in cultures

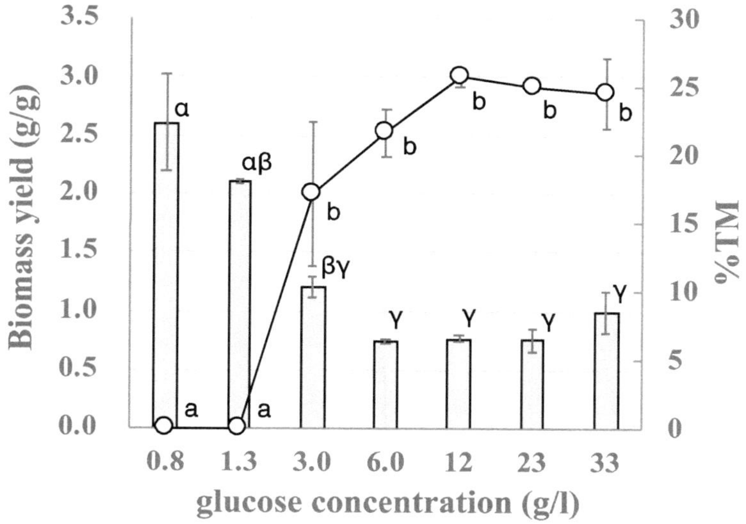

Fig. 2.6: Biomass yield (growth/glucose consumption) and %TM (theoretical maximum of ethanol) from second-day aerobic cultures of *P. sordida* YK-624 incubated in media containing various initial concentrations of glucose (Mori et al. 2019). Gray bars and open circles indicate biomass yields and %TM, respectively. Copyright (2019) Springer Nature.

containing different glucose concentrations were increased according to higher glucose concentrations, up to 6 g/L glucose (Fig. 2.6). At higher glucose concentrations, these values tended to decrease. Biomass yields (growth/glucose consumption) decreased until the initial glucose concentration increased to 6 g/L. On the other hand, ethanol conversion efficiency showed the opposite trend in biomass yields. In differential expression analysis of key enzymes at the pyruvate branch point between mycelia grown under ethanol producing and non-producing conditions, no significant difference was observed (Diaz-Ruiz et al. 2011). Based on these results, it was indicated that the regulatory mechanism of aerobic ethanol production in *P. sordida* YK-624 is different from that of Crabtree-positive yeasts (Dashko et al. 2014, Hagman et al. 2014). Aerobic production of ethanol by this white-rot fungus might result in overflow of respiration catabolism, and overflowing glucose and/or its metabolites is then utilized for ethanol production. Thus, expression of key enzymes at the pyruvate branch point was not dependent on glucose concentration, growth was saturated at higher glucose concentrations, and ethanol production improved in the range of glucose concentrations at which growth was saturated, but there remains a possibility of different regulation mechanisms under anaerobic fermentation conditions.

4. Fermentative Production other than Ethanol

White-rot fungi have a set of usable abilities for bioconversion via ethanol production from lignocellulosic biomass. In addition to ethanol, there are no noticeable reports on the production of valuable chemicals by white-rot fungus, other than the production of some organic acids (Galkin et al. 1998, Takao 1965). Thus, we attempted to introduce other fermentation productivity to white-rot fungus.

4.1 Xylitol

Xylitol is a naturally occurring sugar alcohol in plants such as birch, plum, and strawberry. Artificially, xylitol is produced from xylose, which is a hydrolyzed product of xylan extracted from raw lignocellulosic materials. Then, xylose is reduced into xylitol by a catalytic reaction. Xylitol has 40% fewer calories and similar levels of sweetness as sucrose. This sugar alcohol is metabolized independently of insulin, and is adsorbed more slowly than sugar. In addition, it shows anti-cariogenic effects and several other bioactivities. Because of this range of functions, xylitol is utilized as a supplement in the manufacture of products for diabetes management, body weight management, and dental care.

Generally, xylose is reduced to xylitol by an NAD(P)H-dependent xylose reductase in yeasts and fungi. An $NADP^+$-dependent xylitol dehydrogenase then oxidizes xylitol to xylulose (Moysés et al. 2016). Xylulose enters the pentose phosphate pathway via conversion to xylulose-5P by xylulokinase. As with other fungi and yeasts, *P. sordida* YK-624 can utilize xylose as a sole carbon source for growth. *P. sordida* YK-624 is also able to accumulate small amounts of xylitol during cultivation if grown in culture supplemented with xylose as a sole carbon source (Fig. 2.7). However, if the amount of xylitol accumulation is low, the conversion rate is only 4% of initial xylose. Therefore, we introduced a xylose reductase (XR) gene from *P. chrysosporium* into *P. sordida* auxotrophic mutant to improve xylitol productivity (Hirabayashi et al. 2015). The gene was successfully introduced into *P. sordida* YK-624; an XR gene-introducing transformant showed significantly higher XR intracellular activity, and the xylose conversion efficiency to xylitol was also greatly improved to 24% of initial xylose (Fig. 2.7). Based on these results, it was concluded that the recombinant expression of a gene involved in fermentative production other than ethanol is effective for improving or adding the ability to produce valuable chemicals in white-rot fungi.

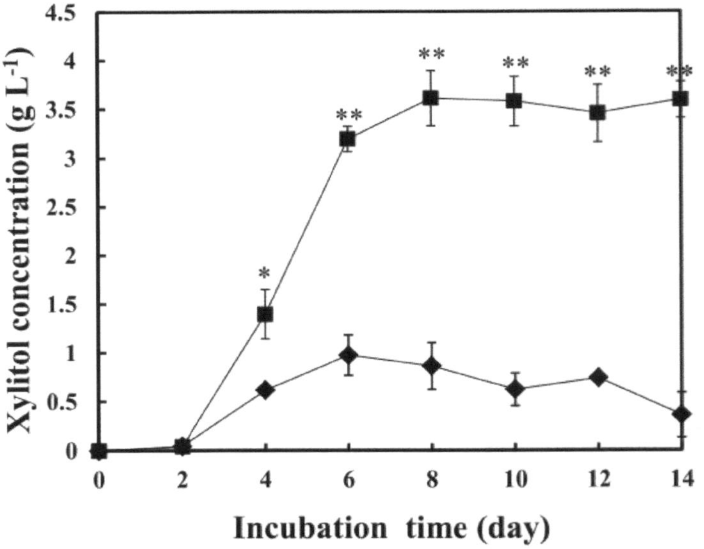

Fig. 2.7: Time courses of xylitol production by wild type (diamonds) and XR gene-introducing transformant (squares) (Hirabayashi et al. 2015) in nitrogen-limited liquid culture supplemented with 1.5% xylose as a sole carbon source. Copyright (2015) Elsevier.

4.2 Lactic Acid

Lactic acid is widely utilized as an additive or raw material in the food, pharmaceutical, and chemical industries. In particular, this acid can be polymerized to produce polylactic acid, a featured renewable and biodegradable polymer. Lactic acid is produced by fermentation from both bacteria and fungi. In bacteria, generally used lactic acid bacteria such as *Lactobacillus* species are utilized. These bacteria produce lactic acid at levels near the theoretical maximum from saccharides via pyruvate (Wang et al. 2015). Among fungi including yeasts, *Rhizopus oryzae* is the best-known lactic acid-producing fungus. This fungus can produce lactic acid in a fermentation process with saccharides and hydrolysate of lignocellulosic biomass at high yield (Wang et al. 2015). However, as with ethanol fermentation, the process of lactic acid fermentation from lignocellulosic also is associated with negative effects in the saccharification stage (Gomes et al. 2015).

Therefore, we attempted the molecular breeding of lactic acid-producing white-rot fungi by heterologous gene expression of NAD-dependent lactate dehydrogenase (LDH), a key enzyme in lactic acid fermentation, to *P. sordida* YK-624 (Mori et al. 2016a). First, we constructed a heterologous *ldh*-expression plasmid, which controlled the expression of LDH obtained from *Bifidobacterium longum* (*Blldh*) under the *gpd* promoter, and introduced it into *P. sordida* YK-624. The resulting transformants successfully produced lactic acid under semi-aerobic liquid culture containing glucose as a carbon source. However, the transformant still produced high amounts of ethanol. This indicates that the majority of pyruvate is utilized for ethanol production in preference to lactic acid, and that lactic acid is derived from the remaining pyruvate. Next, the inhibition of pyruvate decarboxylase (PDC) expression was performed by using the RNA interference method to suppress the ethanol production pathway. As a result, ethanol production was reduced to almost half, and lactic acid production was improved two-fold. In addition, by addition of calcium bicarbonate to the culture for creation of stable lactate-calcium complex, 1.5-fold higher lactic acid production was observed when compared with no addition of calcium bicarbonate. Finally, we attempted successive delignification and lactic acid fermentation with a single organism and single container using our *pdc*-suppressed *ldh*-expression transformant. The outline of experiment is shown in Fig. 2.8A. After preincubation of the transformant in beech wood meal for mycelial growth and delignification, calcium bicarbonate containing glucose-free liquid medium with or without cellulase was added to the delignified wood meal. The culture was then incubated under semi-aerobic conditions. Following fermentation without cellulase, lactic acid was clearly observed in the culture fluid by HPLC analysis (Fig. 2.8B). Following fermentation with the addition of exogenous cellulase, lactic acid and ethanol accumulation in the fluid increased five-fold (Fig. 2.8B). This lactic acid production level was 64% of that obtained in liquid medium containing glucose as a carbon source. On the other hand, a certain amount of glucose remained in the culture fluid after lactic acid fermentation. From these results, we demonstrated the direct production of lactic acid by fermentation from woody biomass using a single microorganism, and confirmed that transgenic white-rot fungi has a potential use for development of simple/easy applications in a wood biorefinery.

This study also highlights two major problems with wood biorefineries using white-rot fungi. Firstly, the saccharification activity of white-rot fungi is poor, and so it must be enhanced for commercial use. Secondly, consumption rates of saccharides are weak because considerable amounts of glucose remained in the culture fluid after fermentation

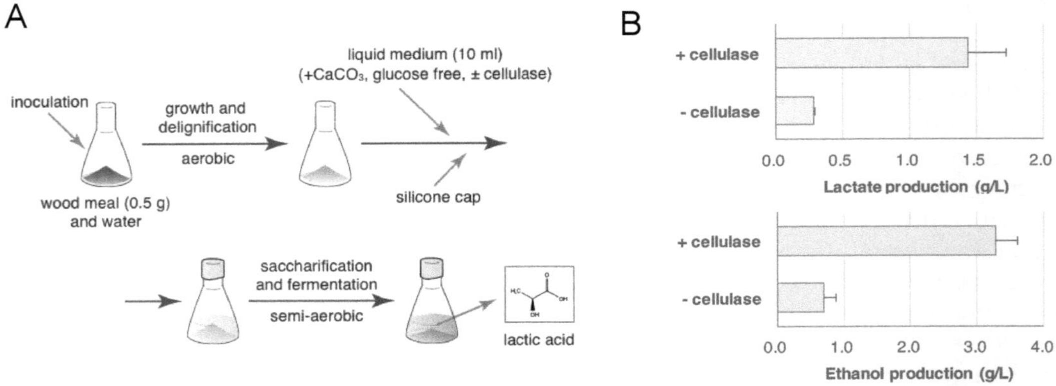

Fig. 2.8: Direct lactate production from wood meal by using *pdc*-suppressed *ldh*-expression transformant (Mori et al. 2016a). (A) Scheme of direct lactate fermentation method from wood meal using transgenic white-rot fungus. (B) Production of lactate and ethanol from wood meal. Copyright (2016) Elsevier.

with cellulase. Therefore, the rate for saccharide uptake is a key factor for improvement of fermentation efficiency.

5. Conclusion

In this chaper, we aimed at the construction of lignocellulosic biorefinery technology that allows all processes to be carried out by a single microorganism. For delignification pretreatment, homologous expression of some genes that encode enzymes related to lignin metabolism was performed in order to improve ligninolytic efficiency of the white-rot fungus *P. sordida* YK-624, which naturally retains superior ligninolytic activity. We attempted to improve ethanol productivity and to grant exogenous fermentation activity to produce other valuable chemicals. As a result, homologous expression of the respective genes studied in this chapter slightly but significantly improved the delignification and fermentation efficiency of *P. sordida* YK-624. Therefore, several enzymes involved in the lignin metabolism and fermentation of white-rot fungi have the potential to transgenically improve fermentation efficiency from lignocellulosic biomass. We also discovered an aerobic ethanol fermentation mechanism for the white-rot fungus *P. sordida* YK-624, which results from overflow of respiratory metabolism, while the effects of single transformation of each gene limit both delignification and fermentation. Accordingly, it will be necessary to attempt new breeding approaches in order to generate new high-functional white-rot fungus that shows improved activity for multiple enzymes, and this work is currently underway in our laboratory.

References

Ander, P., K.E. Eriksson and H. Yu. 1983. Vanillic acid metabolism by *Sporotrichum pulverulentum*: evidence for demethoxylation before ring-cleavage. Arch. Microbiol. 136: 1–6.

Arimoto, M., K. Yamagishi, J. Wang, K. Tanaka, T. Miyoshi, I. Kamei et al. 2015. Molecular breeding of lignin-degrading brown-rot fungus *Gloeophyllum trabeum* by homologous expression of laccase gene. AMB Express 5: 81.

Bjurhager, I., A.M. Olsson, B. Zhang, L. Gerber, M. Kumar, L.A. Berglund et al. 2010. Ultrastructure and mechanical properties of populus wood with reduced lignin content caused by transgenic down-regulation of cinnamate 4-hydroxylase. Biomacromolecules 11: 2359–2365.

Bonnarme, P. and T.W. Jeffries. 1990. Mn(II) Regulation of Lignin peroxidases and manganese-dependent peroxidases from lignin-degrading white rot gungi. Appl. Environ. Microbiol. 56: 210–217.

Brown, J.A., J.K. Glenn and M.H. Gold. 1990. Manganese regulates expression of manganese peroxidase by *Phanerochaete chrysosporium*. J. Bacteriol. 172: 3125–3130.

Cyert, M.S. and C.C. Philpott. 2013. Regulation of cation balance in *Saccharomyces cerevisiae*. Genetics 193: 677–713.

Dashko, S., N. Zhou, C. Compagno and J. Piškur. 2014. Why, when, and how did yeast evolve alcoholic fermentation? FEMS Yeast Res. 14: 826–832.

Dashtban, M., H. Schraft, T.A. Syed and W. Qin. 2010. Fungal biodegradation and enzymatic modification of lignin. Int. J. Biochem. Mol. Biol. 1: 36–50.

Diaz-Ruiz, R., M. Rigoulet and A. Devin. 2011. The Warburg and Crabtree effects: On the origin of cancer cell energy metabolism and of yeast glucose repression. Biochim. Biophys. Acta—Bioenerg. 1807: 568–576.

Fackler, K. and M. Schwanninger. 2010. Polysaccharide degradation and lignin modification during brown rot of spruce wood: A polarised Fourier transform near infrared study. J. Near Infrared Spectrosc. 18: 403–416.

Galkin, S., T. Vares, M. Kalsi and A. Hatakka. 1998. Production of organic acids by different white-rot fungi as detected using capillary zone electrophoresis. Biotechnol. Tech. 12: 267–271.

Gomes, D., A.C. Rodrigues, L. Domingues and M. Gama. 2015. Cellulase recycling in biorefineries—is it possible? Appl. Microbiol. Biotechnol. 99: 4131–4143.

Gómez-Toribio, V., A.B. García-Martín, M.J. Martínez, Á.T. Martínez and F. Guillén. 2009. Induction of extracellular hydroxyl radical production by white-rot fungi through quinone redox cycling. Appl. Environ. Microbiol. 75: 3944–3953.

Guillén, F., Á.T. Martinez and M.J. Martínez. 1990. Production of hydrogen peroxide by aryl-alcohol oxidase from the ligninolytic fungus *Pleurotus eryngii*. Appl. Microbiol. Biotechnol. 32: 465–469.

Hagman, A., T. Säll and J. Piškur. 2014. Analysis of the yeast short-term Crabtree effect and its origin. FEBS J. 281: 4805–4814.

Hasunuma, T., F. Okazaki, N. Okai, K.Y. Hara, J. Ishii and A. Kondo. 2013. A review of enzymes and microbes for lignocellulosic biorefinery and the possibility of their application to consolidated bioprocessing technology. Bioresour. Technol. 135: 513–522.

Hirabayashi, S., J. Wang, H. Kawagishi and H. Hirai. 2015. Improving xylitol production through recombinant expression of xylose reductase in the white-rot fungus *Phanerochaete sordida* YK-624. J. Biosci. Bioeng. 120: 6–8.

Hirai, H., R. Kondo and K. Sakai. 1994. Screening of lignin-degrading fungi and their ligninolytic enzyme activities during biological bleaching of kraft pulp. Mokuzai Gakkaishi 40: 980–986.

Horisawa, S., H. Ando, O. Ariga and Y. Sakuma. 2015. Direct ethanol production from cellulosic materials by consolidated biological processing using the wood rot fungus *Schizophyllum commune*. Bioresour. Technol. 197: 37–41.

Ishchuk, O.P., A.Y. Voronovsky, O.V. Stasyk, G.Z. Gayda, M.V. Gonchar, C.A. Abbas et al. 2008. Overexpression of pyruvate decarboxylase in the yeast *Hansenula polymorpha* results in increased ethanol yield in high-temperature fermentation of xylose. FEMS Yeast Res. 8: 1164–1174.

Kamei, I., Y. Hirota and S. Meguro. 2012a. Integrated delignification and simultaneous saccharification and fermentation of hard wood by a white-rot fungus, *Phlebia* sp. MG-60. Bioresour. Technol. 126: 137–141.

Kamei, I., Y. Hirota, T. Mori, H. Hirai, S. Meguro and R. Kondo. 2012b. Direct ethanol production from cellulosic materials by the hypersaline-tolerant white-rot fungus *Phlebia* sp. MG-60. Bioresour. Technol. 112: 137–142.

Kamm, B. and M. Kamm. 2004. Principles of biorefineries. Appl. Microbiol. Biotechnol. 64: 137–145.

Kenealy, W.R. and D.M. Dietrich. 2004. Growth and fermentation responses of *Phanerochaete chrysosporium* to O2 limitation. Enzyme Microb. Technol. 34: 490–498.

Kersten, P.J. and T.K. Kirk. 1987. Involvement of a new enzyme, glyoxal oxidase, in extracellular H_2O_2 production by *Phanerochaete chrysosporium*. J. Bacteriol. 169: 2195–2201.

Mäkinen, M.A., N. Risulainen, H. Mattila and T.K. Lundell. 2018. Transcription of lignocellulose-decomposition associated genes, enzyme activities and production of ethanol upon bioconversion of waste substrate by *Phlebia radiata.* Appl. Microbiol. Biotechnol. 102: 5657–5672.

Mattila, H., J. Kuuskeri and T.K. Lundell. 2017. Single-step, single-organism bioethanol production and bioconversion of lignocellulose waste materials by phlebioid fungal species. Bioresour. Technol. 225: 254–261.

Mori, T., H. Kako, T. Sumiya, H. Kawagishi and H. Hirai. 2016a. Direct lactic acid production from beech wood by transgenic white-rot fungus *Phanerochaete sordida* YK-624. J. Biotechnol. 239: 83–89.

Mori, T., G. Koyama, H. Kawagishi and H. Hirai. 2016b. Effects of homologous expression of 1,4-benzoquinone reductase and homogentisate 1,2-dioxygenase genes on wood decay in hyper-lignin-degrading fungus *Phanerochaete sordida* YK-624. Curr. Microbiol. 73: 512–518.

Mori, T., Y. Nagai, H. Kawagishi and H. Hirai. 2018. Functional characterization of the manganese transporter *smf2* homologue gene, *PsMnt*, of *Phanerochaete sordida* YK-624 via homologous overexpression. FEMS Microbiol. Lett. 365: 16–21.

Mori, T., O. Kondo, H. Kawagishi and H. Hirai. 2019. Effects of glucose concentration on ethanol fermentation of white-rot fungus *Phanerochaete sordida* YK-624 under aerobic conditions. Curr. Microbiol. 76: 263–269.

Moysés, D.N., V.C.B. Reis, J.R.M. de Almeida, L.M.P. de Moraes and F.A.G. Torres. 2016. Xylose fermentation by *Saccharomyces cerevisiae*: Challenges and prospects. Int. J. Mol. Sci. 17: 207.

Naik, S.N., V.V. Goud, R.K. Rout and A.K. Dalai. 2010. Production of first and second generation biofuels: A comprehensive review. Renew. Sust. Energ. Rev. 14: 578–597.

Nutt, A., A. Salumets, G. Henriksson, V. Sild and G. Johansson. 1997. Conversion of O_2 species by cellobiose dehydrogenase (cellobiose oxidase) and glucose oxidase—A comparison. Biotechnol. Lett. 19: 379–383.

Okamoto, K., K. Imashiro, Y. Akizawa, A. Onimura, M. Yoneda, Y. Nitta et al. 2010. Production of ethanol by the white-rot basidiomycetes *Peniophora cinerea* and *Trametes suaveolens*. Biotechnol. Lett. 32: 909–913.

Okamoto, K., Y. Nitta, N. Maekawa and H. Yanase. 2011. Direct ethanol production from starch, wheat bran and rice straw by the white rot fungus *Trametes hirsuta.* Enzyme Microb. Technol. 48: 273–277.

Okamoto, K., A. Uchii, R. Kanawaku and H. Yanase. 2014. Bioconversion of xylose, hexoses and biomass to ethanol by a new isolate of the white rot basidiomycete *Trametes versicolor*. SpringerPlus 3: 121.

Rasmussen, M.L., P. Shrestha, S.K. Khanal, A.L. Pometto and J. (Hans) van Leeuwen. 2010. Sequential saccharification of corn fiber and ethanol production by the brown rot fungus *Gloeophyllum trabeum.* Bioresour. Technol. 101: 3526–3533.

Reddi, A.R., L.T. Jensen and V.C. Culotta. 2010. Manganese homeostasis in *Saccharomyces cerevisiae*. Chem. Rev. 109: 4722–4732.

Ruiz-Dueñas, F.J. and Á.T. Martínez. 2009. Microbial degradation of lignin: How a bulky recalcitrant polymer is efficiently recycled in nature and how we can take advantage of this. Microb. Biotechnol. 2: 164–177.

Schworm, H.D., K.B. Jedele, E. Holinski, K. Hörtnagel, G. Rudolph, K.P. Boergen et al. 1996. Purification of acetaldehyde dehydrogenase and alcohol dehydrogenases from *Thermoanaerobacter ethanolicus* 39E characterization of the secondary-alcohol dehydrogenase (2° Adh) as bifunctional alcohol dehydrogenase–acetyl-CoA reductive thioesterase. Clin. Genet. 50: 393–397.

Sindhu, R., P. Binod and A. Pandey. 2016. Biological pretreatment of lignocellulosic biomass—An overview. Bioresour. Technol. 199: 76–82.

Sugiura, T., T. Mori, I. Kamei, H. Hirai, H. Kawagishi and R. Kondo. 2012. Improvement of ligninolytic properties in the hyper lignin-degrading fungus *Phanerochaete sordida* YK-624 using a novel gene promoter. FEMS Microbiol. Lett. 331: 81–88.

Suzuki, M.R., C.G. Hunt, C.J. Houtman, Z.D. Dalebroux and K.E. Hammel. 2006. Fungal hydroquinones contribute to brown rot of wood. Environ. Microbiol. 8: 2214–2223.

Takao, S. 1965. Organic acid production by Basidiomycetes, I. Screening of acid-producing strains. Appl. Microbiol. 13: 732–737.

Voelker, S.L., B. Lachenbruch, F.C. Meinzer and S.H. Strauss. 2011. Reduced wood stiffness and strength, and altered stem form, in young antisense 4CL transgenic poplars with reduced lignin contents. New Phytol. 189: 1096–1109.

Wang, J., S. Hirabayashi, T. Mori, H. Kawagishi and H. Hirai. 2016a. Improvement of ethanol production by recombinant expression of pyruvate decarboxylase in the white-rot fungus *Phanerochaete sordida* YK-624. J. Biosci. Bioeng. 122: 17–21.

Wang, J., T. Suzuki, H. Dohra, S. Takigami, H. Kako, A. Soga et al. 2016b. Analysis of ethanol fermentation mechanism of ethanol producing white-rot fungus *Phlebia* sp. MG-60 by RNA-seq. BMC Genomics 17: 616.

Wang, Y., Y. Tashiro and K. Sonomoto. 2015. Fermentative production of lactic acid from renewable materials: Recent achievements, prospects, and limits. J. Biosci. Bioeng. 119: 10–18.

Yamada, Y., J. Wang, H. Kawagishi and H. Hirai. 2014. Improvement of ligninolytic properties by recombinant expression of glyoxal oxidase gene in hyper lignin-degrading fungus *Phanerochaete sordida* YK-624. Biosci. Biotechnol. Biochem. 78: 2128–33.

Yamagishi, K., T. Kimura, S. Oita, T. Sugiura and H. Hirai. 2007. Transformation by complementation of a uracil auxotroph of the hyper lignin-degrading basidiomycete *Phanerochaete sordida* YK-624. Appl. Microbiol. Biotechnol. 76: 1079–1091.

Chapter 3

Energy Conversion of Biomass using Supercritical and Subcritical Fluid

Takeshi Sako[1] and *Idzumi Okajima*[2,3,*]

1. Introduction

To divert from a fossil fuel-dependent society and to reduce CO_2 emissions, it is important to develop alternative energy resources to petroleum oil. Biomass and waste biomass are promising renewable energy sources and investigations regarding its utilization are being enhanced. Using them as biofuels or biochemicals could be effective in reducing fossil fuel consumption and CO_2 emissions. However, there are several problems when using biomass and waste biomass as fuel resources, which are: (a) their high moisture and low calorific value per weight, (b) the large seasonal variations of their composition, and (c) their easy decay leading to a difficult long-term storage, especially with waste biomass. Moreover, it is difficult to directly burn biomass using existing energy production and incineration systems for fuels with high calorific value and low water content. Hence, it is important to develop production techniques for easily available and highly efficient fuels from biomass.

There are several reports of the properties and applications, including several techniques on biomass utilization by supercritical and subcritical water (Arai et al. 2002, Okajima and Sako 2011, Sako 2008, SCEJ 2008, Yoshida 2005). The water characteristics will be introduced below.

[1] Graduate School of Science and Technology, Shizuoka University, 3-5-1 Johoku, Naka-ku, Hamamatsu, Shizuoka 432-8561, Japan.

[2] Applied Chemistry and Biochemical Engineering, Graduate School of Integrated Science and Technology, Shizuoka University, 3-5-1 Johoku, Naka-ku, Hamamatsu, Shizuoka 432-8561, Japan.

[3] Green Energy Research Division, Research Institute of Green Science and Technology, Shizuoka University, 3-5-1 Johoku, Naka-ku, Hamamatsu, Shizuoka 432-8561, Japan.

* Corresponding author: okajima.izumi@shizuoka.ac.jp

2. Characteristics of High-temperature and High-pressure Water

Water is a harmless, environment-friendly, and highly polar general solvent. Its characteristics significantly depend on temperature and pressure. Figure 3.1 shows the phase diagram of water. Here, three types of water are used in utilizing biomass as fuel or energy resource. Supercritical water is located above the critical temperature of 374°C and critical pressure of 22.1 MPa. Subcritical water is located below 374°C and above the saturated vapor pressure at this temperature. High-pressure superheated steam is either located above 374°C and below 22.1 MPa or less than 374°C and its saturated vapor pressure.

The relationship among the temperature, pressure, and density of water (Franck 1987) is shown in Fig. 3.2. Liquid water exists on the right-hand side of the Critical Point (CP) above the convex curve, while steam exists on the left-hand side. Liquid water at point A has a density of approximately 1 g/cm^3, while that of the steam at point B is 0.000017 g/cm^3 at 20°C. Subcritical water at 150–350°C and 0.48–16.5 MPa is widely used for biomass treatment. For instance, the density of subcritical water at 150°C and 0.48 MPa is 0.799 g/cm^3 and that at 350°C and 16.5 MPa is 0.575 g/cm^3. These densities become 0.322 g/cm^3 at the CP, which is the critical density. Above the CP, supercritical water exists. For instance, the density of supercritical water at 450°C and 25 MPa at point F is 0.109 g/cm^3. As previously stated, water density drastically varies at a high-temperature condition. The density of subcritical and supercritical water is located between those of liquid water and steam.

The dielectric constant is among the significantly variable properties of water, which represents the polarity of a solvent. Figure 3.3 shows its temperature and pressure dependency (JSME 1999). The dielectric constant of liquid water at 20°C and 0.1 MPa is

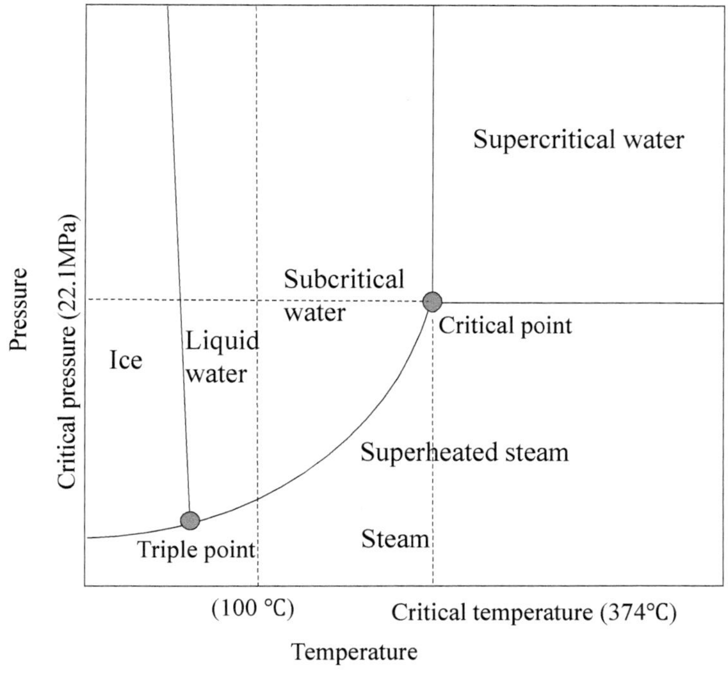

Fig. 3.1: Phase diagram of water.

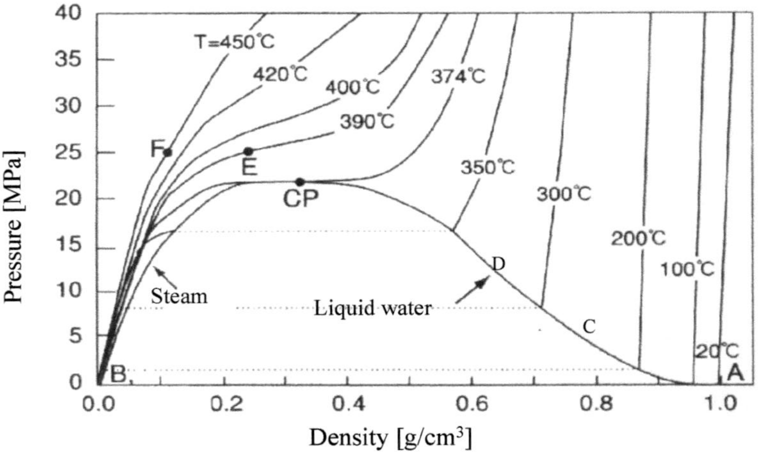

Fig. 3.2: Relation between temperature, pressure, and density of water (Franck 1987, Copyright (1987) Springer Nature).

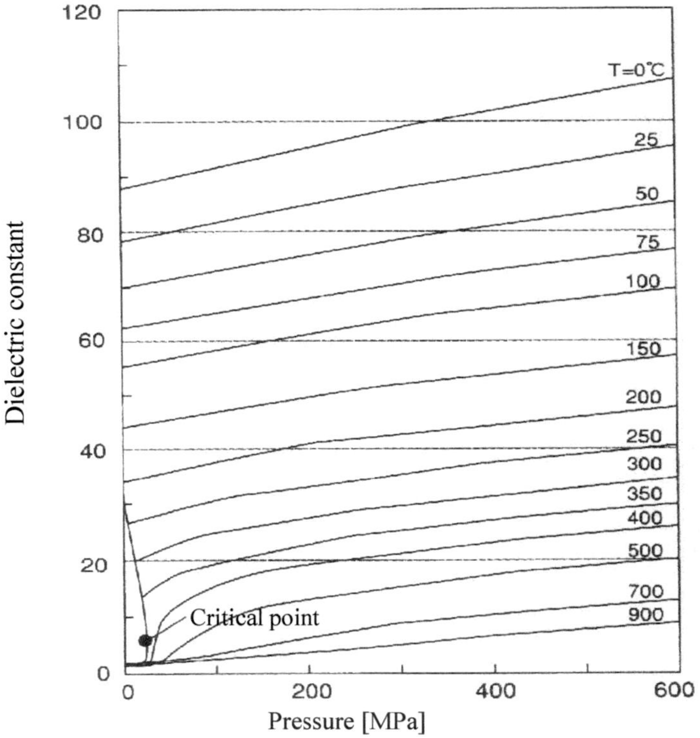

Fig. 3.3: Dependence of dielectric constant of water on temperature and pressure (JSME 1999, Copyright (1999) The Japan Society of Mechanical Engineers).

approximately 80. At this condition, water is a soluble ionic and polar compound, such as salts, alcohols, and acids, and a not dissolved nonpolar compound, such as benzene and toluene. However, the dielectric constant of subcritical water at 250°C and 4.0 MPa is approximately 27 and that of supercritical water at 500°C and 30 MPa is 1.7, which is nearly similar to that of liquid propane at an ambient condition. Hence, water becomes subcritical or supercritical at a high-temperature and high-pressure condition and the solubility of polar and ionic compounds decreases, whereas that of nonpolar organic compounds increases.

The ionic product (K_w) is also a severely variable property of water, which indicates the degree of dissociation of water into protons and hydroxide ions.

$$H_2O = H^+ + OH^- \quad K_W = [H^+][OH^-] \tag{I}$$

The influence of K_w on temperature and pressure (JSME 1999) is shown in Fig. 3.4. The K_w value of saturated liquid water is 10–14 mol^2/kg^2 at 25°C. At 250°C and 4.0 MPa saturated vapor pressure, K_w increases with temperature to a maximum value of approximately 10–11 mol^2/kg^2, which is approximately 1,000 times greater than that at an ambient condition, i.e., there are approximately 32 times more hydroxide ions and protons in subcritical water at 250°C than in liquid water at 25°C. Therefore, these ions act as acid or alkali catalysts and hydrolysis is promoted by subcritical water. Kw rapidly decreases above 300°C and its value becomes 10–15.64 at the CP, which is approximately 2% of that at 25°C. In supercritical water condition, the small K_w decreases with temperature and increases with pressure.

With its varying properties, the water behavior varies from common water at an ambient condition to subcritical water with high reactivity due to high concentration of hydroxide ions and protons to supercritical water with high decomposition.

With these, water in the high-temperature and high-pressure region has unique properties, which is suitable for the degradation of organic compounds, such as waste biomass and plastics.

Next the conversions of biomass to thermal energy, fuel gas, and solid fuel with high-temperature and high-pressure water are explained.

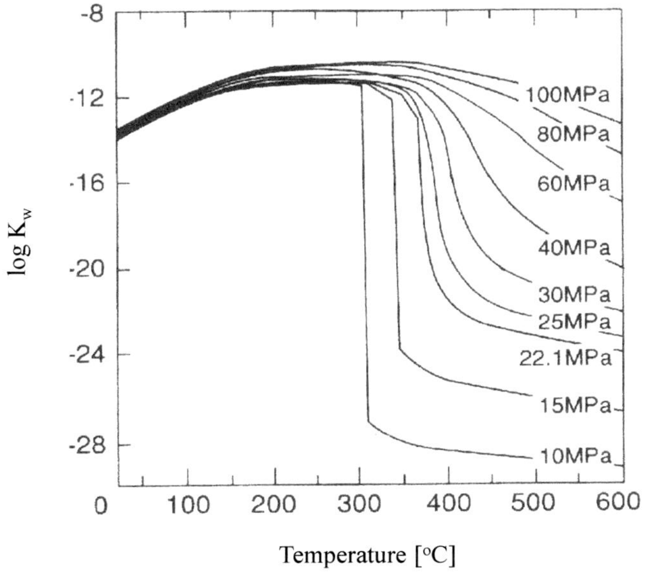

Fig. 3.4: Dependence of ionic product of water on temperature and pressure (JSME 1999, Copyright (1999) The Japan Society of Mechanical Engineers).

3. Treatment and Thermal Energy Recovery from Biomass by High-pressure Superheated Steam Oxidation

There have been reports of waste biomass treatment techniques using supercritical water oxidation (Benjamin and Savage 2005, Cocero 2000, Crain et al. 1993, Ding et al. 1998, Goto et al. 1999, Sako et al. 1997). The main elements of carbon, hydrogen, and nitrogen in waste biomass are decomposed into CO_2, water, and N_2 gas by supercritical water oxidation. However, these treatment conditions require high temperature and pressure above the critical point of water. On the other hand, the oxidation condition could be changed from supercritical water to high-pressure superheated steam without a significant decrease on the degradation power of organic compounds, which is strongly dependent on the temperature than the pressure in the high-temperature and high-pressure region.

A flow-type reactor was used to decompose the sewage sludge, oil cake, and agricultural wastes, such as thinned surplus orange and cut end of welsh onion, with high-pressure superheated steam (Okajima and Sako 2009, Torii et al. 2013). Figure 3.5 shows the flow diagram of the flow-type plant with a maximum waste input of 30 kg/h. This plant has two reactors. The first reactor has an agitator and is connected with pumps. It has an inner diameter of 37 cm, height of 77 cm, and inner volume of 82 L. The second reactor part has three reactors, each with the inner diameter of 10 cm, height of 64 cm, and inner volume of 5 L. It is also packed with MnO_2 particles as the catalyst. Waste biomass slurry and compressed air as oxidants are continuously and alternately supplied into the first reactor by two slurry pumps, a compressor and a gas buster. The supplied waste biomass slurry is decomposed to CO_2, water, low-molecular-weight organics and ammonia. The low-molecular-weight organics and ammonia are decomposed to CO_2, water, and N_2 gas by the MnO_2 catalyst in the second reactors. After treatment, clean hot drainage and exhaust gas are discharged from the plant, with thermal recovery carried out.

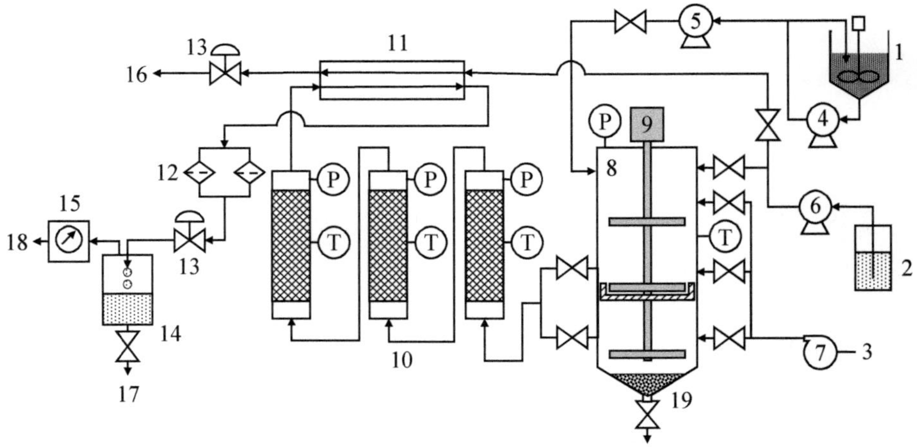

1. Sewage sludge	6. High-pressure pump	11. Double piped-type heat exchanger	16. Hot water
2. Pure water	7. Compressor	12. Filter	17. Clean water
3. Air	8. First-stage reactor	13. Back pressure regulator	18. Clean gas
4. Squeeze pump	9. Agitator	14. Gas-liquid separator	19. Inorganic residue
5. High-pressure slurry pump	10. Second-stage reactors with MnO_2 catalyst	15. Gas meter	

Fig. 3.5: Flow-type two-stage high-pressure superheated steam oxidation bench plant (Torii et al. 2013, Copyright (2013) The Japan Institute of Energy).

Figure 3.6 shows the relationship among the decomposition ratio of carbon in sewage sludge, concentration of water-soluble organic carbon in drainage, and elapsed time for sewage sludge oxidation by high-pressure superheated steam. The concentration of water-soluble organic carbon, which is produced to decompose sewage sludge, is represented by the Total Organic Carbon (TOC) concentration in this section. The TOC concentration is maintained to less than 10 mg/L and a decomposition ratio of approximately 100% is maintained for 100 hours.

Figure 3.7 shows the relationship between the concentration of inorganic nitrogen in the drainage and the elapsed time. The total concentration of nitrogen in the nitric acid, nitrous acid and ammonia was maintained to below 40 mg/L during the experiment.

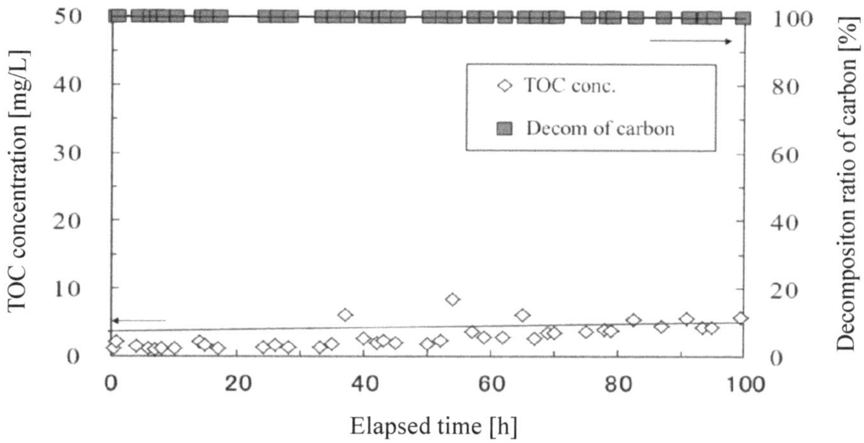

Fig. 3.6: Decomposition ratio of carbon and TOC concentration in two-stage high-pressure superheated steam oxidation (First-stage reactor: 413°C, 8 MPa and 12–17 min, second-stage reactor: 384°C, 8 MPa, 0.2–0.4 min, MnO_2 catalyst, feed rate: 3.0–4.0 kg/h, 1.3 of oxygen supply ratio).

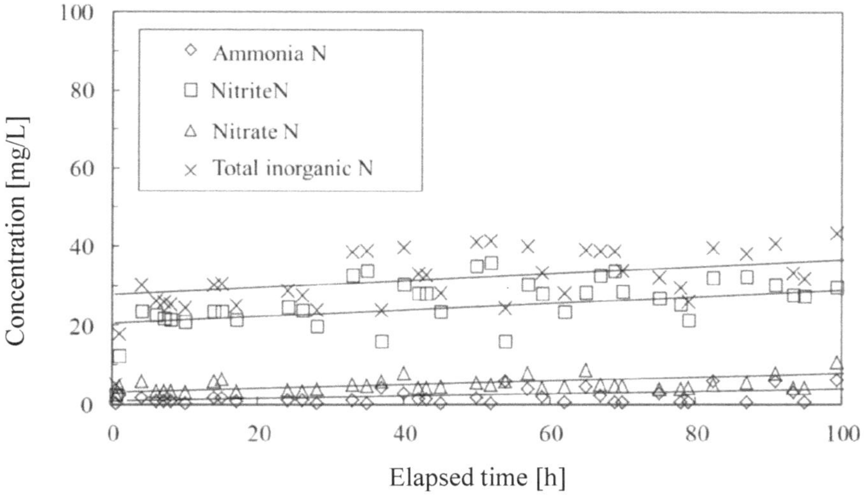

Fig. 3.7: Concentration profile of nitrogen-containing products in two-stage high-pressure superheated steam oxidation (First-stage reactor: 413°C, 8 MPa and 12–17 min, second-stage reactor: 384°C, 8 MPa, 0.2–0.4 min, MnO_2 catalyst, feed rate: 3.0–4.0 kg/h, 1.3 of oxygen supply ratio).

Figure 3.8 shows the sewage sludge before treatment and the drainage water after treatment by high-pressure superheated steam oxidation. As shown in Figs. 3.6 to 3.8, the drainage water has a low impurity concentration and a clear appearance during the high-pressure superheated steam oxidation.

The heat recovery from the high-temperature and high-pressure drainage to the cooling water was conducted using a double-pipe-type heat exchanger attached to the pilot-scale plant. The schematic of the heat exchanger is shown in Fig. 3.9. The temperatures of the drainage and cooling water and the heat recovery efficiency in the heat exchanger are listed in Table 3.1, where Q_1 is the amount of heat absorbed by the cooling water [MJ/h] and Q_2 is the amount of heat released from the drainage water [MJ/h]. The heat recovery efficiency Q_1/Q_2 reached 85–86%. Hence, it will be possible to establish high-pressure superheated steam treatment of waste biomass combined with thermal usage or power generation.

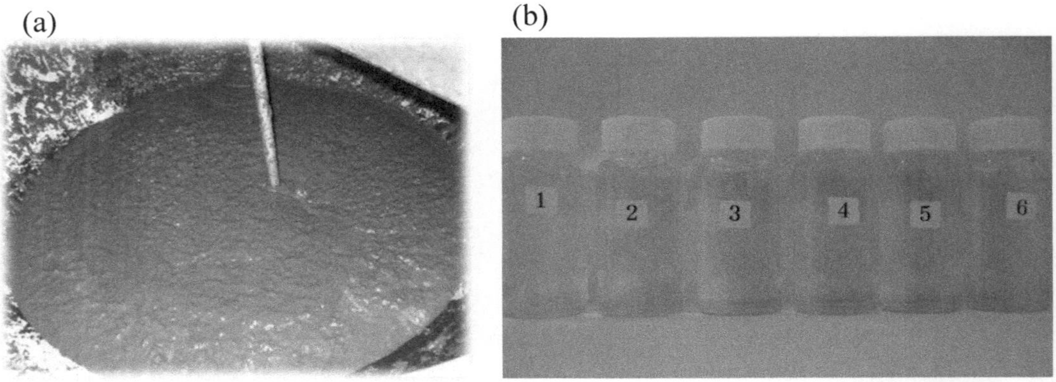

Fig. 3.8: Images of (a) sewage sludge before treatment and (b) drainage water after sewage sludge treatment by high-pressure superheated steam oxidation.

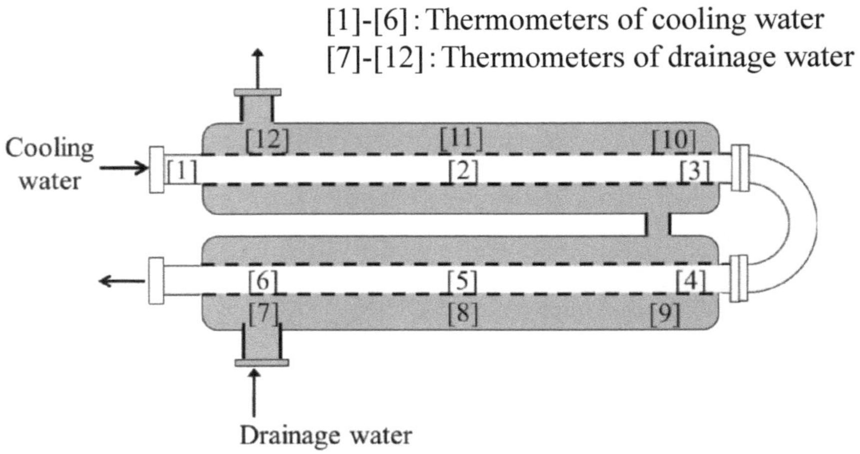

Fig. 3.9: Schematic of heat exchanger attached to pilot-scale plant (Torii et al. 2013, Copyright (2013) The Japan Institute of Energy).

Table 3.1: Temperatures of inlet and outlet of cooling water and drainage water in heat exchanger and heat recovery (Torii et al. 2013).

Cooling water				Drainage water			Heat recovery (Q_1/Q_2) [%]
Flow rate of cooling water [kg/h]	Temperature [°C]		Q_1 [MJ/h]	Temperature [°C]		Q_2 [MJ/h]	
	Inlet	Outlet		Inlet	Outlet		
19	24.3	288.5	22.2	352.4	37.1	31.8	86.4
20	24.1	274.6	21.9	351.7	33.5	32.2	84.9
22	23.9	257.4	22.3	345.7	27.9	27.2	86.4

4. Gasification and Hydrogen Production from Biomass

The gasification of organic waste biomass using subcritical and supercritical water has attracted much attention in the production of hydrogen and methane (Matsumura et al. 2005, Minowa et al. 1998, Schmieder et al. 2000, Xu et al. 1996, Yoshida et al. 2004). Here gasification using high-pressure superheated steam is described (Okajima et al. 2007, Okajima and Sako 2011). Gasifying agents have several advantages such as (a) water, which is an environment-friendly solvent, and is used as a gasifying agent, (b) drying process before gasification is unnecessary, (c) waste biomass can be decomposed to hydrogen and methane, and (d) hydrogen yield is increased because water decomposes to hydrogen. Pig excrement was selected as the main target material in this report.

Reactions (II)–(V) are mainly undergone in waste biomass gasification with high-pressure superheated steam:

$$[C, H, N, O] + H_2O \rightarrow CO + H_2 + H_2O + NH_3 + N_2 \qquad \text{(II)}$$

$$CO + H_2O \rightleftarrows CO_2 + H_2 \qquad \text{(III)}$$

$$CO + 3H_2 \rightleftarrows CH_4 + H_2O \qquad \text{(IV)}$$

$$N_2 + 3H_2 \rightleftarrows 2NH_3 \qquad \text{(V)}$$

where [C, H, N, O] are the carbon, hydrogen, nitrogen, and oxygen atoms in the waste biomass. Reactions (II) and (III), i.e., water-gas reaction and water-gas-shift reaction, respectively, must be promoted, whereas Reactions (IV) and (V) must be inhibited to produce high-yield hydrogen gas. These situations are substantialized at high-temperature and low-pressure conditions by considering chemical equilibrium.

Figure 3.10 shows the batch-type experimental apparatus. The inner volume of the batch reactor is approximately 9 cm^3. Waste biomass sample and alkali catalyst were fed into the reactor. The reactor was then sealed and heated to the reaction temperature in a sand bath. After the reaction, the reactor was cooled as quickly as possible. The volume of the collected gas product was measured and its composition was analyzed.

Figure 3.11 shows the catalytic effects of the gas productivity from the pig excrement using high-pressure superheated steam at 700°C, 10 MPa, and 20-minute reaction time. The Decomposition Efficiency (DE) of biomass is the ratio of carbon atoms in the gaseous products to the introduced carbon atoms including the sample. A 20 wt.% catalyst was added to the organic component in the sample. When a NaOH or KOH catalyst was used,

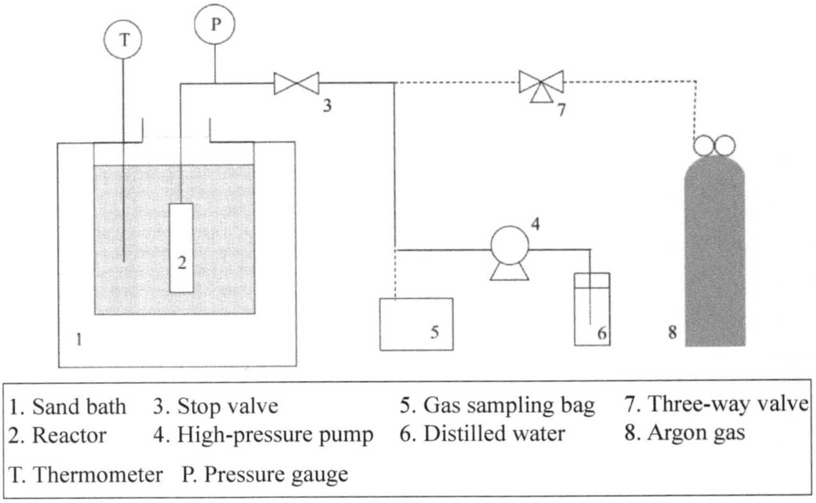

Fig. 3.10: Batch-type experimental setup used for gasification (Okajima et al. 2007, Copyright (2007) The Society of Chemical Engineers, Japan).

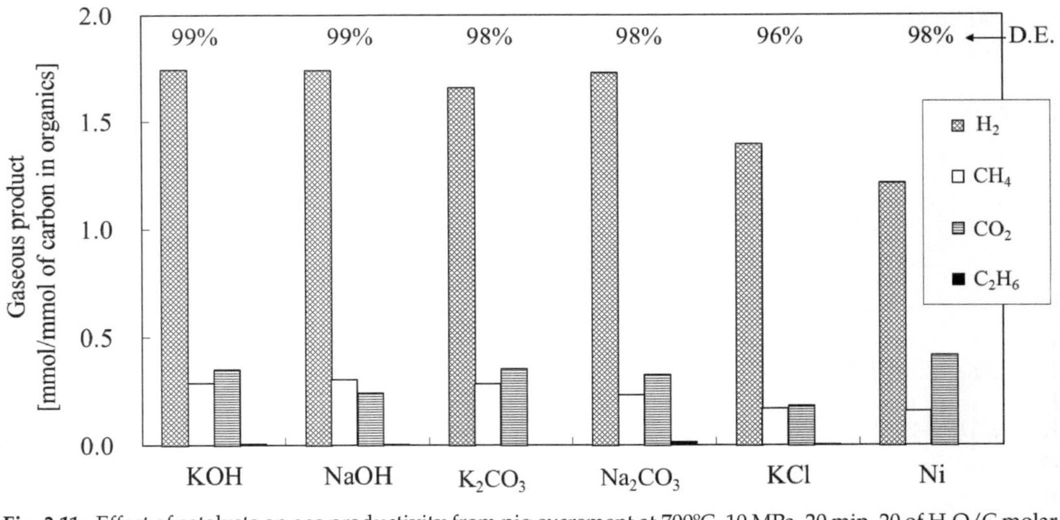

Fig. 3.11: Effect of catalysts on gas productivity from pig excrement at 700°C, 10 MPa, 20 min, 20 of H_2O/C molar ratio and 20 wt% catalyst.

the pig excrement was degraded to 99% and hydrogen gas of 1.75 mmol/mmol carbon was produced in the organic components of the biomass. Using a Na_2CO_3 or K_2CO_3 catalyst, the decomposition efficiency and amount of produced hydrogen gas were slightly less than those with NaOH or KOH catalyst. On the other hand, with a nickel catalyst, the amount of produced hydrogen gas was only 1.22 mmol/mmol carbon in the organics, which is the lowest yield among the catalysts used in the experiments. This is due to the decreased catalytic activity due to the sulfur and chlorine in the pig excrement.

Figure 3.12 shows the effects of the reaction pressure on the gas productivity of the pig excrement at 700°C for 30 minutes using a KOH catalyst. Here, water is high-

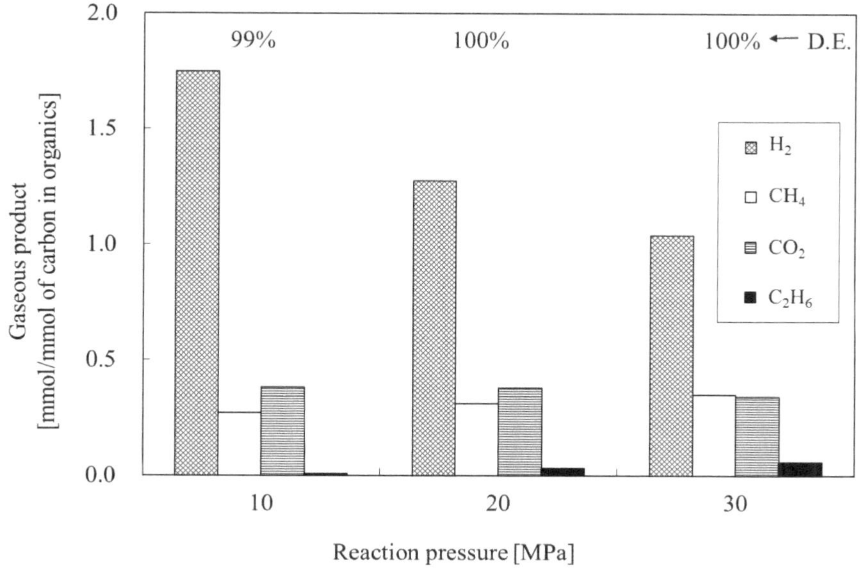

Fig. 3.12: Pressure dependence of gas productivity from pig excrement at 700°C, 30 min, 20 of H_2O/C molar ratio and 20 wt% KOH (Okajima and Sako 2011).

pressure superheated steam at 10–20 MPa and is supercritical water at 30 MPa. The decomposition efficiency was more than 99% at these pressure conditions. As the reaction pressure increased from 10 MPa to 30 MPa, the amount of hydrogen gas produced and the hydrogen gas ratio in the gaseous product decreased from 1.75 mmol/mmol carbon and 73% to 1.04 mmol/mmol carbon and 58%. Moreover, the ratio of methane in the gaseous product increased from 11% at 10 MPa to 20% at 30 MPa because of the promotion of Reaction (IV) with the increase of pressure. High-pressure superheated steam has a large decomposition power similar with supercritical water Moreover, it is more effective for hydrogen production than supercritical water because Reactions (IV) and (V), which consumed hydrogen were enhanced at higher pressure; thereby, strongly inhibiting in high-pressure superheated steam. Furthermore, the amount of hydrogen production was not significantly changed between supercritical water and high-pressure superheated steam. Therefore, high-pressure superheated steam at a lower pressure is more effective for hydrogen gas production than supercritical water at a higher pressure.

Figure 3.13 shows the gas productivity of four types of waste biomass at 700°C, 10 MPa, and 30-minute reaction time with a KOH catalyst. The bean curd refuse (Okara) was generated from the manufacturing process of tofu and soy milk. The strained lees of the distilled spirits are the dehydrated cakes of the residue in distilled spirits production. A large part of which is disposed and a small portion is utilized as a fertilizer, feedstuff, etc. All samples were decomposed more than 96% and 1.99 mmol, 1.95 mmol, 1.93 mmol, and 1.80 mmol of hydrogen gas was produced from the barley husk of distilled spirits, bean curd refuse, strained lees of distilled spirits, and garbage for 1 mmol of carbon atoms in the biomass, respectively. The hydrogen atoms in the biomass per mmol of carbon atoms in the sample were 1.48 mmol, 1.85 mmol, 1.34 mmol, and 1.45 mmol in the barley husk of distilled spirits, bean curd refuse, strained lees of distilled spirits, and garbage, respectively, as listed in Table 3.2. The produced hydrogen gas is larger than the hydrogen

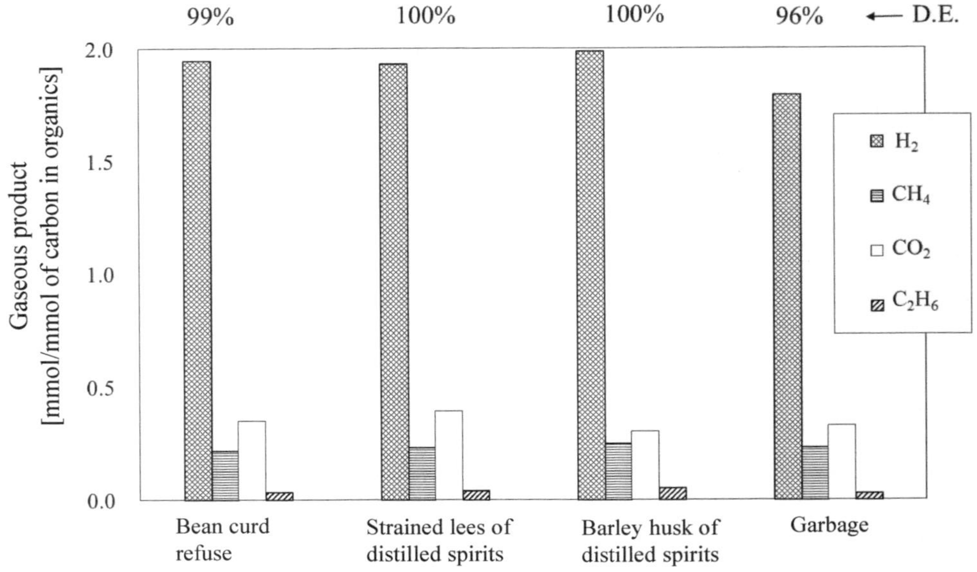

Fig. 3.13: Gas productivity from several types of food wastes at 700°C, 30 min, 20 of H_2O/C molar ratio and 20 wt% KOH (Okajima et al. 2007, Copyright (2007) The Society of Chemical Engineers, Japan).

Table 3.2: Composition of waste biomass.

	Elemental composition organics and inorganics [wt%]					**Amount of hydrogen in the sample [mmol/mmol-C]**
	C	H	N	O	**Other**	
Bean curd refuse	51.8	8.0	4.8	31.5	3.9	1.85
Strained lees of distilled spirits	46.4	5.2	7.1	33.0	8.3	1.34
Barley husk of distilled spirits	49.5	6.1	4.5	34.5	5.4	1.48
Garbage	42.9	5.2	3.4	37.4	10.8	1.45
Pig excrement	43.6	6.3	3.2	39.5	7.4	1.73

atoms including that in the biomass. From these results, it is possible that hydrogen gas is also produced from the decomposition of water itself.

Figure 3.14 shows the influence of the reaction pressure on the yields of ammonia and water-soluble TOC from the bean curd refuse at 700°C for 30 minutes with a KOH catalyst. The ammonia yield increased considerably from 43 to 84% as the reaction pressure increased from 10 MPa to 30 MPa. This was because Reaction (V) of the ammonia production was promoted at a high-pressure condition. In contrast, the yield of TOC decreased gradually with the reaction pressure. These results indicate that high-pressure superheated steam is more preferable than supercritical water for the promotion of hydrogen gas and suppression of ammonia production. In addition, the effect of the reaction temperature on the ammonia yield was investigated at 10 MPa for 30 minutes with a KOH catalyst for the gasification of bean curd refuse. The ammonia yield decreased from 83% at 500°C to 43% at 700°C. Ammonia production was not completely suppressed because hydrogen gas was

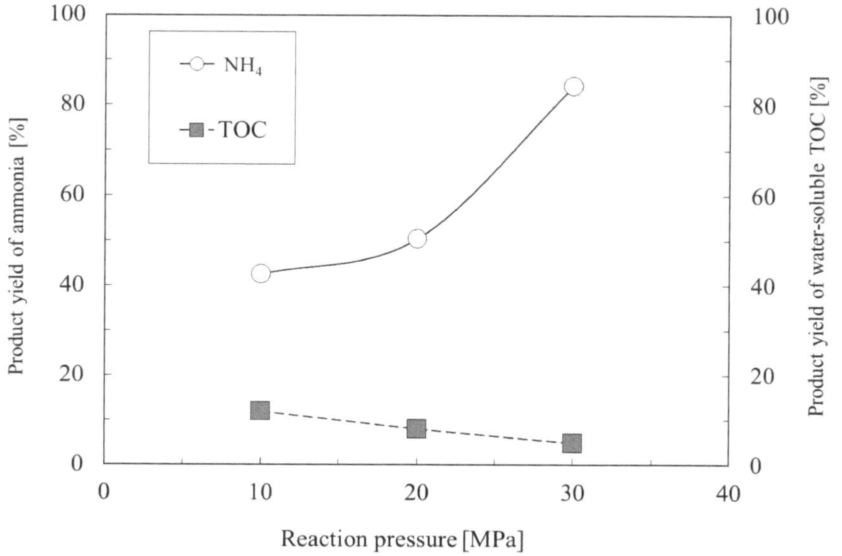

Fig. 3.14: Effect of reaction pressure on byproduct yields from bean cured refuse at 700°C, 20 min, 20 of H_2O/C molar ratio and 20 wt% catalyst (Okajima et al. 2007, Copyright (2007) The Society of Chemical Engineers, Japan).

still produced in the reaction field. Therefore, the amount of ammonia, which is a toxic impurity, significantly decreased at a lower pressure and higher temperature.

From the experimental results, the amount of hydrogen gas from several waste biomasses decreased in the following order: barley husk of distilled spirits > bean curd refuse > strained lees of distilled spirits > garbage > pig excrement. Pig excrement and garbage contains higher chloride than the other biomasses. Therefore, the deactivation of the catalyst was relatively rapid due to the hydrogen product.

5. Production of Powdery Fuel from a Mixture of Waste Biomass and Plastics by Subcritical Water

Mixtures of waste plastics and biomass, such as garbage and agricultural waste, are usually incinerated due to their difficult correct treatment. Recently, biotreatments, such as methane fermentation and composting, have been effectively demonstrated. However, these treatments need separation of waste plastics from the waste biomass and long treatment time.

A composite powdery fuel can be produced from the mixtures of waste biomass and plastics using subcritical water (Okajima et al. 2012a, 2012b). Figure 3.15 shows the principle of this method. The calorific value of the produced composite powdery fuel is 25–30 MJ/kg, which almost similar to that of coal.

Figure 3.16 shows the bench-scale plant of the composite powdery fuel production with subcritical water. The inner volume of this batch-type reactor is approximately 300 L. The horizontal agitator inside the reactor rotates forward and backward in regular intervals over the treatment. 10 kg of biomass, 5 kg of sawdust, and 2 kg of waste plastics were fed into the reactor. The mixture of wastes was agitated in the reactor at a given treatment temperature and pressure for a given treatment time. After treatment, the product was recovered from the reactor, dried, and analyzed.

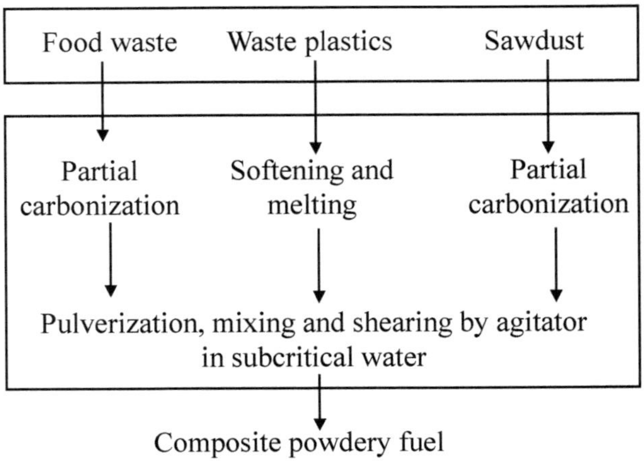

Fig. 3.15: Principle of production of composite powdery fuel with subcritical water (Okajima et al. 2012a, Copyright (2012) The Japan Institute of Energy).

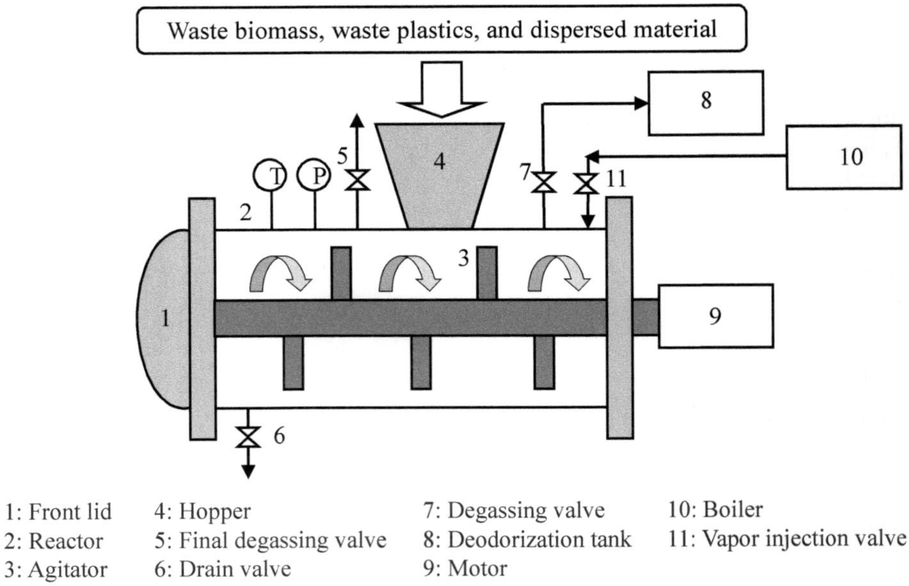

Fig. 3.16: Bench-scale plant of production of composite powdery fuel with subcritical water (Okajima et al. 2012a, Copyright (2012) The Japan Institute of Energy).

A waste mixture of garbage and unsold prepared foods in plastic containers discharged from a food-processing factory or a supermarket was treated. Figure 3.17 shows the dependence of the reaction temperature on the heating value of the powdery product. The heating value increased with the treatment temperature, which was almost equal to that of coal above 200°C. This is because the plastic containers were well dispersed through the powdery product at this condition, which was above the melting point of plastic containers. Figure 3.18 shows images of the powdery product at 205°C and 1.7 MPa, which is black in

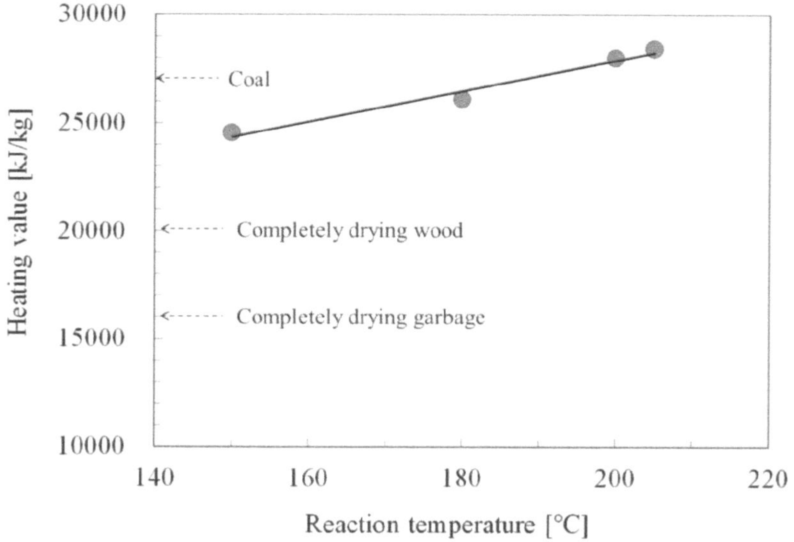

Fig. 3.17: Temperature-dependence of heating value of composite powdery fuel produced by subcritical water at 0.5–1.7 MPa and 30 min (Okajima et al. 2012b, Copyright (2013) JASCO).

Fig. 3.18: Images of composite powdery fuel produced by subcritical water (Okajima et al. 2012b, Copyright (2012) JASCO).

color and less than 5 mm in diameter. This product is composed of a two-layer structure with plastic particles surrounding the partially carbonated biomass. It also contains a very low amount of chlorine, which could be converted to toxic hydrogen chloride and dioxins during the biomass combustion using boilers or stoves. The effect of treatment time on the heating value of the produced powdery fuel was investigated at 205°C and 1.7 MPa, where the heating value was constant at approximately 28 MJ/kg for 10–30 minutes of treatment time.

A combustion test of the produced powdery fuel at 205°C and 1.7 MPa was conducted using a biomass boiler. The powdery fuel was burnt steadily and concentrations of hydrogen chloride, dioxins, sulfur oxide, nitrogen oxide, and dust, including the exhaust gas, were achieved below regulatory values.

Table 3.3: Input and output energies of production of composite fuel using subcritical water (Waste mixture: 6.75 ton, Water content: 50 wt%) (Okajima et al. 2012a).

	Step	**Amount of energy [MJ]**
Input energy	Crushing of waste mixture	137
	Preheating of reactor (25 → 180°C)	1550
	Reheating of reactor (150 → 200°C) and heating of waste mixture (25 → 200°C)	4239
	Compensation of heat release from reactor	28
	Agitation in reactor	128
	Drying (evaporation of water and heating of composite fuel from 50 to 100°C)	11543
Output energy	Heating value of composite fuel (25 MJ/kg × 3037.5 kg)	75938
Energy efficiency	(Output energy)/(Input energy)	4.31

The Energy Profit Ratio (EPR) of the powdery fuel production process was analyzed to estimate its energy efficiency (Goto et al. 2009). EPR is defined as

EPR [–] = Heating value of the produced fuel/Input energy for the fuel production (VI)

The calculation results from the large-scale reactor are listed in Table 3.3, which is a hundred times larger than the bench-scale plant for the experiment. The EPR value of the fuel production process with the large-scale plant was 4.3. Therefore, it could be a new treatment technique for the mixture of waste biomass and plastics.

6. Summary

High-temperature and high-pressure water, such as high-pressure superheated steam, supercritical water and subcritical water, are high-performance environment-friendly solvents for biomass treatment. It is expected establishing a combination of biomass treatment and energy production techniques will contribute to environmental conservation and energy security.

References

Arai, Y., T. Sako and Y. Takebayashi. 2002. Supercritical Fluid—Molecular Interactions, Physical Properties, and New Applications. Springer-Verlag, Germany.

Benjamin, K.M. and P.E. Savage. 2005. Supercritical water oxidation of methylamine. Ind. Eng. Chem. Res. 44: 5318–5324.

Cocero, M.J., E. Alonso, R. Torio, D. Vallelado and F. Fdz-Polanco. 2000. Supercritical water oxidation in a pilot plant of nitrogenous compounds: 2-propanol mixtures in the temperature range 500–750°C. Ind. Eng. Chem. Res. 39: 3707–3716.

Crain, N., S. Tebbal, L. Li and E. Gloyna. 1993. Kinetics and reaction pathways of pyridine oxidation in supercritical water. Ind. Eng. Chem. Res. 32: 2259–2268.

Ding, Z.Y., L. Li, D. Wade and E.F. Gloyna. 1998. Supercritical water oxidation of NH3 over a MnO2/CeO2 catalyst. Ind. Eng. Chem. Res. 37: 1707–1716.

Franck, E.U. 1987. Supercritical studies. p. 338. *In*: M. Bellisent-Funel and G. Neilson (eds.). The Physics and Chemistry of Aqueous Ionic Solutions. Springer Netherlands, Dordrecht, Holland.

Goto, M., T. Nada, A. Kodama and T. Hirose. 1999. Kinetic analysis for destruction of municipal sewage sludge and alcohol distillery wastewater by supercritical water oxidation. Ind. Eng. Chem. Res. 38: 1863–1865.

Goto, T., O. Amano, I. Okajima and T. Sako. 2009. Evaluation method of production techniques based on concept of energy profit ratio. Electronic Journal of the Mottainai Society 3: 1–6 (in Japanese).

[JSME] The Japan Society of Mechanical Engineers. 1999. JSME Steam Tables. Maruzen, Tokyo.

Matsumura, Y., T. Minowa, B. Potic, S.R.A. Kersten, W. Prins, W.P.M. van Swaaij et al. 2005. Biomass gasification in near- and supercritical water: states and prospects. Biomass Bioenerg. 29: 269–292.

Minowa, T., F. Zhen and T. Ogi. 1998. Cellulose decomposition in hot-compressed water with alkali or nickel catalyst. J. Supercrit. Fluids 13: 253–259.

Okajima, I., D. Shimoyama and T. Sako. 2007. Gasification and hydrogen production from food wastes using high pressure superheated steam in the presence of alkali catalyst. J. Chem. Eng. Jpn. 40: 356–364.

Okajima, I. and T. Sako. 2009. 2-Step subcritical water oxidation and its application to energy recovery from waste biomass. Eng. Mater. 57: 42–45.

Okajima, I. and T. Sako. 2011. Gasification and hydrogen production from waste biomass with high pressure superheated steam. Adv. Mater. Res. 222: 305–308.

Okajima, I., N. Kawase, T. Shimizu, K. Tamura, N. Kanno, S. Kawajiri et al. 2012a. Production of composite fuel with high heating value from waste mixture of food and plastic using subcritical water. J. Jpn. Inst. Energy 91: 998–1006 (in Japanese).

Okajima, I., Y. Fujimori, H. Niwa and T. Sako. 2012b. Production of composite fuel with high calorific value from refractory waste mixtures with subcritical water. Jasco Report Special Issue 11: 22–25 (in Japanese).

Saka, S. 2011. The latest trend of next generation technology development for lignocellulosic biomass 2. J. Jpn. Inst. Energy 90: 17–23.

Sako, T., T. Sugeta, K. Otake, M. Sato, M. Tsugumi, T. Hiaki et al. 1997. Decomposition of dioxins in fly ash with supercritical water oxidation. J. Chem. Eng. Jpn. 30: 744–747.

Sako, T., T. Sugeta, K. Otake, C. Kamizawa, M. Okano, A. Negishi et al. 1999. Dechlorination of PCBs with supercritical water hydrolysis. J. Chem. Eng. Jpn. 32: 830–832.

Sako, T. 2008. Development and Application of Supercritical Fluid Technology. CMC Books, Tokyo (in Japanese).

Sako, T. and I. Okajima. 2010. Advanced technologies of energy production from waste biomass using subcritical water. Rev. High Pressure Sci. Tech. 20: 26–32.

Sako, T. and I. Okajima. 2011. Development of utilization techniques of waste biomass with subcritical water. J. Chem. Eng. Japan 75: 772–775.

Sato, O., K. Arai and M. Shirai. 2006. Hydrolysis of poly (ethylene terephthalate) and poly (ethylene 2,6-naphthalene dicarboxylate) using water at high temperature. Catal. Today 111: 297–301.

[SCEJ] The Society of Chemical Engineers Japan. 2008. Introduction to Supercritical Fluid, Maruzen, Tokyo (in Japanese).

Schmieder, H., J. Abeln, N. Boukis, E. Dinjus, A. Kruse, M. Kluth et al. 2000. Hydrothermal gasification of biomass and organic wastes. J. Supercritical Fluids 17: 145–153.

Torii, S., I. Okajima and T. Sako. 2013. Treatment and energy recovery of sewage sludge by high-pressure superheated steam oxidation. J. Jpn. Inst. Energy 92: 945–956 (in Japanese).

Xu, X., Y. Matsumura, J. Setnberg and M.J. Antal Jr. 1996. Carbon-catalyzed gasification of organic feedstocks in supercritical water. Ind. Eng. Chem. Res. 35: 2522–2530.

Yoshida, H. 2005. Waste Refinery by Sub-critical Water. CMC Books, Tokyo (in Japanese).

Yoshida, T., Y. Oshima and Y. Matsumura. 2004. Gasification of biomass model compounds and real biomass in supercritical water. Biomass Bioenerg. 26: 71–78.

Chapter 4

Determination of Diffusion Coefficients in Supercritical Fluid

Chang Yi Kong,[1,2,*] *Toshitaka Funazukuri*[3] and *Koji Miyake*[2]

1. Introduction

In recent years, there is an increasing public awareness of the health, environment and safety hazards associated with the use of conventional organic solvents in the possible solvent contamination of the final products. The high cost due to using harmful organic solvents and the increasingly stringent environmental regulations together with the needs of the new pharmaceutical, food and healthcare industries for ultrapure and high added value products have pointed out the requirements for the development of innovative, advanced and environmentally friendly technologies for the industrial processing. Supercritical Fluid (SF) is an adjustable solvent, which is reached at a temperature and pressure above its critical point and can provide an excellent alternative to the use of the harmful chemical solvents. It can diffuse through solids like a gas, and dissolve materials like a liquid. In addition, close to the critical point, small changes in pressure or temperature result in large changes in density or viscosity, allowing many properties to be tuned. SFs have many advantages, such as good selectivity, facile, near-zero surface tension, low kinetic viscosity, and above the critical temperature there is no phase transition. For this reason, the application of SFs (Byrappa et al. 2008, Knez et al. 2014) has been receiving considerable attention from the scientific community with the aim to develop new technologies to substitute the traditional techniques based on the harmful organic solvents. Table 4.1 presents several fluids investigated as SFs and their critical values.

[1] Green Energy Research Division, Research Institute of Green Science and Technology, Shizuoka University, Hamamatsu 432-8561, Japan.

[2] Department of Applied Chemistry and Biochemical Engineering, Faculty of Engineering, Shizuoka University, Hamamatsu 432-8561, Japan.

[3] Department of Applied Chemistry, Faculty of Science and Engineering, Chuo University, Tokyo 112-8551, Japan.
Emails: tfunazo@kc.chuo-u.ac.jp; miyake.koji@shizuoka.ac.jp

* Corresponding author: kong.changyi@shizuoka.ac.jp

Table 4.1: Critical properties of some commonly used pure solvents (Poling et al. 2001).

Substance (formula)	Critical temperature T_c (K)	Critical pressure P_c (MPa)	Critical density ρ_c (kg/m3)
Carbon dioxide	304.12	7.37	468
Chlorotrifluoromethane ($CClF_3$)	301.84	3.87	579
2,3-Dimethylbutane (C_6H_{14})	499.90	3.13	241
Ethane (C_2H_6)	305.32	4.87	207
Ethanol (C_2H_6O)	513.92	6.15	276
Ethylene (C_2H_4)	282.34	5.04	214
n-Hexane (C_6H_{14})	507.60	3.03	234
Methane (CH_4)	190.56	4.60	163
Propane (C_3H_8)	369.83	4.25	220
2-Propanol (C_3H_8O)	508.30	4.76	273
Sulfur hexafluoride (SF_6)	318.72	3.76	736
Trifluoromethane (CHF_3)	298.97	4.84	526
Water	647.14	22.06	322

Among them, as will be found throughout the present work, the most commonly employed SF is CO_2. It is relatively easy to reach its critical point (CO_2, critical temperature T_c = 304.12 K and critical pressure P_c = 7.37 MPa). Unlike many organic solvents, supercritical CO_2 (SCO_2) is non-flammable. It is inert, non-toxic, widely available and relatively cheap with high purity. The potential application of SCO_2 (Fan et al. 2019, Boock et al. 2019, Kuang et al. 2017, Padrela et al. 2018, Simonov et al. 2018, Zhang et al. 2019) is promising in many fields since it is also a clean, versatile and outstanding solvent, and has high diffusivity combined with the continuously adjustable solvent power/selectivity. In addition, in SCO_2 processes, near-environmental temperatures are employed, thus minimizing heat requirement, thermal and oxidative damage to unstable/labile and easily oxidized products. Another advantage of SCO_2 is that CO_2 is gaseous under standard conditions, which makes products recovery simple, its complete elimination at the end of the process and finally provides solvent free products.

For rational design and efficient operation of the newly proposed SF processes, it is important to fully understand the basic properties in SFs. Diffusion phenomenon is often occurring in a variety of chemical, biological and environmental processes involving mass transfer. Understanding diffusion coefficients of molecules in fluids is needed for fundamental problems. Especially it can play an important role as the rate determining step. Herein, this chapter discusses the techniques for determining the diffusion coefficients in SFs. The measurement methods are described, their advantages and disadvantages are shown and the theories of frequent experimental techniques are discussed in detail. In addition, a brief table of the sources is given for the available diffusion coefficient data as reported in the literature with an emphasis on the results measured in SCO_2.

2. Experimental Methods

As is well known, any mass transfer may include both convection and molecular diffusion (Cussler 2005). This diffusion refers to the net transport of molecular solute in a phase in the absence of convection. In fact, the diffusion can result from pressure gradient, temperature gradients, external force fields and concentration gradients. Only the last type is considered in this chapter. Additionally, in chromatography, the diffusion coefficient refers to the diffusion of a solute in a relatively or very dilute solution. When a small amount of solute component is introduced into a large amount of a fluid, the diffusion occurs by random molecular motion that leads to complete mixing. In this case the diffusion coefficient of the solute in the fluid can be described by the following equation. Note that the Eq. (1) can hold only when the concentration gradient is small.

$$D_{12} = -\left(\frac{dc}{dz}\right)^{-1} J \tag{1}$$

Here, J is the mole flux per unit area of the solute in the fluid, c is the concentration of the diffusion solute in the fluid, z is the spatial length, and D_{12} is often referred to as the infinite dilution binary diffusion coefficient of the solute in the fluid which is constant and independent of c. The dimension of D_{12} is the area per unit time, which is often shown as m^2/s in the SI-unit. The negative sign on the right side of the equation indicates that the concentration of the solute in the fluid decreases with increasing z. Note that the net diffusive transport continues until the concentration gradient disappears. After that, the diffusion process (molecular motion) obviously does not stop but no longer involves net solute transport.

There are many techniques that can be applied to measure diffusion coefficients in SFs. The experimental methods for measuring diffusion coefficients in SFs had been reported (Funazukuri et al. 2004a, Guiochon and Tarafder 2011, Higashi et al. 2001, Kong et al. 2012, Liong et al. 1991a, Medina 2012, Roth 1991, Suárez et al. 1998). The experimental methods may be classified into many types such as chromatographic techniques (the Taylor dispersion, the input-output response, chromatographic impulse response methods), solid dissolution, photon correlation spectroscopy, nuclear magnetic resonance and radioactive tracer methods. Here an overview of the experimental techniques is presented, the advantages and disadvantages are described and some examples of their application are shown.

2.1 Chromatographic Techniques

The chromatographic techniques can be employed to measure the infinite dilution diffusion coefficients.

2.1.1 Taylor Dispersion

The well-known method of measuring diffusion is Taylor dispersion (Aris 1956, Taylor 1953), illustrated schematically in Fig. 4.1.

The apparatus is normally composed of pump, oven, injector, diffusion column, detector, back pressure regulator and flow meter. The high pressure injector is often employed for pulse introducing a solute species into a fully developed laminar flow of SF in a diffusion

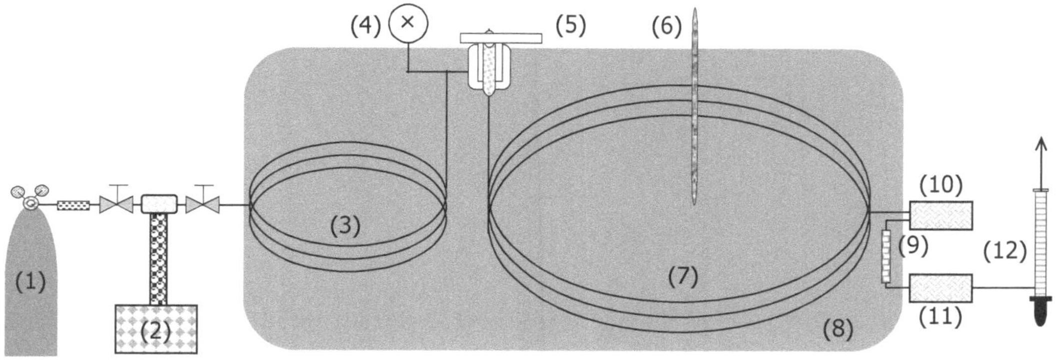

Fig. 4.1: Schematic diagram of the apparatus. (1) solvent; (2) pump; (3) preheating column; (4) pressure gauge; (5) injector; (6) thermometer; (7) diffusion column; (8) constant temperature oven; (9) packed column; (10) detector; (11) back pressure regulator; (12) flow meter.

column. The response of the temporal change in the solute concentration is measured at the end of the column by the non-destructive detectors (such as UV absorption) and the destructive (such as FID).

This method is suitable for determining the D_{12} data for non- or lower polarity compounds and lower molecular weight compounds in SFs. And it has the potential to give accurate results in the region of higher density of SFs. The concentration profile found in this apparatus is that for the decay of the injected pulse. The pulse is dispersed due to the combined action of convection in the axial direction and diffusion both in the axial and radial directions.

The fundamental equations for the tracer concentration $C(r, x, t)$ in the diffusion column together with the initial and boundary conditions are given as Eqs. (2) to (4).

$$\frac{\partial C}{\partial t} = D_{12}\left(\frac{1}{r}\frac{\partial}{\partial r}\left(r\frac{\partial C}{\partial r}\right) + \frac{\partial^2 C}{\partial x^2}\right) - 2u_0\left(1-\left(\frac{r}{R}\right)^2\right)\frac{\partial C}{\partial x} \tag{2}$$

$$C = 0 \quad \text{at} \quad x = +\infty$$

$$C = \frac{m}{\pi R^2}\delta(x) \quad \text{at } t = 0 \tag{3}$$

$$\frac{\partial C}{\partial r} = 0 \quad \text{at } r = 0 \text{ and } R \tag{4}$$

where t is time, r and x are radial and axial variables, u_0 is average fluid velocity, m is injected amount of tracer species, R is column radius, respectively.

In fact, average concentration $c(x, t)$ over the cross section of a column is often employed as Eq. (5).

$$c(x,t) = \frac{2}{R^2}\int_0^R C(r,x,t)rdr \tag{5}$$

The analytical solution $c(t)$ for the cross-sectional average concentration of the tracer at column exit ($x = L$) is given by Eq. (6) with Eq. (7).

$$c(t) = \frac{m/(\pi R^2)}{(4\pi At)^{1/2}} \exp\left\{-\frac{(L-u_0 t)^2}{4At}\right\} \tag{6}$$

where

$$A = D_{12} + \frac{R^2 u_0^2}{48 D_{12}} \tag{7}$$

where L is distance between injection point and detection point, A is the dispersion coefficient, D_{12} is the diffusion coefficient of the solute in the fluid, respectively.

The applicability of the methods can be examined in terms of root-mean-square error ε (defined by Eq. (8)) between measured and calculated concentrations, $c_{exp}(t)$ and $c(t)$, respectively.

$$\varepsilon = \left\{\sum_{i=1}^{N}\left[c(t_i) - c_{exp}(t_i)\right]^2 \Big/ \sum_{i=1}^{N}\left[c_{exp}(t_i)\right]^2\right\}^{1/2} \tag{8}$$

where t_1 and t_N correspond to the times at the frontal and latter 10% peak heights of the experimentally measured concentration $c_{exp}(t)$, respectively. The fitting is considered good when $\varepsilon < 1\%$, and acceptable when $\varepsilon < 2 \sim 3\%$ (Kong et al. 2004). The dispersion coefficient A varies inversely with D_{12}, as shown in Eq. (7). A widely spread pulse means a large A and a small D_{12}. A very sharp pulse indicates small dispersion and hence fast diffusion.

2.1.2 Input-output Response

The experimental apparatus in the input-out response technique (Dahmen et al. 1990a, 1990b, Funazukuri et al. 2002b, Umezawa and Nagashima 1992, Wells et al. 1992) is almost the same as that presented in the Taylor dispersion method except for two detectors and the two columns. The input-out response technique is a modified Taylor dispersion method in which a diffusion column is placed as a pre-column before the diffusion column.

The cross-sectional average concentration $c(t)$ of injected tracer species is described in Eq. (9) when a solvent moves in laminar flow in a circular-cross sectional tubing.

$$A\frac{\partial^2 c}{\partial x^2} - u_0 \frac{\partial c}{\partial x} = \frac{\partial c}{\partial t} \tag{9}$$

where

$$A = D_{12} + \frac{u_0^2 R^2}{48 D_{12}} \tag{10}$$

The initial and the boundary conditions for the input-output response technique become:

$$c = 0 \quad \text{at} \quad t = 0 \tag{11}$$

$$c = c^{I}_{exp}(t) \quad \text{at} \quad x = 0 \tag{12}$$

$$c = \text{finite} \quad \text{at} \quad x = +\infty \tag{13}$$

The solution for Eq. (9) can be obtained by the convolution integral in Eq. (14) with Eq. (15).

$$c(t) = \int_0^t c_{\text{exp}}^{\text{I}}(\xi) f(t-\xi) d\xi \tag{14}$$

where

$$f(t) = \frac{L}{2(\pi A t^3)^{1/2}} \exp\left(\frac{-(L-u_0 t)^2}{4At}\right) \tag{15}$$

Note that $f(t)$ is the Laplace inversion of the transfer function $F(s)$ defined by Eq. (16).

$$F(s) = \frac{\int_0^\infty c_{\text{exp}}^{\text{II}}(t) e^{-st} dt}{\int_0^\infty c_{\text{exp}}^{\text{I}}(t) e^{-st} dt} = \exp\left(\frac{u_0 L(1-(1+4sA/u_0^2)^{1/2})}{2A}\right) \tag{16}$$

where $F(s)$ is equal to the ratio of the Laplace transformation of the output signal $c_{\text{exp}}^{\text{II}}(t)$ to that of the input signal $c_{\text{exp}}^{\text{I}}(t)$, where $c_{\text{exp}}^{\text{I}}(t)$ and $c_{\text{exp}}^{\text{II}}(t)$ are the concentrations measured actually at two different downstream points at the diffusion column distance of L. The output signal $c(t)$ (calculated concentration) can be obtained by Eqs. (14) and (15) with assumed D_{12} or A value. The D_{12} value is determined as to minimize the fitting error defined as Eq. (8). The time changes of the solute concentration ($c_{\text{exp}}^{\text{I}}(t)$ and $c_{\text{exp}}^{\text{II}}(t)$) were measured at the inlet and the outlet of the diffusion column.

Funazukuri et al. (2002b) has employed a polymer coated diffusion column as a separation column set before the diffusion column to separate a solute and a solute-dissolving solvent due to difference in retention factor.

2.1.3 Chromatographic Impulse Response

As well known, the Taylor dispersion method often uses an uncoated diffusion column to measure diffusion coefficients. When a coated diffusion column is employed, it refers to as the Chromatographic Impulse Response (CIR) method (Funazukuri et al. 2000c, Kong et al. 2004). In fact, the CIR method is widely used to measure the solubility, retention factor, partial molar volume and diffusion coefficient for gas, liquid and supercritical fluid chromatographies (Kong et al. 2011, 2014, Funazukuri et al. 2016b, Lauer et al. 1983, Olesik et al. 1987, Roth 1992, 1998).

The apparatus used in the CIR method is almost identical to that used in the Taylor dispersion method shown in Fig. 4.1. However, there is only one difference, namely the use of the coated diffusion column in the CIR method.

The fundamental equations for the tracer concentration $C(r, x, t)$ in the coated diffusion column together with the initial and boundary conditions are given as Eqs. (17) to (19). The analytical solution $c(t)$ for the cross-sectional average concentration (shown in Eq. (5)) of the tracer at column exit is given by Eq. (20) with Eq. (21) (Kong et al. 2004).

$$\frac{\partial C}{\partial t} = D_{12}\left(\frac{1}{r}\frac{\partial}{\partial r}\left(r\frac{\partial C}{\partial r}\right) + \frac{\partial^2 C}{\partial x^2}\right) - 2u_0\left(1-\left(\frac{r}{R}\right)^2\right)\frac{\partial C}{\partial x} \tag{17}$$

$$C = 0 \qquad \text{at} \qquad x = \pm\infty$$

$$C = \frac{m}{\pi R^2} \frac{\delta(x)}{1+k} \qquad \text{at} \quad t = 0 \tag{18}$$

$$\begin{cases} \dfrac{\partial C}{\partial r} = 0 & \text{at} \quad r = 0 \\[2ex] k \dfrac{\partial C}{\partial t} = -\dfrac{2D_{12}}{R} \dfrac{\partial C}{\partial r} & \text{at} \quad r = R \end{cases} \tag{19}$$

$$c(t) = \frac{m}{\pi R^2} \frac{1}{(1+k)(4\pi At)^{1/2}} \exp\left[-\frac{\left(L - u_0 t/(1+k)\right)^2}{4At} \right] \tag{20}$$

where

$$A = \frac{D_{12}}{1+k} + \frac{1+6k+11k^2}{(1+k)^3} \frac{R^2 u_0^2}{48 D_{12}} \tag{21}$$

where k is the retention factor defined as the ratio of a solute amount in the polymer phase to that in the fluid phase.

Lai and Tan (1995) have also employed the CIR method with the moment technique to measure the infinite dilution binary diffusion coefficients of ethyl acetate, toluene, phenol and caffeine in SCO_2. Wakao and Kaguei (1982) have noted that the moment analysis technique can not only evaluate diffusion coefficient correctly, but also cannot directly confirm the validity of the model. However, these shortcomings (Kong et al. 2004) can be overcome by using the curve fitting analysis technique.

One of the advantages of this method is that it can measure multi components. Funazukuri et al. (2007) have succeeded in measuring simultaneously for multiple peaks with imperfect separation by injecting a mixture of cis-4,7,10,13,16,19-docosahexaenoic acid (DHA), cis-5,8,11,14,17-eicosapentaenoic acid (EPA) and α-linolenic acid in SCO_2.

2.2 The Other Techniques

2.2.1 Solid Dissolution

The Solid Dissolution (SD) technique (Higashi et al. 1998, 2005, Knaff and Schlünder 1987) employs a diffusion cell in which the solid solute compressed into a tablet structure. When the SF passes through the diffusion cell, the solute dissolves and diffuses into SF. This method is not necessary to analyze the phase of SF, relatively simple and possible to measure the diffusion coefficients of solid compounds in SF. The diffusion coefficients can be determined by the Eq. (22).

$$D_{12} = \frac{1}{2y} \frac{\rho_2}{\rho_1} \frac{h_E^2 - h_S^2}{\Delta t} \tag{22}$$

where ρ_1 and ρ_2 are the molar densities of the supercritical fluid and the solute, h_S and h_E are the distances from the entrance of diffusion cell to the surface of the compressed solute at the start and end of the experiment, y is the solubility of the solute in the SF, and Δt is the diffusion time. Note that h_S was measured by using a micrometer. h_E was calculated from ρ_2 and the weights of the diffusion cell at the beginning and at the end of the experiments, because the packed mass density changed during depressurization after the experiment. It is assumed in the SD method that the diffusion process is pseudo steady, namely, the rate of solute removed from the cell entrance is larger than the molar change of solute component in the cell space. Further, assuming that the solute concentrations at the diffusion cell entrance and at the solid solute surface are zero and saturated solubility, respectively and that the concentration profile is linear along the axis of diffusion (Knaff and Schlünder 1987, Higashi et al. 1998, 2005).

One disadvantage of this method is that it is limited to solid solutes. Furthermore, it cannot measure the diffusion coefficient for the solid solutes which are liquid under the studied conditions. In addition, the concentration value at saturation is necessary in this technique. This means that it needs solubility data.

2.2.2 Photon Correlation Spectroscopy

Photon correlation spectroscopy with a high pressure scattering cell, which is also referred to as quasielastic light scattering, involves analysis of the spectrum of light scattered by optical heterogeneity (Saad and Gulari 1984). These non-homogeneities are due to the different density regions in fluid. The measurements are based on determining the width of the central or Rayleigh peak in the spectrum of scattered light. It also needs to determine the sample amounts in the cylinders before and after loading. Note that both temperature and concentration fluctuation of solute in SF can contribute to the width at the microscopic scale. When the diffusion coefficient is high under the studied conditions, the scattering elements move further faster, resulting in large fluctuations. If the diffusion coefficient is small, the opposite often occurs. The photon correlation spectroscopy method can measure actually the diffusion coefficients at finite composition range in low density region or in the region near the critical point. However, it is not suitable for measuring diffusion under the condition of infinite dilution. The disadvantages of this method are that the signal-to-noise ratio must be high and the refractive indices of the components in the system must differ by at least 5 to 10%.

2.2.3 Nuclear Magnetic Resonance

The Nuclear Magnetic Resonance (NMR) spectroscopy method is based on the measurements of relaxation time which provide information on the molecular interactions in solution. The measurements are performed with the NMR spectrometer equipped with high pressure NMR diffusion probe and the sample cell, and the diffusion coefficients are determined by using the NMR spin-echo Bessel function analysis technique (Lamb et al. 1989, Xu et al. 2003). The NMR measurements resolve different compounds spectroscopically in a solution based on the different diffusivities of each sample, depending on molecular size and shape. Therefore, this method is suitable for complex samples. Diffusion NMR can be employed to resolve otherwise intractable spectra of mixtures or it can be run to measure microscopic structures. In fact, NMR diffusion measurements provide a way to study the molecular shape, size, aggregate, solution shell, polymerization and chemical bond. For example, research on ligand protein binding is a significant application of NMR diffusion measurements in the drug discovery industry.

While NMR has long been employed for diffusion study, it became truly practical only after recent developments in hardware, pulse sequences and data analysis. Now, this method is employed to measure the diffusion coefficients over a wide range of temperatures (–50 to 350°C) and pressures (0 to 500 MPa). The disadvantage of this method is the need for expensive NMR experimental equipment. Furthermore, it may be difficult to measure the diffusion in solutions containing molecules with similar structures, as the overlap of proton spectra are likely to be very severe.

2.2.4 Radioactive Tracer

The Radioactive Tracer (RT) method (Takahashi and Hongo 1982, Takahashi and Iwasaki 1966) is known to use the radioactive isotopes. It can emit easily detectable radiation and can be used equally well with non-radial counterparts for most purposes. The Fick's second law of diffusion with various boundary conditions is often used in the RT method. It can be classified into four broad types of diaphragm cell, sliding cell, capillary and shear boundary. This depends on the actual experimental layout. Note that the use of radioactive isotopes is a major disadvantage of this method, and another one is the issue of waste disposal after use.

3. Available Data

It is known that the diffusion coefficients are typically on the order of about 10^{-6} to 10^{-4} m^2/s in gases and 10^{-10} to 10^{-9} m^2/s in liquids (Poling et al. 2001). Then, in SF the diffusion coefficient is on the order of about 10^{-9} to 10^{-6} m^2/s. Here, the available diffusion data sources together with the measurement techniques employed and the information of solute, temperature and pressure investigated are shown in Table 4.2. There are 80 reports and 203 compounds, which are collected as much as possible, but the data available only in graphical form is not presented. Note that all diffusion sources shown in Table 4.2 were studied in SCO_2. Most of the diffusion sources presented in Table 4.2 were measured by the Taylor dispersion method and the CIR method. And Catchpole and King (1994), Knaff and Schlünder (1987), Higashi et al. (1998, 2005) have used the SD method; Takahashi and Hongo (1982) have employed the RT method; Lamb et al. (1989), Xu et al. (2003) have used the NMR method; Dahmen et al. (1990a, 1990b), Funazukuri et al. (2002), Umezawa and Nagashima (1992), Wells et al. (1992) have employed the input-output method. As shown in Table 4.2, the diffusion coefficients for non- or low polarity compounds have been measured by the Taylor dispersion method. While those for compounds having polarity or with relatively high molecular weight have been successfully measured by the CIR method. Especially, trinervonin (Kong et al. 2005) studied by the CIR method has a molecular weight of about 1138. This is the highest molecular weight of any compounds ever measured for the diffusion coefficients in SCO_2. The diffusion data measured so far are not enough as yet, especially, in the area near the critical point, in the high temperature (> 100°C) and high pressure (> 40 MPa) region, for the compounds having strong polarity (acid) or having high molecular weight (> 1000).

Table 4.2: Data source, method, solute, temperature and pressure reported on diffusion coefficients in SCO_2.

Data reference (method)	Solute (molecular weight)	*T* (K)	*P* (MPa)	No. of data
Swaid and Schneider (1979) (Taylor dispersion)	Benzene, n-propylbenzene, 1,3,5-trimethylbenzene (78.1–120.2)	308.15–328.15	8.0–16.0	45
Feist and Schneider (1982) (Taylor dispersion)	Benzene, phenol, naphthalene, caffeine (78.1–194.2)	308–333	8.0–16.0	23
Takahashi and Hongo (1982) (RT)	Ethylene (28.1)	298.2–348.2	1.09–20.31	12
Knaff and Schlünder (1987) (SD)	Naphthalene, caffeine (128.2, 194.2)	308.15–333.15	11.8–22.6	16
Sassiat et al. (1987) (Taylor dispersion)	Acetone, benzene, naphthalene, 1,3,5-trimethylbenzene, phenanthrene, pyrene, chrysene (58.1–228.3)	303.15–333.15	11.0–26.5	50
Lamb et al. (1989) (NMR)	Naphthalene (128.2)	308.15–351.15	2.0–100.0	25
Dahmen et al. (1990a) (Input-output)	Cyclopentanone, 3-pentanone, cycloheptanone, 3-heptanone, cyclononanone, 5-nonanone (84.1–142.2)	314	10.0–18.0	48
Dahmen et al. (1990b) (Input-output)	2-Propanone, 3-pentanone, 4-heptanone, adamantanone, 2,4-dimethyl-3-pentanone, 2,2,4,4-tetramethyl-3-pentanone, 2-butanone, 2-pentanone, 2-heptanone, 2-nonanone, 5-nonanone, 6-undecanone, N-4-methoxybenzylidene-4-n-butylaniline, stearic acid, oleic acid, 4-cyano-4′-n-pentoxybiphenyl, Linolenic acid, squalene (58.1–410.7)	313.15–314.5	9.5–18	177
Smith et al. (1990) (Taylor dispersion)	Acridine, phenanthrene, benzoic acid (122.1–179.1)	308.15–328.15	17.3	17
Funazukuri et al. (1991) (Taylor dispersion)	Palmitoleic acid methyl ester, oleic acid methyl ester, elaidic acid methyl ester, linoleic acid methyl ester, γ-linolenic acid methyl ester, eicosapentaenoic acid methyl ester, docosahexaenoic acid methyl ester, nervonic acid methyl ester (268.4–380.7)	308–328	14.0–33.6	26
Liong et al. (1991) (Taylor dispersion)	Behenic acid ethyl ester, DHA ethyl ester, DHA methyl ester, EPA methyl ester, stearic acid ethyl ester (312.5–368.7)	308–318	9.87–21.05	85
Funazukuri et al. (1992) (Taylor dispersion)	Myristoleic acid methyl ester, cis-11-eicosenoic acid methyl ester, erucic acid methyl ester, vitamin K3, vitamin A acetate, vitamin K1, vitamin E, dl-limonene, cis-jasmone, indole, benzene, naphthalene (78.11–450.7)	313	16.0–25.0	24
Liong et al. (1992) (Taylor dispersion)	Palmitic acid ethyl ester (284.5)	308–318	9.67–21.05	17

Table 4.2 contd. ...

...Table 4.2 contd.

Data reference (method)	Solute (molecular weight)	*T* (K)	*P* (MPa)	No. of data
Umezawa and Nagashima (1992) (Input-output)	Acetone, benzene, alkane (C_5H_{12} to $C_{14}H_{30}$) (58.1–198.4)	299.15–308.15	9–10.5	55
Wells et al. (1992) (Input-output)	Phenylacetic acid, vanillin (136.2, 152.2)	308–318	9.67–21.05	16
Suárez et al. (1993) (Taylor dispersion)	Benzene, toluene, ethylbenzene, n-propylbenzene, cumene 78.1–120.2)	313.15–333.15	15.0–35.0	75
Catchpole and King (1994) (SD and Taylor dispersion)	Naphthalene, benzoic acid, oleic acid, α-tocopherol, glycerol trioleate (122.1–885.5)	298.1–318.7	7.1–25.1	16
Lai and Tan (1995) (CIR)	Caffeine, ethyl acetate, phenol, toluene (88.1–194.2)	308–328	7.51–14.46	75
Akgerman et al. (1996) (Taylor dispersion)	Naphthalene, phenanthrene, hexachlorobenzene (128.2–284.8)	308–328	9.64–24.9	45
Funazukuri and Nishimoto (1996) (Taylor dispersion)	Benzene (78.1)	313.2	8.43–30.2	31
Higashi et al. (1998) (SD)	Naphthalene, 2,6-dimethylnaphthalene, 2,7-dimethylnaphthalene (128.2–156.2)	308.20	7.21–20.0	26
Silva and Macedo (1998) (Taylor dispersion)	Diethyl ether, diisopropyl ether, tetrahydrofuran (72.1–102.2)	313.15–333.15	8.11–16.21	45
Ago and Nishiumi (1999) (Taylor dispersion)	Benzene (78.1)	313.15	8–20	57
Funazukkuri and Ishiwata (1999) (Taylor dispersion)	Linoleic acid methyl ester, vitamin K3, indole (117.2– 294.5)	313.20	16.0–25.0	69
Fu et al. (2000) (Taylor dispersion)	Benzoic acid, biphenyl, p-dichlorobenzene (122.1–154.2)	293.15–323.15	7.10–17.10	58
Funazukuri et al. (2000a) (Taylor dispersion)	Acetone (58.1)	308.2–313.2	7.94–40.11	121
Funazukuri et al. (2000b) (Taylor dispersion)	Acetone, 2-butanone, 2-pentanone, 3-pentanone (58.1–86.1)	308.15–328.15	8.33–34.57	143
Funazukuri et al. (2000c) (CIR)	Phenol, α-tocopherol, β-carotene (94.1–536.9)	308.15–328.15	8.72–30.28	107
Funazukuri et al. (2001) (Taylor dispersion)	Benzene (78.1)	308.15–328.15	5.99–34.48	176
González et al. (2001) (Taylor dispersion)	Anisole, 2,4-dimethylphenol, nitrobenzene (108.1–123.1)	313–333	15.0–35.0	15
Yang and Matthews (2001) (Taylor dispersion)	Trifluoroacetylacetone, thenoyltrifluoroacetone, 1,1,1,5,5,5-hexafluoroacetylacetone, copper(II) trifluoroacetylacetonate (154.1–222.2)	308.15–318.15	10.55–22.41	57

Table 4.2 contd. ...

...Table 4.2 contd.

Data reference (method)	Solute (molecular weight)	*T* (K)	*P* (MPa)	No. of data
Coelho et al. (2002) (Taylor dispersion)	Phenol (94.1)	298.15–318.15	9.239–17.036	7
Funazukuri et al. (2002a) (CIR)	Ubiquinone CoQ10 (863.3)	30815–333.15	8.51–30.18	71
Funazukuri et al. (2002b) (Input-output)	Phenol, β-carotene, α-tocopherol, vitamin K3, ubiquinone (94.1–863.3)	308.15–333.15	8.51–30.25	51
González et al. (2002) (Taylor dispersion)	2-Nitroanisole, 1,2-dichlorobenzene, tert-butylbenzene (134.2–153.1)	313–333	15.0–35.0	180
Filho et al. (2003) (Taylor dispersion)	Citral, D-limonene (136.2–152.2)	313.15–333.15	12.0–20.0	30
Funazukuri et al. (2003a) (CIR)	Docosahexaenoic acid, eicosapentaenoic acid, α-linolenic acid (278.4–328.5)	308.15–343.15	8.50–30.14	174
Funazukuri et al. (2003b) (CIR)	Linoleic acid, arachidonic acid (280.5–304.5)	308.15–343.15	8.50–30.29	146
Funazukuri et al. (2003c) (CIR)	β-Carotene, α-tocopherol (430.79, 536.9)	308.15–333.15	8.51–30.31	101
Mantell et al. (2003) (Taylor dispersion)	Malvidin 3,5-diglucoside (655.9)	313.15–333.15	10.0–40.0	32
Matthew and Becnel (2003) (Taylor dispersion)	Methyl orange (327.3)	293.15–313.15	8.8–22.7	87
Xu et al. (2003) (NMR)	Water (18)	283.15–308.15	13.0–30.0	32
Funazukuri et al. (2004b) (CIR)	Oleic acid, oleic acid methyl ester, oleic acid ethyl ester, monoolein, diolein, triolein, elaidic acid, elaidic acid methyl ester, linolenic acid ethyl ester, linoleic acid ethyl ester, elaidic acid ethyl ester, DHA methyl ester, DHA ethyl ester, trilinolenin, trilinolein, trielaidin, triarachidonin (282.5–951.5)	313.21	8.00–30.10	86
Kong et al. (2004) (CIR)	Acetone, phenol (58.1–94.1)	313.15	11.6–28.6	28
Silva et al. (2004) (Taylor dispersion)	α-Pinene, β-pinene (136.2)	313.15–333.15	12.0–20.0	30
Higashi et al. (2005) (SD)	Cetyl alcohol (242.4)	308.20	10.6–18.7	4
Kong et al. (2005) (CIR)	Triarachidonin, trierucin, trinervonin (951.5–1137.9)	308.15–323.15	8.26–30.24	166
Kong et al. (2006a) (CIR)	Benzoic acid, ethanol, methanol, 1-propanol, 2-propanol (32.0–122.1)	313.2	9.5–30.0	78
Kong et al. (2006b) (CIR)	Myristoleic acid, myristoleic acid methyl ester (226.4, 240.4)	313.21–343.15	9.20–30.00	121

Table 4.2 contd. ...

...Table 4.2 contd.

Data reference (method)	Solute (molecular weight)	*T* (K)	*P* (MPa)	No. of data
Kong et al. (2006c) (CIR)	γ-Linolenic acid, γ-linolenic acid methyl ester, γ-linolenic acid ethyl ester (278.4–306.5)	30815–343.15	8.1–30.11	222
Funazukuri et al. (2007) (CIR)	DHA, EPA, α-linolenic acid (278.4–328.5)	313.21	10.83–30.10	18
González et al. (2007a) (Taylor dispersion)	Aniline, mesitylene, styrene (93.1–120.2)	313.16–333.16	15.0–35.0	15
González et al. (2007b) (Taylor dispersion)	Bromobenzene, chlorobenzene, fluorobenzene, iodobenzene (96.1–204.0)	313–333	15.0–35.0	20
Han et al. (2007) (Taylor dispersion)	Arachidonic acid ethyl ester, DHA ethyl ester, EPA ethyl ester (330.5–356.5)	308.15–338.15	8.42–29.95	144
Kong et al. (2007) (CIR)	15-Crown-5, dibenzo-24-crown-8, vitamin K1 (220.3–450.7)	308.15–313.2	7.97–35	74
Nishiumi and Kubota (2007) (Taylor dispersion)	Benzene (78.1)	313.15	2.21–16.71	46
Pizarro et al. (2007) (Taylor dispersion)	5-tert-Butyl-m-xylene, 1,2-diethylbenzene, 1,4-diethylbenzene, phenylacetylene (102.1–162.3)	308–398	15.0–35.0	46
Suárez-Iglesias et al. (2007a) (Taylor dispersion)	Allylbenzene, 2-bromoanisole, 1,3-divinylbenzene, 2-fluoroanisole (118.2–187.0)	313.16–333.16	15.0–35.0	20
Suárez-Iglesias et al. (2007b) (Taylor dispersion)	Benzyl acetate, dibenzyl ether, 2-phenylethyl acetate, 3-phenylpropyl acetate (150.2–198.3)	313.16–333.16	15.0–35.0	20
Kong et al. (2008) (CIR)	Vitamin K_3 (172.2)	308.20	7.80–31.00	69
Lin and Tan (2008) (Taylor dispersion)	p-Chloronitrobenzene (157.6)	323.2–343.2	3.39–10.53	66
Pizarro et al. (2008a) (Taylor dispersion)	n-Butylbenzene, n-pentylbenzene, 1-phenylhexane, 1-phenyloctane, 1-phenyldodecane (134.2–246.4)	308–398	15.0–35.0	50
Pizarro et al. (2008b) (Taylor dispersion)	1-Phenylethanol, 2-phenylethanol, phenylmethanol, 2-phenyl-1-propanol, 3-phenyl-1-propanol (108.1–136.2)	313.16–333.16	15.0–35.0	25
Suárez-Iglesias et al. (2008) (Taylor dispersion)	Benzylacetone, ethyl benzoate, eugenol (148.2–164.2)	313.16–333.16	15.0–35.0	15
Pizarro et al. (2009a) (Taylor dispersion)	2,3-Dimethylaniline, 2,6-dimethylaniline, 2-methylanisole, 4-methylanisole, 3-nitrotoluene (121.2–137.1)	313–333	15.0–35.0	25
Pizarro et al. (2009b) (Taylor dispersion)	2-Ethyltoluene, 3-ethyltoluene, 4-ethyltoluene (120.2)	313–333	15.0–35.0	15

Table 4.2 contd. ...

...*Table 4.2 contd.*

Data reference (method)	Solute (molecular weight)	*T* (K)	*P* (MPa)	No. of data
Dong et al. (2010) (Taylor dispersion)	L-Carvone, L-menthone (150.2, 154.3)	308.15–338.15	15.0–30.0	46
Kong et al. (2010a) (CIR)	$Co(acac)_3$, $Pd(acac)_2$ (356.3, 304.6)	308.15–333.15	8.50–40.0	145
Kong et al. (2010b) (CIR)	Ferrocene, 1,1′-dimethylferrocene (172.5, 214.1)	308.15–323.15	8.02–40.34	126
Lin and Tavlarids (2010) (Taylor dispersion)	Benzene, diesel fuel, 1-methylnaphthalene, 1-hexadecene, toluene, m-xylene (78.1–224.4)	313.15–373.15	10.0–30.0	57
Dong et al. (2011) (Taylor dispersion)	L-Carvone, 1,3-dichlorobenzene, 3-fluorophenol, geraniol (112.1–154.3)	313.15	15.0–24.0	16
Kong et al. (2011) (CIR)	Naphthalene (128.2)	308.15	7.00–40.00	35
Pizarro et al. (2013) (Taylor dispersion)	Isobutylbenzene, *sec*-butylbenzene, 3-methylbutylbenzene (134.2–148.3)	313–333	15.0–35.0	45
Kong et al. (2013a) (CIR)	Platinum(II) acetylacetonate (393.3)	298.15–343.15	8.51–40.0	62
Kong et al. (2013b) (CIR)	Phenylbutazone (308.4)	298.2–343.2	8.50–40.00	104
Kong et al. (2015) (CIR)	Lithium acetylacetonate (106.1)	298.15–333.15	11.00–36.00	73
Kong et al. (2016) (CIR)	Aluminum acetylacetonate (324.3)	300.15–333.15	7.80–40.00	84
Funazukuri et al. (2016a) (Taylor dispersion)	Vitamin K_3 (172.2)	308.2–323.2	7.40–8.06	82
Funazukuri et al. (2016b) (CIR)	Disperse blue 14, disperse orange 11 (237.3, 266.3)	308.2–328.2	20.0–40.0	112
Kong et al. (2017) (CIR)	Chromium(III) acetylacetonate (349.3)	299.15–343.15	8.10–40.00	80

4. Summary

The quantitative knowledge of mass transport phenomena such as diffusion coefficients of solutes in SF is of considerable importance in the design, efficient operation, scale up and simulation of the newly proposed SF processes. In this chapter, the methods of measuring the diffusion coefficients in SF have been discussed. The various compounds measured in SCO_2 have been presented as well. The diffusion data in SCO_2 in the literature were mostly measured by the Taylor dispersion and the CIR methods. It is found that most of the large molecular compounds and polarity compounds have been studied by the CIR method.

Acknowledgments

This research was financially supported by the Ministry of Education, Culture, Sports, Science and Technology in grant-in-aid for scientific research #17K06898.

References

Ago, K.-I. and H. Nishiumi. 1999. Mutual diffusion coefficients of benzene in supercritical carbon dioxide. J. Chem. Eng. Japan 32: 563–568.

Akgerman, A., C. Erkey and M. Orejuela. 1996. Limiting diffusion coefficients of heavy molecular weight organic contaminants in supercritical carbon dioxide. Ind. Eng. Chem. Res. 35: 911–917.

Aris, R. 1956. On the dispersion of a solute in a fluid flowing through a tube. Proc. R. Soc. Lond. A 235: 67–77.

Boock, J.T., A.J.E. Freedman, G.A. Tompsett, S.K. Muse, A.J. Allen, L.A. Jackson et al. 2019. Engineered microbial biofuel production and recovery under supercritical carbon dioxide. Nat. Commun. 10: 587.

Byrappa, K., S. Ohara and T. Adschiri. 2008. Nanoparticles synthesis using supercritical fluid technology—towards biomedical applications. Adv. Drug Deliv. Rev. 60: 299–327.

Catchpole, O.J. and M.B. King. 1994. Measurement and correlation of binary diffusion coefficients in near critical fluids. Ind. Eng. Chem. Res. 33: 1828–1837.

Coelho, L.A.F., J.V. de Oliveira, F.W. Tavares and M.A. Matthews. 2002. Role of attractive forces in self-diffusion and mutual diffusion in dense simple fluids and real substances. Fluid Phase Equilib. 194-197: 1131–1140.

Cussler, E.L. 2005. Diffusion, Mass Transfer in Fluid System. Second edition, Cambridge.

Dahmen, N., A. Dülberg and G.M. Schneider. 1990a. Determination of binary diffusion coefficients in supercritical carbon dioxide with supercritical fluid chromatography (SFC). Ber. Bunsenges. Phys. Chem. 94: 384–386.

Dahmen, N., A. Kordikowski and G.M. Schneider. 1990b. Determination of binary diffusion coefficients of organic compounds in supercritical carbon dioxide by supercritical fluid chromatography. J. Chromatogr. 505: 169–178.

Dong, X., B. Su, H. Xing, Y. Yang and Q. Ren. 2010. Diffusion coefficients of L-menthone and L-carvone in mixtures of carbon dioxide and ethanol. J. Supercrit. Fluids 55: 86–95.

Dong, X., B. Su, H. Xing, Z. Bao, Y. Yang and Q. Ren. 2011. Cosolvent effects on the diffusions of 1,3-dichlorobenzene, L-carvone, geraniol and 3-fluorophenol in supercritical carbon dioxide. J. Supercrit. Fluids 58: 216–225.

Fan, Y., Y.-Q. Zhang, K. Yan and J.-J. Long. 2019. Synthesis of a novel disperse reactive dye involving a versatile bridge group for the sustainable coloration of natural fibers in supercritical carbon dioxide. Adv. Sci. 6: 1801368.

Feist, R. and G.M. Schneider. 1982. Determination of binary diffusion coefficients of benzene, phenol, naphthalene and caffeine in supercritical carbon CO_2 between 308 and 333 K in the pressure range 80 to 160 bar with supercritical fluid chromatography (SFC). Sep. Sci. Techno. 17: 261–270.

Filho, C.A., C.M. Silva, M.B. Quadri and E.A. Macedo. 2003. Tracer diffusion coefficients of citral and D-limonene in supercritical carbon dioxide. Fluid Phase Equilib. 204: 65–73.

Fu, H., L.A.F. Coelho and M.A. Matthews. 2000. Diffusion coefficients of model contaminants in dense CO_2. J. Supercrit. Fluids 18: 141–155.

Funazukuri, T., S. Hachisu and N. Wakao. 1991. Measurements of binary diffusion coefficients of C16–C24 unsaturated fatty acid methyl esters in supercritical carbon dioxide. Ind. Eng. Chem. Res. 30: 1323–1329.

Funazukuri, T., Y. Ishiwata and N. Wakao. 1992. Predictive correlation for binary diffusion coefficients in dense carbon dioxide. AIChE J. 38: 1761–1768.

Funazukuri, T. and N. Nishimoto. 1996. Tracer diffusion coefficients of benzene in dense CO2 at 313.2 K and 8.5–30 MPa. Fluid Phase Equilib. 125: 235–243.

Funazukuri, T. and Y. Ishiwata. 1999. Diffusion coefficients of linoleic acid methyl ester, Vitamin K3 and indole in mixtures of carbon dioxide and n-hexane at 313.2 K, and 16.0 MPa and 25.0 MPa. Fluid Phase Equilib. 164: 117–129.

Funazukuri, T., C.Y. Kong and S. Kagei. 2000a. Binary diffusion coefficients of acetone in carbon dioxide at 308.2 and 313.2 K in the pressure range from 7.9 to 40 MPa. Int. J. Thermophys. 21: 651–669.

Funazukuri, T., C.Y. Kong and S. Kagei. 2000b. Infinite-dilution binary diffusion coefficients of 2-propanone, 2-butanone, 2-pentanone, and 3-pentanone in CO2 by the Taylor dispersion technique from 308.15 to 328.15 K in the pressure range from 8 to 35 MPa. Int. J. Thermophys. 21: 1279–1290.

Funazukuri, T., C.Y. Kong, N. Murooka and S. Kagei. 2000c. Measurements of binary diffusion coefficients and partition ratios for acetone, phenol, α-tocopherol, and β-carotene in supercritical carbon dioxide with a poly(ethylene glycol)-coated capillary column. Ind. Eng. Chem. Res. 39: 4462–4469.

Funazukuri, T., C.Y. Kong and S. Kagei. 2001. Infinite dilution binary diffusion coefficients of benzene in carbon dioxide by the Taylor dispersion technique at temperatures from 308.15 to 328.15 K and pressures from 6 to 30 MPa. Int. J. Thermophys. 22: 1643–1660.

Funazukuri, T., C.Y. Kong and S. Kagei. 2002a. Infinite-dilution binary diffusion coefficient, partition ratio, and partial molar volume for ubiquinone CoQ10 in supercritical carbon dioxide. Ind. Eng. Chem. Res. 41: 2812–2818.

Funazukuri, T., C.Y. Kong and S. Kagei. 2002b. Measurements of binary diffusion coefficients for some low volatile compounds in supercritical carbon dioxide by input-output response technique with two diffusion columns connected in series. Fluid Phase Equilib. 194-197: 1169–1178.

Funazukuri, T., C.Y. Kong and S. Kagei. 2003a. Binary diffusion coefficient, partition ratio, and partial molar volume for docosahexaenoic acid, eicosapentaenoic acid and α-linolenic acid at infinite dilution in supercritical carbon dioxide. Fluid Phase Equilib. 206: 163–178.

Funazukuri, T., C.Y. Kong, T. Kikuchi and S. Kagei. 2003b. Measurements of binary diffusion coefficient and partition ratio at infinite dilution for linoleic acid and arachidonic acid in supercritical carbon dioxide. J. Chem. Eng. Data 48: 684–688.

Funazukuri, T., C.Y. Kong and S. Kagei. 2003c. Binary diffusion coefficients, partition ratios and partial molar volumes at infinite dilution for β-carotene and α-tocopherol in supercritical carbon dioxide. J. Supercrit. Fluids 27: 85–96.

Funazukuri, T., C.Y. Kong and S. Kagei. 2004a. Impulse response techniques to measure binary diffusion coefficients under supercritical conditions. J. Chromatogr. A 1037: 411–429.

Funazukuri, T., C.Y. Kong and S. Kagei. 2004b. Effects of molecular weight and degree of unsaturation on binary diffusion coefficients for lipids in supercritical carbon dioxide. Fluid Phase Equilib. 219: 67–73.

Funazukuri, T., C.Y. Kong and S. Kagei. 2007. Simultaneous determination of binary diffusion coefficients from multiple response curves by chromatographic measurements. J. Chromatogr. A 1150: 105–111.

Funazukuri, T., T. Sugihara, K. Yui, T. Ishii and M. Taguchi. 2016a. Measurement of infinite dilution diffusion coefficients of vitamin K3 in CO_2 expanded methanol. J. Supercrit. Fluids 108: 19–25.

Funazukuri, T., T. Yamasaki, M. Taguchi and C.Y. Kong. 2016b. Measurement of binary diffusion coefficient and solubility estimation for dyes in supercritical carbon dioxide by CIR method. Fluid Phase Equilib. 420: 7–13.

González, L.M., J.L. Bueno and I. Medina. 2001. Determination of binary diffusion coefficients of anisole, 2,4-dimethylphenol, and nitrobenzene in supercritical carbon dioxide. Ind. Eng. Chem. Res. 40: 3711–3716.

González, L.M., J.L. Bueno and I. Medina. 2002. Measurement of diffusion coefficients for 2-nitroanisole, 1,2-dichlorobenzene and tert-butylbenzene in carbon dioxide containing modifiers. J. Supercrit. Fluids 24: 219–229.

González, L.M., O. Suárez-Iglesias, J.L. Bueno, C. Pizarro and I. Medina. 2007a. Limiting binary diffusivities of aniline, styrene, and mesitylene in supercritical carbon dioxide. J. Chem. Eng. Data 52: 1286–1290.

González, L.M., O. Suárez-Iglesias, J.L. Bueno, C. Pizarro and I. Medina. 2007b. Application of the corresponding states principle to the diffusion in CO_2. AIChE J. 53: 3054–3061.

Guiochon, G. and A. Tarafder. 2011. Fundamental challenges and opportunities for preparative supercritical fluid chromatography. J. Chromatogr. A 1218: 1037–1114.

Han, Y., Y. Yang and P. Wu. 2007. Binary diffusion coefficients of arachidonic acid ethyl esthers, cis-5,8,11,14,17-eicosapentaenoic acid ethyl esthers, and cis-4,7,10,13,16,19-docosahexanenoic acid ethyl esthers in supercritical carbon dioxide. J. Chem. Eng. Data 52: 555–559.

Higashi, H., Y. Iwai, Y. Takahashi, H. Uchida and Y. Arai. 1998. Diffusion coefficients of naphthalene and dimethylnaphthalene in supercritical carbon dioxide. Fluid Phase Equilib. 144: 269–278.

Higashi, H., Y. Iwai and Y. Arai. 2001. Solubilities and diffusion coefficients of high boiling compounds in supercritical carbon dioxide. Chem. Eng. Sci. 56: 3027–3044.

Higashi, H., Y. Iwai, T. Oda, K. Murai and Y. Arai. 2005. Diffusion coefficients of cetyl alcohol in supercritical carbon dioxide. J. Chem. Eng. Japan 38: 865–869.

Knaff, G. and E.U. Schlünder. 1987. Diffusion coefficients of naphthalene and caffeine in supercritical carbon dioxde. Chem. Eng. Process 21: 101–105.

Knez, Ž., E. Markočič, M. Leitgeb, M. Primožič, M. Knez Hrnčič and M. Škerget. 2014. Industrial applications of supercritical fluids: A review. Energy 77: 235–243.
Kong, C.Y., T. Funazukuri and S. Kagei. 2004. Chromatographic impulse response technique with curve fitting to measure binary diffusion coefficients and retention factors using polymer-coated capillary columns. J. Chromatogr. A 1035: 177–193.
Kong, C.Y., N.R.W. Withanage, T. Funazukuri and S. Kagei. 2005. Binary diffusion coefficients and retention factors for long-chain triglycerides in supercritical carbon dioxide by the chromatographic impulse response method. J. Chem. Eng. Data 50: 1635–1640.
Kong, C.Y., T. Funazukuri and S. Kagei. 2006a. Binary diffusion coefficients and retention factors for polar compounds in supercritical carbon dioxide by chromatographic impulse response method. J. Supercrit. Fluids 37: 359–366.
Kong, C.Y., M. Mori, T. Funazukuri and S. Kagei. 2006b. Measurements of binary diffusion coefficients, retention factors and partial molar volumes for myristoleic acid and its methyl ester in supercritical carbon dioxide. Anal. Sci. 22: 1431–1436.
Kong, C.Y., N.R.W. Withanage, T. Funazukuri and S. Kagei. 2006c. Binary diffusion coefficients and retention factors for γ-linolenic acid and its methyl and ethyl esters in supercritical carbon dioxide. J. Supercrit. Fluids 37: 63–71.
Kong, C.Y., N. Takahashi, T. Funazukuri and S. Kagei. 2007. Measurements of binary diffusion coefficients and retention factors for dibenzo-24-crown-8 and 15-crown-5 in supercritical carbon dioxide by chromatographic impulse response technique. Fluid Phase Equilib. 257: 223–227.
Kong, C.Y., T. Funazukuri and S. Kagei. 2008. Reliability of binary diffusion coefficients determined from tailing response curves measured by the Taylor dispersion method in the near critical region. J. Supercrit. Fluids 44: 294–300.
Kong, C.Y., Y.Y. Gu, M. Nakamura, T. Funazukuri and S. Kagei. 2010a. Diffusion coefficients of metal acetylacetonates in supercritical carbon dioxide. Fluid Phase Equilib. 297: 162–167.
Kong, C.Y., M. Nakamura, K. Sone, T. Funazukuri and S. Kagei. 2010b. Measurements of binary diffusion coefficients for ferrocene and 1,10-dimethylferrocene in supercritical carbon dioxide. J. Chem. Eng. Data 55: 3095–3100.
Kong, C.Y., K. Sone, T. Sako, T. Funazukuri and S. Kagei. 2011. Solubility determination of organometallic complexes in supercritical carbon dioxide by chromatographic impulse response method. Fluid Phase Equilib. 302: 347–353.
Kong, C.Y., T. Funazukuri, S. Kagei, G. Wang, F. Lu and T. Sako. 2012. Applications of the chromatographic impulse response method in supercritical fluid chromatography. J. Chromatogr. A 1250: 141–156.
Kong, C.Y., T. Siratori, G. Wang, T. Sako and T. Funazukuri. 2013a. Binary diffusion coefficients of platinum(II) acetylacetonate in supercritical carbon dioxide. J. Chem. Eng. Data 58: 2919–2924.
Kong, C.Y., K. Watanabe and T. Funazukuri. 2013b. Diffusion coefficients of phenylbutazone in supercritical CO_2 and in ethanol. J. Chromatogr. A 1279: 92–97.
Kong, C.Y., T. Siratori, T. Funazukuri and G. Wang. 2014. Infinite dilution partial molar volumes of platinum(II) 2,4-pentanedionate in supercritical carbon dioxide. J. Chromatogr. A 1362: 294–300.
Kong, C.Y., Y. Yakumara, T. Funazukuri and S. Kagei. 2015. Measurement and correlation of binary diffusion coefficients of lithium acetylacetonate in supercritical carbon dioxide and in liquid ethanol. J. Supercrit. Fluids 104: 265–271.
Kong, C.Y., K. Watanabe and T. Funazukuri. 2016. Determination and correlation of infinite dilution binary diffusion coefficients for aluminum acetylacetonate in supercritical and liquid fluids. Fluid Phase Equilib. 420: 83–88.
Kong, C.Y., K. Watanabe and T. Funazukuri. 2017. Measurement and correlation of the diffusion coefficients of chromium (III) acetylacetonate at infinite dilution in supercritical carbon dioxide and in liquid ethanol. J. Chem. Thermodynamics 105: 86–93.
Kuang, T., F. Chen, L. Chang, Y. Zhao, D. Fu, X. Gong et al. 2017. Facile preparation of open-cellular porous poly (L-lactic acid) scaffold by supercritical carbon dioxide foaming for potential tissue engineering applications. Chem. Eng. J. 307: 1017–1025.
Lai, C.-C. and C.-S. Tan. 1995. Measurement of molecular diffusion coefficients in supercritical carbon dioxide using a coated capillary column. Ind. Eng. Chem. Res. 34: 674–680.
Lamb, D.M., S.T. Adamy, K.W. Woo and J. Jonas. 1989. Transport and relaxation of naphthalene in supercritical fluids. J. Phys. Chem. 93: 5002–5005.

Lauer, H.H., D. McManigill and R.D. Board. 1983. Mobile-phase transport properties of liquefied gases in near-critical and supercritical fluid chromatography. Anal. Chem. 55: 1370–1375.

Lin, I.-H. and C.-S. Tan. 2008. Measurement of diffusion coefficients of p-chloronitrobenzene in CO2-expanded methanol. J. Supercrit. Fluids 46: 112–117.

Lin, R. and L.L. Tavlarides. 2010. Diffusion coefficients of diesel fuel and surrogate compounds in supercritical carbon dioxide. J. Supercrit. Fluids 52: 47–55.

Liong, K.K., P.A. Wells and N.R. Foster. 1991a. Diffusion in supercritical fluids. J. Supercrit. Fluids 4: 91–108.

Liong, K.K., P.A. Wells and N.R. Foster. 1991b. Diffusion coefficients of long-chain esters in supercritical carbon dioxide. Ind. Eng. Chem. Res. 30: 1329–1335.

Liong, K.K., P.A. Wells and N.R. Foster. 1992. Diffusion of fatty acid esters in supercritical carbon dioxide. Ind. Eng. Chem. Res. 31: 390–399.

Mantell, C., M. Rodríguez and E. Maetínez de la Ossa. 2003. Measurement of the diffusion coefficient of a model food dye (malvidin 3,5-diglucoside) in a high pressure CO2+ methanol system by the chromatographic peak-broadening technique. J. Supercrit. Fluids 25: 57–68.

Matthews, M.A. and J.M. Becnel. 2003. Diffusion coefficients of methyl orange in dense carbon dioxide with the micelle-forming surfactant dehypon Ls-54. J. Chem. Eng. Data 48: 1413–1417.

Medina, I. 2012. Determination of diffusion coefficients for supercritical fluids. J. Chromatogr. A 1250: 124–140.

Nishiumi, H. and T. Kuboda. 2007. Fundamental behavior of benzene–CO_2 mutual diffusion coefficients in the critical region of CO_2. Fluid Phase Equilib. 261: 146–151.

Olesik, S.V., J.L. Steger, N. Kiba, M. Roth and M.V. Novotny. 1987. High-pressure apparatus for physico-chemical measurements by capillary supercritical fluid chromatography. J. Chromatogr. 392: 165–174.

Padrela, L., M.A. Rodrigues, A. Duarte, A.M.A. Dias, M.E.M. Braga and H.C. de Sousa. 2018. Supercritical carbon dioxide-based technologies for the production of drug nanoparticles/nanocrystals—A comprehensive review. Adv. Drug Deliv. Rev. 131: 22–78.

Pizarro, C., O. Suárez-Iglesias, I. Medina and J.L. Bueno. 2007. Using supercritical fluid chromatography to determine diffusion coefficients of 1,2-diethylbenzene, 1,4-diethylbenzene, 5-tert-butyl-m-xylene and phenylacetylene in supercritical carbon dioxide. J. Chromatogr. A 1167: 202–209.

Pizarro, C., O. Suárez-Iglesias, I. Medina and J.L. Bueno. 2008a. Diffusion coefficients of n-butylbenzene, n-pentylbenzene, 1-phenylhexane, 1-phenyloctane, and 1-phenyldodecane in supercritical carbon dioxide. Ind. Eng. Chem. Res. 47: 6783–6789.

Pizarro, C., O. Suárez-Iglesias, I. Medina and J.L. Bueno. 2008b. Molecular diffusion coefficients of phenylmethanol, 1-phenylethanol, 2-phenylethanol, 2-phenyl-1-propanol, and 3-phenyl-1-propanol in supercritical carbon dioxide. J. Supercrit. Fluids 43: 469–476.

Pizarro, C., O. Suárez-Iglesias, I. Medina and J.L. Bueno. 2009a. Binary diffusion coefficients for 2,3-dimethylaniline, 2,6-dimethylaniline, 2-methylanisole, 4-methylanisole and 3-nitrotoluene in supercritical carbon dioxide. J. Supercrit. Fluids 48: 1–8.

Pizarro, C., O. Suárez-Iglesias, I. Medina and J.L. Bueno. 2009b. Binary diffusion coefficients of 2-ethyltoluene 3-ethyltoluene, and 4-ethyltoluene in supercritical carbon dioxide. J. Chem. Eng. Data 54: 1467–1471.

Pizarro, C., O. Suárez-Iglesias, I. Medina and J.L. Bueno. 2013. Diffusion coefficients of isobutylbenzene, sec-butylbenzene, and 3-methylbutylbenzene in supercritical carbon dioxide. J. Chem. Eng. Data 58: 2001–2007.

Poling, B.E., J.M. Prausnitz and J.P. O'connell. 2001. The Properties of Gases and Liquids, 5th Edition.

Roth, M. 1991. Diffusion and thermodynamic measurements by supercritical fluid chromatography. J. Microcolumn Separations 3: 173–184.

Roth, M. 1992. Influence of TiO_2 surface on 1,2-chlorine shift in β-chlorine-substituted radicals as studied by radiation chemistry and photocatalysis. J. Phys. Chem. 96: 8522–8556.

Roth, M. 1998. Partial molar properties from solute retention in supercritical fluid chromatography: thermodynamic framework, advantages, and limitations. Fluid Phase Equilib. 148: 189–199.

Saad, H. and E. Gulari. 1984. Diffusion of liquid hydrocarbons in supercritical CO_2 by photon correlation spectroscopy. Ber. Bunsenges. Phys. Chem. 88: 834–837.

Sassiat, P.R., P. Mourier, M.H. Caude and R.H. Rosset. 1987. Measurement of diffusion coefficients in supercritical carbon dioxide and correlation with the equation of Wilke and Chang. Anal. Chem. 59: 1164–1170.

Silva, C.M. and E.A. Macedo. 1998. Diffusion coefficients of esters in supercritical carbon dioxide. Ind. Eng. Chem. Res. 37: 1490–1498.

Silva, C.M., C.A. Filho, M.B. Quadri and E.A. Macedo. 2004. Binary diffusion coefficients of α-pinene and β-pinene in supercritical carbon dioxide. J. Supercrit. Fluids 32: 167–175.

Simonov, A.S., M.S. Kondratenko, I.V. Elmanovich, V.E. Sizov, E.P. Kharitonova, S.S. Abramchuk et al. 2018. Modification of Nafion with silica nanoparticles in supercritical carbon dioxide for electrochemical applications. J. Membr. Sci. 564: 106–114.

Smith, S.A., V. Shenai and M.A. Matthews. 1990. Diffusion in supercritical mixture: CO_2+cosolvent+solute. J. Supercrit. Fluids 3: 175–179.

Suárez, J.J., J.L. Bueno and I. Medina. 1993. Determination of binary diffusion coefficients of benzene and derivatives in supercritical carbon dioxide. Chem. Eng. Sci. 48: 2419–2427.

Suárez, J.J., I. Medina and J.L. Bueno. 1998. Diffusion coefficients in supercritical fluids: available data and graphical correlations. Fluid Phase Equilib. 153: 167–212.

Suárez-Iglesias, O., I. Medina, C. Pizarro and J.L. Bueno. 2007a. Diffusion coefficients of 2-fluoroanisole, 2-bromoanisole, allylbenzene and 1,3-divinylbenzene at infinite dilution in supercritical carbon dioxide. Fluid Phase Equilib. 260: 279–286.

Suárez-Iglesias, O., I. Medina, C. Pizarro and J.L. Bueno. 2007b. Diffusion of benzyl acetate, 2-phenylethyl acetate, 3-phenylpropyl acetate, and dibenzyl ether in mixtures of carbon dioxide and ethanol. Ind. Eng. Chem. Res. 46: 3810–3819.

Suárez-Iglesias, O., I. Medina, C. Pizarro and J.L. Bueno. 2008. Limiting diffusion coefficients of ethyl benzoate, benzylacetone, and eugenol in carbon dioxide at supercritical conditions. J. Chem. Eng. Data 53: 779–784.

Swaid, I. and G.M. Schneider. 1979. Determination of binary diffusion coefficients of benzene and some alkylbenzenes in supercritical CO_2 between 308 and 328 K in the pressure range 80 to 160 bar with supercritical fluid chromatography (SFC). Ber. Bunsenges. Phys. Chem. 83: 969–974.

Takahashi, S. and H. Iwasaki. 1966. The diffusion of gases at high pressures. I. The self-diffusion coefficient of carbon dioxide. Bull. Chem. Soc. Japan 39: 2105–2109.

Takahashi, S. and M. Hongo. 1982. Diffusion coefficients of gases at high pressures in the CO_2-C_2H_4 system. J. Chem. Eng. Japan 15: 57–59.

Taylor, G. 1953. Dispersion of soluble matter in solvent flowing slowly through a tube. 1953. Proc. R. Soc. Lond. A 219: 186–203.

Umezawa, S. and A. Nagashima. 1992. Measurement of the diffusion coefficients of acetone, benzene, and alkane in supercritical CO_2 by the Taylor dispersion method. J. Supercrit. Fluids 5: 242–250.

Wakao, N. and S. Kaguei. 1982. Heat and Mass Transfer in Packed Beds. Gordon & Breach. Pub., New York.

Wells, T., N.R. Foster and R.P. Chaplin. 1992. Diffusion of phenylacetic acid and vanillin in supercritical carbon dioxide. Ind. Eng. Chem. Res. 31: 927–934.

Xu, B., K. Nagashima, J.M. DeSimone and C.S. Johnson, Jr. 2003. Diffusion of water in liquid and supercritical carbon dioxide: An NMR study. J. Phys. Chem. A 107: 1–3.

Yang, X.N. and M.A. Matthews. 2001. Diffusion of chelating agents in supercritical CO_2 and a predictive approach for diffusion coefficients. J. Chem. Eng. Data 46: 588–595.

Zhang, J., Y. Qiao, X. Zu, H. Zheng, S. Gao and L. Zheng. 2019. Cleaner strategy for the scouring and bleaching of flax rove with enzymes in supercritical carbon dioxide. J. Clean Prod. 210: 759–766.

Chapter 5

Catalytic Green Organic Synthesis in Unique Reaction Environments

Nobuyuki Mase[1,*] and *Kohei Sato*[2]

1. Introduction

Since the beginning of the 19th century, organic synthetic chemistry has evolved constantly and extensively, making a significant contribution to the development of new research fields (Chen 2018, Nicolaou 2018). Although technological innovation of equipment used for organic synthesis and structural analysis is rapidly developing, certain aspects of the laboratory have remained unchanged for many years. One example is glassware that is used for reaction, distillation, and storage. Organic solvents with various physical properties such as methanol, diethyl ether, ethyl acetate, dichloromethane, and dimethylformamide have also been used for over 100 years as a medium for organic transformation of bond formation/cleavage reactions. In other words, organic synthetic chemistry is a fusion of traditional and cutting-edge technologies, and also continues to evolve constantly. We have been conducting research into catalytic organic synthesis in unique reaction environments, such as water, supercritical fluid, fine bubbles, microwave, and tube flow, which are not commonly used in organic synthesis (Fig. 5.1). Here the prospects for the above research topics are outlined and described.

[1] Green Energy Research Division, Research Institute of Green Science and Technology, Shizuoka University, 3-5-1 Johoku, Naka-ku, Hamamatsu, Shizuoka 432-8561, Japan.

[2] Division of Applied Chemistry and Biochemical Engineering, College of Engineering, Academic Institute, Shizuoka University, 3-5-1 Johoku, Naka-ku, Hamamatsu, Shizuoka 432-8561, Japan.
Email: sato.kohei@shizuoka.ac.jp

* Corresponding author: mase.nobuyuki@shizuoka.ac.jp

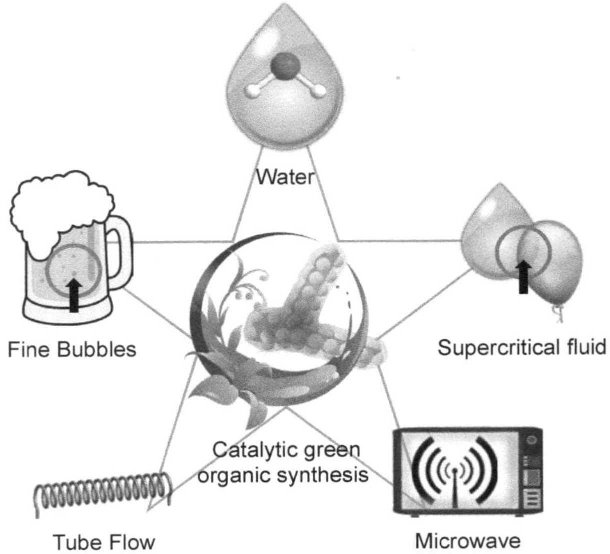

Fig. 5.1: Catalytic green organic synthesis in unique reaction environments.

2. Organocatalysis in Water: Development of Environmentally Friendly Asymmetric Synthesis

2.1 *Water in Organic Synthesis*

"Water" within a cell plays an important role, not only as a solvent for various biological substances, but also as a reactant in daily life activities. On the other hand, in organic synthetic chemistry in flasks, "water" functions as a reagent, an activator, and a solvent, and its importance has increased manifold in the 21st century (Kobayashi 2012). In the Diels-Alder reaction, reported by Breslow (Rideout and Breslow 1980), reactivity in water (H_2O + LiCl) is approximately 1800 times faster than that in isooctane, which has triggered the use of "water" as a solvent in organic synthesis. Because hydrophobic interactions work most strongly in water, hydrophobic substrates tend to aggregate "in water", and the Diels-Alder reaction is accelerated. On the other hand, the theoretical consideration of "on water" (Narayan et al. 2005) has been reported by Marcus group (Jung and Marcus 2007). The interface between the oil droplets of the substrate and the water phase, i.e., the water molecules at the interface, is similar to the surface of hexagonal crystal ice, and the hydroxyl group regularly exists at the interface. Therefore, a large reaction acceleration effect was observed ($k_{surface}/k_{neat} = 1.5 \times 10^5$, $k_{surface}/k_{homogeneous} = 600$) by activating the substrate by hydrogen bonding of these surface hydroxyl groups. However, applicable substrates and catalysts are limited due to the underdevelopment of this research field. In particular, highly stereoselective catalytic organic transformation is a challenging research topic. If we can skillfully use "water", which can be handled easily, safely and cheaply in organic synthetic chemistry, we can propose more efficient and practical methods from the viewpoint of green sustainable chemistry than conventional methods using organic solvents. We aimed to develop a catalyst that functions in water and establish environmentally friendly asymmetric synthesis.

2.2 Organocatalysis

Many of the catalysts developed so far have been inspired by natural enzymes as a model. Approximately one-third of known enzymes require metal ions to express catalytic activity, and "organometallic complex catalysts" have been greatly developed in organic synthesis. Meanwhile, as a model of non-metallic enzymes, the "organocatalyst" (Dalko 2013, Rovis 2016, Torres 2013) has attracted significant attention in the 21st century as a "third catalyst", next to "enzyme" and "organometallic complex catalyst". Organocatalysts are attractive from the viewpoint of environmentally friendly and elemental strategy, as they are easy to handle, inexpensive, and promote organic reactions without the use of metals with high residual toxicity. In particular, organocatalysts based on aldolase enzymes have been investigated explosively as they catalyze various nucleophilic and electrophilic reactions to carbonyl compounds (List 2012, Maruoka 2012). We have focused on the fact that the native enzyme aldolase achieves ideal stereoselective carbon-carbon bond formation/cleavage under biological reaction conditions, i.e., in the presence of large amounts of water (Fig. 5.2).

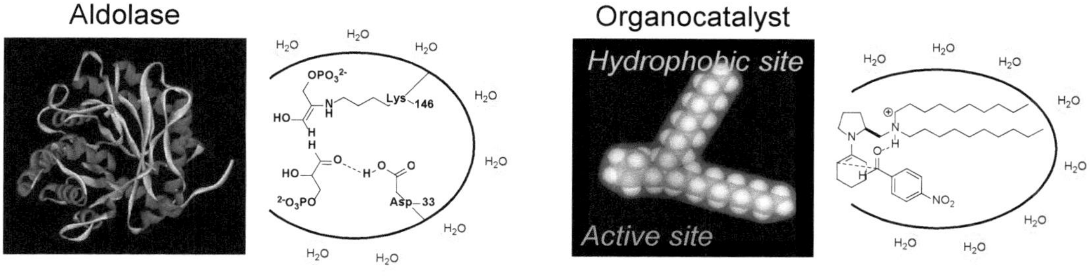

Fig. 5.2: Design of organocatalyst.

2.3 Organocatalytic Direct Asymmetric Reactions in Water

Study of stereoselective aldol reaction using water as reaction medium in the presence of newly designed and synthesized organocatalysts with a long chain alkyl group resulted in a yield of 99% and the stereoselectivity of 99% enantiomeric excess (ee) (Mase et al. 2006, 2009, Mase and Barbas 2010). Furthermore, in a conventional study in organic solvents, a large excess amount of donor (~ 30 equivalents) was required, but in the reaction system in water, the reaction proceeded efficiently, even when one equivalent of donor was used, and the product was obtained in 98% yield with 92% ee. Furthermore, when this catalyst system was applied to the Michael reaction, the polymerization reaction could be suppressed by using brine instead of water, giving the Michael adduct in 99% yield with 97% ee (Mase et al. 2006). In conventional methods, extraction and column chromatography were necessary to isolate the product, but in our reaction system, the brine was removed from the reaction mixture, and the syn-adduct was obtained by recrystallization in 73% yield with 99% ee. The catalyst and water could be recovered and reused in both the aldol reaction and Michael reaction, and asymmetric synthesis could be carried out as much as possible without waste (Fig. 5.3).

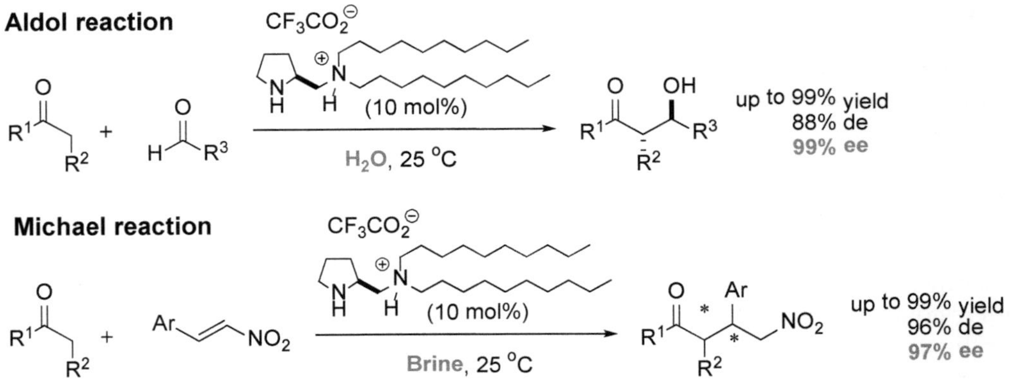

Fig. 5.3: Organocatalysis in water.

3. Organocatalysis in Supercritical Carbon Dioxide: Development of High-purity Polylactide Synthesis

3.1 Polylactide

The advantage of no metal contamination in the product by the organocatalytic method is also powerful in polymer synthesis. For example, linear polylactide (PLA), which is a carbon-neutral bio-based polymer, has high biodegradability and biocompatibility, and is expected to be applied to the medical and pharmaceutical fields. Ring-opening polymerization of lactide, which is a representative synthetic method of PLA, is carried out by bulk polymerization in the presence of a metal catalyst, such as tin (II) 2-ethylhexanoate (Mehta et al. 2005). However, there are concerns about the decrease in strength due to the presence of the residual monomer and toxicity caused by residual metals. Although tin is an element with low toxicity, the Ministry of Health, Labor and Welfare of Japan regulates that "the content of tin in soft drinks should not exceed 150 ppm" (https://www.mhlw.go.jp/hourei/doc/tsuchi/T180809I0040.pdf, https://www.mhlw.go.jp/hourei/doc/tsuchi/T181018I0050.pdf). Organocatalysis has been reported as an alternative to metal complex catalysis, which is solution polymerization with low reactivity in toxic halogenated organic solvents, such as chloroform and dichloromethane (Kamber et al. 2007, Nederberg et al. 2001). To solve these problems, we aimed to establish a PLA synthesis method by utilizing the superior features of "organocatalyst" and "supercritical carbon dioxide ($scCO_2$)".

3.2 Supercritical Carbon Dioxide: CO_2 Plasticizing Polymerization Method

In addition to the three classical states of solid, liquid, and gas, the state above the critical temperature (Tc) and the critical pressure (Pc) is called supercritical fluid (SCF), which is a characteristic fluid combining the property of liquid with excellent solubility and the property of gas with excellent diffusibility (Brunner 2010). The solubility and diffusibility of an SCF can be continuously changed by controlling the temperature and pressure, and gas recovery is easy. In particular, $scCO_2$ is inexpensive, safe, nonflammable, has low surface tension and low viscosity, and its critical temperature is relatively mild (Tc = 31.1°C, Pc = 7.38 MPa). Although $scCO_2$ has various advantages, the carbon atom of CO_2 is sandwiched between the electronegative oxygen atoms, and the central carbon atom is in the electron-deficient state. Therefore, CO_2 shows Lewis acidity. As a result, CO_2 reacts

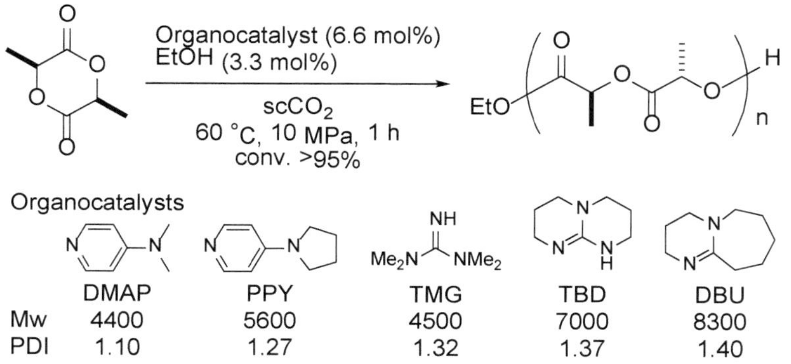

Fig. 5.4: Organocatalysis in $scCO_2$: high-purity PLA synthesis.

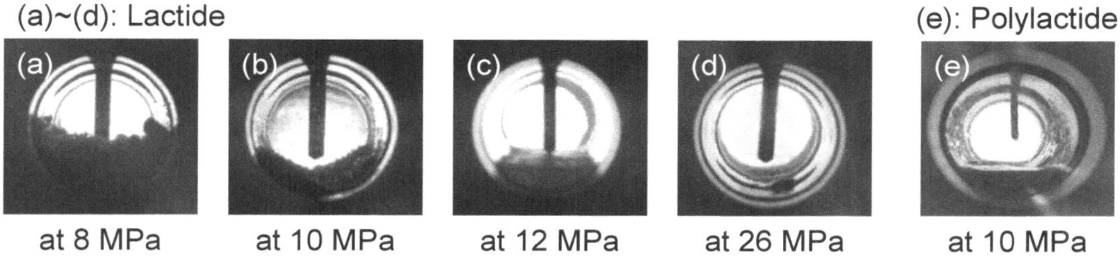

Fig. 5.5: Phase changes in $scCO_2$ at 60°C.

with various nucleophiles, and the anionic polymerization reaction in $scCO_2$ is considered challenging. In PLA synthesis, using a basic or nucleophilic organocatalyst, it was expected that the reactivity would be decreased by the formation of an anion or zwitterion intermediate. However, in the ring-opening polymerization of L-lactide using ethanol as a reaction initiator in the presence of an organocatalyst such as DMAP, the polymerization proceeded rapidly (1 h) and quantitatively (> 95% conversion rate) in $scCO_2$. We succeeded in preparing high-purity PLA without metals, organic solvents, or residual monomers (Fig. 5.4) (Mase et al. 2018). To elucidate the factors providing high reactivity, the behavior of L-lactide and PLA in $scCO_2$ was observed, and we found that both melted in $scCO_2$ (Fig. 5.5). In $scCO_2$, the polymerization reaction would proceed under high concentration, similar to bulk polymerization, and under homogeneous conditions similar to solution polymerization. This method was named the CO_2 plasticizing polymerization method (CPP method).

3.3 Polymerization and Granulation in One-pot

Because PLA is generally obtained in a hard-massive form, it is processed into pellets. Granular PLA was obtained under CPP conditions using a silicone surfactant, which is inexpensive and safer than conventional fluorochemical surfactants. In other words, PLA synthesis and granulation were achieved by one operation. As it is easy to remove from the reaction vessel by granulation, simplification of the post-treatment process is expected. Micelle formation of the surfactant and monomer probably caused the formation of granular PLA, but details are unknown.

3.4 High Purification of PLA by $scCO_2$ Extraction

Because the polymer PLA has low solubility in $scCO_2$, the $scCO_2$-soluble organocatalyst can be removed from the reaction mixture by $scCO_2$ extraction. By examining the extraction conditions, the extraction efficiency was improved under low-temperature and high-pressure conditions. More than 95% DMAP was removed from the mixture of PLA and DMAP under conditions of 40°C, 20 MPa, and 5 hours, and high purification of PLA was achieved.

3.5 Control of Epimerization by Brønsted Acid Catalyst

The optical purity of PLA greatly influences its physical and chemical properties. For example, PLA with low optical purity is amorphous, while PLA with high optical purity is crystalline. Furthermore, in terms of biodegradability, the presence of a non-natural (*R*)-configuration reduces the rate of degradation by microorganisms. It is necessary to control stereoregularity by suppressing the epimerization of PLA, not only in academic research but also in industrial PLA production. L-lactide as a monomer is available with high enantiopurity (> 99.5% ee), but the enantioselectivity decreases to 88.0% ee in the conventional DMAP-catalyzed PLA synthesis in chloroform. On the other hand, with the use of a Brønsted acid catalyst, such as Tf_2NH under CPP conditions, no epimerization was observed (> 99.5% ee).

3.6 High-molecular-weight PLA

PLA with high optical purity, quality and productivity was achieved. To develop the CPP method for the production of general polymer materials, the synthesis of PLA, with a theoretical molecular weight of more than 10,000, was investigated, but the molecular weight decreased and became lesser than the theoretical value. The formation of cyclic polylactide (*c*PLA) through intramolecular cyclization reaction was presumed to be a cause of molecular weight decrease. We found that the sterically bulky and highly nucleophilic 9-azajurolidine as an organocatalyst affected the molecular weight increase of PLA (M_n: 32000 g/mol, M_w: 72000 g/mol).

3.7 Cyclic Polylactide Synthesis

On the other hand, *c*PLA, which is a structural isomer of PLA, is expected as a new functional material because of its different topology, low biodegradability and low viscosity from those of linear PLA. However, in conventional *c*PLA synthesis using an organocatalyst and a metal complex catalyst, control of high-yield and optical purity has not been achieved. However, the polymerization reaction proceeded smoothly under CPP conditions, and a solid *c*PLA with high enantiopurity was obtained (Mase et al. 2018). The purity of the synthesized highly enantiopure *c*PLA stereocomplex is higher than the reported value (Shin et al. 2012), despite its smaller molecular weight, and therefore, development of new materials in the future is expected.

As mentioned above, $scCO_2$ melts lactide below its melting point, and the organocatalyst is sensitive to lactide monomers but insensitive to CO_2. Therefore, synthesis of high-purity linear- and cyclic-PLA containing no metals, organic solvents, or residual monomers was developed (Fig. 5.6).

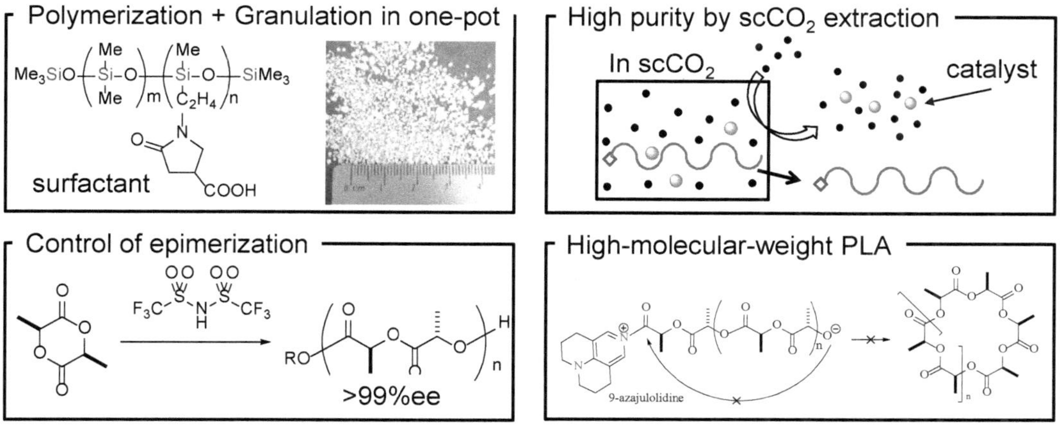

Fig. 5.6: High-purity PLA synthesis by CO_2 plasticizing polymerization method (CPP method).

4. Organic Synthesis Using Fine Bubbles: Development of Gas-Liquid Phase Reaction Without Autoclave

4.1 Fine Bubbles

Gas solubility in the liquid phase decreases as the temperature of the solution increases and it increases as the temperature decreases. On the other hand, the chemical reactivity increases as the temperature increases and decreases as the temperature decreases. To combine these two seemingly contradictory properties, it is common to carry out gas-liquid reaction while vigorously mechanically stirring in a pressure-resistant vessel, which can be used under high-temperature and high-pressure conditions. This high-pressure reaction system has not changed for more than a century. To avoid high-pressure conditions, a bubbling system has been alternatively developed. Bubbles are divided by a fine filter, but the size of the bubbles is limited. Furthermore, to obtain products with higher conversion, excess gas supply has been required. As a result, the reaction involving the gas phase is clean and simple, but the scope of application in "manufacturing" was limited because of the initial cost of the reactor and the cost of maintenance. "Reversal of thinking" is necessary to change the reaction mode, which remained unchanged for more than a century, and thus, we have reached the following idea "Disperse the gas phase in the liquid phase at the micro- to nano-level, instead of forcibly dissolving the gas phase". We introduce the next-generation gas-liquid and gas-liquid-solid reaction process using Fine Bubbles (FBs) (Alheshibri et al. 2016, Tsuge 2014), which are different from conventional bubbles as they do not require high pressure (Fig. 5.7).

FBs have a diameter of less than 100 µm and contain visible microbubbles (MB) with a diameter of 1–100 µm and invisible colloidal Ultra-Fine Bubbles (UFB) with a diameter less than 1 µm. Nanobubbles (NB) and ultrafine bubbles (UFB) are synonyms. In 2013, ISO Committee TC 281 was established to promote international standardization of fine bubbles. FB have features such as (1) retention in the system for a long time because of a slow rise rate, (2) electronic repulsion due to negative charge on the bubble surface, and (3) enhancement of dissolution rate by self-pressurization, developed in fields such as environmental engineering, agriculture, aquaculture and medical care. As a result, many studies have been conducted on FB in water, but research on FB in organic solvents has been extremely limited. In particular, the effect of FB on gas-related chemical reactions is not known. Recently, interesting research results on the self-pressurizing effect have been reported by White (German 2016). The Young-Laplace equation ($\Delta P = 2\gamma/r$), which is widely used in surface chemistry, can also be applied to the nanoscale. When smaller bubbles are formed, gas solubilization into the liquid phase is accelerated owing to the increase in the Laplace internal pressure. Therefore, if the UFB can be efficiently produced in various liquid phases, it is possible to construct a high-pressure micro-space inside the liquid phase, even under normal pressure. Therefore, even if the dissolved gas is consumed by the reaction, the dissolved gas is always saturated-dissolved by MB → UFB → dissolution, and thus, a highly efficient reaction system could be constructed (Fig. 5.8). To discuss the possibility of forming a unique reaction field by UFB, the existence of UFB should be confirmed. Bubbles are different from solid particles, as their size varies dynamically, and their scattering intensity is weak. Visible MB can be observed with home cameras and videos, but it is difficult to measure nano-sized UFB by the general Dynamic Light Scattering method (DLS). However, the Brownian motion of particles scattered by laser irradiation is tracked individually in real time, and from the Stokes-Einstein equation, the particle-size distribution and particle number concentration in the suspension can be measured by Nano Tracking Analysis (NTA) (Filipe et al. 2010, Tian et al. 2016). NTA has been applied to the measurement of UFB in water since the late 2000s. For example, when

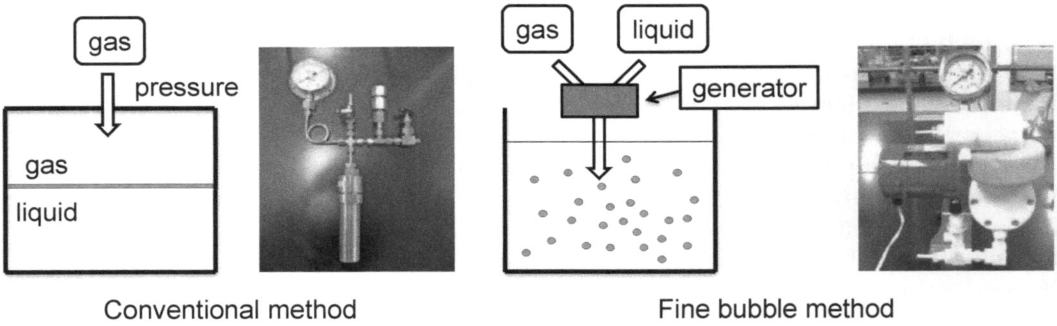

Fig. 5.7: Comparison of gas-liquid reaction methods.

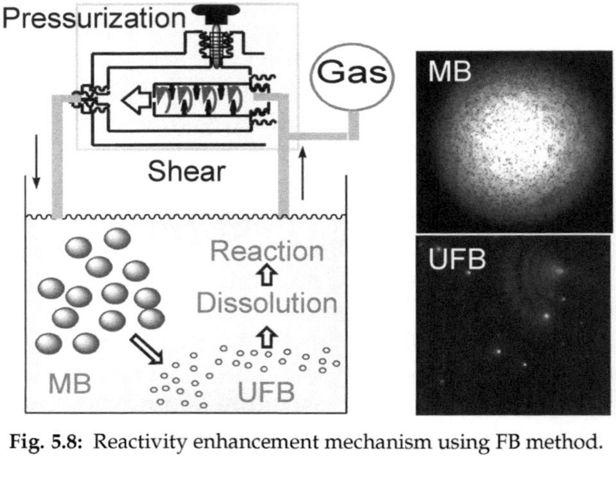

Fig. 5.8: Reactivity enhancement mechanism using FB method.

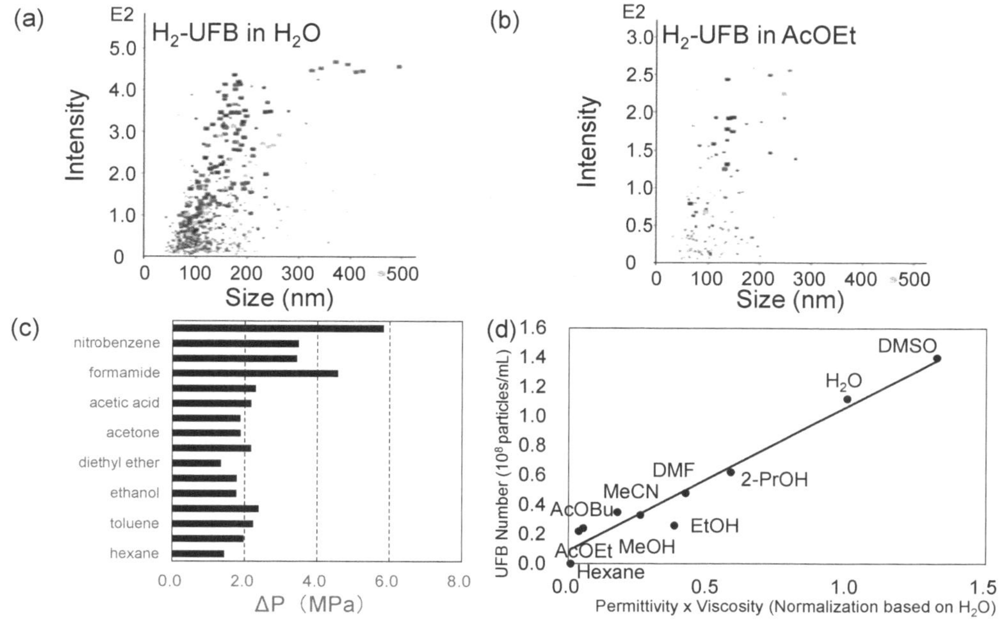

Fig. 5.9: (a), (b) Hydrogen-UFB in water and organic solvent (AcOEt), (c) Laplace internal pressure of 50 nm UFB, (d) relationship between number of air-UFB and permittivity × viscosity.

hydrogen-UFB is produced using the fine bubble generator MA3FS (ASUPU), the UFB number concentration is 1.35×10^8 particles/mL with an average particle diameter of 148.5 ± 6.4 nm (Fig. 5.9a). Using ethyl acetate as a representative organic solvent, UFB, with an average particle diameter of 134.1 ± 10.8 nm, was measured at a concentration of 2.17×10^7 particles/mL (Fig. 5.9b). Since air bubbles with a diameter of about 50 nm were also observed, ΔP of bubbles with a diameter of 50 nm was calculated according to the Young-Laplace equation (Fig. 5.9c). More than 1 MPa of ΔP in various organic solvents is calculated, therefore, a similar effect to that of the pressure vessel can be expected. UFB in the organic solvent were measured, and UFB were observed in a solvent other than hexane. Plotting the permittivity × viscosity on the horizontal axis and the number of UFB on the vertical axis, we found that the number of UFB is proportional to the permittivity × viscosity (Fig. 5.9d).

4.2 Aerobic Oxidation Reaction of Alcohol by FB Method

We attempted to develop an organic synthetic process through the environmentally friendly FB method to serve as a fundamental technology for creating useful molecules and materials. The oxidation reaction of alcohol is an industrially important reaction, but it is inappropriate for industrialization because of the generation of hazardous waste in stoichiometric Jones oxidation, Swern oxidation, and Dess-Martin oxidation. In TEMPO oxidation, which is a representative catalytic oxidation method, sodium hypochlorite is often used as a reoxidant for TEMPO, but volumetric efficiency against productivity decreases because of the low concentration of an aqueous sodium hypochlorite solution (~ 12%). In an effort to overcome these problems, catalytic air oxidation reactions accompanied by formation of water as a by-product has attracted attention. However, there is always

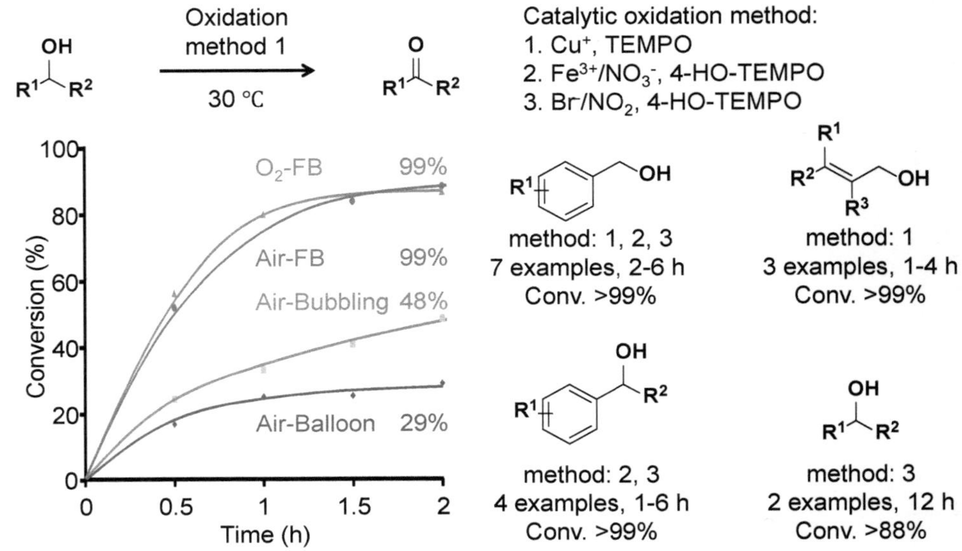

Fig. 5.10: Aerobic oxidation reaction of alcohol by FB method.

a danger of explosion because the oxidation reaction is carried out in flammable organic solvent with combustion-supporting oxygen. On the other hand, in the FB method, since the amount of oxygen can be controlled to a minimum, the suppression of explosion is expected. Air (oxygen) oxidation of primary and secondary alcohols using TEMPO as a catalyst was chosen as a model reaction and the advantages of FB method were evaluated (Fig. 5.10).

The FB, bubbling (supplying gas through ceramic filter), and balloon (supplying gas in balloon at normal pressure) methods were compared. In the Cu^+/TEMPO catalyst system, the FB method exhibited about twice as much reactivity as the bubbling method, achieving 99% conversion. In the FB method, all the substrates were consumed, and similar reactivity was observed with air rather than oxygen, which is industrially advantageous. In the Cu^+/TEMPO catalyst system, the oxidation reaction of secondary alcohol is difficult because of the steric hindrance of TEMPO itself. The 4-HO-TEMPO and $Fe(NO_3)_3/9H_2O$ catalyst system was suitable for the oxidation of aromatic secondary alcohols, and the oxidation reaction proceeded quantitatively. This catalyst system can also be applied to the oxidation of aromatic primary alcohols. On the other hand, contamination with residual metals can be an issue in fine chemical synthesis of pharmaceuticals, agrichemicals, and electronic materials. Catalyst systems containing metal atoms cannot overcome this problem. However, the metal-free organocatalytic oxidation reaction of primary and secondary aromatic alcohols, and aliphatic alcohols was achieved by using mixed FB of nitrogen dioxide and oxygen in the Br^-/NO_2/TEMPO catalyst system. The oxidation reaction of alcohol was accelerated by oxygen or air FB as the final oxidizing agent, and various alcohols were oxidized by selecting the catalyst system (Mase et al. 2011).

4.3 Catalytic Hydrogenation Reaction by FB Method

Hydrogenation is widely used in industry. For example, aniline, an intermediate in the production of various chemicals such as pigments, drugs, and pesticides, is produced at approximately 3 million tons/year by hydrogenation of nitrobenzene. In catalytic hydrogenation, a catalyst supported on a solid phase is often used. In the gas-liquid-solid phase reaction, component A in the gas phase migrates to the liquid phase and then to the solid phase on the catalyst surface, where it reacts with component B in the liquid phase. Because the concentration of component B is much higher than that of component A, the reaction rate of the gas-liquid-solid reaction is often a function of the dissolved concentration of gaseous component A. The dissolved concentration of component A in the liquid phase increases rapidly with FB, and thus, the efficiency of the gas-liquid-solid reaction should be improved. A heterogeneous catalyst (Pd/Al_2O_3) with high abrasion resistance, which is easily separated from the reaction mixture, was used as a catalyst for the model hydrogenation of styrene. Compared to the balloon method and the bubbling method, the reaction rate was improved by 10 times under FB conditions. The FB method is applicable to the hydrogenation of internal or polysubstituted olefins/alkynes, and the reaction proceeds quantitatively (Fig. 5.11, left). Hydrogenation on the scale of 10 g (about 100 mmol) was also achieved (Mase et al. 2013). Similar to the hydrogenation of multiple bonds, the catalytic hydrogenation reaction rate of nitrobenzene was improved by more than 20 times. Various nitroarenes with electron-donating, electron-withdrawing, and heteroaryl groups were easily hydrogenated with hydrogen FB, and synthesis of various aniline derivatives was achieved (Fig. 5.11, right) (Mase et al. 2017). This method can also be applied to the synthesis of fine chemicals, such as liquid crystal intermediates and aramid fiber monomers (Fig. 5.12). In the further development of the three-phase system, the FB method is effective not only in the gas-liquid-solid reaction but also in gas-liquid-liquid phase reaction, i.e., under phase transfer conditions.

4.4 Atmospheric Pressure One-Pot Hydrogen Peroxide Synthesis by FB Method

Hydrogen peroxide (H_2O_2) has higher oxidation-reduction potential than oxygen, and it is drawing attention as an environmentally friendly oxidizing agent on account of the formation of water as a sole by-product. The industrial synthesis of H_2O_2 is a multistep method called the anthraquinone method, which is a process of hydrogenating anthraquinone and synthesizing H_2O_2 by autoxidation of the hydrogenated anthrahydroquinone,

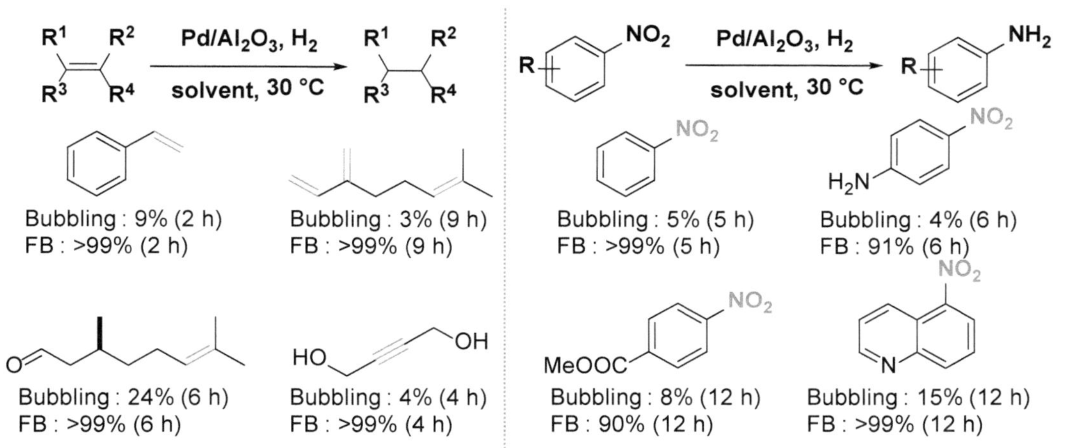

Fig. 5.11: Catalytic hydrogenation reaction by FB method.

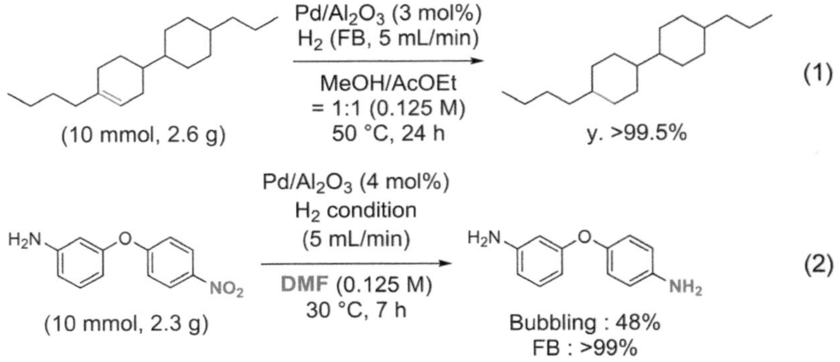

Fig. 5.12: Fine chemical synthesis by FB catalytic hydrogenation.

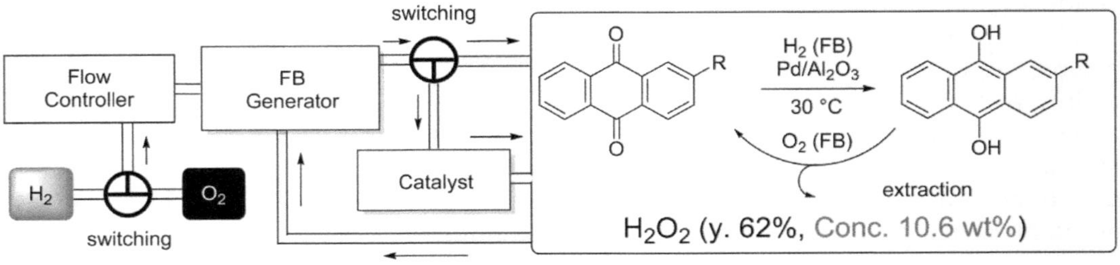

Fig. 5.13: Atmospheric pressure one-pot hydrogen peroxide synthesis by FB method.

followed by purification by extracting H_2O_2. Although the synthesis of large amounts of H_2O_2 is desirable to reduce costs, the risk of storing and transporting large amounts of high-concentration explosive H_2O_2 should be avoided. We aimed to develop a quick method for synthesizing H_2O_2 at room temperature and atmospheric pressure (Fig. 5.13). Under FB conditions, different gases can be introduced into the reaction mixture by switching flow paths. H_2-hydrogenation of anthraquinone, followed by air-oxidation of anthrahydroquinone, can be carried out continuously in the same vessel. Inside the FB generator, a static mixer with a function as an extractor is installed. Two continuous gas-liquid phase reactions and extraction can be carried out only by operation of the switching valve of the reaction system. As a result, 3.1 wt% of H_2O_2 was synthesized under normal temperature and pressure conditions, and the synthesis process was simplified a great deal. Investigating the oxidation reaction using synthesized H_2O_2, the oxidation of secondary alcohol, oxidation of sulfide, and epoxidation reaction of cyclooctene proceeded quantitatively, achieving on-site synthesis. Currently, we are working to further increase the H_2O_2 concentration and achieve a concentration of over 10%.

4.5 Photooxidation by FB Method

Singlet oxygen ($^1O_2$) is easily generated by the UV irradiation of triplet oxygen ($^3O_2$) in the presence of a photosensitizer and has strong oxidizing ability due to the vacant orbital. Because the reaction reagent is oxygen itself, a large amount of dissolved oxygen is necessary to improve reactivity. We investigated the photooxidation reaction using the FB method. Generally, it is challenging to conduct a gas-liquid photoreaction in a pressure-

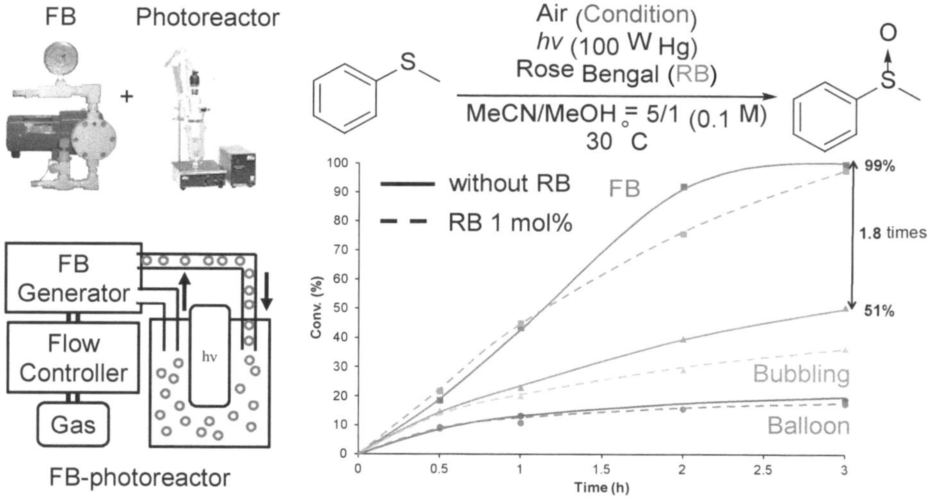

Fig. 5.14: Sensitizer-free photooxidation of sulfide by FB method.

resistant reactor, but by combining an FB generator with a photoreactor, it can be performed at normal temperature and pressure (Fig. 5.14, left). To apply the FB + UV irradiation method to the photooxidative dehydrogenation of amines followed by homocoupling reaction, O_2-FB was generated in the presence of a catalytic amount (0.01 mol% = 100 ppm) of photosensitizer (TPP: tetraphenylporphyrin). The reaction rate was three times faster than the bubbling method. In addition, the amount of photosensitizer could be reduced to 10 ppm. Similarly, the oxidative aromatization of biomass-derived γ-terpinene was applicable to the FB + UV irradiation method. In addition, because this method can be applied not only to oxidative dehydrogenation but also to the oxygen addition reaction, sulfoxide was quantitatively obtained by the photooxidation reaction of sulfide (Fig. 5.14, right). Generally, it is necessary to add a photosensitizer to the $^1O_2$-mediated reaction, but this oxidation reaction gave sulfoxide selectively even without any photosensitizer. Therefore, the desired sulfoxide was easily obtained by stopping the irradiation and gas supply after complete oxidation and then simply removing the solvent.

4.6 FB-flow Method

FB-mediated organic synthesis, introduced so far, is a batch system using circulating FB. In the flow method, when the gas supply is large, it becomes a pipe flow. The reactivity is high but productivity decreases; thus, recovery of the exhaust gas becomes difficult. It is said that the slug flow, which is a gas-liquid alternating flow, is most efficient, and the constriction of the narrow gas-liquid interval is important. We proposed the FB flow method as a new alternative method (Fig. 5.15). Recently, better results than slug flow were obtained from the viewpoint of yield and chemoselectivity.

FB is a technology developed in Japan, with a substantial application field. The development of general equipment has evolved from small to large scale, but the development of an FB generator has developed from large to small scale. Application of FB-mediated organic synthesis to industrialization is considered promising. Furthermore, this method can be implemented by installing an FB generator in an existing reaction

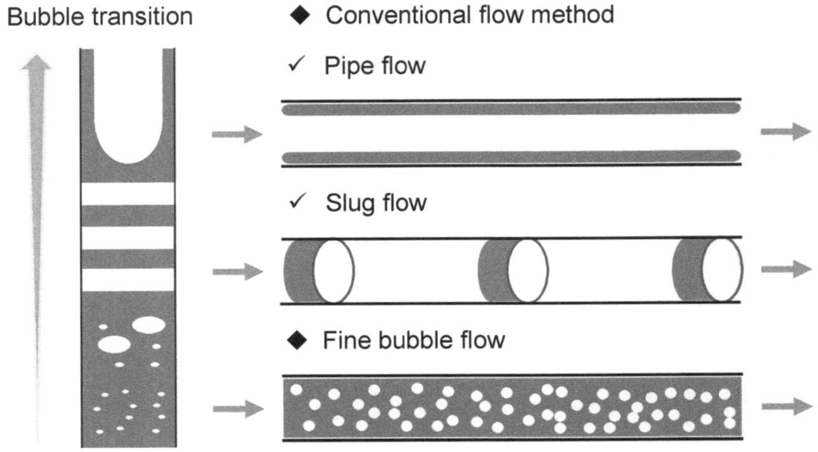

Fig. 5.15: Application of FB method to flow reaction.

vessel, which is advantageous in terms of cost. This research is a "quiet revolution" study, but we believe that the FB method will produce a huge ripple effect.

5. Organic Synthesis Using Microwave-Assisted Flow Reactor: Rapid Optimization of Reaction Conditions

5.1 μW Flow Method: Desktop Plant

In recent chemical and pharmaceutical industries, it has been desirable to shift from mass production/disposal processes to small, diverse, highly efficient and sustainable manufacturing processes. Microwave (μW) heating is known as one of the most highly efficient material manufacturing methods and is widely used for organic synthesis because of advantages such as uniform and rapid heating and reaction acceleration effects (Tierney and Lidstrom 2007). However, because of the limitation of the penetration depth of microwaves, it was difficult to increase production scale by using large reaction vessels. These problems can be solved by implementing a flow system using pipes instead of batch systems using conventional flasks. We aimed to construct a large-scale flow system that meets the requirements of pharmaceutical products (~ 1 kg/d) and special chemicals (~ 10 kg/d). The problem with the conventional μW flow method is the difficulty in fixing the relative positional relationship with the sample, because the electric field distribution of the microwave region is complicated. We thought that these problems could be solved by using a "μW resonant cavity flow system" designed and developed by SAIDA·FDS.

Since protection and deprotection occupy approximately 20% of the pharmaceutical synthesis process, efficient and quantitative methods are required (Carey et al. 2006). Acetylation of alcohol was investigated as a model reaction of scale-up flow synthesis (Fig. 5.16). When benzyl alcohol and acetic anhydride were used as model substrates, the acetylation reaction proceeded quantitatively by microwave heating under no-catalyst/solvent-free conditions. Acetylation of various primary, secondary and phenolic alcohols proceeded quantitatively. In addition, this process was also applied to tertiary alcohols, which did not proceed in high yield under batch conditions. Since the composition of the contents changes as the reaction progresses, the absorption of microwaves continuously

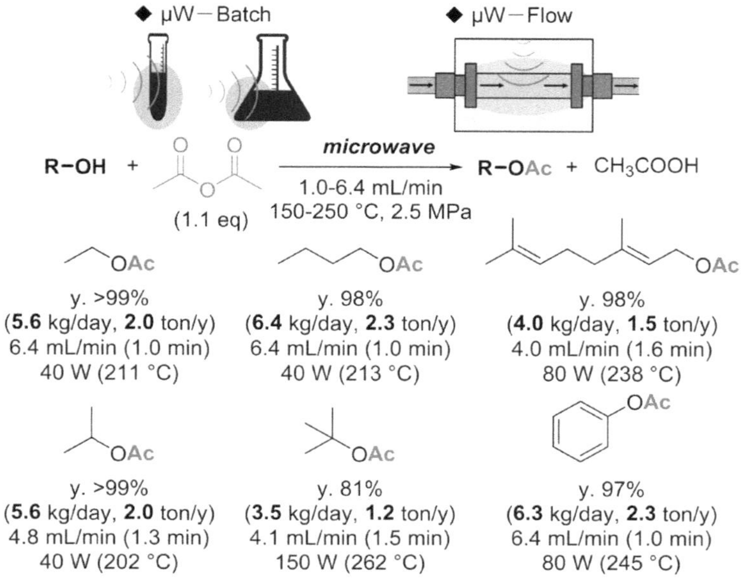

Fig. 5.16: Application to μW flow method.

changes. However, by using this device, which can automatically set the optimum frequency at all times, productivities of 3.5–6.4 kg/day and 1.2–2.3 ton/year were achieved with a residence (reaction) time of 1.0–1.6 minutes. This system can be installed in the laboratory bench. As annual production reaches the ton level in the one-flow path, it is expected to be put to practical use as a desktop plant.

5.2 Rapid Condition Optimization Using Flow Reaction System

Because the reaction conditions can be changed strictly and easily, the flow method is suitable for identifying the optimal reaction conditions. Furthermore, by using in-line monitoring, it is possible to analyze the progress or abnormality of the reaction in real time. Focusing on the features of these flow methods, we aimed to develop a condition examination and analysis method that can promptly optimize reaction conditions (Fig. 5.17) (Vámosi et al. 2019). First, reaction conditions were set based on the design of experiment and optimized. For model reactions, Fischer indole synthesis, which is widely used for alkaloids and pharmaceutical synthesis, was chosen. The upper and lower limits of the irradiation power and flow rate of the microwave flow system were equally divided into three, and overall, nine points were optimized to eliminate the possibility of biased optimizing within a specific range. After analyzing nine samples using flow IR and GC, the points giving the best yield were divided into four, and four more experiments were carried out. A three-dimensional approximate surface was created from 13 experimental data and a 14th experiment was conducted according to the expected optimum condition. As a result of actually examining the predicted optimum conditions, Fischer indole product was obtained at an IR yield of 98% (GC yield > 99%) under the conditions of 6.2 mL/min, 250°C. This result exceeded the previously reported yield of 83% (Yokozawa et al. 2015). Within 140 minutes, optimization was achieved in a total of 14 experiments. In addition, this method is also applicable to the acetylation of tertiary alcohols, even though substrates

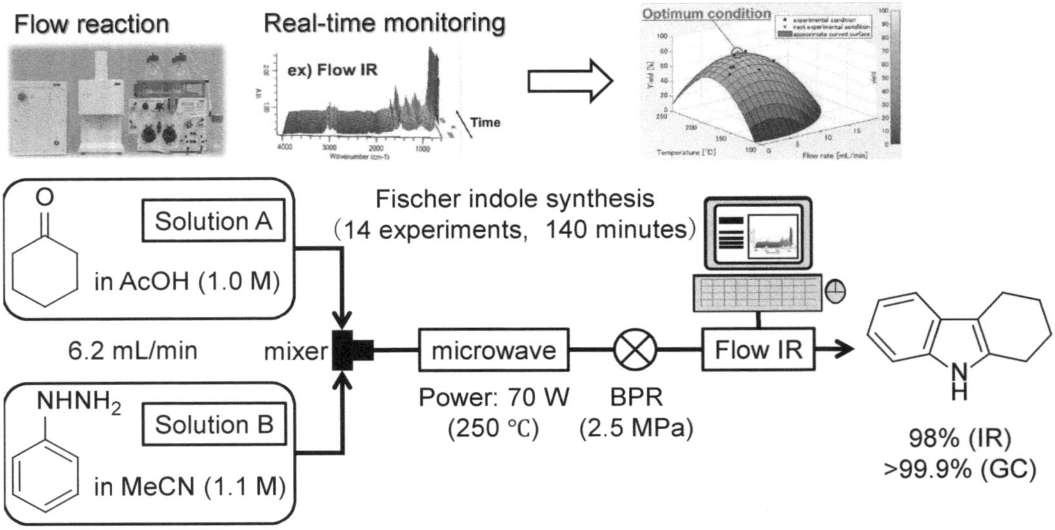

Fig. 5.17: Rapid condition optimization using flow reaction system.

and products may decompose during the reaction. The optimization method developed in this research is called the "9 + 4 + 1 method". This optimization method is expected to be applied not only to model substrates but also to other various substrates. Furthermore, it is expected to be applied to optimum condition identification for low-yield batch reaction.

6. Conclusion

We discussed catalytic organic reaction in water, supercritical fluid, fine bubbles, microwave and tube flow, but water is a common medium in biological reactions, and supercritical fluids are also common on the Earth. In addition, ultra-fine bubbles, which have been difficult to visualize until recently, are clearly present, now that they can be generally visualized. The process of challenging and then solving a previously unresolved problem leads to the general adoption of solutions and, ultimately the advancement of culture. It is my hope that this study will help to facilitate this process.

Acknowledgments

This work was supported in part from the Ministry of Education, Culture, Sports, Science and Technology (MEXT), the Japan Society for the Promotion of Science (JSPS), and the Suzuki Foundation.

References

Alheshibri, M., J. Qian, M. Jehannin and V.S.J. Craig. 2016. A history of nanobubbles. Langmuir 32: 11086–11100.

Brunner, G. 2010. Applications of supercritical fluids. Annu. Rev. Chem. Biomol. Eng. 1: 321–342.

Carey, J.S., D. Laffan, C. Thomson and M.T. Williams. 2006. Analysis of the reactions used for the preparation of drug candidate molecules. Org. Biomol. Chem. 4: 2337–2347.

Chen, D.Y.-K. 2018. A personal perspective on organic synthesis: Past, present, and future. Isr. J. Chem. 58: 85–93.

Dalko, P.I. 2013. Comprehensive Enantioselective Organocatalysis: Catalysts, Reactions and Applications. Wiley-VCH Verlag GbmH & Co. KGaA.

Filipe, V., A. Hawe and W. Jiskoot. 2010. Critical evaluation of nanoparticle tracking analysis (NTA) by nanosight for the measurement of nanoparticles and protein aggregates. Pharm. Res. 27: 796–810.

German, S.R., M.A. Edwards, Q. Chen and H.S. White. 2016. Laplace pressure of individual H2 nanobubbles from pressure–addition electrochemistry. Nano Lett. 16: 6691–6694.

https://www.mhlw.go.jp/hourei/doc/tsuchi/T180809I0040.pdf.

https://www.mhlw.go.jp/hourei/doc/tsuchi/T181018I0050.pdf.

Jung, Y.S. and R.A. Marcus. 2007. On the theory of organic catalysis on water. J. Am. Chem. Soc. 129: 5492–5502.

Kamber, N.E., W. Jeong, R.M. Waymouth, R.C. Pratt, B.G.G. Lohmeijer and J.L. Hedrick. 2007. Organocatalytic ring-opening polymerization. Chem. Rev. 107: 5813–5840.

Kobayashi, S. 2012. Science of Synthesis: Water in Organic Synthesis. Georg Thieme Verlag.

List, B. 2012. Science of Synthesis: Asymmetric Organocatalysis 1, Lewis Base and Acid Catalysts. Georg Thieme Verlag.

Maruoka, K. 2012. Science of Synthesis: Asymmetric Organocatalysis 2, Broensted Base and Acid Catalysts, and Additional Topics. Georg Thieme Verlag.

Mase, N., K. Watanabe, H. Yoda, K. Takabe, F. Tanaka and C.F. Barbas, III. 2006. Organocatalytic direct Michael reaction of ketones and aldehydes with β-nitrostyrene in brine. J. Am. Chem. Soc. 128: 4966–4967.

Mase, N., Y. Nakai, N. Ohara, H. Yoda, K. Takabe, F. Tanaka et al. 2006. Organocatalytic direct asymmetric aldol reactions in water. J. Am. Chem. Soc. 128: 734–735.

Mase, N., N. Noshiro, A. Mokuya and K. Takabe. 2009. Effect of long chain fatty acids on organocatalytic aqueous direct aldol reactions. Adv. Synth. Catal. 351: 2791–2796.

Mase, N. and C.F. Barbas, III 2010. In water, on water, and by water: mimicking nature's aldolases with organocatalysis and water. Org. Biomol. Chem. 8: 4043–4050.

Mase, N., T. Mizumori and Y. Tatemoto. 2011. Aerobic copper/TEMPO-catalyzed oxidation of primary alcohols to aldehydes using a microbubble strategy to increase gas concentration in liquid phase reactions. Chem. Commun. 47: 2086–2088.

Mase, N., S. Isomura, M. Toda and N. Watanabe. 2013. Micro and nanobubble based strategy for gas-liquid-solid multiphase reactions: Palladium-catalysed hydrogenation of carbon-carbon unsaturated bonds. Syn. Lett. 24: 2225–2228.

Mase, N., Y. Nishina, S. Isomura, K. Sato, T. Narumi and N. Watanabe. 2017. Fine-bubble-based strategy for the palladium-catalyzed hydrogenation of nitro groups: Measurement of ultrafine bubbles in organic solvents. Syn. Lett. 28: 2184–2188.

Mase, N., S. Moniruzzaman, J. Mori, F. Ishizuka, S. Kumazawa, K. Yamamoto et al. 2018. Organocatalytic ring-opening polymerization of L-lactide in supercritical carbon dioxide under plasticizing conditions. Tetrahedron Lett. 59: 4392–4396.

Mase, N., S. Moniruzzaman, Y. Yamamoto, K. Nakaya, T. Sato and T. Narumi. 2018. Organocatalytic stereoselective cyclic polylactide synthesis in supercritical carbon dioxide under plasticizing conditions. Polymers 10: 713.

Mehta, R., V. Kumar, H. Bhunia and S.N. Upadhyay. 2005. Synthesis of poly(lactic acid): A review. J. Macromol. Sci. Polym. Rev. C 45: 325–349.

Narayan, S., J. Muldoon, M.G. Finn, V.V. Fokin, H.C. Kolb and K.B. Sharpless. 2005. "On water": Unique reactivity of organic compounds in aqueous suspension. Angew. Chem. Int. Ed. 44: 3275–3279.

Nederberg, F., E.F. Connor, M. Moller, T. Glauser and J.L. Hedrick. 2001. New paradigms for organic catalysts: The first organocatalytic living polymerization. Angew. Chem. Int. Ed. 40: 2712–2715.

Nicolaou, K.C. 2018. The emergence and evolution of organic synthesis and why it is important to sustain it as an advancing art and science for its own Sake. Isr. J. Chem. 58: 104–113.

Rideout, D.C. and R. Breslow. 1980. Hydrophobic acceleration of diels-alder reactions. J. Am. Chem. Soc. 102: 7816–7817.

Rovis, T. 2016. Handbook of Reagents for Organic Synthesis: Reagents for Organocatalysis. John Wiley & Sons Ltd.

Shin, E.J., A.E. Jones and R.M. Waymouth. 2012. Stereocomplexation in cyclic and linear polylactide blends. Macromolecules 45: 595–598.

Tian, X.S., M.R. Nejadnik, D. Baunsgaard, A. Henriksen, C. Rischel and W. Jiskoot. 2016. A comprehensive evaluation of nanoparticle tracking analysis (nanosight) for characterization of proteinaceous submicron particles. J. Pharm. Sci. 105: 3366–3375.

Tierney, J.P. and P. Lidstrom. 2007. Microwave Assisted Organic Synthesis. Blackwell Publishers.

Torres, R.R. 2013. Stereoselective Organocatalysis: Bond Formation Methodologies and Activation Modes. John Wiley & Sons, Inc.

Tsuge, H. 2014. Micro- and Nanobubbles: Fundamentals and Applications. CRC Press.

Vámosi, P., K. Matsuo, T. Masuda, K. Sato, T. Narumi, K. Takeda et al. 2019. Rapid optimization of reaction conditions based on comprehensive reaction analysis using a continuous flow microwave reactor. Chem. Rec. 19: 77–84.

Yokozawa, S., N. Ohneda, K. Muramatsu, T. Okamoto, H. Odajima, T. Ikawa et al. 2015. Development of a highly efficient single-mode microwave applicator with a resonant cavity and its application to continuous flow syntheses. RSC Adv. 5: 10204–10210.

Chapter 6

Sequestration of Carbon Dioxide in Deep Portions of Ocean through a Gas-lift Effect

Takayuki Saito

1. Introduction

Human's history has been of increases in the population and in the energy consumption directly relating to the population. Figure 6.1 shows a solid evidence of this relationship. Getting much more amount of free-spending energy enabled human's rapid growth in the population. Why do they need such huge amounts of energy in order to exist and to keep their act? This question is one of the bases of Research Institute Green Science and Technology. The motivation of the members of Green Energy Research Division is to struggle with this difficult question.

We, the members of Green Energy Research Division, started to discuss the relationship between input energy to the earth and primary mammals. Figure 6.2 shows the relationship between the habitat density of plant-eating mammals and their weight (Damuth 1981). This relationship is expressed by the following simple equation,

$$\log ND = -0.75 (\log W) + 4.23 \quad (1)$$

To put this briefly, this fact indicates that the habitat density of larger plant-eating mammals becomes smaller. For instance, the habitat density of a plant-eating mammal of 60 kg (similar to human's average weight) is one to ten; i.e., there is a limit to the habitat density of plant-eating mammals and flesh-eating mammals. From Eq. (1), the habitat density of a plant-eating mammal of 62 kg is calculated at 2.9/km^2. This fact indicates that the habitat density of human beings (a species of animals) that are positioned in halfway situation between plant-eating mammals and flesh-eating mammals is limited and is less than 2.9. Actually, the habitat density of all mammals is expressed by

$$N_D = 55 \times w^{-0.90} \quad (2)$$

From this equation, the habitat density of human beings is calculated at 1.3/km^2.

Green Energy Research Division, Research Institute of Green Science and Technology, Shizuoka University, 3-5-1 Johoku, Naka-ku, Hamamatsu, Shizuoka 432-8011, Japan.
Email: saito.takayuki@shizuoka.ac.jp

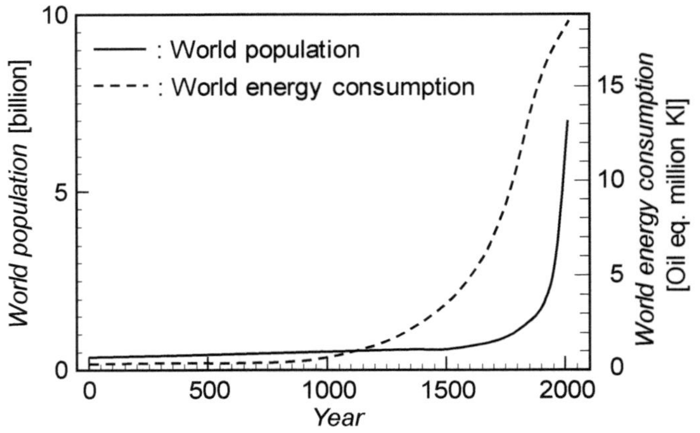

Fig. 6.1: Increases in world population and world energy consumption. World population: modified based on UN, World Population Prospects (2010). World energy consumption: modified based on Energy Transitions: History, Requirements, Prospects.

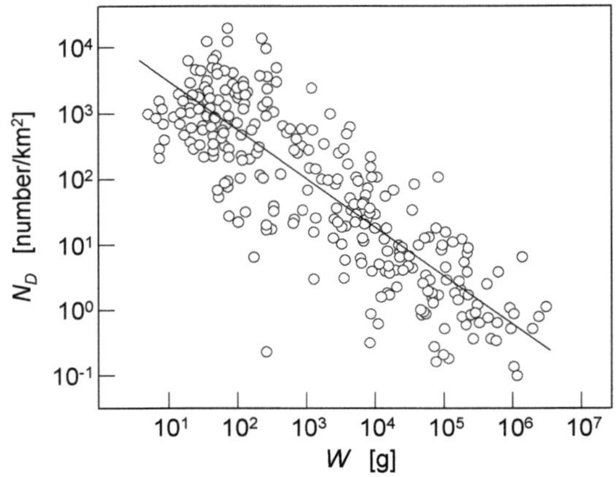

Fig. 6.2: Relationship between habitat density of plant-eating mammals and their weight (Damuth 1981). Copyright (1981) Springer Nature.

As of 2017, the world population was 7550 million and the population density was 58/km2. This value of 58 is 45-fold of 1.3. Required energy for human beings to exist is about 8.4 × 106 J/day. A human consumes 2.52 × 108 J/day. Thus, the human being consumes 30-fold energy of its existence. The humans obtain this amount of energy mainly from fossil fuel, atomic energy and hydraulic energy. This excessive energy consumption keeps such high population density of humans.

2. CO_2 Sequestration at Deep Portions of the Ocean by Gas-lift System with Shallow Injection and Deep Releasing

2.1 Introduction

To mitigate global warming, the best option is to develop and to use CO_2-free new energy; the second-best one is to restrain or to stop consumption of fossil fuel such as oil, coal and natural gas; the third-best one is to save energy; the fourth-best one is to improve energy efficiency. These are rational, however, as stated above humans need enormous amount of energy to maintain themselves. Under the present circumstances of no new energy taking the place of the fossil fuel, best mixing of these options is reasonable. Furthermore, according to progress status of the global warming, emergency evacuation procedures such as underground sequestration of CO_2 or ocean sequestration of CO_2 is required, for the specific purpose of reducing estimated peak of the atmospheric CO_2 concentration. According to IPCC report 2013 (Intergovernmental Panel on Climate Change 2013), annually about 32.6 Gt of CO_2 is discharged from humans' activities, and about 45% of the CO_2 remains in the atmosphere for several tens of years to several hundred years, in consideration of these, the emergency evacuation procedures should be developed in advance. Here deep ocean sequestration of CO_2 through a gas-lift pump system called Gas Lift Advanced Dissolution system (Kajishima et al. 1995, 1997, Kosugi et al. 2001, Niwa et al. 2005, Saito et al. 1999, 2000, 2001) of CO_2 (GLAD), are described and discussed on the basis of experimental and computational results done in RIGST.

2.2 *Outline of Ocean Sequestration of CO_2*

Figure 6.3 shows carbon circulation according to IPCC report 2013 (Intergovernmental Panel on Climate Change 2013), Annual discharge of CO_2 from the human's activities is 32.6 $GtCO_2$ on an average. 14.7 $GtCO_2$ stays behind in the atmosphere, and the 9.5 $GtCO_2$ is absorbed in land (plants, animals, soil, etc.), and the 8.4 $GtCO_2$ is absorbed in the ocean. Before the Industrial Revolution, natural CO_2 transferred from the atmosphere and land to the ocean is estimated at 3.3 $GtCO_2$; the 2.6 $GtCO_2$ is transferred from the ocean to the atmosphere and the 0.7 $GtCO_2$/year is fixed on the deep ocean floors by biological pump. Before the Industrial Revolution, the amount of CO_2 discharged from humans' activities was small and was put in the natural carbon circulation smoothly. Carbon circulation was balanced for long term before the Industrial Revolution. However, late during the 20th century and early in the 21th century, 14.7 $GtCO_2$ stays behind in the atmosphere disproportionally.

Even now, the ocean is an enormous reservoir (i.e., an enormous absorbent source) for various kinds of substances. Furthermore, the ocean has three stably-stratified layers: surface layer, intermediate layer and deep layer. This layer structure keeps a long-term (a period of several hundred years) ocean general circulation, and works as a high-performance damper to delay the peak of atmospheric CO_2 concentration. The ocean sequestration of CO_2 utilizes the above characteristics of the ocean. Hoffert et al. (1979) reported the effects of the ocean sequestration of CO_2 as shown in Fig. 6.4. As shown in this figure, the ocean sequestration of CO_2 is a mere buffer to avoid the rapid increase in the atmospheric CO_2 concentration which the earth had not experienced. Deeper injection is more efficient, as the damper against the increase in the atmospheric CO_2 concentration.

As described above, the ocean sequestration of CO_2 is one of the emergency evacuation options; therefore, the necessary conditions are summarized as follows;

1) technically, highly feasible and short development period,
2) long term sequestration, at least several hundred years more,
3) low secondary environmental impact,
4) large net reduction of CO_2 in the atmosphere,
5) low cost.

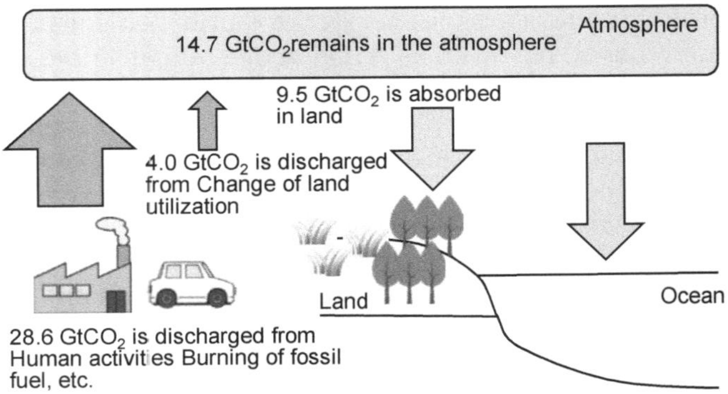

Fig. 6.3: Carbon circulation. Modified based on IPCC Report 2013.

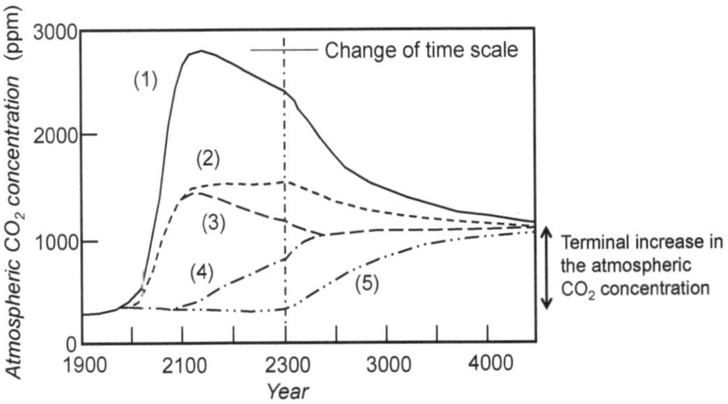

Fig. 6.4: The effects of the ocean sequestration of CO_2, estimated by Hoffert et al. (1979). (1) 100% of CO_2 from humans' activities is discharged in the atmosphere. (2) 50% of the CO_2 is injected at a depth of 1500 m. (3) 50% of the CO_2 is injected at a depth of 4000 m. (4) 100% of the CO_2 is injected at a depth of 1500 m. (5) 100% of the CO_2 is injected at a depth of 4000 m. Copyright (1979) Springer Nature.

2.3 Typical Options of the Ocean Sequestration of CO2, and Analysis of them

Thus far, technically feasible options the ocean sequestration of CO_2 are categorized into three types: (a) liquid CO_2 sequestration (: pool of liquid CO_2) on the deep ocean floor, (b) direct release of liquid CO_2 into intermediate water, and (c) direct injection of gaseous CO_2 into shallow water, as illustrated in Fig. 6.5 (Saito et al. 2000). According to a paper (Saito et al. 2000) published in Environmental Science and Technology, these three are outlined as follows;

(a) Liquid CO_2 sequestration on the deep ocean floor

Liquid CO_2 is pumped from a transport ship and dumped on the deep ocean floor deeper than 3,400 m, at which the Liquid CO_2 density is larger than that of the surrounding seawater, through a pipeline. CO_2-gas-hydrate covers the surface of the liquid CO_2 stored like a pond at a pan. The CO_2-gas-hydrate is stable under deep-ocean conditions of temperature and pressure. This property enables long-term isolation of CO_2. Note that the seawater near the CO_2-gas-hydrate is supersaturated by CO_2; i.e., this indicates the seawater is acidized.

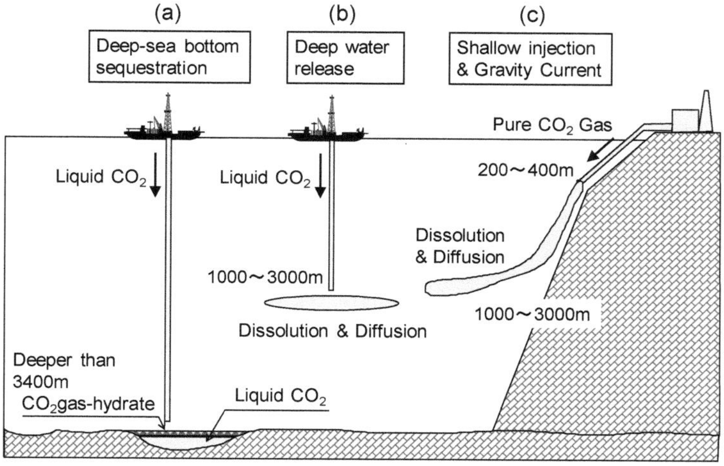

Fig. 6.5: Category of ocean sequestration of CO_2 (Saito et al. 2000). Copyright (2000) American Chemical Society.

Separation of CO_2 from exhausted gas of thermal power plant, and liquefaction of the CO_2 consume 10–30% of energy generated at a power plant (Department of Energy USA 1993). This immediately means cost rise, and decrease in the net CO_2 reduction, as well as direct disposal of energy into the ocean.

(b) Direct release of liquid CO_2 into intermediate water

The liquid CO_2 is directly released into intermediate-depth waters of 1,000 m to 3,000 m through a pipeline from a moving transporter ship. Under these conditions of depth and pressure, the CO_2-gas-hydrate covers the surfaces of the dispersed CO_2 droplets. This effect enables gradual dissolution of CO_2 into the surrounding seawater, and transportation of the CO_2-concentration by intermediate currents. Note that the seawater near the CO_2-gas-hydrate is supersaturated by CO_2; i.e., this indicates the seawater is acidized. The density of the liquid CO_2 under intermediate water conditions is smaller than that of the surrounding seawater. This indicates ascent of the CO_2 droplets.

From extra energy consumption, and decrease in the net CO_2 reduction, this option has the same situation as option (a).

(c) Direct injection of gaseous CO_2 into shallow water

Gaseous CO_2 separated at a thermal power plant is pumped and directly released into shallow waters (i.e., surface water) of 200 m to 400 m from an offshore pipeline laid on the continental shelf. The CO_2 bubbles dissolve into the shallow waters while they ascend. Since the initial injection depth is small, liquefaction of CO_2 is not required. Hence, cost and energy required for sequestration are considerably improved. CO_2-rich seawater sinks to the deep ocean due to a small density difference between the CO_2-rich seawater and ambient seawater. Note that upward currents owing to the bubbles' buoyancy occur, and transportation of the CO_2-rich seawater owing to turbulence of ambient ocean currents occurs, before the CO_2-rich seawater reaches intermediate depth waters. Furthermore, there are thermocline and pycnocline in the ocean. These disturb settling of the CO_2-rich seawater. In addition, in surface water, organism species are most diverse and the number of the organism is the largest.

2.4 Principle of the GLAD System

To overcome these difficulties of the typical options described above, we invented the GLAD system and have been improving its energy efficiency and environmental acceptability, on the basis of fundamental researches in multiphase flows and measurement science. Here, the GLAD system (Kajishima et al. 1995, 1997, Kosugi et al. 2001, Niwa et al. 2005, Saito et al. 1999, 2000, 2001) is outlined. It is an ingeniously practical application of a gas lift pump, which dissolves gaseous CO_2 at shallow portions of 300–400 m and transports the seawater including the CO_2 into deep portions of 3000–5000 m by a gas-lift pumping effect. Figure 6.6 (Kajishima et al. 1995, 1997, Saito et al. 1999, 2000) and Fig. 6.7 (Kosugi et al. 2001, Niwa et al. 2005, Saito et al. 2001) diagram two types of the GLAD system.

A prototype of the GLAD system shown in Fig. 6.6 is composed mainly of (a) inverse-J-shape pipes built at a shelf edge, (b) drain pipes laid on a continental slope, (c) floats equipped with insoluble gases separators, (d) underwater pipeline and (e) manifold laid on a continental shelf, and (f) compressor built at a thermal power plant. The inverse-J-shape pipes are suspended from the floats. CO_2 gas with a concentration of 80 to 90% coarsely separated from exhaust gas from thermal power plant pump is pumped by the compressor and transported through the underwater pipeline to a manifold. The gas including CO_2 is injected into risers of the inverse-J-shape pipes at a depth of 300 m to 400 m. The bubbles including CO_2 gas ascend in the risers, and at the same time the CO_2 included in the bubbles dissolve to seawater. The buoyancy of the bubbles and the decrease (owing to inclusion of the bubbles) in superficial density in the risers generate a gas-lift pumping effect. CO_2 dissolution into seawater is completed in the risers, i.e., seawater including CO_2 is perfectly isolated from the shallow water. The insoluble gases of mostly N_2 and less O_2 are separated and discharged by the separators, without pressure drop of the air-lift pumping; the pumping effect is thus preserved. The pumping effect is strong enough to overcome a friction loss in the downcomers and drain pipes and also to transport the seawater including CO_2 into deep portions. The density of the seawater including CO_2 is slightly larger than that of ambient seawater; in the downcomers and drain pipes, the gravitational force thus acts on the seawater including CO_2, additionally.

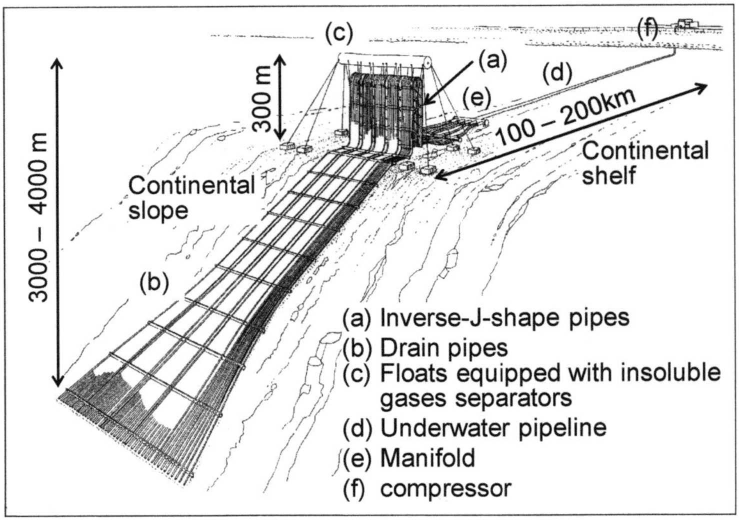

Fig. 6.6: Prototype of the GLAD system (Saito et al. 2000). Copyright (2000) American Chemical Society.

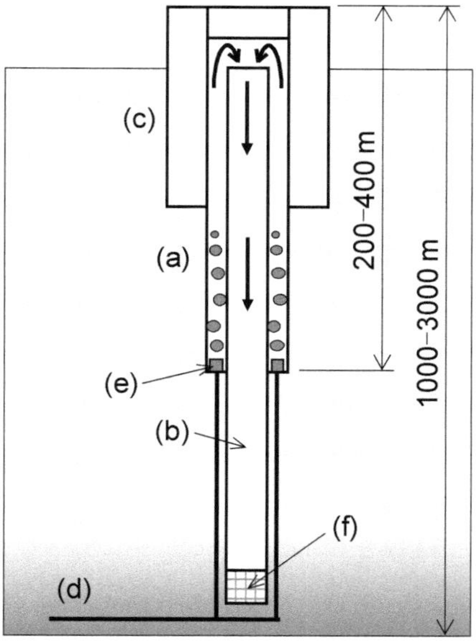

Fig. 6.7: Improved type of the GLAD system. SPAR-type GLAD. (a) Riser (dissolution pipe), (b) Downcomer (drain pipe), (c) Float, (d) Underwater pipeline, (e) Bubble generator, (f) Diffuser.

For the purpose of reducing the ocean current drag force acting on the suspended pipes as well as of improvement of the stability and safety of the system, the GLAD system was redesigned as shown in Fig. 6.7. This new type was named SPAR-type GLAD; SPAR is a type of floating oil platform used in very deep waters such as the US Gulf of Mexico, offshore Malaysia and Norway. A riser (dissolution pipe) (a) and a downcomer (drain pipe) (b) constitute partially a concentric pipe. These pipes are suspended by float (c). Gas including CO_2 is pumped from a thermal power plant through an underwater pipeline (d), and injected into the riser through a diffuser (f). In the same way as the prototype GLAD system, the gas-lift pumping effect is generated in the riser. The upper end of the riser is open to the atmosphere; therefore, without the gas separator, the insoluble gases of mostly N_2 and less O_2 are separated and discharged to the atmosphere, naturally. The seawater including CO_2 flows toward the deep portions in the downcomer by hydraulic head.

Both GLAD systems do not need liquefaction and highly pure CO_2; hence, energy efficacy and cost for a total system of CO_2 sequestration in the deep ocean are improved dramatically. Furthermore, since the pressure at the gas injection depth is low, amount of CO_2 dissolution is a little, from the Henry's law. Immediately, a degree of acidity is mild.

2.5 Results of Fundamental Researches on Bubble Dissolution to the Surrounding Liquid

2.5.1 Mass Transfer from a Single CO_2 Bubble to the Surrounding Water

Perfect dissolution of the CO_2 of the bubbles to the seawater in the risers/riser of 200 m to 300 m and satisfactory gas-lift pumping power has to be simultaneously achieved. A gas-liquid two-phase flow takes patterns of bubbly flow, slug flow, froth flow and annular dispersed flow in response to the gas and liquid flow rate (Ohnuki and Akimoto 1996)

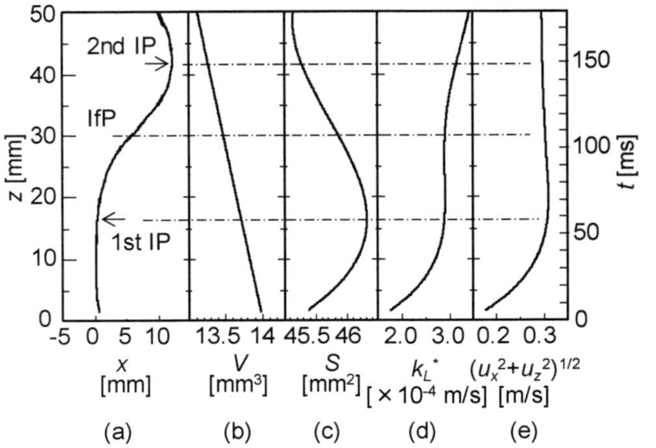

Fig. 6.8: Properties of a single CO2 bubble in pure water (Saito and Toriu 2015). Trajectory of the bubble centroid motion, Bubble volume, surface area, and instantaneous average mass transfer coefficient kL*. (a) Trajectory of the bubble centroid motion in a x–z plane; (b) Bubble volume V; (c) Bubble surface area S; (d) Instantaneous average mass-transfer coefficient kL*; (e) Bubble velocity (ux2 + uz2)1/2. 1st IP: First inversion point, IfP: Inflection point, 2nd IP: Second inversion point. Copyright (2015) Elsevier.

A bubbly flow is the best for the GLAD system. Hence, mass transfer from a single CO_2 bubble to the surrounding water is a key.

Saito's group (Hanyu and Saito 2010, Morikawa et al. 2008, Saito and Toriu 2015, Sakakibara et al. 2007, Sone et al. 2008, Tsuchiya et al. 2003, 1999, 2001) experimentally carried out a series of researches in the above mass transfer and in interaction of the bubble surface motion and the surrounding liquid motion in consideration of the bubble's gravity-center motion. Properties of a single CO_2 bubble of 2.9 mm in initial equivalent diameter are shown in Fig. 6.8 (Saito et al. 2015). This bubble size is good for initial size for the GLAD system. This bubble ascends zigzagging with dissolution into the surrounding water. Its mass transfer coefficient ranges from 1.8×10^{-4} to 3.4×10^{-4}.

These values of mass transfer coefficient are large enough to completely dissolve CO_2 included in the bubbles into seawater in an ascent distance of 200–300 m.

2.5.2 Influences of Bubble Surface Contamination

Various substances included in seawater contaminate a bubble surface, inevitably. This contamination influences the bubble's centroid motion, bubble's surface motion and mass transfer. Saito et al. (Huang and Saito 2015, 2017a, 2017b, Saito et al. 2010, Tsuchiya et al. 1997) investigated influences of the bubble surface contamination on these systematically, by using a single CO_2 bubble of 2.9 mm in equivalent diameter.

Saito et al. simulated bubble surface contamination by using one-pentanol (mass concentration of 500 ppm). They reported that, between a pure system (: pure water) and a contaminated system (: 500-ppm-onepentanol solution), there are the significant differences in the bubble's centroid motion, velocity, mass transfer coefficient and wakes. Figure 6.9 (Huang and Saito 2017b) shows them. The amplitude of the zigzagging motion attenuated in the contaminated system. The velocity in the contaminated system periodically changed and was smaller by about 30% than that in the pure system. Furthermore, the mass transfer coefficient in the contaminated system became about one tenth of that in the pure system.

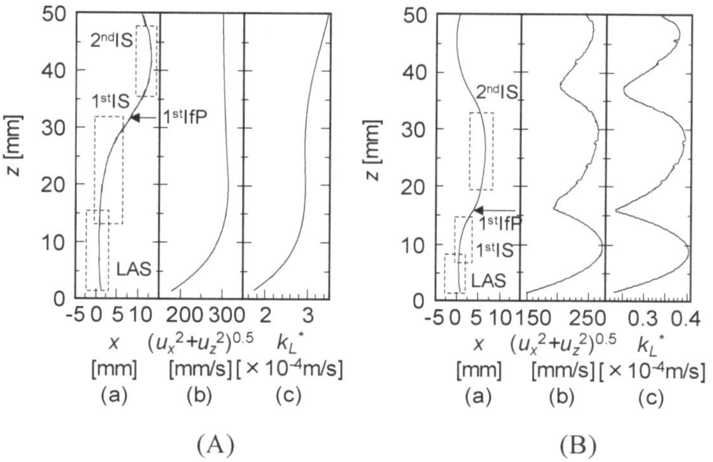

Fig. 6.9: Bubble's trajectory (a), velocity (b), average instantaneous mass transfer coefficient (c) in purified water (A) and contaminated water (B).

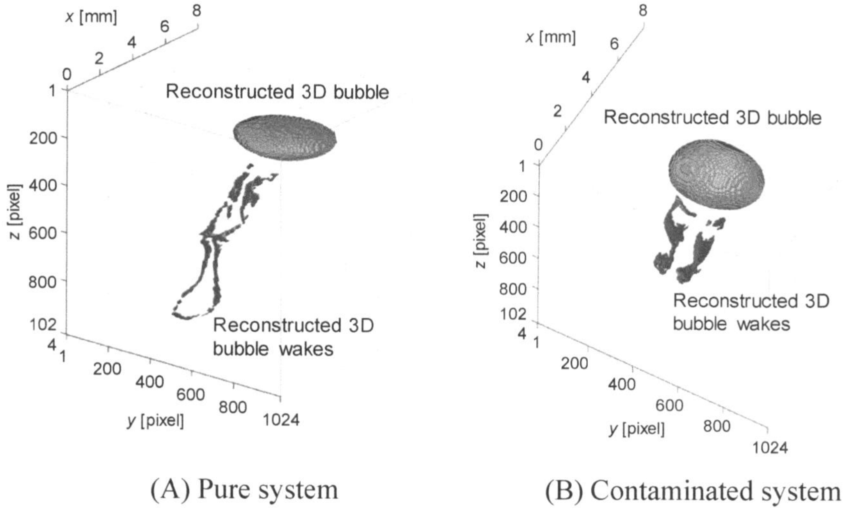

Fig. 6.10: Three-dimensional babble wakes (Huang and Saito 2017a). Copyright (2017) Elsevier.

The three-dimensional bubble wake structure was reconstructed from results measured by stereo PIV (Particle Image Velocimeter) system and LIF/HPTS (Laser Induced fluorescence with 8-Hydroxypyrene-1, 3, 6-Trisulfonic Acid, Trisodium Sal) method (Huang and Saito 2017a). As shown in Fig. 6.10, the bubble's three-dimension in the contaminated system was different from that in the pure system. In the pure system the wakes exhibited a horseshoe-like shape, on the other hand, in the contaminated system the wakes were out of a horseshoe-like shape.

Marangoni convection generated on the contaminated bubble surface played a great role against the above differences between the clean bubble surface and contaminated bubble surface (Huang and Saito 2017b).

2.5.3 Characteristics of Dispersed-bubble Flows in a Large-diameter Pipe

Generally, multiphase flows possess multi scales from the molecular state to continuum state. Scaling-up in engineering design needs large-scale structure of the flows. Saito et al. carried out a series of researches to reveal the large-scale structure of turbulent bubbly flows (Higuchi and Saito 2010, Mudde and Saito 2001, Nagami and Saito 2013, 2014). Using an optical fiber described next, we revealed a large-scale structure of a bubble swarm in a column of 380 mm in inner diameter. The bubble swarm possessed long period fluctuation in spatial number density near the gas injection (Higuchi and Saito 2010); however, the fluctuation gradually attenuated in the upper region of the column.

We revealed that the liquid phase turbulence structure of a bubbly pipe flow was superposition of the turbulence structure of the liquid flow and the bubble-induced turbulence structure (Mudde and Saito 2001, Nagami and Saito 2013, 2014).

2.5.4 Measurement Instruments

In order to perform the above experiments, Saito et al. developed an optical fiber probe to simultaneously measure bubble chord length, bubble velocity and bubble number density (Mizushima and Saito 2012, Mizushima et al. 2013, Mizushima and Saito 2019, Saito et al. 2009, 2018, Sakamoto and Saito 2012a, Sakamoto and Saito 2012b). A single tip optical fiber probe is shown in Fig. 6.11. The primary sensing part is ground into a wedge shape, and the secondary sensing part is microfabricated by a femtosecond laser beam. Its optical system and data acquisition system are shown in Fig. 6.12.

Optical fiber probe is an intrusive measurement; however, it successfully detects bubbles even if the number density is high. Furthermore, this probe is almost free of electrical noise and intrinsic response time limitations, and is very practical. Laser beams propagated through the fiber reflect at the sensing parts, and some of the reflected beams are propagated back again through the same fiber. After reflection of a beam splitter, the return beams are detected by optical sensor. The returned beams' intensity depends on the difference in the refractive indices between the gas and liquid phase around the sensing parts; hence, the output signals are basically a type of on-off signal. From these on-off signals, various quantities, such as the velocities, sizes, number density, volumetric and number density fluxes and interfacial area density, are calculated.

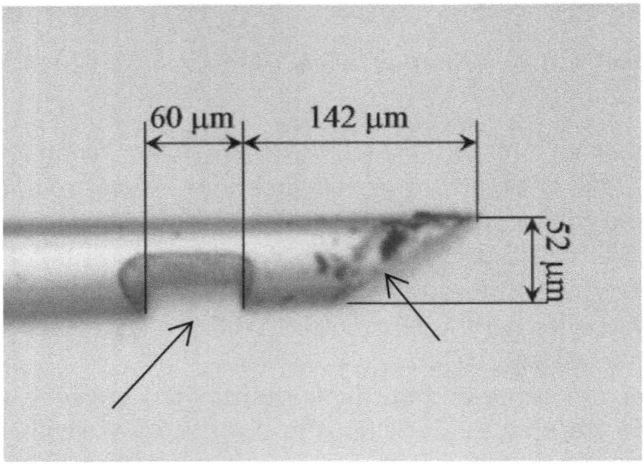

Fig. 6.11: Close-up photograph of an optical fiber probe developed by Saito et al. (Saito et al. 2009).

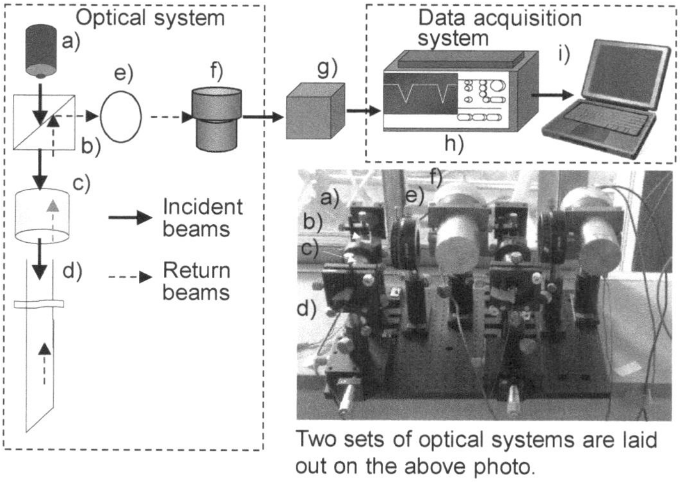

Fig. 6.12: Typical optical and data acquisition systems, and the configuration (Saito et al. 2018). (a) Laser diode, (b) Half mirror, (c) Objective lens, (d) OFP, (e) Polarization element, (f) Electromagnetic shield (photomultiplier is contained in this shield), (g) Amplifier, (h) A/D converter, (i) Personal computer.

2.6 *Large-scale Gas-lift Pump*

A large-scale gas-lift pump was experimentally investigated for the other specific purpose of lifting manganese nodules (Saito et al. 2004, 2005). The results were very useful for development of the GLAD system. They are briefly summarized below.

The large-scale setup is illustrated in Fig. 6.13. A shaft (a) of 1.5 m in the inner diameter and 200 m in depth was built in the underground. The shaft was filled with a tap water. A gas-lift pump made of high-tension steel with 151 mm of inner diameter and 212.6 m of total length was set in the shaft. Air (almost indissoluble gas) injection depth was set at 71 m, 100 m, 131 m and 184 m. Static pressures and pressure difference inside the gas-lift pipe were measured as well as an air flow rate, and a water flow rate at the upper stream of the gas injection; these factors were essential to discuss the GLAD system.

Its lifting performance of pumped water volume and the static pressure (corresponding with discharge pressure of a mechanical pump) inside the gas-lift pipe is plotted in Fig. 6.14 and 6.15, respectively. An effective lifting height is 12.6 m; is defined as a distance between the water surface of the shaft and the outlet of the gas-lift pump. The ratio, L_1/L_2, of L_1 (a length between a gas injection and the water level) to L_2 (an effective lifting height) is called a submerged ratio, and is a good index against a pump load. A larger submergence ratio means a smaller pump load. The submerged ratio becomes 1.0 in the prototype GLAD system, and 0.8–0.9 in the SPAR-type GLAD system, respectively.

As shown in Fig. 6.14, the lifted water volume ranged from 1.2 m^3/min to 5.1 m^3/min. The lifted water volume got larger with increase in submergence ratio. A flow pattern near the gas injection was bubbly flow, and slug flow appeared in most part in the gas-lift pump. Furthermore, from profiles of the static pressure shown in Fig. 6.15, static pressure was larger than the outside static pressure; this difference between the inside static pressure and the outside static pressure corresponds to discharge pressure of a mechanical pump. The pumping performance of the large-scale gas-lift pump is satisfactory to transport a liquid over long distance.

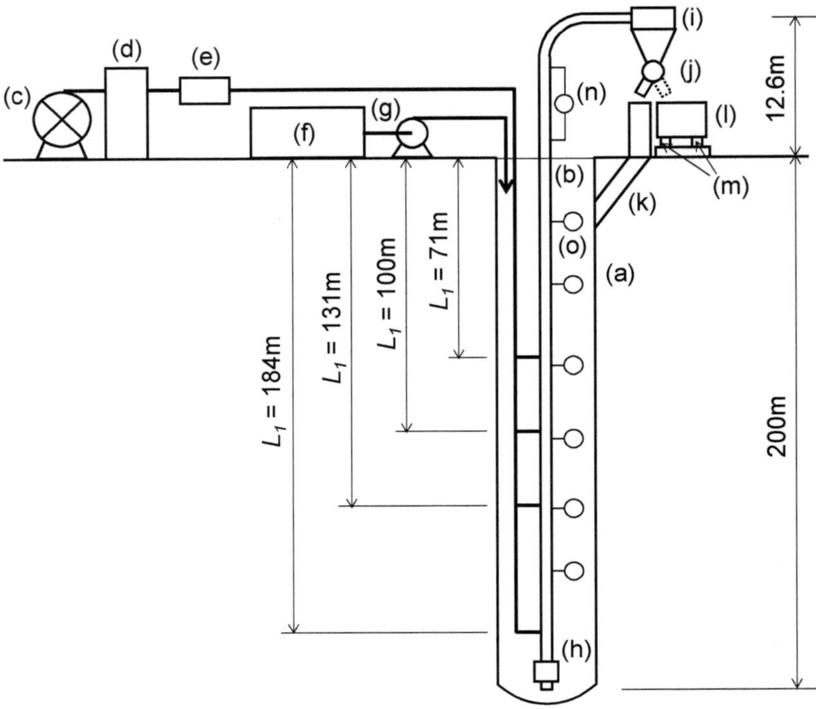

Fig. 6.13: Schematic diagram of 200-m gas-lift pump. (a) Shaft, (b) Gas-lift pipe, (c) Compressor, (d) Cooling tower, (e) Mass flow controller, (f) Water supply tank, (g) Pump, (h) Electromagnetic flow meter, (i) Gas-water separator, (j) Swing pipe, (k) Return pipe, (l) Measure tank, (13) Load cell, (14) Differential pressure transducer, (15) Pressure transducer.

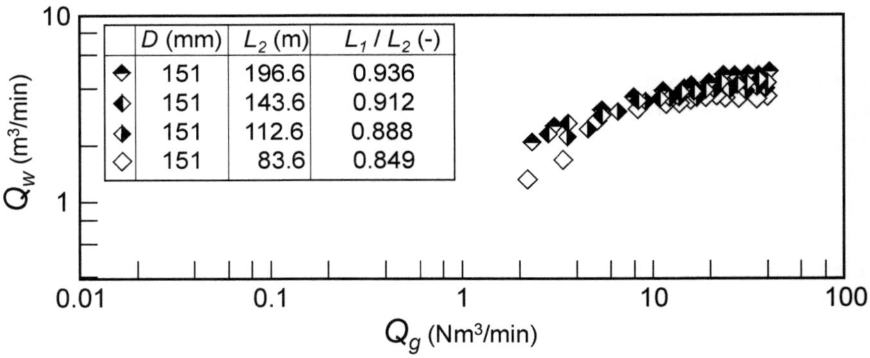

Fig. 6.14: Pumping performance of the large-scale gas-lift pump.

2.7 Computational Simulation

We made numerical models (Kajishima and Saito 1996, Saito et al. 2000) from gas-liquid two-phase flow composed of CO_2 gas, air and seawater, as well as of CO_2 gas, air and water, on the basis of results of fundamental researches. The numerical models and governing equations are described below.

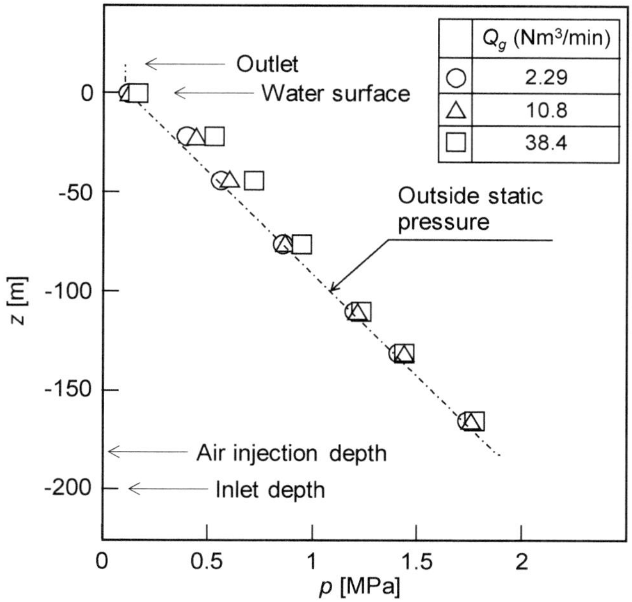

Fig. 6.15: Profiles of the static pressure in the large-scale gas-lift pump.

The mass and momentum conservations are represented as follows,

$$\frac{\partial \rho_M}{\partial t} + \frac{\partial(\rho_M v_M)}{\partial z} = qG, \tag{1}$$

$$\frac{\partial(\rho_M v_M)}{\partial t} + \frac{\partial(\alpha \rho_M v_G^2)}{\partial z} + \frac{\partial(1-\alpha)\rho_L v_L^2}{\partial z} = -\frac{\partial P}{\partial z} - \rho_M g - \frac{4\tau_W}{D_P}. \tag{2}$$

Here, $\rho_M = \alpha\rho_G + (1 - \alpha)$ and $\rho_M v_M = \alpha\rho_G v_G + (1 - \alpha)\,\rho_L v_L$. ρ [kg/m^3] is the density, v [m/s] is the velocity, P [Pa] is the static pressure and α [–] is the void fraction. τ_W [Pa] is the friction stress at the pipe wall, and D_p [m] is the diameter of the riser. The subscripts G, L and M denote gas phase, liquid phase and mixture, respectively.

In addition to Eqs. (1) and (2), the conservation equations for the mass and the number the bubbles to evaluate the void fraction α and the bubble diameter d_B [m] are expressed as follows,

$$\frac{\partial(\alpha\rho_G)}{\partial t} + \frac{\partial(\alpha\rho_G v_G)}{\partial z} = q_G - \Gamma_G \tag{3}$$

$$\frac{\partial n_B}{\partial t} + \frac{\partial(n_B v_G)}{\partial z} = \frac{q_G}{m_{G0}} \tag{4}$$

$$\Gamma_G = 4\pi d_B^{\,2} n_B k_L M_G (C_s - C_\infty) \tag{5}$$

$$d_B = (3\alpha/4\rho\pi n_B) \tag{6}$$

We carried out computations of characteristics of the GLAD system, on the basis of the above experimental results and numerical models. Figure 6.16 shows a comparison between the experimental results (Saito et al. 2004, 2005) of lifted water volume in the large-scale gas-lift pump (gas injection depth: 184 m) and the computational results (Kajishima et al. 1996). These results, i.e., gas-lift pump without gas dissolution, were consistent with each other. Hence, our numerical models and computation methods were satisfactory to analyze the GLAD system.

Next, we computationally analyzed a small-scale experimental setup of the GLAD system; riser inner diameter: 25 mm, riser length: 7.69 m. The comparison between experiments and computations are shown in Fig. 6.17. Our numerical models and computation well reproduce the experimental results.

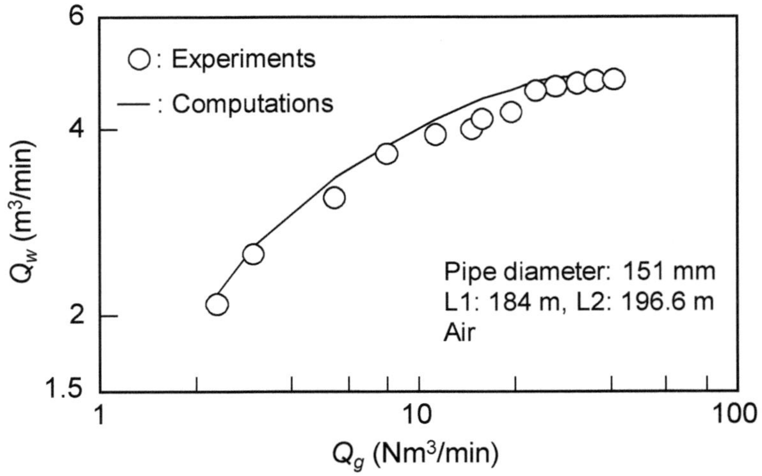

Fig. 6.16: Comparison between experimental and computational results Large-scale gas-lift pump (air).

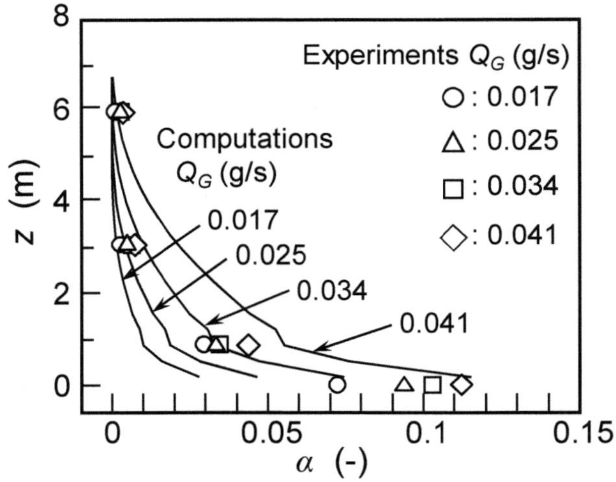

Fig. 6.17: Comparison between experimental and computational results. z represents the distance from the gas injection, α represents the void fraction. α of 0 means complete dissolution of CO_2 included in bubbles into the surrounding water.

2.8 Discussion of Environmental Acceptability

We have investigated the environmental acceptability of the GLAD system based on computations including following: the performance of CO_2 dissolution in a laboratory-scale small apparatus (Saito et al. 1999, 2000); confirmatory experiments on the environmental performance (Yamada and Saito 2012); confirmation of the pump-up performance in a large-scale apparatus (Saito et al. 2004, 2005); determination of the surrounding liquid motion of a bubble (Hanyu et al. 2010, Morikawa et al. 2008, Saito et al. 2015, Sakakibara et al. 2007, Sone et al. 2008, Tsuchiya et al. 1999, 2001, 2003); consideration of the relation between the bubble motion and the surface contamination (Huang and Saito 2015, 2017a, 2017b); examination of turbulent bubbly flows in large-diameter vessels (Higuchi and Saito 2010, Mudde and Saito 2001, Nagami and Saito 2013, 2014).

In a model case of the prototype GLAD system (a drainpipe: 1000 mm in the inner diameter and 20 km in length; a riser: 1000 mm in the inner diameter and 400 m in length; gas-injection depth: 380 m), typical computational results (Saito et al. 2011) for pH level in the drainpipe are plotted in Fig. 6.18. The pH in the drainpipe can be controlled to remain higher than 5.5, provided Q_g is less than 0.1 kg/s (= 360 kg/h); where Q_g is the mass flow rate of the supplied gas with 80% CO_2 concentration. Since the discharged seawater including CO_2 is transported and spread into the surrounding deep water due to deep water currents, the secondary environmental impact is expected to be mitigated in the short term. Advanced researches are required from the viewpoint of ocean biology and ocean environment.

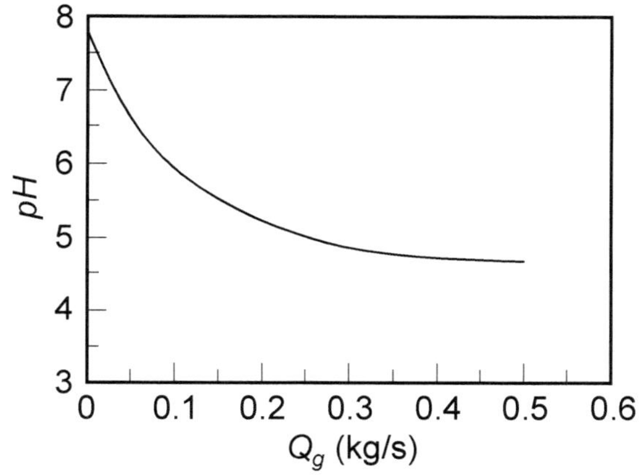

Fig. 6.18: pH value of a CO_2 solution in the drainpipe of the prototype GLAD system via numerical simulation.

3. Conclusions

Here we have discussed CO_2 sequestration at deep portions of the ocean by gas-lift system with shallow injection and deep releasing; this method named GLAD system (Gas Lift Advanced Dissolution system) has been carefully developed on the basis of experimental and computational researches. The GLAD system possesses hopeful and practical characteristics. However, ocean disposal of CO_2 is an emergency evacuation procedure. We would like to stress, in order to mitigate global warming, that the best option is to develop and to use CO_2-free new energy; the second-best one is to restrain or to stop consumption of fossil fuel such as oil, coal and natural gas; the third-best one is to save energy; the fourth-best one is to improve energy efficiency.

References

Damuth, J. 1981. Population density and body size in mammals. Nature 290: 699–700.
Department of Energy USA. 1993. DOE Report, DOE/ER-30194.
Hanyu, K. and T. Saito. 2010. Dynamical mass transfer process of a CO2 bubble measured by using LIF/HPTS visualization and photoelectric probing. Can. J. of Chem. Eng. 88: 551–560.
Higuchi, M. and T. Saito. 2010. Quantitative characterizations of long-period fluctuations in a large-diameter bubble column based on point-wise void fraction measurements. Chem. Eng. J. 160: 284–292.
Hoffert, M.I., Y.C. Wey, A.J. Callegari and W.S. Broecker. 1979. Atmospheric response to deep-sea injections of fossil-fuel carbon dioxide. Climatic Change 2: 53–68.
Huang, J. and T. Saito. 2015. Influence of bubble-surface contamination on instantaneous mass transfer. Chem. Eng. & Technol. 38: 1947–1954.
Huang, J. and T. Saito. 2017a. Discussion about the differences in mass transfer, bubble motion and surrounding liquid motion between a contaminated system and a clean system based on consideration of three-dimensional wake structure obtained from LIF visualization. Chem. Eng. Sci. 170: 105–115.
Huang, J. and T. Saito. 2017b. Influences of gas-liquid interface contamination on bubble motions, bubble wakes, and instantaneous mass transfer. Chem. Eng. Sci. 157: 182–199.
Intergovernmental Panel on Climate Change. 2013. Climate Change 2013: The Physical Science Basis.
Kajishima, T., T. Saito, R. Nagaosa and H. Hatano. 1995. A gas-lift system for CO2 release into shallow seawater. Energy Conversion & Management 36: 467–470.
Kajishima, T. and T. Saito. 1996. Numerical simulation of unsteady flow in air-lift pump. The Japan Society of Mechanical Engineers, Int. J. Series B 39: 525–532.
Kajishima, T., T. Saito and R. Nagaosa. 1997. GLAD: A gas lift method for CO2 disposal into the Ocean. Energy 22: 257–262.
Kosugi, S., K. Niwa, T. Saito, T. Kajishima and K. Hamaogi. 2001. Design factors in gas-lift advanced dissolution (GLAD) system for CO2 sequestration into the ocean. Chem. Eng. Sci. 56: 6205–6210.
Mizushima, Y. and T. Saito. 2012. Detection method of a position pierced by a single-tip optical fiber probe in bubble measurement. Meas. Sci. & Technol. 23: 085308.
Mizushima, Y., A. Sakamoto and T. Saito. 2013. Measurement technique of bubble velocity and diameter in a bubble column via Single-Tip Optical-fiber Probing with judgment of the pierced position and angle. Chem. Eng. Sci. 100: 98–104.
Mizushima, Y. and T. Saito. 2019. Improving the accuracy of droplet measurement by optical fiber probe using 3D ray-tracing simulation. J. of Chem. Eng. of Japan 52: 171–178.
Morikawa, K., S. Urano and T. Saito. 2008. Turbulence modulation induced by bubble swarm in oscillating-grid Turbulence. Journal of Power and Energy Systems, The Japan Society of Mechanical Engineers 2: 330–339.
Mudde, R.F. and T. Saito. 2001. Hydrodynamical similarities between bubble column and bubbly pipe flow. J. of Fluid Mech. 437: 203–228.
Nagami, Y. and T. Saito. 2013. Measurement of modulation induced by the interaction between bubble motion and liquid-phase motion in decaying turbulence formed by an oscillating-grid. Particuology 11: 158–169.
Nagami, Y. and T. Saito. 2014. An experimental study of the modulation of the bubble motion by gas-liquid-phase interaction in oscillating-grid decaying turbulence. Flow, Turbulence and Combustion 92: 147–174.
Niwa, K., S. Kosugi, T. Saito, T. Kajishima and K. Hamaogi. 2005. Correlations among the design factors of the CO2 ocean sequestration system, GLAD. Energy 30: 2308–2317.
Ohnuki, A. and H. Akimoto. 1996. An experimental study on developing air-water two-phase flow along a large vertical pipe: effect of air injection method. Int. J. of Multiphase Flow 22: 1143–1154.
Saito, T., T. Kajishima, K. Tsuchiya and S. Kosugi. 1999. Mass transfer and structure of bubbly flows in a system of CO2 disposal into the ocean by gas-lift column. Chem. Eng. Sci. 54: 4945–4951.
Saito, T., T. Kajishima and R. Nagaosa. 2000. CO2 sequestration at sea by gas-lift system of shallow injection and deep releasing. Environ. Sci. & Technol. 34: 4140–4145.

Saito, T., S. Kosugi, T. Kajishima and K. Tsuchiya. 2001. Characteristics and performance of a deep ocean disposal system for low-purity CO2 gas via a gas lift effect. Energy & Fuels 15: 285–292.

Saito, T., T. Kajishima and K. Tsuchiya. 2004. Pumping characteristics of a large-scale gas-lift system. Experimental Thermal & Fluid Sci. 28: 479–488.

Saito, T., K. Tsuchiya and T. Kajishima. 2005. Local and global scale structure of bubbly flows in glad (gas lift advance dissolution) system. Experimental Thermal & Fluid Sci. 29: 305–313.

Saito, T., K. Matsuda, Y. Ozawa, S. Oishi and S. Aoshima. 2009. Measurement of tiny-droplets using a newly developed optical fiber probe microfabricated by femtosecond pulse laser. Meas. Sci. & Technol. 20: 114002.

Saito, T., K. Sakakibara, Y. Miyamoto and M. Yamada. 2010. A Study of surfactant effects on the liquid-phase motion around a zigzagging-ascent bubble using a recursive cross correlation PIV. Chem. Eng. J. 158: 39–50.

Saito, T., T. Kajishima and K. Tsuchiya. 2011. Deep ocean CO2 sequestration via GLAD (Gas-Lift Advanced Dissolution) system. J. of Environ. & Eng. 6: 412–415.

Saito, T. and M. Toriu. 2015. Effects of a bubble and the surrounding liquid motions on the instantaneous mass transfer across the gas-liquid interface. Chem. Eng. J. 265: 164–175.

Saito, T., K. Sato, A. Nihei and H. Muramatsu. 2018. Improvement of optical fiber probing in multiphase systems, and the possibility of practical application in chemical engineering processes. J. of Chem. Eng. of Japan 51: 331–341.

Sakakibara, K., M. Yamada, Y. Miyamoto and T. Saito. 2007. Measurement of the surrounding liquid motion of a single rising bubble using a Dual-Camera PIV system. Flow Meas. & Instrument. 18: 1612–1618.

Sakamoto, A. and T. Saito. 2012a. Computational analysis of responses of a wedge-shaped optical fiber probe in bubble measurement. Rev. of Sci. Instru. 83: 075107.

Sakamoto, A. and T. Saito. 2012b. Robust algorithms for quantifying noisy signals of optical fiber probes employed in industrial-scale practical bubbly flows. Int. J. of Multiphase Flow 41: 77–90.

Sone, D., K. Sakakibara, M. Yamada and T. Saito. 2008. Bubble motion and its surrounding liquid motion through the collision of a pair of bubbles. Journal of Power and Energy Systems, The Japan Society of Mech. Engs. 2: 306–317.

Tsuchiya, K., H. Mikasa and T. Saito. 1997. Absorption dynamics of CO2 bubbles in a pressurized liquid flowing downward and its simulation in seawater. Chem. Eng. Sci. 52: 4119–4126.

Tsuchiya, K., M. Koshihata, T. Tomida and T. Saito. 1999. Effect of satellite bubbles on dynamics of gas absorption from a CO2 bubble into a downward-flowing liquid. Korean J. of Chem. Eng. 16: 635–639.

Tsuchiya, K., T. Saito, T. Kajishima and S. Kosugi. 2001. Coupling between mass transfer from dissolving bubbles and formation of bubble-surface wave. Chem. Eng. Sci. 56: 6411–6417.

Tsuchiya, K., T. Ishida, T. Saito and T. Kajishima. 2003. Dynamics of interfacial mass transfer in a gas-dispersed system. Can. J. Chem. Eng. 81: 647–654.

Yamada, M. and T. Saito. 2012. A newly developed photoelectric optical fiber probe for simultaneous measurements of a CO2 bubble chord length, velocity, and void fraction and the local CO2 concentration in the surrounding liquid. Flow Meas. & Instrument. 27: 8–19.

Chapter 7

Nutrient Concentrations in Natural Waters from Mt. Fuji to the Suruga Bay

Rumi Sohrin

1. Introduction

Primary production, a biological transformation of inorganic carbon to organic carbon, is a key process for sustaining our lives over the increase in carbon emission and a population expansion since the industrial revolution. Now humans use 11 to 24% of net primary production on the land surface for food, fuels, and fiber (Schlesinger and Bernhardt 2013). On a global scale, about 50% of primary production is processed by marine algae, and about 30% of the anthropogenic CO_2 emitted since 1750 has been absorbed into the ocean, owing to the air–sea disequilibrium of CO_2 concentrations driven by algal primary production (IPCC 2013). Primary production requires elements in inorganic form–not only carbon, hydrogen and oxygen that are expressed in the equation of photosynthesis–but also others like nitrogen, phosphorus, silicon and heavy metals. The latter elements are collectively designated as "nutrients" in environmental science.

This chapter presents fundamental knowledge about nutrients and a brief review of the highest elevation in Japan (Mt. Fuji) and the deepest (Suruga Bay) terrains located in Shizuoka prefecture, central Japan, where the Research Institute of Green Science and Technology (RIGST) at Shizuoka University is located. The environment encompassing Mt. Fuji and the Suruga Bay is like a miniature earth and is a good place for understanding the linkage between land and ocean via nutrient transport over a water cycle. This aspect is highlighted in recent findings of inorganic nutrients dynamics by the author's group as to the natural waters associated with Mt. Fuji and the Suruga Bay. The goal of this chapter is to demonstrate the benefits of terrestrial nutrients loaded to the Suruga Bay.

Green Biology Research Division, Research Institute of Green Science and Technology (RIGST), Shizuoka University, 836 Ohya, Surugaku, Shizuoka, 422-8529 Japan.
Email: sohrin.rumi@shizuoka.ac.jp

2. Nutrients

Nutrients are essential chemical compounds for all organisms as sources of elements and energy. Organic nutrients are vital carbon sources for heterotrophs (e.g., animals), and some specific compounds like vitamins are also essential for autotrophs (e.g., plants). Inorganic nutrients are required not only as sources of elements other than carbon by autotrophs, but also as energy sources by chemoautotrophic microbes and as co-factors by all organisms. Since organic nutrients are produced by autotrophs via a process designated as primary production, inorganic nutrients are primordial components sustaining an ecosystem.

2.1 Major Nutrients

Nutrients are divided by their concentrations in environments into macro nutrients (Ca^{2+}, K^+, Na^+, Mg^{2+}, Cl^-, > 95% in total), micro nutrients (≥ 1 ppm other than macro nutrients), and trace nutrients (< 1 ppm), and are also distinguished in their relative requirements by organisms. Large requirements are the major [inorganic] nutrients of nitrate (NO_3^-), nitrite (NO_2^-), ammonium (NH_4^+), phosphate (PO_4^{3-}), and silicic acid ($Si(OH)_4$) which are determined by the colorimetric method. Phosphate is also designated as Soluble Reactive Phosphorus (SRP) or Dissolved Inorganic Phosphorus (DIP), but all terms refer the phosphorus compounds reactive to acidified molybdate and hydrogen phosphate (HPO_4^{2-}) as the major component of PO_4^{3-} in fresh water and seawater (Millero 2013). $Si(OH)_4$ is thought to account for 94.6% of soluble silicon compounds in seawater and is the only silicon compound incorporated into constituents of living organisms represented by diatoms (algae) and sponge (animal) (Exley and Sjöberg 2014). The sum of concentrations of NO_3^-, NO_2^- and NH_4^+ is defined as Dissolved Inorganic Nitrogen (DIN). The molar ratio of DIN to PO_4^{3-} (N : P) has an effect on algal growth because algae uptake major nutrients at similar ratios to their cellular ratios. This favorable condition promotes their growth (Geider and Roche 2002). Algal cellular N : P ratios converge to 16 : 1 across various aquatic environments. This convergence is known as the Redfield ratio (Redfield 1958), though the ratios vary among algal species and physiological states (Ho et al. 2003).

2.2 Trace Metals

Trace metals are representatives of trace nutrients. Iron (Fe) is the most-studied trace metal as it is contained in various enzymes (Table 7.1). Low Fe concentration limits primary production in the Southern Ocean, equatorial Pacific Ocean and Alaskan Gyre, which are known as High-Nutrient-Low-Chlorophyll (HNLC) regions. HNLC regions attract the attention of geoengineering owing to their potential of CO_2 sequestration from the atmosphere by inducing primary production with artificial Fe input from ships, which is called ocean iron fertilization (Singh and Ahluwalia 2013). In nature, Fe is exported from land to the ocean by rivers, aerosols, and Submarine Groundwater Discharge (SGD). Upon reaching the oxygenated ocean surface, Fe^{2+} is easily oxidized to Fe^{3+} and precipitates in the form of $Fe(OH)_3$, which algae cannot uptake. Besides Fe, many trace metals like cadmium (Cd), cobalt (Co), copper (Cu), manganese (Mn), molybdenum (Mo), nickel (Ni), and zinc (Zn) are essential for primary production (Table 7.1). Trace metals occur in various forms (various valence of ions in the form of free ion/oxide/hydroxide/complex with organic ligand), but the dissolved form is best-measured as a proxy of a total amount of bioavailable trace metals. Dissolved trace metals are determined with filtrates through each of < 0.1, 0.2

Table 7.1: Biological functions of trace metals.

Metal	Function
Cd	Dissociation of bicarbonate
Co	Dissociation of bicarbonate
	Active center of vitamin B_{12}
Cu	Denitrification
	Ammonium oxidation
Fe	Nitrogen fixation
	Denitrification
	Nitrification
	Ammonium oxidation
	Electron transport (photosynthesis, respiration)
	Replication and repair of DNA
Mn	Oxygen evolution from water
	Scavenge of superoxides and hydroxy radicals
Mo	Nitrogen fixation
	Denitrification
Ni	Dissociation of urea
Zn	Dissociation of bicarbonate
	Dephosphorylation
	hydrolysis of protein

or 0.45 µm-pore-size filters by inductively coupled mass spectrometry (ICP-MS) or atomic absorption spectrometry after storage with acid at room temperature for several months to ensure complete recovery of the soluble form (Löscher et al. 1997). Desalination and organic matter removal are necessary for seawater samples and organic-rich samples, respectively, before analysis, to avoid interference of major cations (e.g., Na^+) and carbon. Air dust is the most critical contaminant source of trace metals. It is noteworthy that the most accurate measurement of trace metals in seawater has been achieved in the last 30 years by progress in clean technique: measures before were one to three orders of magnitude higher than the current values (Nakayama et al. 1989).

3. Mt. Fuji and Kakita River—Greatest Volume of Pure Natural Water in Japan

3.1 Overview

Mt. Fuji is the highest mountain in Japan (elevation, 3,776 m) and is located between Shizuoka and Yamanashi prefectures (Fig. 7.1). Mt. Fuji is an active volcano and consists of four volcanoes which were active in the past; Pre-Komitake, Komitake, Ko-Fuji and Shin-Fuji Volcanoes, ordering from old (base) to young (top). Mt. Fuji has ~ 2.2 billion

Fig. 7.1: Location of the study sites of Kakita River, Kano River and the Suruga Bay. Land and ocean images were obtained from the Red Relief Image Map by Asia Air Survey Co., Ltd., and R. Schlitzer, Ocean Data View, https://odv.awi.de, 2019, respectively. Bar shows water depth.

tons of annual precipitation of rain and snow, most of which is supplied from the Suruga Bay and the Sagami Bay locating in the South (Yasuhara et al. 2007). The daily average of ~ 4.5 million tons of precipitation is recharged (penetrates into the aquifer which is a geological underground layer of water-bearing rock), mainly at the altitude of 1100–2700 m (Yasuhara et al. 2007). The major aquifer is basaltic lava flowing with many cracks, which are called "clinkers," on the surface of Shin-Fuji Volcano (Tsuchi 2007). On the eruption of Shin-Fuji Volcano, ejected, low viscous basaltic magma flowed far down the side of Ko-Fuji Volcano, and the surface and bottom parts were rapidly cooled, hardened and then crushed by stress from the flowing hot inner fluid. These reactions made clinkers. Precipitates immediately penetrated through voids of pyroclastic materials (e.g., dust, ash, lapillus, and bomb) deposited on the Shin-Fuji Volcano lava flow and piles of clinkers beneath. The covering of permeable clinkers is one of the reasons why Mt. Fuji has no lakes and steadily flowing rivers (Tsuchi 2007). Precipitates cannot recharge into the densely solidified inner part. Therefore, groundwater stored in voids of piled clinkers (unconfined aquifer) is pressurized by dense rock (confined aquifer), moves down toward the base, and then discharges at the end of the lava flow. The volume of discharges from Mt. Fuji is estimated at 6.55 million tons a day in total (Yasuhara et al. 2007).

There are many springs at the lower end of the Shin-Fuji Volcano lava flows (Tsuchi 2007). The largest volume of spring water forms the Kakita River located at the end of the largest Shin-Fuji Volcano lava flow, named Mishima lava flow. Kakita River receives more than 1 million tons of water a day from springs, which is the greatest volume of spring water in Japan (Numazu Office of River and National Highway, http://www.cbr.mlit.go.jp). Kakita River is only 1.2 km in length and flows into Kano River, which originates in the Izu Peninsula and is the fourth largest river in volume running into the Suruga Bay (4.11 million tons a day; database of Ise Bay environment, www.isewan-db.go.jp).

3.2 Nutrient Dynamics in Kakita River

Our laboratory started the study on Kakita River and Kano River in 2016. Major nutrients and nine elements of trace metals were determined, the latter being selected by taking into account their biological function and expected concentration. Previous studies and our data show that Kakita River spring water is very clean, with extremely low phytoplankton biomass (Table 7.2). Chlorophyll *a* (Chl. *a*) concentration ranged from 0.004 to 0.092 μg L^{-1} in Kakita River spring water in 2016–2018, which is comparable to that in the deep (≥ 200 m) seawater of the Suruga Bay. The concentration of Chl. *a* is an index of algal biomass because Chl. *a* is a major photosynthetic pigment possessed by most algae. However, we found higher concentrations of NO_3^- (44–53 μmol L^{-1}), PO_4^{3-} (2.7–2.9 μmol L^{-1}), $Si(OH)_4$ (649–666 μmol L^{-1}), vanadium (V) (620–710 nmol L^{-1}), and chromium (Cr) (5–12 nmol L^{-1}) in Kakita River spring water than those in Kano River (Table 7.2). Low Chl. *a* concentration in Kakita River spring water is mostly attributed to long residence time occurring underground (~ 10 years; Mahara et al. 1993) while primary production is halted. In general, Fe and Mn are the dominant trace metals in groundwater (e.g., Groschen et al. 2008), but V accounts for > 96% of the total of measured nine trace metals in Kakita River spring water. High concentrations of V were reported for groundwaters recharged in Mt. Fuji back in 1968 (Okabe and Morinaga 1968) and those recently reported by National Institute of Advanced Industrial Science and Technology (AIST) (AIST 2016). Many studies suggest that high V concentrations of Mt. Fuji groundwater are attributed to higher content of V in basalt than in other rocks (Ando et al. 1989). Concentrations of trace metals other than V in Mt. Fuji groundwater are reported only on the side of Yamanashi prefecture (Uchiyama et al. 2014) but not on the side of Shizuoka prefecture, which might be partly due to low concentrations near the detection limit (Table 7.2). It is highly likely that low concentrations of Fe (< 2 nmol L^{-1}) and Mn (< 0.3 to 3 nmol L^{-1}) in Kakita River spring water relates to high content of dissolved O_2 (Ino 1990) resulting from active ventilation across clinkers (Tsuchi 2007). Oxidized Fe and Mn would have precipitated into the aquifer or they would be removed by filtration with 0.2 μm-pore size filter before analysis. Oxidative water also keeps Cr concentration high since the decrease in Cr concentration in groundwater has been reported along the progress of reductive condition (Raddatz et al. 2011). High concentrations of V and Cr are not of benefit to primary production as they are toxic and not essential for algae and plant growth, though some algae accumulate Cr inside their cells by active transport (Volland et al. 2012). Two vanadic enzymes have been reported thus far, V-nitrogenase catalyzing oxidation of N_2 to NH_3 and V-haloperoxidase by which algae reduce H_2O_2 and produce halogenated organic compounds (Fournier et al. 2014). Bromo organic compounds produced by V-bromoperoxigenase reaction exhibit alleopathy by which *Synechococcus* sp. (Cyanobacteria) protect themselves from grazers and conquer bacteria (Johnson et al. 2011). Suppliers of drinking mineral waters sell the Mt. Fuji groundwater as V-rich natural water. The best-known medicinal benefit of V is an insulin-like effect to reduce blood sugar levels with diabetes (Gupta and Gupta 2014). A daily dosage of 0.2–1 mg of V as vanadyl sulfate is recommended (Balch and Balch 2000), which is equivalent to the V content in 6–30 L of Kakita River spring water. Regarding toxicity of V in humans, some studies report no apparent detrimental effects of a dose of 4.5 and 9 mg V a day for 6 to 16 months, while induction of diarrhea was demonstrated by intakes of 5–218 mg V a day over 6 to 10 weeks (Anke 2004).

Down the stream of Kakita River, Chl. *a* increased to 0.1–0.6 μg L^{-1}, but the level was still lower than that at the upstream sites of Kano River to the confluent point with Kakita River (0.47–1.1 μg L^{-1}). Chl. *a* in Kano River between the confluent point and river mouth

Table 7.2: Range and average (in parenthesis) of concentrations of chlorophyll *a* (Chl. *a*) and inorganic nutrients in Kakita and Kano rivers and the Suruga Bay. Average was calculated by assuming of the values below detection limit = 0. ND: not determined.

System	Site (depth)	Period	Chl. *a* (μg L^{-1})	Dissolved trace metal (nmol L^{-1})								Major nutrients (μmol L^{-1})				
				Fe	Zn	Cu	Mn	Pb	Cd	Cr	V	NO_3^-	NO_2^-	NH_4^+	PO_4^{3-}	$Si(OH)_4$
Kakita River spring water	Kakita River park, well 1 (0 m)	Dec 2016 Jul 2017 Dec 2017 Mar 2018	0.004–0.092 (0.04)	< 2	< 1–7.6 (3.0)	< 1–7.1 (2.4)	< 0.3–2.7 (1.0)	0.3–1.2 (0.6)	< 0.05–0.43 (0.15)	5.0–12.4 (7.4)	623–709 (669)	44.5–52.9 (48.3)	< 0.02–0.22 (0.09)	< 0.01–0.23 (0.06)	2.7–2.9 (2.8)	649–666 (660)
middle of Kakita River	Kakita bridge [upstrem of the confluent point] (0 m)	Nov 2016 Dec 2016 Jan 2017 Jul 2017	0.09–0.28 (0.16)	< 2–52 (22)	< 1–12 (5.5)	< 1–4.3 (1.7)	< 0.3–9.5 (4.2)	< 0.1–0.66 (0.39)	0.06–0.28 (0.18)	3.2–5.9 (4.7)	484–661 (607)	29.5–48.8 (40.2)	0.06–0.39 (0.18)	< 0.01–0.75 (0.43)	1.8–2.6 (2.2)	675 (675)
Kakita River & Kano River	confluent point (0 m)	Nov 2016 Dec 2016 Jan 2017 Jul 2017	0.08–0.42 (0.22)	9.6–106 (41)	5.1–65 (24)	< 1–9.0 (4.0)	< 0.3–19 (7.8)	< 0.1–1.1 (0.55)	< 0.05–0.39 (0.11)	5.7–9.2 (7.2)	569–671 (629)	38.6–56.0 (46.6)	< 0.02–0.94 (0.41)	< 0.01–1.2 (0.61)	2.0–3.3 (2.5)	596–647 (622)
middle of Kano River	Joyama and Kamishima bridge [upstream of the confluent point] (0 m)	Dec 2016 Jan 2017 Jul 2017 Aug 2017 Mar 2018	0.47–1.13 (0.83)	85–219 (139)	< 1–22 (7.6)	< 1–5.2 (1.8)	2.1–28 (20)	< 0.1–1.1 (0.52)	< 0.05–0.23 (0.14)	< 0.2–1.4 (0.8)	44–68 (54)	13.6–36.7 (26.6)	0.12–0.51 (0.32)	< 0.01–0.54 (0.19)	0.39–0.73 (0.56)	368–447 (416)
end of Kano River	river mouth (0 m)	Nov 2016	0.48–0.59 (0.54)	61.1–91.2 (76.2)	14–24 (19)	3.5–6.6 (5.0)	< 0.3–3.4 (1.7)	< 0.1–0.53 (0.27)	< 0.05	3.7–4.3 (4.0)	300–308 (304)	53.3–56.2 (54.8)	0.35–0.75 (0.55)	0.2 (0.2)	1.9 (1.9)	ND[a]
Suruga Bay[b]	siphon-sampling station (397 m)	Oct 2009 Apr 2010	ND	1.2–1.6 (1.4)	5.9–7.1 (6.5)	0.9 (0.9)	0.25–0.29 (0.27)	ND	0.94–0.97 (0.96)	ND	ND	32.9	0.07	0.07	2.3	72.6

[a] ND: not determined.
[b] Suzuki and Sohrin (2015).

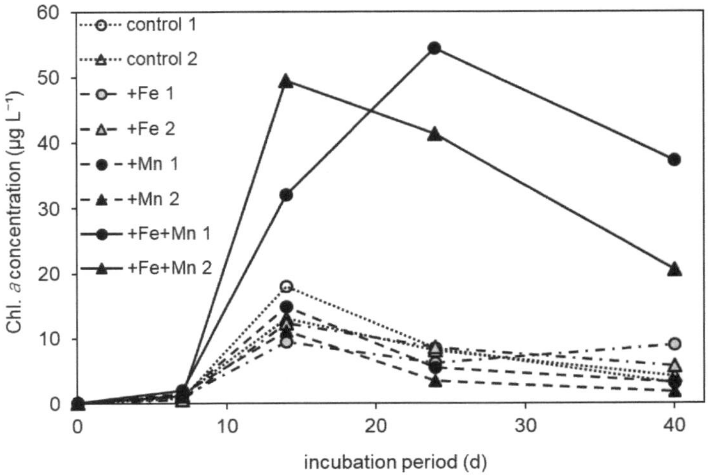

Fig. 7.2: Changes in Chl. *a* concentration over the course of the incubation experiment with Kakita River spring water collected in July, 2018. Expected concentrations at the start are follows: control, 0 nmol Fe L^{-1}/0 nmol Mn L^{-1}; +Fe, 180 nmol Fe L^{-1}/0 nmol Mn L^{-1}; +Mn, 0 nmol Fe L^{-1}/50 nmol Mn L^{-1}; +Fe+Mn, 180 nmol Fe L^{-1}/50 nmol Mn L^{-1}. Samples were maintained in duplicate bottle (1 and 2) at 15.9°C under 14 h : 10 h light-dark cycle.

was comparable (0.5–0.6 µg L^{-1}) to that in the downstream of Kakita River, suggesting that the diluted algal population in Kano River with Kakita River water enters the Suruga Bay without significant growth. By conducting incubation experiments with algae in Kakita River spring water after being enriched with Fe and/or Mn, it is confirmed that primary production in Kakita River spring water is potentially under co-limitation by Fe and Mn, mostly a type of "simultaneous co-limitation" or "serial co-limitation" with Mn as a principal limiting factor over Fe (Fig. 7.2). Simultaneous co-limitation is defined as stimulation of primary production by the combined nutrient addition (in this case, +Fe+Mn), but not by the single nutrient addition (+Fe or +Mn), whereas serial co-limitation is defined as stimulation of primary production by addition of only one nutrient (+Mn, not +Fe) and even stronger stimulation by the combined addition of multiple nutrients (+Fe+Mn) (Burson et al. 2016). However, algal response is slow, and it takes a few days for initiation of growth. *In situ* population would likely enter the Suruga Bay before starting growth. This diagnosis is consistent with insignificant algal growth in the downstream of Kano River to the confluent point with Kakita River. *In situ* primary production also might be suppressed by water movement driven by high flow velocity (Feipeng et al. 2013). Suppression of primary production infers that Kakita River and Kano River deliver major nutrients to the Suruga Bay without significant algal consumption.

4. Suruga Bay—Steepest, Deep Terrain in the Global Ocean

4.1 Overview

4.1.1 Topography

The Suruga Bay located in Shizuoka prefecture opens to the Pacific Ocean (Fig. 7.1). The Suruga Bay is the deepest bay (maximum depth, ~ 2,500 m) in Japan and one of the deepest seas that have "bay" in their names in the world. Based on my survey on Google Earth, the

Suruga Bay likely places as the second deepest bay after the Bay of Bengal of India (max. depth, 4694 m; surface area, 2,172,000 km^2), followed by the Baffin Bay of the Arctic Circle (2,136 m; 689,000 km^2). Compared to other "deep bays" in the world, quite smaller Suruga Bay (surface area, 2,300 km^2) is characterized by a narrower continental shelf and steeper slope. Typical width of the continental shelf is from ~ 1 km in the northern part (shore side) to ~ 10 km in the southern part (bay month), with an exception of the widest sections (8 ~ 10 km) off the Uchiura Bay (off the base of Izu Peninsula) (Misawa et al. 2007), which are much narrower than the global average (~ 60 km; Pinet 2014). Mouths of large rivers (Kano River, Fuji River, and Abe River) located in the northern part, and a lot of terrestrial materials like nutrients may pass across the narrow continental shelf, reaching the oceanic province (area of the sea beyond the continental shelf). The gradient of the continental slope is about 150/1000 (8.5 degree) in the northern part (AIST 2016), which is steeper than the global average (4 degree; Pinet 2014). Beyond the continental slope, a narrow channel extends along the north-south direction in the center of the bay, from the mouth of the Fuji River (north) to the bay mouth (south), which is named the Suruga Trough, otherwise known as the Suruga Bay Central Channel (Shiba 2016). The Suruga Trough is a boundary of the Eurasia Plate and the Philippine Sea Plate; the latter subducts beneath the former. The Suruga Trough is the northwestern most part of the Nankai Trough, whose underlying fault is a source of Nankai megathrust earthquakes. The most recent large earthquake occurring in the Suruga Bay is the 2009 Suruga Bay Earthquake, which occurred in August, 2009, with a magnitude of 6.5 (Japan Meteorological Agency 2009).

4.1.2 Biota

Steep terrains of the Suruga Bay can link the land tightly to the ecosystem of the oceanic province not only in the epipelagic zone (0–100 m) but also in the mesopelagic (100–1000 m) and bathypelagic (1000–4000 m) zones and contribute to high biodiversity in the Suruga Bay. The number of described fish (Pisces) in the Suruga Bay is about 1,200 species out of 3,790 species in the Japanese exclusive economic zone, with 38% as deep-water fishes that dwell in the depths > 200 m (Fujikura et al. 2010, Fukui 2015).

Sakura-ebi shrimp (*Lucensosergia lucens Hansen*, 1922; former name, *Sergia lucens*; Vereshchaka et al. 2014) is probably the most famous organism dwelling in the Suruga Bay. Sakura-ebi is a 40–50 mm-long decapod and is a commercially important species usually offered as dried shrimp in markets. The Suruga Bay is the type locality of sakura-ebi, and the dense aggregation can be found only in the Suruga Bay and along the coast of Taiwan (Omori et al. 1988), with significant genetical difference between the populations (Imai et al. 2013). Sakura-ebi are also found in the adjacent waters of the Tokyo Bay and the Sagami Bay as well as off Borneo, off the Philippines, and off New Guinea (Vereshchaka 2000). Sakura-ebi is a benthopelagic species that swims near the seafloor during the day, mostly along the 200 m isobath near the edge of the continental shelf, to escape from predators, and then migrates to 20–100 m after dusk to feed (Omori 1969). Our preliminary study on sakura-ebi suggests an increase in the relative amount of *cis*-astaxanthin over *trans*-astaxanthin by diel vertical migration (Sohrin et al. 2017). Astaxanthin is a reddish pigment with an anti-oxidant effect, and *cis*-astaxanthin is considered more effective than *trans*-astaxanthin (Liu and Osawa 2007). There are two main fishery grounds of sakura-ebi in the Suruga Bay: off the Fuji River (northern part) and off the Oi River (western part). The fishing is restricted in spring (the lucrative season from middle March to beginning of June) and autumn (from end of October to end of December) to protect spawning shrimps, and the volume of the caught shrimps has been limited since 1977. However, the catch decreased after 2008 from an annual value of ≥ 2000 tons to the record low of 312 tons in the spring of 2018, which is less than half of the average value in recent years. The decrease in catch is considered to relate to overfishing of spawning females. The autumn fishing of 2018 was canceled to protect juveniles under 35 mm in size, whose females would spawn the next summer, and the following spring fishing was prohibited in the spawning site off the Fuji River. Water temperature is known to have a great influence on the number of eggs spawned, with positive correlation with the thickness of the preferential temperature zone (18–25°C) at the depths of active spawning (0–50 m) (Abe and Tsukui 2014). High temperature decreases the ratio of spawned females to matured females (Doi et al. 2018). Global Sea Surface Temperatures (SST) were markedly high in 2014–2018, all placing in the top five of the high SST since1891 (Global Climate Report–Annual 2018, https://www.ncdc.noaa.gov). The influence of global warming is revealed on the reproduction and life-cycle of sakura-ebi in the Suruga Bay.

Some researchers consider that SGD maintains the preferred water temperature for the spawning of sakura-ebi and delivers inorganic/organic nutrients to diatom, which is the expected dominant prey of the juveniles (Omori 2015). AIST surveyed the bottom geology of the Suruga Bay in 2013–2016 and found a few SGD sites off Tagonoura Port (northern continental shelf) by photographs taken from a remotely operated vehicle (ROV). Ochiai (1970) calculated the volume of SGD in the Suruga Bay from Mishima lava flow as 0.12 million tons a day. Concentrations of nutrients in SGD have not been determined in the Suruga Bay. ROV is a good sampling device to collect SGD with minimizing seawater mixing, but frequent sampling with ROV is difficult due to its high running cost. However, frequent sampling is important to unveil the influence of the nutrient supply by SGD on the ecosystem (Sugimoto et al. 2017). A sampling device satisfying the requirement is a peristaltic pump by which my group took seawater containing volcanic gas from Teishi Knoll with a summit crater of ~ 200 m across and ~ 30 m deep on the sea floor of 122 m below sea level (Notsu et al. 2014).

4.1.3 Water Mass Structure

The Suruga Bay receives not only terrestrial waters but also plenty of open waters with different origins at the distinctive depths from the Pacific Ocean. In oceanography, all seawater on earth is considered to occur at the surface at a certain point in time. If the surface seawater becomes denser (by cooling, evaporation, sea ice formation, etc.), or is disturbed by wind-driven turbulence, it sinks down to the interior with holding temperature and salinity obtained at the surface. The place where seawater downwelling occurs is designated as "origin," and the water parcels with a homogeneous feature inside are called "water mass." Water mass with different origins can be distinguished by difference in physicochemical features (temperature, salinity, density, concentrations of dissolved oxygen and nutrients, etc.). In the Suruga Bay, five distinctive water masses are defined from surface to the depth as (A) Coastal Water near the bay head, influenced by river water; (B) Offshore Water characterized by a salinity maximum and exhibiting the characteristics of Kuroshio Water; (C) Surface Water, a mixture of the Coastal Water and the Offshore Water at 0–100 m depth; (D) Intermediate Water characterized by a salinity minimum at 200–1200 m; (E) Pacific Deep Water at > 1200 m (Nakamura 1982). Regarding Intermediate Water, salinity minima found in the mesopelagic zone of the Pacific Ocean have consistent density value and are identified as the North Pacific Intermediate Water, whose origin is considered the Okhotsk Sea (Yasuda 1997). The Suruga Bay has a siphon-sampling system, which takes Intermediate Water from 397 m depth and supplies it directly to tanks in the underground basement of a land station of Suruga-Bay Deep Sea Water Aquaculture Research Center, Shizuoka Prefectural Research Institute of Fishery in Yaizu City. Sampled seawater is used for research and provided to manufacturers of seafood products and others. Details of the system are described in Sohrin et al. (2011) and others. It should be noted that the system planned to take seawater from 687 m depth was damaged by the 2009 Suruga Bay Earthquake, and, currently, the system is considered to be taking seawater from 270 m depth, which is more influenced by surface water. The siphon-sampling system also takes seawater from 20 m depth. These sampling systems are a powerful tool to promote the study of biogeochemical oceanography as they provide seawater with different origins constantly.

4.2 Nutrient Concentrations in the Oceanic Province of the Suruga Bay

The project named "Suruga Bay Research for Understanding Marine Ecosystems" (SURUME) was launched in 2014 by researchers at RIGST, Tokai University, and Ehime University. The goal of this project is to figure out temporal changes in biomass and diversity of organisms at lower trophic-level dwellings in the Suruga Bay over seasonal to decadal scales and the factors contributing to them. Seawater and plankton are collected once a month at the nearshore station (SR1; water depth, 1000 m) and offshore station (SR3; 1500 m) in the oceanic province nearby the Suruga Trough (Fig. 7.1). Trace metals are not determined in SURUME because of insufficient equipment on the research ship to meet clean techniques. Regarding trace metals in the Suruga Bay, my group determined concentrations of Fe, Zn, Mn, Cu, and Cd in 397 m-seawater taken by the siphon-sampling system (Suzuki and Sohrin 2015) (Table 7.2). The concentration of Cd in the intermediate water was higher than in Kakita River and Kano River, while there was no significant difference for other trace metals.

The concentrations of major nutrients in 2014–2017 exhibited noticeable seasonal changes in the uppermost 200 m with highest in winter and lowest in summer (Fig. 7.3). This condition is driven by a seasonal change of water column stability: the cold and wind wave situation in winter promotes intrusion of deeper water with high nutrient concentrations into the surface layer via convection, whereas the intrusion is avoided by developed stratification in summer, resulting in a shortage of nutrient supply for algal demand. Algal bloom occurred in spring and late fall mainly within the upper 50 m, and its magnitude (maximum Chl. *a* concentration and duration) varied between years and stations. Concentrations of NO_2^- and NH_4^+ increased in the upper 150 m in spring (Fig. 7.3). This increase reveals a large supply of organic nitrogen by primary production and active microbial decomposition of it. Depths of high NO_2^- concentration extended to deeper levels at the offshore station (down to 150 m, SR3) than at the nearshore station (< 100 m, SR1), which suggests low activity of nitrification (i.e., microbial oxidation of NO_2^- to NO_3^-) below the subsurface in the offshore area, presumably due to low production of organic nitrogen.

Over the course of algal bloom, N : P ratio in the surface layer noticeably increased from 14 (winter) to > 20 (spring) and then decreased to < 10 (summer), reflecting the transition from PO_4^{3-} depletion in spring to NO_3^- depletion in summer (Fig. 7.3). The winter surface N : P ratio is comparable to the average at 500 m over an entire period (14.2), reflecting nutrient export from the depth to the surface. The deep N : P ratio is also comparable to the average in the mesopelagic and bathypelagic zones of the Pacific Ocean (~ 14, Fanning 1992), but they are lower than the canonical Redfield ratio (16). The lower N : P ratio in the deep Pacific waters is thought to result from N-removal via active denitrification (i.e., microbial reduction of inorganic N to N_2 gas) in the eastern tropical Pacific (Sarmiento and

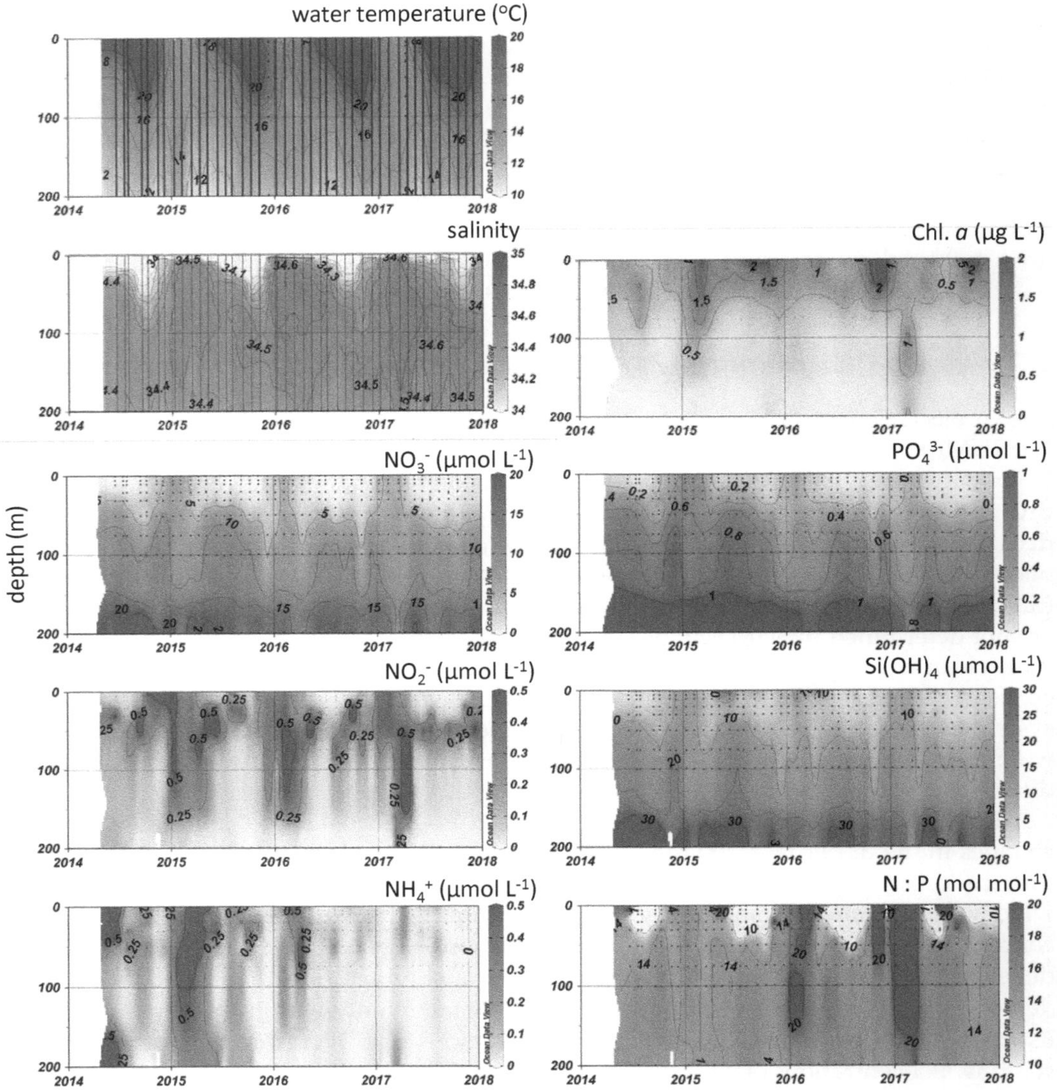

Fig. 7.3: Water temperature, salinity, concentrations of Chl. *a* and major nutrients, and the molar ratio of sum of NO_3^-, NO_2^-, and NH_4^+ to PO_4^{3-} (N : P) at the nearshore station SR1 in the Suruga Bay in the uppermost 200 m from July, 2014 to December, 2018. Dot shows the sampling depth.

Gruber 2006) or more intensive regeneration of inorganic P over N from sinking particles with high content of bioavailable P, like fecal pellets (Fanning 1992). Many literatures have reported depletion of the surface N : P ratio in summer, but the reports on prior elevation are quite limited, only with regard to off the Scotland (Painter et al. 2017). They (Painter et al. 2017) suggest a reason for the elevation as excess PO_4^{3-} consumption over NO_3^- by actively growing algae (Geider and Roche 2002).

$Si(OH)_4$ is the only major nutrient that was detected at any depth over an entire period. We found a concomitant increase in $Si(OH)_4$ concentration and a decrease in salinity in the surface water in summer, 2015 (Fig. 7.4), which is evidence of nutrient export from

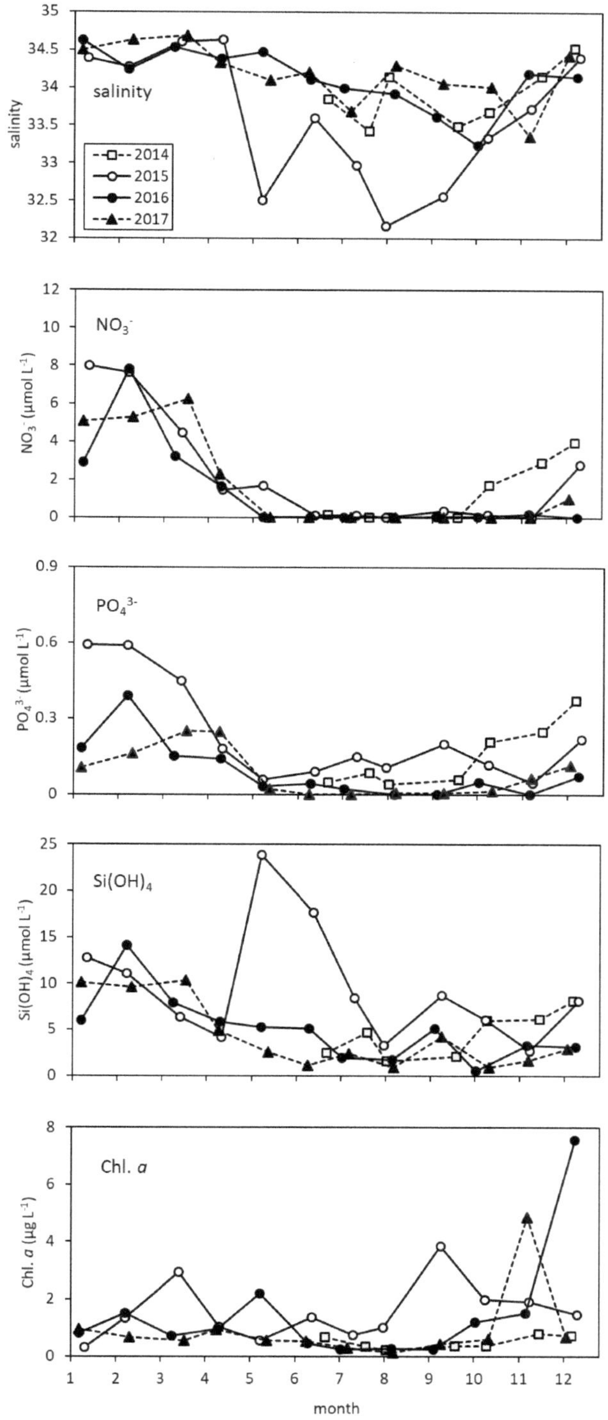

Fig. 7.4: Seasonal and interannual changes of salinity and concentrations in Chl. a, NO_3^-, PO_4^{3-}, and $Si(OH)_4$ at the offshore station SR3 in the Suruga Bay from August, 2014 to December, 2018. No data in January and February, 2015 due to bad weather.

land to the oceanic province of the Suruga Bay because terrestrial waters have a generally high concentration of $Si(OH)_4$, as shown with Kakita River and Kano River. Summertime surface concentrations of NO_3^- and PO_4^{3-} did not increase in 2015, but they were higher in 2015 than in the other years. These findings suggest fluvial input of NO_3^- and PO_4^{3-} with their large consumption before reaching the oceanic province. Given the decrease in surface salinity every summer and higher summertime surface Chl. *a* concentration in 2015 than in the other years, it is highly likely that fluvial runoff mitigates depleted or quasi-depleted conditions of the surface NO_3^- and PO_4^{3-} in the Suruga Bay in summer. A previous study shows dilution of major nutrients in offshore water by an occasional increase in intrusion of Kuroshio water (Iwata et al. 2005). Thus, terrestrial nutrients are important to sustain biological production of the Suruga Bay.

5. Conclusion and Future Direction

The monitoring of natural waters associated with the Suruga Bay and Mt. Fuji was launched in 2014 and 2016, respectively, to confirm their nutritional link. Notably high concentrations of V (670 nmol L^{-1}) and Cr (7 nmol L^{-1}) and low concentrations of Fe (< 2 nmol L^{-1}) and Mn (1 nmol L^{-1}) are a unique feature of Kakita River spring water originating from Mt. Fuji groundwater. Results of incubation experiments with Kakita River spring water suggest that low concentrations of Fe and Mn suppress primary production in Kakita River and downstream of its main river, Kano River, and may allow transport of high concentrations of major nutrients to the Suruga Bay. The uppermost 200 m of the oceanic province of the Suruga Bay exhibits noticeable seasonal changes in major nutrient concentrations with N-depletion in summer. Over the course of algal bloom, N : P ratio in the surface layer increased from 14 (winter) to > 20 (spring) and decreased to < 10 (summer). A concomitant decrease in salinity and increase in $Si(OH)_4$ in the surface water of summer 2015 as well as summertime low surface salinity suggest that surface N-depletion is more or less mitigated by fluvial nutrient input.

The Kuroshio large meander, a phenomenon of the Kuroshio Current taking a meander path to the south, started in August 2017, and is going on as of April 2019. Previous studies have shown increased intrusion of Kuroshio water into the Suruga Bay during the Kuroshio large meander (Kutsuwada et al. 2007) and dilution of major nutrients in offshore water of the Suruga Bay by an increase in intrusion of Kuroshio water (Iwata et al. 2005). Further accumulation of nutrient data by SURUME will unveil the influence of the Kuroshio large meander on the ecosystem of the Suruga Bay.

Acknowledgements

I thank Drs. K. Kato, K. Nagaosa, S. Sato (Shizuoka University), Y. Funahashi, and M. Ota (Japan Riverfront Research Center), and Mr. Sugisawa (Numazu Office of River and National Highway, Ministry of Land, Infrastructure, Transport and Tourism) for their kind help to take the samples of Kakita and Kano Rivers. Special thanks are given to the captain and the crew of Hokuto (Tokai University), Drs. J. Nishikawa, H. Matsuura, T. Yoshikawa (Tokai University), and Y. Obayashi (Ehime University), and all the participants in the SURUME cruises for their support. I am grateful Ms. Nagakura, Ms. Saiki, Mr. Okemoto, Mr. Ozawa, Mr. Mogi, and Mr. Nogami (Shizuoka University) for the analyses of astaxanthin and nutrients. This work was supported by River Foundation, KELEK, and Institute of Oceanic Research and Development, Tokai University.

References

Abe, M. and F. Tsukui. 2014. Effect of water temperature on the amount of sergestid shrimp, *Sergia lucens*, spawning. Bull. Shizuoka Pref. Res. Inst. Fish 46: 93–96 (in Japanese with English abstract).

AIST. 2016. Accomplishment report on survey, proof and technical development of marine geological environment (in Japanese with English abstract).

Ando, A., H. Kamioka, S. Terashima and S. Itoh. 1989. 1988 values for GSJ rock reference samples, "Igneous rock series". Geochem. J. 23: 143–148.

Anke, M. 2004. Vanadium—An element both essential and toxic to plants, animals and humans. Anales de la Real Academia Nacional de Farmacia 70: 961–999.

Balch, P.A. and J.F. Balch. 2000. Prescription for Nutritional Healing. 3rd edition. Avery, New York.

Burson, A., M. Stomp, L. Akil, C.P.D. Brussaard and J. Huisman. 2016. Unbalanced reduction of nutrient loads has created an offshore gradient from phosphorus to nitrogen limitation in the North Sea. Limnol. Oceanogr. 61: 869–888.

Doi, W., A. Nakahiro, H. Washiyama and N. Suzuki. 2018. Spawning rate estimation and histological characteristics of post-spawning ovaries of the sergestid shrimp *Lucensosergia lucens* based on shipboard spawning experiments. Aquaculture Sci. 66: 11–16 (in Japanese with English abstract).

Exley, C. and S. Sjöberg. 2014. Silicon species in seawater. Spectrochim. Acta, Part A 117: 820–821.

Fanning, K.A. 1992. Nutrient provinces in the sea: concentration ratios, reaction rate ratios, and ideal covariation. J. Geophys. Res. 97: 5693–5712.

Feipeng, L., Z. Haiping, Z. Yiping, X. Yihua and C. Ling. 2013. Effect of flow velocity on phytoplankton biomass and composition in a freshwater lake. Sci. Total Environ. 447: 64–71.

Fournier, J.-B., E. Rebuffet, L. Delage, R. Grijol, L. Meslet-Cladière, J. Rzonca et al. 2014. The vanadium iodoperoxidase from the marine Flavobacteriaceae species *Zobellia galactanivorans* reveals novel molecular and evolutionary features of halide specificity in the vanadium haloperoxidase enzyme family. Appl. Environ. Microbiol. 80: 7561–7573.

Fujikura, K., D. Lindsay, H. Kitazato, S. Nishida and Y. Shirayama. 2010. Marine biodiversity in Japanese waters. PLOS ONE 5: e11836.

Fukui, A. 2015. Mysterious deep-sea fishes. pp. 138–164. *In*: School of Marine Science and Technology, Tokai University (ed.). The Deep Sea—The Suruga Bay, the Deepest Bay in Japan. Shizuoka Shimbun, Shizuoka, Japan (in Japanese).

Geider, A.J. and J.L. Roche. 2002. Redfield revisited: variability of C:N:P in marine microalgae and its biochemical basis. Eur. J. Phycol. 37: 1–17.

Groschen, G.E., T.L. Arnold, W.S. Morrow and K.L. Warner. 2008. Occurrence and distribution of iron, manganese, and selected trace elements in ground water in the glacial aquifer system of the Northern United States. Geological Survey Scientific Investigations Report 2009–5006, U.S. Geological Survey. Reston, VA, USA.

Gupta, U.C. and S.C. Gupta. 2014. Sources and deficiency diseases of mineral nutrients in human health and nutrition: A review. Pedosphere 24: 13–38.

Ho, T.-Y., A. Quigg, Z.V. Finkel, A.J. Milligan, K. Wyman, P.G. Falkowski et al. 2003. The elemental composition of some marine phytoplankton. J. Phycol. 39: 1145–1159.

Imai, H., Y. Hanamura and J.-H. Cheng. 2013. Genetic and morphological differentiation in the Sakura shrimp (*Sergia lucens*) between Japanese and Taiwanese populations. Contributions to Zoology 82: 123–130.

Ino, M. 1990. Visiting exquisite water, 10: Kakita River spring water. J. Groundwater Hydrol. 32: 111–118 (in Japanese).

IPCC. 2013. The physical science basis. Contribution of Working Group I to the fifth assessment report of the Intergovernmental Panel on Climate Change. *In*: T.F. Stocker, D. Qin, G.-K. Plattner, M.M.B. Tignor, S.K. Allen, J. Boschung et al. (eds.). Cambridge University Press, Cambridge, UK and New York, NY, USA.

Iwata, T., Y. Shinomura, Y. Natori, Y. Igarashi, R. Sohrin and Y. Suzuki. 2005. Relationship between salinity and nutrients in the subsurface layer in the Suruga Bay J. Oceanogr. 61: 721–732.

Japan Meteorological Agency. 2009. Monthly Report on Earthquakes and Volcanoes in Japan, August 2009 (in Japanese).

Johnson, T.L., B. Palenik and B. Brahamsha. 2011. Characterization of a functional vanadium-dependent bromoperoxidase in the marine cyanobacterium *Synechococcus* sp. CC93111. J. Phycol. 47: 792–801.

Kutsuwada, K., Masahiro Tanikawa, Y. Hagiwara and T. Katsumata. 2007. Long-term variability of upper oceanic condition in Suruga Bay and its surrounding area. Oceanography in Japan 16: 277–290 (in Japanese with English abstract).

Liu, X. and T. Osawa. 2007. *Cis* astaxanthin and especially 9-*cis* astaxanthin exhibits a higher antioxidant activity *in vitro* compared to the all-trans isomer. Biochem. Biophys. Res. Commun. 357: 187–193.

Löscher, B.M., H.J.W. De Baar, J.T.M. De Jong, C. Veth and F. Dehairs. 1997. The distribution of Fe in the antarctic circumpolar current. Deep Sea Res. Part II 44: 143–187.

Mahara, Y., T. Igarashi and T. Yasuharu. 1993. Groundwater ages of confined aquifer in Mishima lava flow, Shizuoka. J. Groundwater Hydrol. 35: 201–215 (in Japanese with English abstract).

Millero, F.L. 2013. Chemical Oceanography, 4th edition. CRC Press, Boca Raton, FL, USA.

Misawa, Y., M. Kadota and S. Matsui. 2007. Submarine topography, geology of the eastern part of the Suruga Bay and basement complex of the Izu Peninsula. Bulletin of institute of Oceanic Research and Development, Tokai University 28: 1–12 (in Japanese with English abstract).

Nakamura, Y. 1982. Oceanographic feature of Suruga Bay from view point of fisheries oceanography. Bulletin of Shizuoka Prefectural Fisheries Experiment Station 17: 1–153 (in Japanese with English abstract).

Nakayama, E., Y. Suzuki, K. Fujiwara and Y. Kitano. 1989. Chemical analyses of seawater for trace elements recent progress in Japan on clean sampling chemical speciation of trace elements. Anal. Sci. 5: 129–139.

Notsu, K., R. Sohrin, H. Wada, T. Tsuboi, H. Sumino, T. Mori et al. 2014. Leakage of magmatic–hydrothermal volatiles from a crater bottom formed by a submarine eruption in 1989 at Teishi Knoll, Japan. J. Volcanol. Geotherm. Res. 270: 90–98.

Ochiai, T. 1970. Study on fissure water in Mishima lava flow-Estimation of flow velocity distribution of groundwater, porosity and amount of groundwater flow in the section of lava flow. The journal of the Japanese Association of Groundwater Hydrology 16-17: 7–16 (in Japanese with English abstract).

Okabe, S. and T. Morinaga. 1968. Distribution of Vanadium in the Kuroshio and Adjacent Regions. Journal of the Chemical Society of Japan 89: 284–287 (in Japanese).

Omori, M. 1969. The biology of a sergestid shrimp *Sergestes lucens* Hansen. Bull. Ocean Res. Inst. Univ. Tokyo 4: 1–83.

Omori, M., Y. Ukishima and F. Muranaka. 1988. New record of occurrence of *Sergia lncens* (Hansen) (Crustacea, Sergestidae) off Tung-kang, Taiwan, with special reference to phylogeny and distribution of the species. Journal of the Oceanographical Society of Japan 44: 261–267 (in Japanese with English abstract.

Omori, M. 2015. Today's fishing of sakura-ebi shrimp in the Suruga Bay. Ocean Newsletter 349 (in Japanese with English abstract).

Painter, S.C., S.E. Hartman, C. Kivimäe, L.A. Salt, N.M. Clargo, C.J. Daniels et al. 2017. The elemental stoichiometry (C, Si, N, P) of the Hebrides Shelf and its role in carbon export. Prog. Oceanogr. 159: 154–177.

Pinet, P.R. 2014. Invitation to Oceanography, 6th edition. Jones & Bartlett Learning, Burlington, MA, USA.

Raddatz, A.L., T.M. Johnson and T.L. McLing. 2011. Cr stable isotopes in Snake River Plain aquifer groundwater: evidence for natural reduction of dissolved Cr (VI). Environ. Sci. Technol. 45: 502–507.

Redfield, A.C. 1958. The biological control of chemical factors in the environment. Am. Sci. 46: 205–221.

Sarmiento, J.L. and N. Gruber. 2006. Ocean Biogeochemical Dynamics. Princeton University Press, Princeton, NJ, USA.

Schlesinger, W.H. and E.S. Bernhardt. 2013. Biogeochemistry, 3rd edition: An Analysis of Global Change. Academic Press, Waltham, MA, USA.

Shiba, M. 2016. How was the Suruga Bay formed? Journal of Fossil Research 49: 3–12 (in Japanese with English abstract).

Singh, U.B. and A.S. Ahluwalia. 2013. Microalgae: a promising tool for carbon sequestration. Mitigation and Adaptation Strategies for Global Change 18: 73–95.

Sohrin, M. Isaji, Y. Obara, S. Agostini, Y. Suzuki, Y. Hiroe et al. 2011. Distribution of *Synechococcus* in the dark ocean. Aquat. Microb. Ecol. 64: 1–14.

Sohrin, R., C. Nagakura, J. Nishikawa, H. Matsuura, T. Yoshikawa and Y. Obayashi. 2017. Astaxanthin of Malacostraca in the Suruga Bay. Fall meeting in 2019 of the Oceanographic Society of Japan. https://jdream3.com/(in Japanese).
Sugimoto, R. et al. 2017. Phytoplankton primary productivity around submarine groundwater discharge in nearshore coasts. Mar. Ecol. Prog. Ser. 563: 25–33.
Suzuki, M. and R. Sohrin. 2015. Preparation of a large volume of low-trace metal seawater for trace metal manipulation experiments. Geosci. Repts. Shizuoka Univ. 42: 25–36 (in Japanese with English abstract).
Tsuchi, R. 2007. Groundwater and springs of Fuji volcano, Japan. pp. 375–387. *In*: S. Aramaki, T. Fujii, S. Nakada and N. Miyaji (eds.). Fuji Volcano. Yamanashi Institute of Environmental Science, Yamahashi, Japan (in Japanese with English abstract).
Uchiyama, T., K. Otagiri, T. Sano, H. Kobayashi, T. Nakamura, T. Hasegawa et al. 2014. Hydrologic science research for the management and utilization of ground water resources in the northern piedmont area of Mount Fuji. Y-CROST Research Report 9: 67–85 (in Japanese with English abstract).
Vereshchaka, A.L. 2000. Revision of the genus *Sergia* (Decapoda: Dendrobranchiata: Sergestidae): taxonomy and distribution. Galathea Report 18: 69–207.
Vereshchaka, A.L., J. Olesen and A.A. Lunina. 2014. Global diversity and phylogeny of pelagic shrimps of the former genera *Sergestes* and *Sergia* (Crustacea, Dendrobranchiata, Sergestidae), with definition of eight New Genera. PLOS ONE 9: e112057.
Volland, S., C. Lütz, B. Michalke and U. Lütz-Meindl. 2012. Intracellular chromium localization and cell physiological response in the unicellular alga *Micrasterias*. Aquat. Toxicol. 109: 59–69.
Yasuda, I. 1997. The origin of the North Pacific Intermediate Water. J. Geophys. Res. 102: 893–909.
Yasuhara, M., K. Kazahaya and A. Marui. 2007. An isotopic study on where, when, and how groundwater is recharged in Fuji volcano, central Japan. pp. 389–405. *In*: S. Aramaki, T. Fujii, S. Nakada and N. Miyaji (eds.). Fuji Volcano. Yamanashi Institute of Environmental Science, Yamahashi, Japan (in Japanese with English abstract).

Chapter 8

The Diversity of Cyanobacterial Photoresponsive Systems

Keita Miyake,[1] *Keiji Fushimi*[1,2] and *Rei Narikawa*[1,2,3,*]

1. Introduction

Living organisms possess systems that sense and respond to external environmental changes such as light, temperature and chemicals in order to optimize living activities even under fluctuating environments. Because photosynthetic organisms utilize light as an energy source, they have developed highly organized systems to sense and respond to light signals. The photosynthetic organisms are highly diversified in their photosynthetic pigments and apparatuses, and so they have various preferences to light color and intensity. In this context, the photosynthetic organisms have various photoreceptors to sense information on light color and intensity of the living environment, to transduce the light signals into enzymatic and transcriptional activities, and to finally optimize their cellular and morphological processes. In the case of the terrestrial plants, phytochrome photoreceptors reversibly sense red and far-red lights to control photomorphological responses such as shade avoidance, flowering, and light germination (Chen and Chory 2011). Furthermore, phototropin and cryptochrome photoreceptors sense blue light and its intensity to control photomorphological responses such as the circadian clock, chloroplast photo-relocation movement, phototropism, and stomatal opening (Christie 2007, Liu et al. 2016). In the case of eukaryotic algae and cyanobacteria, it has been elucidated that various photoresponsive phenomena are regulated by diverse photoreceptors partly utilizing photoreceptors common to the terrestrial plants (Duanmu et al. 2017, Wiltbank and Kehoe 2019, Fushimi and Narikawa 2019). In this chapter, we focus on cyanobacterial photoresponsive systems to briefly outline the photoreceptors and the photoresponsive phenomena regulated by these photoreceptors.

[1] Graduate School of Integrated Science and Technology, Shizuoka University, 836 Ohya, Suruga, Shizuoka 422-8529, Japan.

[2] Core Research for Evolutional Science and Technology, Japan Science and Technology Agency, 4-1-8 Honcho, Kawaguchi, Saitama 332-0012, Japan.

[3] Research Institute of Green Science and Technology, Shizuoka University, 836 Ohya, Suruga, Shizuoka 422-8529, Japan.

* Corresponding author: narikawa.rei@shizuoka.ac.jp

2. Cyanobacteria

Cyanobacteria are prokaryotic organisms that perform oxygen-evolving photosynthesis and their origin is the same as the ancestry of the plant chloroplasts. Many cyanobacterial species have been isolated from various environments such as lakes, oceans, lands, hot springs and polar sites, and their cell shapes are also highly diversified: unicellular cocci (Fig. 8.1A), unicellular bacilli (Fig. 8.1B), multicellular filament (Fig. 8.1C) branching filament, and so on. Furthermore, some cyanobacteria differentiate specialized cells such

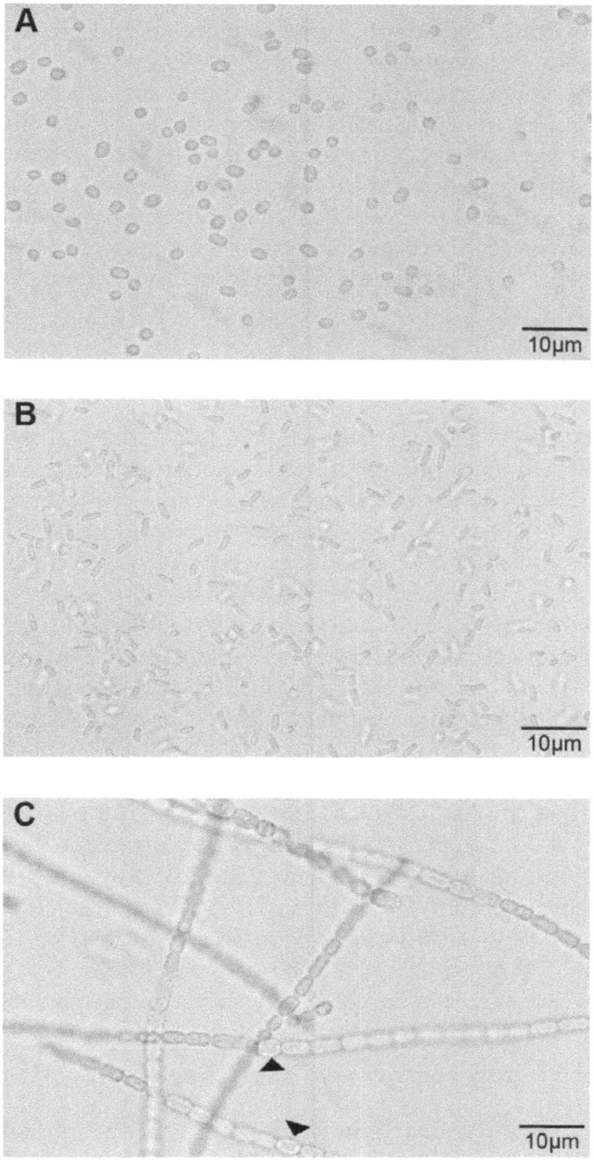

Fig. 8.1: Photographs of various cyanobacteria. (A) *Acaryochloris marina* MBIC 11017. (B) *Synechococcus elongatus* PCC 7942. (C) *Anabaena* sp. PCC 7120. Arrowheads indicate heterocysts.

as heterocysts (Fig. 8.1C), nitrogen-fixing cells, akinetes, dormant cells and hormogonia, motile cells. Because of the recent progress of genome sequencing technology, genome information of more than 400 cyanobacterial species are now available (Prabha and Singh 2019). In good agreement with the physiological and ecological diversity of cyanobacteria as described above, their genomic contents are also diversified in their qualities and quantities. The genome sizes range from less than 2 Mbp to 10 Mbp, and gene compositions are also diverse (Hess 2011, Yerrapragada et al. 2015). In particular, genes related to signal transduction proteins tend to be abundant in terrestrial and freshwater cyanobacteria, but not in marine cyanobacteria (Prabha and Singh 2019). In the marine environment, cyanobacteria form niches by the sea depth, and the fluctuation of environmental factors under the adapted environment should be small (Ting et al. 2002). Under such an environment, signal transduction genes may not be much necessary. On the other hand, the freshwater and land environments are not so deep to form niches. Under such environments, heterogenous photosynthetic organisms compete for light and nutrition sources, and so the genes for signal transduction should be abundant. In fact, photoreceptor genes showed such a tendency, although a few exceptions are present. Among the freshwater and land cyanobacteria, genetic transformation and modification of some species have been established, and molecular mechanisms of the photosynthetic and photoresponsive systems have been studied for a long time by using these species. We will introduce these systems next.

3. Cyanobacterial Photosynthetic System

Oxygen-evolving photosynthesis performed by land plants, eukaryotic algae, and cyanobacteria consists of two photosynthetic reaction centers, photosystem II (PSII) and photosystem I (PSI). The manganese cluster of PSII photooxidizes water molecule to generate chemical energy and reducing power. In addition to the photosystems, cyanobacteria possess light-harvesting antenna complex, phycobilisome (PBS), which is a supercomplex consisting of rod and core substructures (Grossman et al. 1993, Watanabe and Ikeuchi 2013). Although the PBS mainly transfers absorbed light energy to PSII, the PBS shows the state transition to transfer the light energy to PSI but not PSII in response to changes in light quality and quantity (Murata 2009). The PBS is composed of chromophore-binding phycobiliproteins and linker proteins. Phycobiliproteins such as allophycocyanin (λ max = 650 nm), phycocyanin (PC, λ max = 620 nm), phycoerythrocyanin (λ max = 575 nm), and phycoerythrin (PE, λ max = 560 nm) have been identified (Grossman et al. 1993, Watanabe and Ikeuchi 2013). Phycobiliproteins absorbing shorter wavelength light quality are located distal from the reaction centers and transfer the energy to the other phycobiliproteins absorbing longer wavelength light quality step by step, and finally transfers to the reaction centers. In addition to cyanobacteria, PBS is also present in glaucophytes and rhodophytes. Although the main role of the PBS is to harvest the light energy under normal light conditions, the PBS possesses another role as a storage material to be degraded under the nutrition depleted conditions (Grossman et al. 1993).

4. Cyanobacterial Photoreceptors and their Regulatory Mechanisms

It has been reported that cyanobacteria have many diverse photoreceptors (Losi and Gärtner 2008). Among them, linear tetrapyrrole-binding photoreceptors play central roles in many photo-acclimation processes (Fushimi and Narikawa 2019). The linear tetrapyrrole is a generic name for various chromophores synthesized by various enzymes starting from a chromophore, biliverdin, which is a cleavage product from heme. The linear tetrapyrrole is also known as bilin (bile pigment), because it has been originally found as a degradation product of heme in animals. The lengths of the conjugated systems of the linear tetrapyrrole chromophores vary depending on their structures, and their absorbing wavelengths differ accordingly (Fig. 8.2A) (Dammeyer and Frankenberg-Dinkel 2008). The conjugated systems are highlighted by gray colors, and the conjugated systems are in the order of biliverdin (BV) > phytochromobilin (PΦB) > phycocyanobilin (PCB) > phycoviolobilin (PVB). The longer conjugated system results in the longer wavelength light absorption. The linear tetrapyrroles are pigments that are used not only for light perception but also for light harvesting for photosynthesis in cyanobacteria and some eukaryotic algae.

Linear tetrapyrrole-binding photoreceptors in cyanobacteria are categorized into phytochromes and cyanobacteriochromes (CBCRs) (Rockwell and Lagarias 2010, Fushimi and Narikawa 2019). The cyanobacterial phytochromes have high homology with the plant phytochromes and function as photoreceptors to reversibly sense red and far-red lights also in cyanobacteria (Hughes et al. 1997, Yeh et al. 1997). Although physiological roles of the cyanobacterial phytochromes have long been unknown, their roles have recently been discovered in the cyanobacteria harboring chlorophyll *f* (Chl. *f*), which absorbs longer wavelength far-red light than chlorophyll *a* (Chl. *a*) (Gan et al. 2014a, 2014b, Zhao et al. 2015). The cyanobacteria harboring Chl. *f* accumulate only Chl. *a* under the normal white light conditions, but accumulate Chl. *f* in addition to Chl. *a* under the far-red light conditions, enabling efficient photosynthesis even under the far-red light conditions (Gan et al. 2014a, 2014b). It has been elucidated that this photo-acclimation process is regulated at the transcriptional level by the phytochrome called RfpA and the downstream factors RfpB and RfpC (Zhao et al. 2015).

While most phytochromes exhibit red/far-red reversible photoconversions, the CBCRs are highly diversified to show various photoconversions covering the ultraviolet-to-visible spectrum (Fushimi and Narikawa 2019). Many photoreceptors showing various photoconversions such as red/green, red/blue, green/blue, green/teal and blue/teal

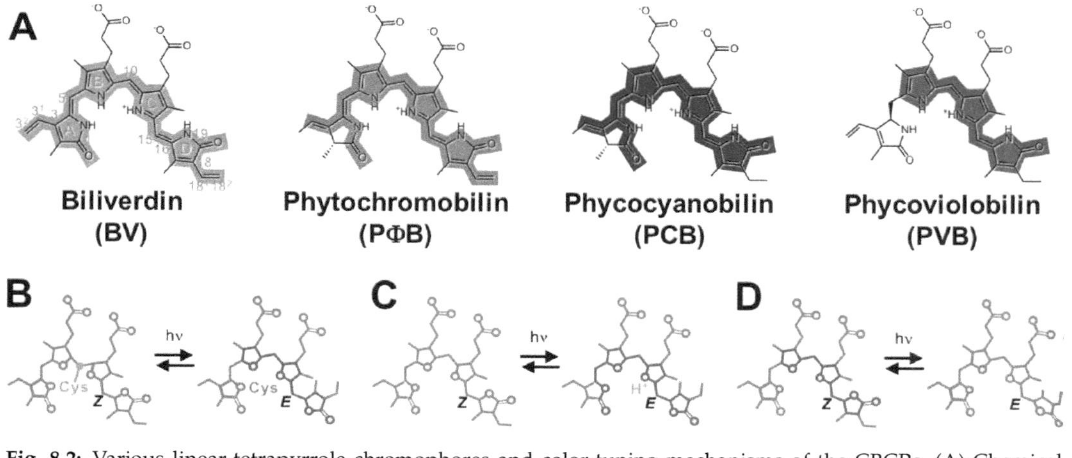

Fig. 8.2: Various linear-tetrapyrrole chromophores and color-tuning mechanisms of the CBCRs. (A) Chemical structures of the various linear-tetrapyrrole chromophores. The conjugated regions are highlighted by gray colors. (B-D) Color tuning mechanisms of the CBCRs: reversible covalent bond formation (B), reversible protonation (C), and reversible distortion of the ring D (D).

ones have been identified so far, and their color-tuning mechanisms have been elucidated based on the structural and spectral studies (Enomoto et al. 2012, Fushimi et al. 2016a, 2016b, Hirose et al. 2008, Ishizuka et al. 2006, Ma et al. 2012, Narikawa et al. 2008a, 2008b, 2014, 2015a, 2015b, Rockwell 2015b, Rockwell et al. 2008, 2011, 2012a, 2015a, 2016, 2017, Yoshihara et al. 2004). PVB and PCB have been identified as chromophores to bind the CBCRs (Ishizuka et al. 2007, Narikawa et al. 2008b). Since PVB and PCB themselves absorb green and red-light regions, respectively, photoreceptors incorporating these chromophores also sense shorter and longer wavelength light regions, respectively. Both the phytochromes and CBCRs are known to show rotation of the double bond between the C and D rings of the chromophore, Z/E isomerization, as a primary reaction triggered by light irradiation. In the case of the CBCRs, it has been clarified that the combination of four color-tuning mechanisms results in diverse spectral properties (Fushimi and Narikawa 2019): binding chromophore species (Ishizuka et al. 2007, Narikawa et al. 2008b, 2015b) (Fig. 8.2A), reversible covalent bond formation (Ishizuka et al. 2011, Narikawa et al. 2014, Rockwell et al. 2008, 2011, 2017) (Fig. 8.2B), reversible protonation (Hirose et al. 2013) (Fig. 8.2C) and reversible distortion of the ring D (Lim et al. 2018, Rockwell et al. 2012b, 2014, Wiebeler et al. 2018) (Fig. 8.2D).

Several photo-acclimation processes regulated by these CBCRs have been revealed (Enomoto et al. 2014, 2015, Hirose et al. 2010, Kehoe and Grossman 1996, Narikawa et al. 2011, Savakis et al. 2012, Song et al. 2011, Yang et al. 2018, Yoshihara et al. 2000, Wiltbank and Kehoe 2016). We here briefly introduce two photo-acclimation processes, complementary chromatic acclimation of the photosynthetic antennae complexes and light-dependent cell aggregation.

4.1 Chromatic Acclimation

In addition to the photosynthetic reaction centers, the photosynthetic organisms develop light-harvesting antennae complexes absorbing light energy and transferring the energy into the reaction centers. As mentioned above, cyanobacteria possess light-harvesting antenna complex, PBS, to harvest green-to-orange light region which cannot be efficiently absorbed by the chlorophylls (Grossman et al. 1993, Watanabe and Ikeuchi 2013). Among various cyanobacteria, some cyanobacteria possess the PE absorbing green light and the PC absorbing orange-to-red light and alter the ratio of the PE to the PC in response to the ambient light environment. Namely, the PC but not the PE are highly accumulated under the red light environment, whereas the PE but not PC highly are accumulated under the green light environment in order to efficiently perform photosynthesis (Hirose et al. 2010, Kehoe and Grossman 1996). In these organisms, the CBCR photoreceptors (RcaE and CcaS) show reversible photoconversion between a red-absorbing form and a green-absorbing form to sense ratio of the two light colors, alter the autophosphorylation activity in response to the light color, and finally regulate gene expression of the PC and PE components (Hirose et al. 2008, 2010, 2013).

4.2 Light Dependent Cell Aggregation

Thermophilic cyanobacteria isolated from hot springs show cell aggregation composed of a large number of cells under low-temperature light conditions (Kawano et al. 2011). Because dark reactions are rate-limiting in the photosynthetic process under the low-temperature light conditions, cyanobacteria under such conditions are thought to be more susceptible to photoinhibition than normal conditions. In this context, the formation of the cell aggregates under the low-temperature light conditions is regarded as an active photoacclimation process avoiding excess light, because light intensity should be reduced at the cells deeper within the cell aggregates by self-shading. This aggregation is readily resolved by cellulase addition, indicating that cellulose accumulation is an essential factor for cell aggregation. The cellulose accumulation and cell aggregation were inhibited by disruption of cellulose synthase and secretion genes (Kawano et al. 2011, Maeda et al. 2018). Furthermore, it has been revealed that this cell aggregation occurs under blue light but not under green light. Detailed biochemical and physiological studies have revealed that CBCR photoreceptors SesA, SesB, and SesC cooperatively work to regulate the blue light-dependent cell aggregation by modifying cellular c-di-GMP content (Enomoto et al. 2014, 2015).

5. Unique Cyanobacterium Acaryochloris marina

As mentioned above, there are various species within the cyanobacterial phylum, and it has been clarified that each species has developed a highly organized photoresponsive strategy coping with the adapted environment. In order to further discover unique photoreceptors and photoresponsive phenomena, we are currently focusing on a quite unique cyanobacterium, *Acaryochloris marina* MBIC 11017 (Fig. 8.1A). *A. marina* MBIC 11017 had coexisted within the capsule of the sea squirt colony, *Lissoclinum patella*, inhabiting Palau, and separated from the capsule (Miyashita et al. 1996). Although most cyanobacteria utilize the Chl. *a* as the photosynthetic reaction center pigment (Fig. 8.3A), *A. marina* MBIC 11017 exceptionally utilizes Chl. *d* as the reaction center pigment (Fig. 8.3B), and 95% or more of the chlorophylls are occupied by the Chl. *d* (Chen et al. 2009, Murakami et al. 2004). The Chl. *d* absorbs far-red light, which is a longer wavelength light than that absorbed by the Chl. *a* (Fig. 8.3C) (Miyashita et al. 1996). Since the light quality used for photosynthesis is red-shifted in comparison with the other cyanobacteria, we hypothesized that there would exist photoreceptors to sense longer wavelength light quality. Therefore, we searched candidate genes for the CBCRs from the genome sequence of *A. marina* MBIC 11017 and biochemically and spectrally analyzed them (Fushimi et al. 2016a, Hasegawa et al. 2018, Narikawa et al. 2014, 2015a, 2015b). As a result, in addition to PCB, some CBCRs covalently bound not only PCB but also BV, which is a precursor of PCB and absorbs about 50 nm red-shifted light quality in comparison with PCB (Fushimi et al. 2016a, Narikawa et al. 2015b). In good agreement with the red-shifted property of BV, these photoreceptors showed reversible photoconversion between far-red-absorbing forms and orange-absorbing forms upon BV binding, whereas these photoreceptors showed reversible photoconversion between red-absorbing forms and green-absorbing forms upon PCB binding. These red-shifted properties may be physiologically important for the Chl. *d*-utilizing *A. marina* MBIC 11017.

Although *A. marina* has the PBS as the light-harvesting antennae complex as well as the other cyanobacteria, a little knowledge about its structure and function has been accumulated. It has been revealed that the PBS of *A. marina* MBIC1107 is composed of only the rod substructure, lacking the core substructure (Chen et al. 2009). Comparing the cell absorption spectrum with the model cyanobacterium *Synechocystis* sp. PCC 6803, absorption of the photosynthetic reaction center is red-shifted, but the absorption corresponding to

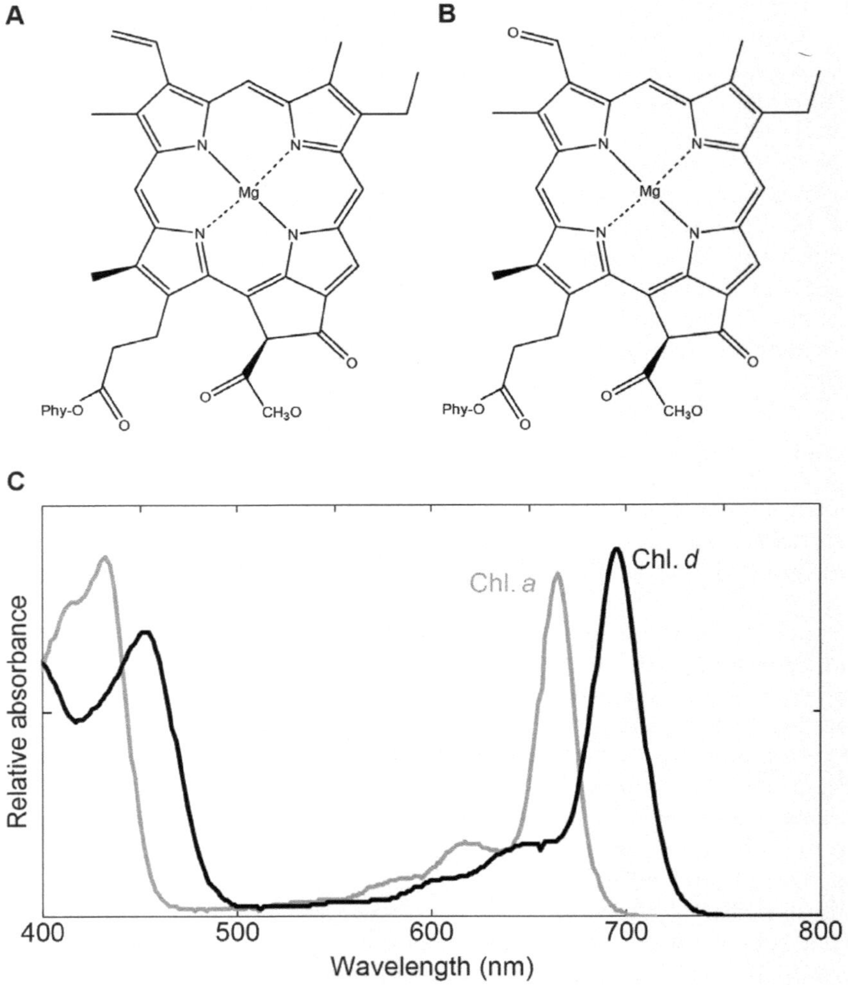

Fig. 8.3: Chemical structures and absorbance spectra of Chl. *a* and *d*. (A) Chemical structure of the Chl. *a*. (B) Chemical structure of the Chl. *d*. (C) Absorption spectra of the Chl. *a* (gray line) and Chl. *d* (black line).

the PBS is very similar (Fig. 8.4). Most of the energy transfer mechanism from the PBS to the reaction center in *A. marina* MBIC 11017 is still unknown. According to our recent study mentioned above, it has been suggested that *A. marina* MBIC 11017 may sense the ratio of the far-red light to the orange light. The orange light corresponds to that absorbed by the PBS, while the far-red light corresponds to that absorbed by the reaction center. In this context, the photoreceptors to sense the ratio of the far-red light to the orange light may be involved in the photo-acclimation processes of these photosynthetic apparatuses. Currently, we are analyzing transcriptional and proteomic profiles of *A. marina* MBIC 11017 under the far-red and orange monochromic light conditions. By promoting the transcriptomic and proteomic analyses, we hope to unravel the unique photoresponsive strategy of *A. marina* MBIC 11017.

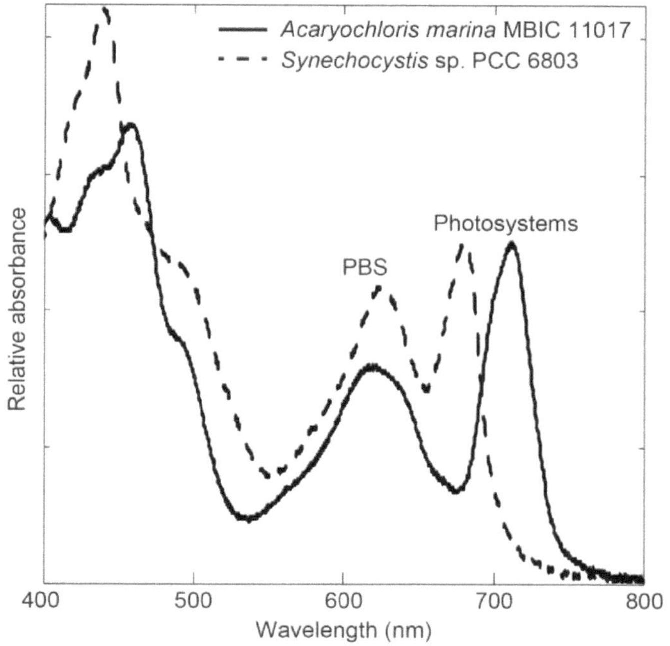

Fig. 8.4: Cellular absorbance spectra of model cyanobacterium *Synechocystis* sp. PCC 6803 (dashed line) and *Acaryochloris marina* MBIC 11017 (solid line).

6. Summary

As described in the introduction, various cyanobacterial species exist in various environments. Until now, most studies to analyze the molecular mechanisms of photoresponsive systems have been focused on species whose cultivation and genetic modification have been established. In recent years, novel regulatory mechanisms for photosynthesis have been discovered from cyanobacteria which are non-model and newly isolated cyanobacteria (Gan et al. 2014a, 2014b, Sanfilippo et al. 2016). Because of the remarkable progress on genome editing and DNA sequencing technologies, the threshold to deal with non-model organisms should be further reduced in the near future. In this context, it is largely expected that the generality and diversity of highly developed photoresponsive strategies of various cyanobacteria will be revealed in more detail. We also would like to approach this important issue by focusing on the unique cyanobacterium *A. marina* MBIC 11017.

References

Chen, M., M. Floetenmeyer and T.S. Bibby. 2009. Supramolecular organization of phycobiliproteins in the chlorophyll *d*-containing cyanobacterium *Acaryochloris marina*. FEBS Lett. 583: 2535–2539.

Chen, M. and J. Chory. 2011. Phytochrome signaling mechanisms and the control of plant development. Trends Cell Biol. 21: 664–671.

Christie, J.M. 2007. Phototropin blue-light receptors. Annu. Rev. Plant Biol. 58: 21–45.

Dammeyer, T. and N. Frankenberg-Dinkel. 2008. Function and distribution of bilin biosynthesis enzymes in photosynthetic organisms. Photochem. Photobiol. Sci. 7: 1121–1130.

Duanmu, D., N.C. Rockwell and J.C. Lagarias. 2017. Algal light sensing and photoacclimation in aquatic environments. Plant Cell Environ. 40: 2558–2570.

Enomoto, G., Y. Hirose, R. Narikawa and M. Ikeuchi. 2012. Thiol-based photocycle of the blue and teal light-sensing cyanobacteriochrome Tlr1999. Biochemistry 51: 3050–3058.

Enomoto, G., R. Nomura, T. Shimada, Ni-Ni-Win, R. Narikawa and M. Ikeuchi. 2014. Cyanobacteriochrome SesA is a diguanylate cyclase that induces cell aggregation in *Thermosynechococcus*. J. Biol. Chem. 289: 24801–24809.

Enomoto, G., Ni-Ni-Win, R. Narikawa and M. Ikeuchi. 2015. Three cyanobacteriochromes work together to form a light color-sensitive input system for c-di-GMP signaling of cell aggregation. Proc. Natl. Acad. Sci. U.S.A. 112: 8082–8087.

Fushimi, K., T. Nakajima, Y. Aono, T. Yamamoto, Ni-Ni-Win, M. Ikeuchi et al. 2016a. Photoconversion and fluorescence properties of a red/green-type cyanobacteriochrome AM1_C0023g2 that binds not only phycocyanobilin but also biliverdin. Front. Microbiol. 7: 588.

Fushimi, K., N.C. Rockwell, G. Enomoto, Ni-Ni-Win, S.S. Martin, F. Gan et al. 2016b. Cyanobacteriochrome photoreceptors lacking the canonical Cys residue. Biochemistry 55: 6981–6995.

Fushimi, K. and R. Narikawa. 2019. Cyanobacteriochromes: photoreceptors covering the entire UV-to-visible spectrum. Curr. Opin. Struct. Biol. 57: 39–46.

Gan, F., G. Shen and D.A. Bryant. 2014a. Occurrence of far-red light photoacclimation (FaRLiP) in diverse cyanobacteria. Life 5: 4–24.

Gan, F., S. Zhang, N.C. Rockwell, S.S. Martin, J.C. Lagarias and D.A. Bryant. 2014b. Extensive remodeling of a cyanobacterial photosynthetic apparatus in far-red light. Science 345: 1312–1317.

Grossman, A.R., M.R. Schaefer, G.G. Chiang and J.L. Collier. 1993. The phycobilisome, a light-harvesting complex responsive to environmental conditions. Microbiol. Rev. 57: 725–749.

Hasegawa, M., K. Fushimi, K. Miyake, T. Nakajima, Y. Oikawa, G. Enomoto et al. 2018. Molecular characterization of DXCF cyanobacteriochromes from the cyanobacterium *Acaryochloris marina* identifies a blue-light power sensor. J. Biol. Chem. 293: 1713–1727.

Hess, W.R. 2011. Cyanobacterial genomics for ecology and biotechnology. Curr. Opin. Microbiol. 14: 608–614.

Hirose, Y., T. Shimada, R. Narikawa, M. Katayama and M. Ikeuchi. 2008. Cyanobacteriochrome CcaS is the green light receptor that induces the expression of phycobilisome linker protein. Proc. Natl. Acad. Sci. U.S.A. 105: 9528–9533.

Hirose, Y., R. Narikawa, M. Katayama and M. Ikeuchi. 2010. Cyanobacteriochrome CcaS regulates phycoerythrin accumulation in *Nostoc punctiforme*, a group II chromatic adapter. Proc. Natl. Acad. Sci. U.S.A. 107: 8854–8859.

Hirose, Y., N.C. Rockwell, K. Nishiyama, R. Narikawa, Y. Ukaji, K. Inomata et al. 2013. Green/red cyanobacteriochromes regulate complementary chromatic acclimation via a protochromic photocycle. Proc. Natl. Acad. Sci. U.S.A. 110: 4974–4979.

Hughes, J., T. Lamparter, F. Mittmann, E. Hartmann, W. Gärtner, A. Wilde et al. 1997. A prokaryotic phytochrome. Nature 386: 663.

Ishizuka, T., T. Shimada, K. Okajima, S. Yoshihara, Y. Ochiai, M. Katayama et al. 2006. Characterization of cyanobacteriochrome TePixJ from a thermophilic cyanobacterium *Thermosynechococcus elongatus* strain BP-1. Plant Cell Physiol. 47: 1251–1261.

Ishizuka, T., R. Narikawa, T. Kohchi, M. Katayama and M. Ikeuchi. 2007. Cyanobacteriochrome TePixJ of *Thermosynechococcus elongatus* harbors phycoviolobilin as a chromophore. Plant Cell Physiol. 48: 1385–1390.

Ishizuka, T., A. Kamiya, H. Suzuki, R. Narikawa, T. Noguchi, T. Kohchi et al. 2011. The cyanobacteriochrome, TePixJ, isomerizes its own chromophore by converting phycocyanobilin to phycoviolobilin. Biochemistry 50: 953–961.

Kawano, Y., T. Saotome, Y. Ochiai, M. Katayama, R. Narikawa and M. Ikeuchi. 2011. Cellulose accumulation and a cellulose synthase gene are responsible for cell aggregation in the cyanobacterium *Thermosynechococcus vulcanus* RKN. Plant Cell Physiol. 52: 957–966.

Kehoe, D.M. and A.R. Grossman. 1996. Similarity of a chromatic adaptation sensor to phytochrome and ethylene receptors. Science 273: 1409–1412.

Lim, S., Q. Yu, S.M. Gottlieb, C.W. Chang, N.C. Rockwell, S.S. Martin et al. 2018. Correlating structural and photochemical heterogeneity in cyanobacteriochrome NpR6012g4. Proc. Natl. Acad. Sci. U.S.A. 115: 4387–4392.

Liu, B., Z. Yang, A. Gomez, B. Liu, C. Lin and Y. Oka. 2016. Signaling mechanisms of plant cryptochromes in *Arabidopsis thaliana*. J. Plant Res. 129: 137–148.

Losi, A. and W. Gärtner. 2008. Bacterial bilin- and flavin-binding photoreceptors. Photochem. Photobiol. Sci. 7: 1168–1178.

Maeda, K., J. Tamura, Y. Okuda, R. Narikawa, T. Midorikawa and M. Ikeuchi. 2018. Genetic identification of factors for extracellular cellulose accumulation in the thermophilic cyanobacterium *Thermosynechococcus vulcanus*: proposal of a novel tripartite secretion system. Mol. Microbiol. 109: 121–134.

Ma, Q., H-H. Hua, Y. Chen, B-B. Liu, A.L. Krämer, H. Scheer et al. 2012. A rising tide of blue-absorbing biliprotein photoreceptors: characterization of seven such bilin-binding GAF domains in *Nostoc* sp. PCC7120. FEBS J. 279: 4095–4108.

Miyashita, H., H. Ikemoto, N. Kurano, K. Adachi, M. Chihara and S. Miyachi. 1996. Chlorophyll *d* as a major pigment. Nature 383: 402.

Murakami, A., H. Miyashita, M. Iseki, K. Adachi and M. Mimuro. 2004. Chlorophyll *d* in an epiphytic cyanobacterium of red algae. Science 303: 1633.

Murata, N. 2009. The discovery of state transitions in photosynthesis 40 years ago. Photosyn. Res. 99: 155–160.

Narikawa, R., Y. Fukushima, T. Ishizuka, S. Itoh and M. Ikeuchi. 2008a. A novel photoactive GAF domain of cyanobacteriochrome AnPixJ that shows reversible green/red photoconversion. J. Mol. Biol. 380: 844–855.

Narikawa, R., T. Kohchi and M. Ikeuchi. 2008b. Characterization of the photoactive GAF domain of the CikA homolog (SyCikA, Slr1969) of the cyanobacterium *Synechocystis* sp. PCC 6803. Photochem. Photobiol. Sci. 7: 1253–1259.

Narikawa, R., F. Suzuki, S. Yoshihara, S.-I. Higashi, M. Watanabe and M. Ikeuchi. 2011. Novel photosensory two-component system (PixA-NixB-NixC) involved in the regulation of positive and negative phototaxis of cyanobacterium *Synechocystis* sp. PCC 6803. Plant Cell Physiol. 52: 2214–2224.

Narikawa, R., G. Enomoto, Ni-Ni-Win, K. Fushimi and M. Ikeuchi. 2014. A new type of dual-Cys cyanobacteriochrome GAF domain found in cyanobacterium *Acaryochloris marina*, which has an unusual red/blue reversible photoconversion cycle. Biochemistry 53: 5051–5059.

Narikawa, R., K. Fushimi, Ni-Ni-Win and M. Ikeuchi. 2015a. Red-shifted red/green-type cyanobacteriochrome AM1_1870g3 from the chlorophyll *d*-bearing cyanobacterium *Acaryochloris marina*. Biochem. Biophys. Res. Commun. 461: 390–395.

Narikawa, R., T. Nakajima, Y. Aono, K. Fushimi, G. Enomoto, Ni-Ni-Win et al. 2015b. A biliverdin-binding cyanobacteriochrome from the chlorophyll d-bearing cyanobacterium *Acaryochloris marina*. Sci. Rep. 5: 7950.

Prabha, R. and D.P. Singh. 2019. Cyanobacterial phylogenetic analysis based on phylogenomics approaches render evolutionary diversification and adaptation: an overview of representative orders. 3 Biotech. 9: 87.

Rockwell, N.C., S.L. Njuguna, L. Roberts, E. Castillo, V.L. Parson, S. Dwojak et al. 2008. A second conserved GAF domain cysteine is required for the blue/green photoreversibility of cyanobacteriochrome Tlr0924 from *Thermosynechococcus elongatus*. Biochemistry 47: 7304–7316.

Rockwell, N.C. and J.C. Lagarias. 2010. A brief history of phytochromes. ChemPhysChem. 11: 1172–1180.

Rockwell, N.C., S.S. Martin, K. Feoktistova and J.C. Lagarias. 2011. Diverse two-cysteine photocycles in phytochromes and cyanobacteriochromes. Proc. Natl. Acad. Sci. U.S.A. 108: 11854–11859.

Rockwell, N.C., S.S. Martin, A.G. Gulevich and J.C. Lagarias. 2012a. Phycoviolobilin formation and spectral tuning in the DXCF cyanobacteriochrome subfamily. Biochemistry 51: 1449–1463.

Rockwell, N.C., S.S. Martin and J.C. Lagarias. 2012b. Mechanistic insight into the photosensory versatility of DXCF cyanobacteriochromes. Biochemistry 51: 3576–3585.

Rockwell, N.C., S.S. Martin, A.G. Gulevich and J.C. Lagarias. 2014. Conserved phenylalanine residues are required for blue-shifting of cyanobacteriochrome photoproducts. Biochemistry 53: 3118–3130.

Rockwell, N.C., S.S. Martin, F. Gan, D.A. Bryant and J.C. Lagarias. 2015a. NpR3784 is the prototype for a distinctive group of red/green cyanobacteriochromes using alternative Phe residues for photoproduct tuning. Photochem. Photobiol. Sci. 14: 258–269.

Rockwell, N.C., S.S. Martin and J.C. Lagarias. 2015b. Identification of DXCF cyanobacteriochrome lineages with predictable photocycles. Photochem. Photobiol. Sci. 14: 929–941.

Rockwell, N.C., S.S. Martin and J.C. Lagarias. 2016. Identification of cyanobacteriochromes detecting far-red light. Biochemistry 55: 3907–3919.

Rockwell, N.C., S.S. Martin and J.C. Lagarias. 2017. There and back again: loss and reacquisition of two-Cys photocycles in cyanobacteriochromes. Photochem. Photobiol. 93: 741–754.

Sanfilippo, J.E., A.A. Nguyen, J.A. Karty, A. Shukla, W.M. Schluchter, L. Garczarek et al. 2016. Self-regulating genomic island encoding tandem regulators confers chromatic acclimation to marine *Synechococcus*. Proc. Natl. Acad. Sci. U.S.A. 113: 6077–6082.

Savakis, P., S. De Causmaecker, V. Angerer, U. Ruppert, K. Anders, L.-O. Essen et al. 2012. Light-induced alteration of c-di-GMP level controls motility of *Synechocystis* sp. PCC 6803. Mol. Microbiol. 85: 239–251.

Song, J.-Y., H.S. Cho, J.-I. Cho, J.-S. Jeon, J.C. Lagarias and Y.-I. Park. 2011. Near-UV cyanobacteriochrome signaling system elicits negative phototaxis in the cyanobacterium *Synechocystis* sp. PCC 6803. Proc. Natl. Acad. Sci. U.S.A. 108: 10780–10785.

Ting, C.S., G. Rocap, J. King and S.W. Chisholm. 2002. Cyanobacterial photosynthesis in the oceans: the origins and significance of divergent light-harvesting strategies. Trends Microbiol. 10: 134–142.

Watanabe, M. and M. Ikeuchi. 2013. Phycobilisome: architecture of a light-harvesting supercomplex. Photosyn. Res. 116: 265–276.

Wiebeler, C., R.A. Gopalakrishna, W. Gärtner and I. Schapiro. 2018. The effective conjugation length is responsible for the red/green spectral tuning in the cyanobacteriochrome Slr1393g3. Angew. Chem. Int. Ed. Engl. In press.

Wiltbank, L.B. and D.M. Kehoe. 2016. Two cyanobacterial photoreceptors regulate photosynthetic light harvesting by sensing teal, green, yellow, and red light. MBio. 7: e02130–15.

Wiltbank, L.B. and D.M. Kehoe. 2019. Diverse light responses of cyanobacteria mediated by phytochrome superfamily photoreceptors. Nat. Rev. Microbiol. 17: 37–50.

Yang, Y., V. Lam, M. Adomako, R. Simkovsky, A. Jakob, N.C. Rockwell et al. 2018. Phototaxis in a wild isolate of the cyanobacterium *Synechococcus elongatus*. Proc. Natl. Acad. Sci. U.S.A. 115: E12378–E12387.

Yeh, K.C., S.H. Wu, J.T. Murphy and J.C. Lagarias. 1997. A cyanobacterial phytochrome two-component light sensory system. Science 277: 1505–1508.

Yerrapragada, S., A. Shukla, K. Hallsworth-Pepin, K. Choi, A. Wollam, S. Clifton et al. 2015. Extreme sensory complexity encoded in the 10-megabase draft genome sequence of the chromatically acclimating cyanobacterium *Tolypothrix* sp. PCC 7601. Genome Announc. 3: e00355-15.

Yoshihara, S., F. Suzuki, H. Fujita, X.X. Geng and M. Ikeuchi. 2000. Novel putative photoreceptor and regulatory genes Required for the positive phototactic movement of the unicellular motile cyanobacterium *Synechocystis* sp. PCC 6803. Plant Cell Physiol. 41: 1299–1304.

Yoshihara, S., M. Katayama, S. Geng and M. Ikeuchi. 2004. Cyanobacterial phytochrome-like PixJ1 holoprotein shows novel reversible photoconversion between blue- and green-absorbing forms. Plant Cell Physiol. 45: 1729–1737.

Zhao, C., F. Gan, G. Shen and D.A. Bryant. 2015. RfpA, RfpB, and RfpC are the master control elements of far-red light photoacclimation (FaRLiP). Front. Microbiol. 6: 1303.

Chapter 9

Hyperspectral Remote Sensing of Plant Water Status and Plant Water Use under Drought Stress

Quan Wang[1,2,*] and *Jia Jin*[1]

1. Introduction

Significant impacts of global change on various ecosystems have received worldwide attention in recent decades (Jin and Wang 2016a, Li et al. 2018, Stocker et al. 2014). As reported by the Intergovernmental Panel on Climate Change (IPCC), the globally averaged land and ocean surface temperature increased linearly up to ca. 0.72°C for the period from 1951 to 2012 (IPCC 2014). Such a warming trend would amplify drought-induced stress and increase mortality in forests further (Young et al. 2017) since the drought has been projected to accompany more frequently with the rising temperature and thus would impact significantly on socio-economic activities and environments (Gao and Zhang 2016).

Water availability is essential for plant growth (Pasqualotto et al. 2018). It controls plant water status and determines a series of plant physiological processes as well (Cao et al. 2015). Among all available physiological parameters indicating plant water status, the leaf scale water content is one of the fundamental parameters for assessing plants' physiological status, especially on detecting plant drought stress (Pasqualotto et al. 2018). Similarly, for plant water use, transpiration is a critical component among manifold physical processes and biological cycles (Katul et al. 2012, Marshall et al. 2016) and is another pivotal parameter for indicating vegetation physiological status (Jin et al. 2019b). A clear understanding of both parameters is therefore important for elucidating the response of plants to different water conditions (Jin et al. 2019a).

[1] Faculty of Agriculture, Shizuoka University, Shizuoka 422-8529, Japan.
[2] Research Institute of Green Science and Technology, Shizuoka University, Shizuoka 422-8529, Japan.
* Corresponding author: wang.quan@shizuoka.ac.jp

Unfortunately, our understanding is severely limited by the availability of reliable field sampling/monitoring data of plant water status and plant water use in current, which is far beyond the requirement for assessing the impacts of global change. Traditionally, the plant water content is calculated from the change of weight of newly harvested leaves after a drying process (Pasqualotto et al. 2018), while leaf scale transpiration is always estimated from the increased vapor contents inside a chamber closely attached to a leaf surface (Wang and Jin 2015). On the other hand, even though canopy scale transpiration measurements can be conducted continuously using sap-flow measurement techniques (Verstraeten et al. 2008), intensive installation and maintaining costs are nevertheless preventing the approaches from being applied to large scale monitoring. To date, direct measurements of plant water content or transpiration are in general time-consuming, labor-extensive, expensive and not feasible at large spatial scales.

Alternatively, hyperspectral remote sensing, providing information of targets with high spectral and spatial resolutions, has shown to be an effective tool to identify ecosystem water use conditions (Ben-Dor et al. 2013, El-Hendawy et al. 2017, Goetz 2009). During the past decades, to assess plant water status accurately using reflected spectra has been one of main goals of hyperspectral remote sensing (Furuuchi et al. 2013, Pasqualotto et al. 2018), even though photosynthetic pigments are by far the most popularly estimated plant parameters because of their close relations with radiative absorption (Gitelson et al. 2002, Gitelson et al. 2006, Jin and Wang 2019a, Kira et al. 2015, Main et al. 2011, Sims and Gamon 2002, Sonobe and Wang 2017, Sonobe and Wang 2018, Ustin et al. 2009, Yi et al. 2014, Yu et al. 2016). Since water also shows specific absorption patterns in the wavelength domain from 400 nm to 2500 nm, the plant water content has naturally been another commonly estimated parameter from hyperspectral remote sensing (Pu et al. 2003, Rodríguez-Pérez et al. 2007, Sims and Gamon 2003, Zhang et al. 2012). Comparatively, the plant water use (transpiration) has far less been investigated through hyperspectral reflectance. Furthermore, currently available studies on plant water use from remote sensing data relied heavily on energy balance approaches (e.g., Allen et al. 2007, Bastiaanssen et al. 1998, Su 2002), making all of them confronting with serious biases and errors resulting from partitioning ET into transpiration and soil evaporation, since the ratio of transpiration to total evapotranspiration is affected by various factors and may vary greatly under different environments (Gowda et al. 2008, Marshall et al. 2016, Wang and Dickinson 2012). As a recent developing, the Two-Source Energy Balance model could provide separate evaluations on respective heat fluxes in soil and vegetation (Kustas and Norman 1999, Norman et al. 1995), but the supplemental data it requires can hardly be fulfilled (Timmermans et al. 2007, Yao et al. 2010). What is the most important, such kind of models could not provide straightforward links for fluxes and reflected hyperspectral information (Devitt et al. 2011, Loukas et al. 2005). To date, only a very few hyperspectral remote sensing models can provide a direct assessment of plant transpiration (Jin et al. 2019a, Jin and Wang 2016b, Marino et al. 2014, Marshall et al. 2016, Sun et al. 2014, Wang and Jin 2015).

In this chapter, we report our recent progress on applying remotely sensed hyperspectral information to trace plant water status and plant water use in both mesic and xeric temperate ecosystems. The datasets used for calibration and validation for temperate mesic ecosystems, mainly composed from deciduous species, were based on leaf samples collected from typical temperate deciduous forests in the Naeba Mountain (138° 40′ E, 36° 51′ N) in Japan, a site where beech (*Fagus crenata*) is popularly distributed (Wang and Li 2012b). This mountain region has plenty of precipitation near 2000 mm year^{-1} with a high proportion of snow falling (Wang et al. 2008). Mean annual air temperature at 550-, 900- and 1500-m was 10.0, 9.3 and 5.6°C, respectively (Wang et al. 2008). On the other side, the field data of xeric ecosystems were taken in the Integrated Remote Sensing Experimental Site (IRSES, 87° 54′ E, 44 °25′ N), located at the southern border of the Gurbantünggüt Desert in China. The drought-resistant shrub, *Haloxylon ammodendron*, is the dominant species here (Huang and Li 2017). An extreme drought temperate climate is prevalent in this region, which typically has annual precipitation from 80 mm to 160 mm but the corresponding pan evaporation may reach 2000 mm (Jin et al. 2019b, Li and Wang 2012, Zheng and Wang 2014), approximately. The region experiences a vast range of annual air temperature change, with the year-round mean near 6.6°C but the lowest and the highest temperature of –40°C and 40°C, respectively (Jin and Wang 2016b). Species from the two contrasting ecosystems were included in this study, targeting for generating robust indices or models applicable to varying drought conditions.

The traditional way for retrieving plant variables from hyperspectral information is to set up relationships between the reflectance signals and the biophysical or biochemical traits (Dorigo et al. 2007), based primarily on synchronous spectral, biophysical and biochemical data under varying conditions (either in the field or in the laboratory). In this chapter, we used several different datasets including the leaf water content and reflected spectral data that synchronously measured for deciduous species at different phenological stages, simultaneously measured leaf water content and reflectance of deciduous species from lab-controlled dehydration experiments, and multi-scale field-measured synchronous transpiration and reflectance of the desert plant *Haloxylon ammodendron*, as well as simulated data from radiative transfer models including PROSPECT (Feret et al. 2008) and SCOPE (van der Tol et al. 2009). Compared with field-measured datasets, the lab-controlled dataset has covered a number of extreme cases (e.g., leaves with low water contents) that are hardly obtained in the field. Similarly, simulation datasets could contain vast ranges of plant reflectance spectra and biophysical/biochemical variables compared with purely monitoring in the field, and has commonly been used for developing not only universal but also consistent hyperspectral vegetation indices for different variables (Féret et al. 2011, Haboudane et al. 2004, Li and Wang 2013, le Maire et al. 2004, le Maire et al. 2008, Wang and Li 2012b).

Developed approaches of hyperspectral remote sensing for retrieving plant traits can typically be grouped into two categories, namely, physically based inversions and empirically/statistically based methods. Both approaches are widely used to retrieve different biophysical and biochemical variables from hyperspectral information (Dorigo et al. 2007, Verrelst et al. 2015, Wang et al. 2018). Inversion approaches take full advantage of physically based Radiative Transfer Models (RTMs) but confront inherently with the famous "ill-posed" problem, along with the technical difficulties in inversion processes (Combal et al. 2003, Li and Wang 2011, Wang et al. 2007, Wang and Li 2012a, Yebra and Chuvieco 2009). In comparison, empirically/statistically based algorithms are simple but effective, leading to being the most popular in current remote sensing applications (Verrelst et al. 2015). Among various empirical/statistical approaches, Vegetation Indices (VIs) are evidently the oldest, simplest, most popularly applied as well (Wang et al. 2018). Even so, the Partial Least Squares Regression (PLSR) has also been widely used in optical remote sensing applications. Comparatively, the PLSR is less restrictive and can be used even for the cases when there are fewer samples than predictor variables (Huang et al. 2004), particularly orientating for analyzing multi-collinear spectral datasets (Jin and Wang 2019a). In this chapter, we reported our recent studies on tracing plant water status and plant water use using hyperspectral reflectance, mainly based on the two prevalent statistical approaches (VIs and PLSR).

2. Synchronous Datasets and Analytical Approaches

2.1 The Build-up of Simultaneous Datasets

Synchronous information of hyperspectral reflectance and plant water status or plant water use is a prerequisite for revealing potential underlying linkages between them. Accordingly, simultaneously obtained data were composited at first hand for both mesic and xeric ecosystems via either direct measurements or model outputs at different scales.

2.1.1 Direct Measurements: Leaf Scale Data Collected from Field Monitoring vs. from Lab-based Controlled Experiments

a) Field monitoring

All leaf samples were collected following the protocol of the detached branch technique and applied to both mesic and xeric ecosystems (Cao et al. 2015). Specifically designed for non-*in situ* conditions, the technique requires that the branches with the targeted leaf samples are cut before dawn and need to be recut under water to ensure the reliability of the measurements (Foley et al. 2006). After that, all samples need to be stored under humid, dim, and cool conditions before reflectance measurements are taken. All leaf samples collected were mature leaves.

Leaf reflectance spectra covering the wavelength region from 350 nm to 2500 nm were measured using a leaf clip attached ASD field spectroradiometer (ASD FR, USA). In the meantime, an electronic balance was used to take leaf weight immediately after reflectance measurement to ensure both measurements were synchronous to the maximum.

Leaf samples of mesic ecosystems were collected from the four tower plots (termed as 550 m, 900 m X1, 990 m X5, and 1500 m henceforth) in the Naeba Mountain, Japan in 2007. Every two weeks during the start (leaf flushing) and the end (leaf shedding) of the growth period, as well as once per month during the summer, synchronous field measurements were conducted (Wang and Li 2012b), which finally resulted to a total of 220 measurements of beech leaves being archived. Furthermore, in 2008 a total sum of 83 leaf scale measurements was obtained from the five layers at different heights (20, 18, 16, 14, and 12 m, respectively) of the *F. crenata* canopy at the 900-m X5 site. In addition, 45 leaves from 15 popular species alongside beech near the 900-m site were collected in August of 2008 (Wang and Li 2012a). Consequently, the field-measured dataset of mesic ecosystems contains hundreds of leaf samples from different species, from multiple layers of canopies (sunlit and sun-shaded), and from different growing conditions and growth stages. Beside leaf reflectance, pigments and leaf weight, leaf-scale gas exchange was also taken for each leaf sample. A portable photosynthesis system LI-6400 (LI-COR, Inc., USA) was used to examine leaf scale gas exchange properties.

Fieldworks for xeric ecosystems were carried out at the IRSES located 20 km away from the Fukang Station of Desert Ecology of the Chinese Academy of Sciences (87° 54′ 13″ E, 44° 25′ 56″ N) (Jin et al. 2019b). This area is prevalent with an arid continental climate, typically having dry hot summer and cold winter. The soil type was sandy. Even though the vegetation coverage reached ca. 15% according to the inventory data in 2011, no surface runoff is detectable in this desert plot (Zheng and Wang 2015).

Leaf scale transpiration rate (Tr) was gauged using an HCM-1000 portable gas exchange system (HCM-1000, Waltz, Germany). Once per month from June to August in 2013 and 2014, diurnal course measurements were performed. Both leaves from the upper and lower canopy layers were selected for the hourly step tracing from 8:00 to 18:00 local time. Meanwhile, leaf reflectance spectra from 400 to 2500 nm were recorded using a leaf clip equipped field spectroradiometer (ASD FR, USA) at the same pace. Averaged spectra from five repeats were used for further analysis. As every transpiration measurement lasted for approximately 10 minutes, the leaf scale spectra were taken from nearby leaves instead.

b) Lab-controlled experiments

Plant response to extreme drought conditions may not be available if solely based on field monitoring, especially for mesic ecosystems. To ensure building up a comprehensive dataset enclosing diverse drought conditions as possible, we have further conducted leaf dehydration experiments in the laboratory for deciduous species including *Liquidambar styraciflua*, *Prunus* x yedoensis, *Zelkova serrata*, *Idesia polycarpa*, and *Fagus crenata*.

During the experiments, reflectance spectra and weights of fresh leaves were measured at once before they were dehydrated under natural conditions. Leaf reflectance spectra and weights were then recorded once per hour for the initial 5 hours but at a much slower pace after that during the entire dehydration experiment (ca. 24 hours) before all samples were air-dried to stable weights. At last, the samples were dried in an oven (70°C) for 72 hours before being reweighed. These experiments finally produced 224 data pairs in total.

2.1.2 Direct Measurements: Canopy Scale Dataset

Canopy-scale hyperspectral reflectance and plant water use (transpiration) were taken following a similar protocol for both mesic and xeric ecosystems, which has estimated transpiration from sap flow measurements and has measured canopy scale reflectance over towers using ASD spectroradiometer (ASD FR, USA).

Sap flow was monitored using the Granier type of sensors (Granier 1985). All sensors were installed into the tree stems up to a depth of 20-mm in both ecosystems and were northerly set to prevent heating from direct sunshine and to minimize temperature variations caused by other environmental factors. They were further sealed with silicone caulk to protect them from being affected by precipitation and moisture as well. In addition, white foils were wrapped outside to avoid the effects of ground longwave radiation (Zheng and Wang 2015). Sap flux density (m^3 m^{-2} s^{-1}) was calculated following Lu et al. (2004).

Canopy hyperspectral reflectance was taken using the ASD field spectroradiometer. For each sample tree, canopy measurements were conducted on clear days once per month during the growing seasons. All reflectance spectra were taken using a bare fiber having a 25° FOV (Field-Of-View) attached to the spectroradiometer at the nadir direction. The measuring height was about 1-m above the tree. The final canopy spectra used for further analysis were the averaged output from the five repeats taken at every time. Finally, the transpiration rate when canopy scale hyperspectral reflectance was taken was extracted to form the canopy scale simultaneous dataset.

2.1.3 Simulation Datasets

The representativeness of a dataset for training, the propriety of selected wavelengths and the independent validation will eventually determine the success of using statistical approaches to estimate plant parameters from hyperspectral information (Féret et al. 2011). It is always ideal to use a comprehensive database including as many different cases as possible to identify an efficient and robust statistical model (le Maire et al. 2008). Unfortunately, this is not reachable if using field-measured data only. As a recent development, combining both field-measured data and simulation data generated from RTMs is getting popular to address the problem (Féret et al. 2011, Haboudane et al. 2004, Li and Wang 2013, le Maire et al. 2004, le Maire et al. 2008, Wang and Li 2012b). Following the approach, we have produced simulation datasets from the leaf scale reflectance model PROSPECT (Feret et al. 2008, Jacquemoud and Baret 1990) and the integrated model of SCOPE (van der Tol et al. 2009).

For generating the leaf scale simulation dataset, we calibrated the PROSPECT-4 model using field-measured spectra randomly picked out from the deciduous species dataset (Wang and Li 2012a), following the calibration strategy recommended by Feret et al. (2008). The PROSPECT-4 model simulates leaf scale hemispherical reflectance and transmittance for the entire wavelength domain ranging from 400 nm to 2500 nm (le Maire et al. 2004). A total amount of 10,000 spectra were simulated from different combinations of input parameters, which were treated as following a uniform distribution for each. More details about the simulation dataset can be found in Wang and Li (2012a).

Similarly, synchronous canopy-scale spectra and transpiration were simulated using the SCOPE model. Extensive calibrations on the two specific modules of SCOPE, namely the radioactive transfer module and the energy balance module, were performed using the synchronous field data of the desert plant *Haloxylon ammodendron* before generating the simulation dataset. The detailed calibration procedure was given elsewhere in Jin and Wang (2019b), in which a time-series simulation to produce the synthetic dataset (n = 2204) was presented. The SCOPE model was driven using locally recorded meteorological data along with the timely varied chlorophyll contents that linearly interpolated from field-measured data, in order to exclude unrealistic combinations to the maximum.

2.2 Analytical Approaches

2.2.1 Spectra Forms Used for Statistical Analysis: Original Reflectance and Derivative Spectra

Although original reflectance is the most commonly used spectral mode in hyperspectral remote sensing, it is not always effective in reducing noise (Tsai and Philpot 1998). In most cases, derivative spectrum analysis can reduce background effects and has been demonstrated as a powerful option for quantifying plant pigments or physical parameters from hyperspectral data (Imanishi et al. 2004). Accordingly, in addition to the original reflectance spectra, the first-order derivative spectra were also used in this study. The first-order derivative spectra were calculated as:

$$d_\lambda = \left(\frac{dR}{d\lambda}\right)_\lambda = \frac{R_{\lambda+1} - R_\lambda}{\Delta\lambda} \tag{1}$$

where d_λ denotes the first-order derivative at the band λ, while $R_{\lambda+1}$ and R_λ refer to the reflectance values at the two continuous bands of $\lambda+1$ and λ, respectively. $\Delta\lambda$ represents the wavelength increment.

2.2.2 Hyperspectral Indices

Vegetation indices are by far the most popularly applied statistical approach for retrieving plant parameters (Jin and Wang 2016b, Verrelst et al. 2015). During the past 40 years, a number of hyperspectral indices in different types and in different band combinations have been successfully used for tracing biophysical and biochemical parameters (e.g., Haboudane et al. 2004, Main et al. 2011, le Maire et al. 2004, Wang and Li 2012a, Zarco-Tejada et al. 2005, Zhao et al. 2005), including those for plant water status (Gao 1996, Cao et al. 2015, Cao and Wang 2017, Chen et al. 2005, Cheng et al. 2006, Hunt and Rock 1989, Pasqualotto et al. 2018, Peñuelas et al. 1997, Sims and Gamon 2003, Sun et al. 2014, Wang and Li 2012b, Zhang et al. 2012) but only a few for plant water use (Jin and Wang 2016b, Marino et al. 2014, Marshall et al. 2016, Sun et al. 2014).

The best hyperspectral indices were identified for the five different types of indices that were developed using either the original reflectance spectra or the first-derivative spectra. They are: (1) simple Reflectance (R) or simple derivative spectra (dR); (2) Simple Ratio (SR); (3) Simple Difference (SD); (4) Normalized Difference (ND); and (5) Double-Difference (DDn). In detail:

$$R(\lambda 1) = R_{\lambda 1} \tag{2}$$

$$SR(\lambda 1, \lambda 2) = \frac{R_{\lambda 1}}{R_{\lambda 2}} \tag{3}$$

$$SD(\lambda 1, \lambda 2) = R_{\lambda 1} - R_{\lambda 2} \tag{4}$$

$$ND(\lambda 1, \lambda 2) = \frac{R_{\lambda 1} - R_{\lambda 2}}{R_{\lambda 1} + R_{\lambda 2}} \tag{5}$$

$$DDn(\lambda 1, \Delta) = 2R_{\lambda} - R_{\lambda - \Delta} - R_{\lambda + \Delta} \tag{6}$$

where R_{λ} represents the reflectance value at the given wavelength λ, and a prefix "d" denotes the indices calculated using the first-derivative spectra, e.g., dSR, dND, dDDn, etc.

2.2.3 Partial Least Squares Regression

Unlike VIs using only a few spectral bands, the Partial Least Squares Regression (PLSR) makes full use of all available bands and is hence good for analyzing multi-collinear spectral datasets (Jin and Wang 2019b, Kawamura et al. 2010). Several previous studies have asserted that the PLSR outperformed other traditional statistical approaches including VIs, stepwise and multiple linear regressions, or principal component regression, on retrieving various leaf biochemical contents (Dorigo et al. 2007, Hansen and Schjoerring 2003, Huang et al. 2004, Yi et al. 2014). However, one of the inherent challenges of the approach is the risk of overfitting due to the high dependency or correlation between different bands (Jin and Wang 2019a, Thenkabail et al. 2012). Performing waveband selection is, therefore , inevitable in order to enhance the performance of PLSR analysis eventually (Kawamura et al. 2010, Mehmood et al. 2012). Accordingly, the stepwise band selection and cross-validation were conducted in this study in order to ensure the robustness of the calibrated PLSR model (Jin and Wang 2019b). Previous studies suggest that the simplest and the most practical method on searching is the stepwise selection, in which variables are stepwisely added or removed from a multilinear model according to their statistical significance (Andersen and Bro 2010, Dorigo et al. 2007, Norgaard et al. 2000, Schmitt and Ployhart 1999). Furthermore, the principal components used for the PLSR models were determined using cross-validation (Chen et al. 2004).

3. Hyperspectral Remote Sensing of Plant Water Status and Plant Water Use: Applications

3.1 Retrieval of Leaf Water Status from Hyperspectral Indices

Searching the best index for assessing the leaf water status (represented by EWT) of deciduous species suggested that the DDn type of indices performed best among the aforementioned five types of indices if using the original reflectance spectra since this type of indices had the smallest RMSE (0.0011 g/cm^2). Among the type of indices, the DDn (1530, 525) was the best for the simulation dataset (RMSE of 0.0010 g/cm^2) (Wang and Li 2012a). For the indices developed from the derivative spectra, the dSR type of indices, especially the dSR (1510, 1560), was the best for estimating EWT of deciduous species (Wang and Li 2012b). However, this type of indices was less efficient on tracing EWT of the desert plant *Haloxylon ammodendron*. Instead, the identified best indices for the desert plant *Haloxylon ammodendron* were SR (2280, 2300) and ND (2280, 2300), although their RMSEs were much larger than those of deciduous species.

All five types of indices calculated either from the original reflectance spectra or from the first-order derivative spectra were screened for their potential to trace the change of EWT during leaf dehydration. For those indices developed from the original reflectance spectra, the most commonly used bands were found near the two central wavelengths of 1400 nm and 1900 nm. Furthermore, the best index was SR (1670, 1880) (R^2 = 0.95, RMSE = 0.0009 g/cm^2), which performed identically well with the validation dataset (Cao et al. 2015). Whereas, for the first-order derivative spectral indices, the most used wavelengths were located well within the domains from 1300 to 1600 nm and from 1750 to 1900 nm. Among them, the best index for estimating EWT was found to be dSR (1530, 1895), which had an R^2 of 0.96 and an RMSE of 0.0009 g/cm^2. This index performed ever better on validation (R^2 = 0.97) (Cao et al. 2015).

3.2 Retrieval of Leaf Transpiration from Hyperspectral Indices

All indices developed from original reflectance spectra performed really poor on estimating leaf scale transpiration. Among them, the DDn type indices at the wavelength of 2075 nm with a Δ of 60 nm was the relative efficient index for tracing transpiration. Its R^2 reached 0.26 and had an RMSE of 2.46 μmol/g/s. The DDn type index with band combinations around the wavelengths 2200 nm and 2400 nm and with the Δ less than 200 nm showed with more efficacy among the entire wavelength region.

In comparison, derivative spectra-based indices performed much better on tracing leaf transpiration. For instance, the simple derivative spectra index (dR_{2210}) had a dramatic improvement if compared with the R type index of the original spectra (with the R^2 of

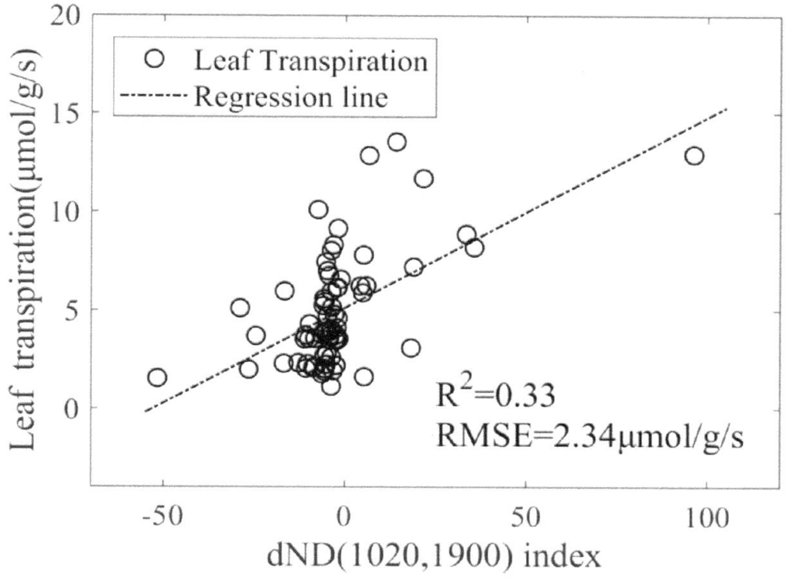

Fig. 9.1: Scatter diagrams of field measured leaf transpiration and the best index dND (1020, 1900) identified from derivative spectra.

0.17 vs. 0.05). Furthermore, the dDDn (1770, 675) had an R^2 of 0.27 with leaf transpiration. Similarly, the simple difference index calculated using the wavelength combination of 1465 nm and 2030 nm, and the simple ratio index dSR(1480, 1980), had the R^2 of 0.29 and 0.28, respectively. However, the best index among all types of derivative spectral indices was the normalized difference dND(1020,1900), which had an R^2 of 0.33 with leaf transpiration (Fig. 9.1).

3.3 Retrieval of Leaf Transpiration with PLSR Models

The PLSR models developed using either the original reflectance spectra or the first-order derivative spectra were examined for their abilities on estimating leaf water use (represented by transpiration). It is clear from the results that the PLSR models developed from the original reflectance spectra could not have traced leaf transpiration very satisfactorily, even though they were statistically significant. Nevertheless, the R^2 of the best one hit 0.50 with an RMSE of 2.02 μmol/g/s ($p < 0.001$), suggesting the PLSR model may produce good results even based on the original reflectance spectra, and the leaf transpiration can be captured to a certain level using the original reflectance.

Unlike the result of using the original reflectance spectra, the best PLSR model using the specific bands of 2210, 2215, 1980, 2205, 2000, 2140, 435, 2375 and 2345 nm of the first-order spectra closely traced the leaf transpiration of desert plant *Haloxylon ammodendron*. The R^2 of this PLSR model even hit 0.71 and its RMSE was 1.53 μmol/g/s (Fig. 9.2), clearly showing the advantage of using the first-order derivative spectra in PLSR analysis. Apparently, the first-order derivative spectra based PLSR model could have predicted leaf Tr much better than those PLSR models developed from the original reflectance spectra (Wang and Jin 2015).

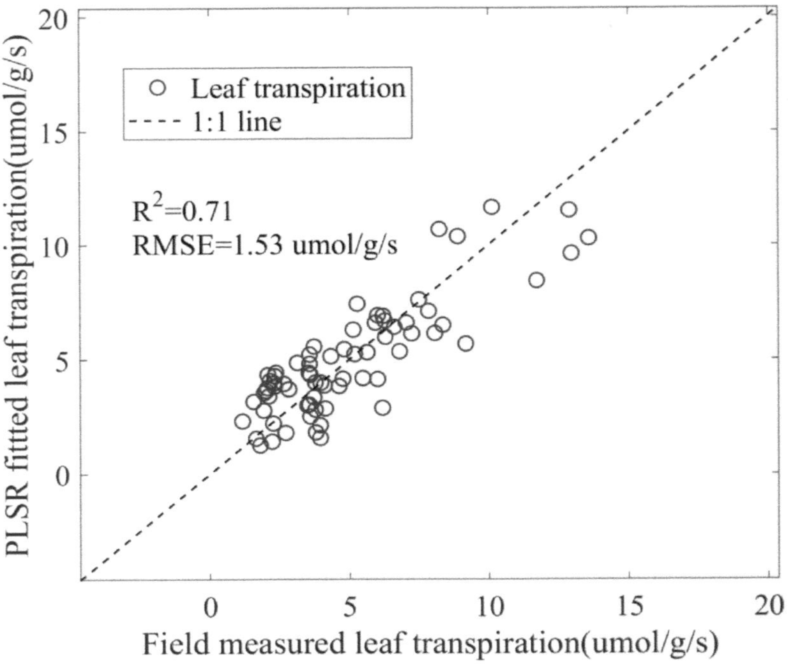

Fig. 9.2: The scatter diagram of observed and the best PLSR fitted leaf scale transpiration using the first-order derivative spectra.

3.4 Retrieval of Canopy-Scale Transpiration from Hyperspectral Indices

Statistical analysis revealed that SD (2315, 2370) was the most effective one among all indices developed using the original reflectance spectra for retrieving canopy scale transpiration. The R^2 and RMSE of this index for estimating transpiration of the xeric ecosystem were 0.42 and 1.8×10^{-5} m^3/m^2/s, respectively (Jin and Wang 2016b).

In comparison, much better indices were obtained if using the derivative spectra. The best type of indices was found to be the dSR type, and the dSR (660, 1040) was the best one among the type of indices to trace canopy scale transpiration, which had an R^2 of 0.54 and an RMSE of 1.6×10^{-5} m^3/m^2/s (Jin and Wang 2016b). Linear regression of the index and canopy transpiration was illustrated in Fig. 9.3.

However, the identified indices performed inconsistently with the simulation and the field-measured dataset, obviously. Results indicated that the five types of indices had near identical accuracies ($R^2 \approx 0.46$) for the simulation dataset, among which the indices dSR (970, 1335), dSD (1435, 1580), dDDn (1250, 550) and dND (970, 1330) were identified as the best one for each individual type (Jin et al. 2019a). Unfortunately, these indices did not function well with the dataset measured in the field. As a step further, we calibrated the indices using a composite dataset by integrating both the simulation and the field-measured data but validated using the simulation and the field-measured data separately. For this case, the best performed SR (1525, 2150) had the R^2 of 0.46 and 0.33 for the field-measured dataset and the simulation dataset, respectively (Jin et al. 2019a). Similarly, the R^2 values of ND (1425, 2145), the best ND-type index identified, were 0.44 and 0.36, respectively (Jin et al. 2019a). Furthermore, the previously developed derivative hyperspectral index from

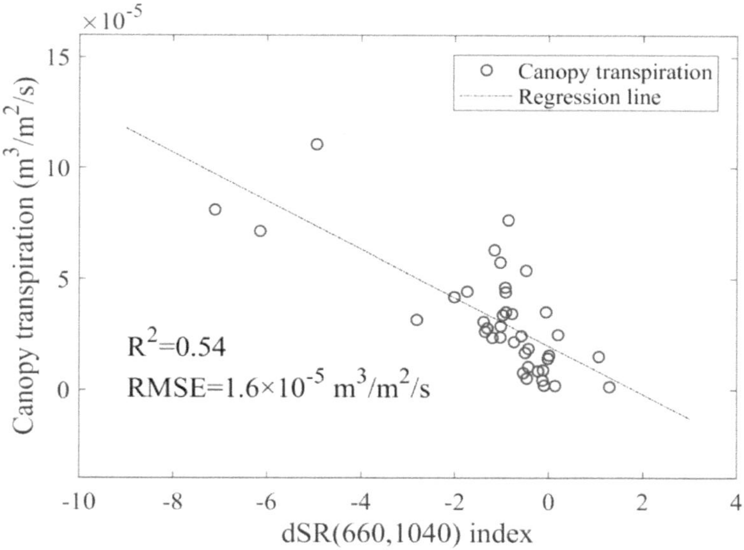

Fig. 9.3: Scatter diagram of field measured canopy transpiration and the dSR (660, 1040).

the field-measured data, the dSR (640, 1040), kept relatively good performance with the simulation dataset (R^2 = 0.32).

3.5 Retrieval of Canopy-Scale Transpiration from PLSR

PLSR models developed from either the original reflectance spectra or the first-order derivative spectra performed better than any types of hyperspectral indices on estimating canopy scale transpiration. The calibrated PLSR model based on seven bands (760, 2370, 2315, 2160, 2385, 2305 and 2170 nm) of the original reflectance spectra can estimate canopy transpiration of the desert plant *Haloxylon ammodendron* with an R^2 of 0.75 and an RMSE of 1.21 × 10^{-5} $m^3/m^2/s$. Furthermore, the best PLSR model developed from the first-order derivative spectra was even more efficient, since the PLSR model using the specific bands of 670, 1195, 2190, 915, 2425, 2365 and 2205 nm of the first-order spectra could have traced canopy transpiration of *Haloxylon ammodendron* at a level of R^2 = 0.78 (Fig. 9.4). The RMSE was also dropped to 1.13 × 10^{-5} $m^3/m^2/s$, clearly showing the advantage of using the first-order derivative spectra in PLSR analysis.

4. Discussion and Future Developments

4.1 General Applicable Indices or PLSRs and Common Hyperspectral Bands Used for Plant Water Status and Plant Use Estimation

Although statistical methods are simple on linking plant properties with hyperspectral information, the success of one statistical model largely depends upon the specific dataset used for model calibration. Clearly, the robustness of the relationships would depend much on the representativeness of the specific dataset (Li and Wang 2011). Previous studies archived dozens of indices ranging from the SR to the ND types for detecting water status,

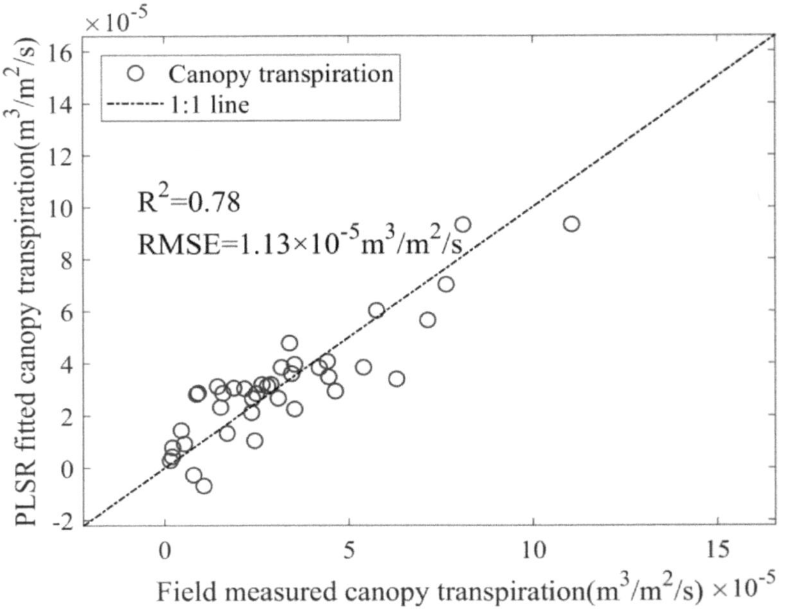

Fig. 9.4: The scatter diagram of observed and estimated canopy transpiration based on the PLSR model.

e.g., the MSI (R_{1599}/R_{819}) by Hunt and Rock (1989), the NDWI (($R_{860} - R_{1240}$)/($R_{860} + R_{1240}$)) by Gao (1996), and the WI (R_{900}/R_{970}) by Peñuelas et al. (1997). In this study, searching the best index for leaf water status (represented by EWT) resulted in different indices for different datasets, clearly suggesting vegetation indices developed from a particular database could be inapplicable for other datasets. The truth in the meantime indicated that the success of retrieving leaf water status was affected by many other factors such as different plant species, different growing conditions or different growth stages, etc., as also pointed out by le Maire et al. (2008).

Nevertheless, sensitive bands for tracing leaf water status could have been revealed from the composite data covering both mesic and xeric ecosystems and leaf dehydration experiment as well. Our leaf dehydration experiments suggested that the values of original reflectance increased greatly with the loss of leaf water content, especially in the wavelength domains from 350 nm to 700 nm and from 1300 nm to 2500 nm (Cao et al. 2015). The most sensitive wavelengths, nevertheless, were within the wavelength domain of 1300–2500 nm, similar to those reported by previous studies (Carter 1991, Danson et al. 1992, Hunt et al. 1987, Tucker 1980). In addition, the reflectance derivatives within the domain of 1500–1550 nm were found to increase but those within the range from 1950 to 2000 nm were found to decrease with the increase of EWT (Wang and Li 2012b).

Close linear relationships of leaf scale transpiration with the original reflectance spectra from 400 nm to 800 nm and with the two specific domains nearby 1400 nm or 1900 nm have been revealed from the statistical analyses. Alternatively, the bands within the wavelength domain of 1900–2500 nm were the most selected in the derivative spectra based PLSR models for leaf transpiration estimation. Especially, the first-order derivative spectra information at 2435, 2440, 2445, and 2470 nm were essential for developing PLSR models to trace leaf scale water use (Wang and Jin 2015). Additionally, the spectrum derivatives around 520, 700, 1030 and 1540 nm had significant positive correlations with leaf and canopy transpiration, while the spectrum derivatives around 600, 1300 and 1745 nm had significant negative correlations (Jin and Wang 2016b). These bands were therefore had very high possibilities to be the commonly applied hyperspectral bands for estimating plant water use.

4.2 Robustness of Statistical Models and Future Large-Scale Applications

In this chapter, we have applied two generally applied statistical approaches (VIs and PLSRs) for tracing plant water status and plant water use using hyperspectral information. Similar to many previous studies claiming that the PLSR was superior over other statistical approaches including vegetation indices, multiple linear or stepwise regressions, or principal component analysis (Atzberger et al. 2010, Darvishzadeh et al. 2008, Fu et al. 2014, Huang et al. 2004, Ryu et al. 2011, Yi et al. 2014), our results also suggested that the PLSR models performed much better than VIs, especially on plant water use (transpiration) estimation. However, it is clear that the PLSR models always involved more bands than VIs did, facing with a high probability of being overfitted (Kawamura et al. 2010, Jin and Wang 2019b) since there were high collinearities or correlations among different spectral bands. Furthermore, compared with the simple VIs, how to discriminate informative bands for PLSR models to minimize the risk of overfitting remains a dilemma when applying the method for retrieving plant variables from reflected spectra (Jin and Wang 2019b).

It is essential to have a large calibration dataset in order to develop generally applicable statistical models (le Maire et al. 2008, Wang and Li 2012a). Following the way, we have attempted to develop general models for retrieving plant water status and plant water use using large simulation datasets generated from radiative transfer models of PROSPECT (Feret et al. 2008) and SCOPE (van der Tol et al. 2009). However, the identified best indices if only from the simulation datasets performed poorly with the field-based datasets (both EWT and Tr), due primarily to the simulation errors resulted from the inconsistency between the simulation dataset and the field-measured dataset. Furthermore, a simulation dataset by itself alone will not be sufficient for identifying a generally applicable index (Wang and Li 2012a). Thus, for developing robust statistical models, both simulation and field-measured datasets should be integrated.

A next critical step is on how to identify the best index by considering its overall performance for the joint simulation and field-measured data. The strategy adopted in this step is essential because the samples contained in the two datasets might vary considerably. Unfortunately, there was no consensus yet on the strategy for the dilemma (le Maire et al. 2004, le Maire et al. 2008, Sonobe and Wang 2017, Wang and Li 2012b). Furthermore, how to separate data into calibration and validation also needs more attention to ensure the robustness of the built-up model (Richter et al. 2012).

In addition, in spite of 30 years research on leaf reflectance spectroscopy, it remains a great challenge on selecting informative spectral bands for estimating plant variables from hyperspectral information if solely based on statistical approaches, since there are high dependencies or correlations at different bands (Jin and Wang 2019a, Kira et al. 2015). For most cases, wavelengths used by empirical models were purely determined by statistics. Unfortunately, their underlying physical and biological mechanisms are either neglected intentionally or only poorly examined, making it very hard to maintain the robustness of empirical models that are needed for general applications (Féret et al. 2011, Jin and Wang 2018). In future, it is essential to clarify the biophysical, biochemical, and physiological mechanisms of wavelengths used by the statistical models, via either control experiments or modeling activities, in order to develop generally applicable statistical models (Féret et al. 2017, Jin and Wang 2018).

References

Allen, R.G., M. Tasumi and R. Trezza. 2007. Satellite-based energy balance for mapping evapotranspiration with internalized calibration (METRIC)—Model. J. Irrig. Drainage Eng. 133(4): 380–394.

Andersen, C.M. and R. Bro. 2010. Variable selection in regression—a tutorial. J. Chemom. 24(11-12): 728–737.

Atzberger, C., M. Guérif, F. Baret and W. Werner. 2010. Comparative analysis of three chemometric techniques for the spectroradiometric assessment of canopy chlorophyll content in winter wheat. Comput. Electron. Agric. 73(2): 165–173.

Bastiaanssen, W.G.M., M. Menenti, R.A. Feddes and A.A.M. Holtslag. 1998. A remote sensing surface energy balance algorithm for land (SEBAL). 1. Formulation. J. Hydrol. 212-213: 198–212.

Ben-Dor, E., D. Schläpfer, A.J. Plaza and T. Malthus. 2013. Hyperspectral remote sensing. pp. 413–456. *In*: M. Wendisch and J.-L. Brenguier (eds.). Airborne Measurements for Environmental Research. Wiley-VCH Verlag GmbH & Co. KGaA, Weinheim, Germany.

Cao, Z., Q. Wang and C. Zheng. 2015. Best hyperspectral indices for tracing leaf water status as determined from leaf dehydration experiments. Ecol. Indic. 54: 96–107.

Cao, Z. and Q. Wang. 2017. Retrieval of leaf fuel moisture contents from hyperspectral indices developed from dehydration experiments. Eur. J. Remote Sens. 50(1): 18–28.

Carter, G.A. 1991. Primary and secondary effects of water content on the spectral reflectance of leaves. Am. J. Bot. 78(7): 916–924.

Chen, D., J. Huang and T.J. Jackson. 2005. Vegetation water content estimation for corn and soybeans using spectral indices derived from MODIS near- and short-wave infrared bands. Remote Sens. Environ. 98(2-3): 225–236.

Chen, S., X. Hong, C.J. Harris and P.M. Sharkey. 2004. Sparse modeling using orthogonal forward regression with PRESS statistic and regularization. IEEE Trans. Syst. Man Cy. B 34(2): 898–911.

Cheng, Y.-B., P.J. Zarco-Tejada, D. Riaño, C.A. Rueda and S.L. Ustin. 2006. Estimating vegetation water content with hyperspectral data for different canopy scenarios: Relationships between AVIRIS and MODIS indexes. Remote Sens. Environ. 105(4): 354–366.

Combal, B., F. Baret, M. Weiss, A. Trubuil, D. Macé, A. Pragnère et al. 2003. Retrieval of canopy biophysical variables from bidirectional reflectance using prior information to solve the ill-posed inverse problem. Remote Sens. Environ. 84(1): 1–15.

Danson, F.M., M.D. Steven, T.J. Malthus and J.A. Clark. 1992. High-spectral resolution data for determining leaf water content. Int. J. Remote Sens. 13(3): 461–470.

Darvishzadeh, R., A. Skidmore, M. Schlerf, C. Atzberger, F. Corsi and M. Cho. 2008. LAI and chlorophyll estimation for a heterogeneous grassland using hyperspectral measurements. ISPRS J. Photogramm. 63(4): 409–426.

Devitt, D., L. Fenstermaker, M. Young, B. Conrad, M. Baghzouz and B. Bird. 2011. Evapotranspiration of mixed shrub communities in phreatophytic zones of the Great Basin region of Nevada (USA). Ecohydrology 4(6): 807–822.

Dorigo, W.A., R. Zurita-Milla, A.J.W. de Wit, J. Brazile, R. Singh and M.E. Schaepman. 2007. A review on reflective remote sensing and data assimilation techniques for enhanced agroecosystem modeling. Int. J. Appl. Earth Obs. Geoinf. 9(2): 165–193.

El-Hendawy, S., N. Al-Suhaibani, W. Hassan, M. Tahir and U. Schmidhalter. 2017. Hyperspectral reflectance sensing to assess the growth and photosynthetic properties of wheat cultivars exposed to different irrigation rates in an irrigated arid region. PLOS ONE 12(8): e0183262.

Feret, J.-B., C. François, G.P. Asner, A.A. Gitelson, R.E. Martin, L.P.R. Bidel et al. 2008. PROSPECT-4 and 5: Advances in the leaf optical properties model separating photosynthetic pigments. Remote Sens. Environ. 112(6): 3030–3043.

Féret, J.-B., C. François, A. Gitelson, G.P. Asner, K.M. Barry, C. Panigada et al. 2011. Optimizing spectral indices and chemometric analysis of leaf chemical properties using radiative transfer modeling. Remote Sens. Environ. 115(10): 2742–2750.

Féret, J.B., A.A. Gitelson, S.D. Noble and S. Jacquemoud. 2017. PROSPECT-D: Towards modeling leaf optical properties through a complete lifecycle. Remote Sens. Environ. 193: 204–215.

Foley, S., B. Rivard, G.A. Sanchez-Azofeifa and J. Calvo. 2006. Foliar spectral properties following leaf clipping and implications for handling techniques. Remote Sens. Environ. 103(3): 265–275.

Fu, Y., G. Yang, J. Wang, X. Song and H. Feng. 2014. Winter wheat biomass estimation based on spectral indices, band depth analysis and partial least squares regression using hyperspectral measurements. Comput. Electron. Agric. 100: 51–59.

Furuuchi, H., M.W. Jenkins, R.S. Senock, J.L.J. Houpis and J.C. Pushnik. 2013. Estimating plant crown transpiration and water use efficiency by vegetative reflectance indices associated with chlorophyll fluorescence. Open J. Ecol. 3(3): 122–132.

Gao, B. 1996. NDWI—a normalized difference water index for remote sensing of vegetation liquid water from space. Remote Sens. Environ. 58(3): 257–266.

Gao, L. and Y. Zhang. 2016. Spatio-temporal variation of hydrological drought under climate change during the period 1960–2013 in the Hexi Corridor, China. J. Arid Land 8(2): 157–171.

Gitelson, A.A., Y. Zur, O.B. Chivkunova and M.N. Merzlyak. 2002. Assessing carotenoid content in plant leaves with reflectance spectroscopy. Photochem. Photobiol. 75(3): 272–281.

Gitelson, A.A., G.P. Keydan and M.N. Merzlyak. 2006. Three-band model for noninvasive estimation of chlorophyll, carotenoids, and anthocyanin contents in higher plant leaves. Geophys. Res. Lett. 33(11): L11402.

Goetz, A.F.H. 2009. Three decades of hyperspectral remote sensing of the Earth: A personal view. Remote Sens. Environ. 113: S5–S16.

Gowda, P.H., J.L. Chavez, P.D. Colaizzi, S.R. Evett, T.A. Howell and J.A. Tolk. 2008. ET mapping for agricultural water management: present status and challenges. Irrig. Sci. 26(3): 223–237.

Granier, A. 1985. A new method of sap flow measurement in tree stems. Ann. Sci. For. 42(2): 193–200.

Haboudane, D., J.R. Miller, E. Pattey, P.J. Zarco-Tejada and I.B. Strachan. 2004. Hyperspectral vegetation indices and novel algorithms for predicting green LAI of crop canopies: Modeling and validation in the context of precision agriculture. Remote Sens. Environ. 90(3): 337–352.

Hansen, P.M. and J.K. Schjoerring. 2003. Reflectance measurement of canopy biomass and nitrogen status in wheat crops using normalized difference vegetation indices and partial least squares regression. Remote Sens. Environ. 86(4): 542–553.

Huang, G. and Y. Li. 2017. Photodegradation effects are related to precipitation amount, precipitation frequency and litter traits in a desert ecosystem. Soil Biol. Biochem. 115(Supplement C): 383–392.

Huang, Z., B.J. Turner, S.J. Dury, I.R. Wallis and W.J. Foley. 2004. Estimating foliage nitrogen concentration from HYMAP data using continuum removal analysis. Remote Sens. Environ. 93(1-2): 18–29.

Hunt, E.R., B.N. Rock and P.S. Nobel. 1987. Measurement of leaf relative water content by infrared reflectance. Remote Sens. Environ. 22(3): 429–435.

Hunt, E.R. and B.N. Rock. 1989. Detection of changes in leaf water content using near- and middle–infrared reflectances. Remote Sens. Environ. 30(1): 43–54.

Imanishi, J., K. Sugimoto and Y. Morimoto. 2004. Detecting drought status and LAI of two Quercus species canopies using derivative spectra. Comput. Electron. Agric. 43(2): 109–129.

IPCC. 2014. Climate Change 2014: Impacts, Adaptation and Vulnerability. Cambridge University Press UK.

Jacquemoud, S. and F. Baret. 1990. PROSPECT: A model of leaf optical properties spectra. Remote Sens. Environ. 34(2): 75–91.

Jin, J. and Q. Wang. 2016a. Assessing ecological vulnerability in western China based on time-integrated NDVI data. J. Arid Land 8(4): 533–545.

Jin, J. and Q. Wang. 2016b. Hyperspectral indices based on first derivative spectra closely trace canopy transpiration in a desert plant. Ecol. Inform. 35: 1–8.

Jin, J. and Q. Wang. 2018. Informative bands used by efficient hyperspectral indices to predict leaf biochemical contents are determined by their relative absorptions. Int. J. Appl. Earth Obs. Geoinf. 73: 616–626.

Jin, J. and Q. Wang. 2019a. Selection of informative spectral bands for PLS models to estimate foliar chlorophyll content using hyperspectral reflectance. IEEE Trans. Geosci. Remote Sens. 57(5): 3064–3072.

Jin, J. and Q. Wang. 2019b. Evaluation of informative bands used in different PLS regressions for estimating leaf biochemical contents from hyperspectral reflectance. Remote Sens. 11(2): 197.

Jin, J., Q. Wang and J. Wang. 2019a. Combing both simulated and field-measured data to develop robust hyperspectral indices for tracing canopy transpiration in drought-tolerant plant. Environ. Monit. Assess. 191: 13.

Jin, J., Q. Wang, J. Wang and D. Otieno. 2019b. Tracing water and energy fluxes and reflectance in an arid ecosystem using the integrated model SCOPE. J. Environ. Manage. 231: 1082–1090.

Katul, G.G., R. Oren, S. Manzoni, C. Higgins and M.B. Parlange. 2012. Evapotranspiration: A process driving mass transport and energy exchange in the soil-plant-atmosphere-climate system. Rev. Geophys. 50(3): RG3002.

Kawamura, K., N. Watanabe, S. Sakanoue, H.J. Lee, Y. Inoue and S. Odagawa. 2010. Testing genetic algorithm as a tool to select relevant wavebands from field hyperspectral data for estimating pasture mass and quality in a mixed sown pasture using partial least squares regression. Grassl. Sci. 56(4): 205–216.

Kira, O., R. Linker and A. Gitelson. 2015. Non-destructive estimation of foliar chlorophyll and carotenoid contents: Focus on informative spectral bands. Int. J. Appl. Earth Obs. Geoinf. 38: 251–260.

Kustas, W.P. and J.M. Norman. 1999. Evaluation of soil and vegetation heat flux predictions using a simple two-source model with radiometric temperatures for partial canopy cover. Agric. For. Meteorol. 94(1): 13–29.

le Maire, G., C. François and E. Dufrêne. 2004. Towards universal broad leaf chlorophyll indices using PROSPECT simulated database and hyperspectral reflectance measurements. Remote Sens. Environ. 89(1): 1–28.

le Maire, G., C. François, K. Soudani, D. Berveiller, J.-Y. Pontailler, N. Bréda et al. 2008. Calibration and validation of hyperspectral indices for the estimation of broadleaved forest leaf chlorophyll content, leaf mass per area, leaf area index and leaf canopy biomass. Remote Sens. Environ. 112(10): 3846–3864.

Li, P. and Q. Wang. 2011. Retrieval of leaf biochemical parameters using PROSPECT inversion: A new approach for alleviating Ill-posed problems. IEEE Trans. Geosci. Remote Sens. 49(7): 2499–2506.

Li, P. and Q. Wang. 2012. Retrieval of chlorophyll for assimilating branches of a typical desert plant through inversed radiative transfer models. Int. J. Remote Sens. 34(7): 2402–2416.

Li, P. and Q. Wang. 2013. Developing and validating novel hyperspectral indices for leaf area index estimation: Effect of canopy vertical heterogeneity. Ecol. Indic. 32: 123–130.

Li, W., P. Ciais, Y. Wang, Y. Yin, S. Peng, Z. Zhu et al. 2018. Recent changes in global photosynthesis and terrestrial ecosystem respiration constrained from multiple observations. Geophys. Res. Lett. 45(2): 1058–1068.

Loukas, A., L. Vasiliades, C. Domenikiotis and N.R. Dalezios. 2005. Basin-wide actual evapotranspiration estimation using NOAA/AVHRR satellite data. Phys. Chem. Earth 30(1-3): 69–79.

Lu, P., L. Urban and P. Zhao. 2004. Granier's thermal dissipation probe (TDP) method for measuring sap flow in trees: theory and practice. Acta Bot. Sin. 46(6): 631–646.

Main, R., M.A. Cho, R. Mathieu, M.M. O'Kennedy, A. Ramoelo and S. Koch. 2011. An investigation into robust spectral indices for leaf chlorophyll estimation. ISPRS J. Photogramm. 66(6): 751–761.

Marino, G., E. Pallozzi, C. Cocozza, R. Tognetti, A. Giovannelli, C. Cantini et al. 2014. Assessing gas exchange, sap flow and water relations using tree canopy spectral reflectance indices in irrigated and rainfed Olea europaea L. Environ. Exp. Bot. 99: 43–52.

Marshall, M., P. Thenkabail, T. Biggs and K. Post. 2016. Hyperspectral narrowband and multispectral broadband indices for remote sensing of crop evapotranspiration and its components (transpiration and soil evaporation). Agric. For. Meteorol. 218-219: 122–134.

Mehmood, T., K.H. Liland, L. Snipen and S. Sæbø. 2012. A review of variable selection methods in partial least squares regression. Chemometr. Intell. Lab. 118: 62–69.

Norgaard, L., A. Saudland, J. Wagner, J.P. Nielsen, L. Munck and S.B. Engelsen. 2000. Interval partial least-squares regression (iPLS): A comparative chemometric study with an example from near-infrared spectroscopy. Appl. Spectrosc. 54(3): 413–419.

Norman, J.M., W.P. Kustas and K.S. Humes. 1995. Source approach for estimating soil and vegetation energy fluxes in observations of directional radiometric surface temperature. Agric. For. Meteorol. 77(3): 263–293.

Pasqualotto, N., J. Delegido, S. Van Wittenberghe, J. Verrelst, J.P. Rivera and J. Moreno. 2018. Retrieval of canopy water content of different crop types with two new hyperspectral indices: Water absorption area index and depth water index. Int. J. Appl. Earth Obs. Geoinf. 67: 69–78.

Peñuelas, J., J. Pinol, R. Ogaya and I. Filella. 1997. Estimation of plant water concentration by the reflectance water index WI (R900/R970). Int. J. Remote Sens. 18(13): 2869–2875.

Pu, R., S. Ge, N. Kelly and P. Gong. 2003. Spectral absorption features as indicators of water status in coast live oak (Quercus agrifolia) leaves. Int. J. Remote Sens. 24(9): 1799–1810.

Richter, K., C. Atzberger, T.B. Hank and W. Mauser. 2012. Derivation of biophysical variables from Earth observation data: validation and statistical measures. J. Appl. Remote Sens. 6(1): 063557.

Rodríguez-Pérez, J.R., D. Riaño, E. Carlisle, S. Ustin and D.R. Smart. 2007. Evaluation of hyperspectral reflectance indexes to detect grapevine water status in vineyards. Am. J. Enol. Viticult. 58(3): 302–317.

Ryu, C., M. Suguri and M. Umeda. 2011. Multivariate analysis of nitrogen content for rice at the heading stage using reflectance of airborne hyperspectral remote sensing. Field Crops Res. 122(3): 214–224.

Schmitt, N. and R.E. Ployhart. 1999. Estimates of cross-validity for stepwise regression and with predictor selection. J. Appl. Psychol. 84(1): 50.

Sims, D.A. and J.A. Gamon. 2002. Relationships between leaf pigment content and spectral reflectance across a wide range of species, leaf structures and developmental stages. Remote Sens. Environ. 81(2-3): 337–354.

Sims, D.A. and J.A. Gamon. 2003. Estimation of vegetation water content and photosynthetic tissue area from spectral reflectance: a comparison of indices based on liquid water and chlorophyll absorption features. Remote Sens. Environ. 84(4): 526–537.

Sonobe, R. and Q. Wang. 2017. Towards a universal hyperspectral index to assess chlorophyll content in deciduous forests. Remote Sens. 9(3): 191.

Sonobe, R. and Q. Wang. 2018. Nondestructive assessments of carotenoids content of broadleaved plant species using hyperspectral indices. Comput. Electron. Agric. 145: 18–26.

Stocker, T., D. Qin, G.-K. Plattner, M. Tignor, S.K. Allen, J. Boschung et al. 2014. Climate change 2013: The Physical Science Basis. Cambridge University Press Cambridge, UK, and New York.

Su, Z. 2002. The Surface Energy Balance System (SEBS) for estimation of turbulent heat fluxes. Hydrol. Earth Syst. Sci. 6(1): 85–100.

Sun, P., S. Wahbi, T. Tsonev, M. Haworth, S. Liu and M. Centritto. 2014. On the use of leaf spectral indices to assess water status and photosynthetic limitations in Olea europaea L. during water-stress and recovery. PloS One 9(8): e105165.

Thenkabail, P.S., J.G. Lyon and A. Huete. 2012. Hyperspectral Remote Sensing of Vegetation. CRC Press Boca Raton, FL.

Timmermans, W.J., W.P. Kustas, M.C. Anderson and A.N. French. 2007. An intercomparison of the Surface Energy Balance Algorithm for Land (SEBAL) and the Two-Source Energy Balance (TSEB) modeling schemes. Remote Sens. Environ. 108(4): 369–384.

Tsai, F. and W. Philpot. 1998. Derivative analysis of hyperspectral data. Remote Sens. Environ. 66(1): 41–51.

Tucker, C.J. 1980. Remote sensing of leaf water content in the near infrared. Remote Sens. Environ. 10(1): 23–32.

Ustin, S.L., A.A. Gitelson, S. Jacquemoud, M. Schaepman, G.P. Asner, J.A. Gamon et al. 2009. Retrieval of foliar information about plant pigment systems from high resolution spectroscopy. Remote Sens. Environ. 113: S67–S77.

van der Tol, C., W. Verhoef, J. Timmermans, A. Verhoef and Z. Su. 2009. An integrated model of soil-canopy spectral radiances, photosynthesis, fluorescence, temperature and energy balance. Biogeosciences 6(12): 3109–3129.

Verrelst, J., G. Camps-Valls, J. Muñoz-Marí, J.P. Rivera, F. Veroustraete, J.G.P.W. Clevers et al. 2015. Optical remote sensing and the retrieval of terrestrial vegetation bio-geophysical properties—A review. ISPRS J. Photogramm. 108: 273–290.

Verstraeten, W.W., F. Veroustraete and J. Feyen. 2008. Assessment of evapotranspiration and soil moisture content across different scales of observation. Sensors 8(1): 70–117.

Wang, K. and R.E. Dickinson. 2012. A review of global terrestrial evapotranspiration: Observation, modeling, climatology, and climatic variability. Rev. Geophys. 50(2): RG2005.

Wang, Q., A. Iio, J. Tenhunen and Y. Kakubari. 2008. Annual and seasonal variations in photosynthetic capacity of Fagus crenata along an elevation gradient in the Naeba Mountains, Japan. Tree Physiol. 28(2): 277–285.

Wang, Q. and P. Li. 2012a. Hyperspectral indices for estimating leaf biochemical properties in temperate deciduous forests: Comparison of simulated and measured reflectance data sets. Ecol. Indic. 14(1): 56–65.

Wang, Q. and P. Li. 2012b. Identification of robust hyperspectral indices on forest leaf water content using PROSPECT simulated dataset and field reflectance measurements. Hydrol. Process 26(8): 1230–1241.

Wang, Q. and J. Jin. 2015. Leaf transpiration of drought tolerant plant can be captured by hyperspectral reflectance using PLSR analysis. iForest 9: 30–37.

Wang, Q., J. Jin, R. Sonobe and J.M. Chen. 2018. Derivative hyperspectral vegetation indices in characterizing forest biophysical and biochemical quantities. pp. 27–63. *In*: P.S. Thenkabail, J.G. Lyon and A. Huete (eds.). Hyperspectral Indices and Image Classifications for Agriculture and Vegetation. CRC Press, Boca Raton, FL.

Wang, Y., X. Li, Z. Nashed, F. Zhao, H. Yang, Y. Guan et al. 2007. Regularized kernel-based BRDF model inversion method for ill-posed land surface parameter retrieval. Remote Sens. Environ. 111(1): 36–50.

Yao, W., M. Han and S. Xu. 2010. Estimating the regional evapotranspiration in Zhalong wetland with the Two-Source Energy Balance (TSEB) model and Landsat7/ETM+ images. Ecol. Inform. 5(5): 348–358.

Yebra, M. and E. Chuvieco. 2009. Linking ecological information and radiative transfer models to estimate fuel moisture content in the Mediterranean region of Spain: Solving the ill-posed inverse problem. Remote Sens. Environ. 113(11): 2403–2411.

Yi, Q., G. Jiapaer, J. Chen, A. Bao and F. Wang. 2014. Different units of measurement of carotenoids estimation in cotton using hyperspectral indices and partial least square regression. ISPRS J. Photogramm. Remoten Sens. 91: 72–84.

Young, D.J., J.T. Stevens, J.M. Earles, J. Moore, A. Ellis, A.L. Jirka et al. 2017. Long-term climate and competition explain forest mortality patterns under extreme drought. Ecol. Lett. 20(1): 78–86.

Yu, K.-Q., Y.-R. Zhao, F.-L. Zhu, X.-L. Li and Y. He. 2016. Mapping of chlorophyll and SPAD distribution in pepper leaves during leaf senescence using visible and near-infrared hyperspectral imaging. Trans. ASABE 59(1): 13.

Zarco-Tejada, P.J., A. Berjon, R. Lopez-Lozano, J.R. Miller, P. Martin, V. Cachorro et al. 2005. Assessing vineyard condition with hyperspectral indices: Leaf and canopy reflectance simulation in a row-structured discontinuous canopy. Remote Sens. Environ. 99(3): 271–287.

Zhang, L., Z. Zhou, G. Zhang, Y. Meng, B. Chen and Y. Wang. 2012. Monitoring the leaf water content and specific leaf weight of cotton (Gossypium hirsutum L.) in saline soil using leaf spectral reflectance. Eur. J. Agron. 41: 103–117.

Zhao, D.H., J.L. Li and J.G. Qi. 2005. Identification of red and NIR spectral regions and vegetative indices for discrimination of cotton nitrogen stress and growth stage. Comput. Electron. Agric. 48(2): 155–169.

Zheng, C. and Q. Wang. 2014. Water-use response to climate factors at whole tree and branch scale for a dominant desert species in central Asia: Haloxylon ammodendron. Ecohydrology 7(1): 56–63.

Zheng, C. and Q. Wang. 2015. Seasonal and annual variation in transpiration of a dominant desert species, Haloxylon ammodendron, in Central Asia up-scaled from sap flow measurement. Ecohydrology 8(5): 948–960.

Chapter 10

Aroma Glycosides Contribute Plant Chemical Defense

Toshiyuki Ohnishi

1. Introduction

Plants biosynthesize and emit various Volatile Organic Compounds (VOCs), including aromatic esters, green leaf alcohols (C6), phenylpropanoids (C9), monoterpenes (C10), sesquiterpenes (C15), and norisoprenoids (C16). VOCs are highly volatile and mobile in air and are considered as signaling molecules that transmit biological information between plants and other organisms (Arimura et al. 2009). VOCs play physiological roles as repellents against herbivores, anti-microbial agents against fungi and as pollinators. Plants have developed intracellular mechanisms to stably store VOCs so that evaporation and toxicity are reduced such as apoptosis (Izumi et al. 1999, Pattnaik et al. 1997). Among VOC storage mechanisms, hydrophobic or highly reactive VOCs are accumulated locally in specialized storage cells such as glandular trichomes and resin ducts. Glandular trichomes on the surfaces of leaves and stems of tomato (*Solanum lycopersicum*), basil (*Ocimum basilicum*), thyme (*Thymus*) or spearmint (*Mentha spicata*) have been described as specialized storage cells for hydrophobic chemicals, and VOCs in glandular trichomes are quickly emitted into the atmosphere in response to attacks by insect herbivores or mechanical wounding (Yazaki et al. 2017). Alternatively, plants convert VOCs into Glycosidic Bound Volatile compounds (GBVs) to enhance water solubility. GBVs have higher hydrophilicity than VOCs, and can therefore be accumulated stably inside intracellular vacuoles, as shown in apricot, peach, yellow plum (Krammer et al. 1991), grape berries (Günata et al. 1985), kiwi (Young and Paterson 1995), strawberry (Roscher et al. 1997), raspberry (Pabst et al. 1991), and tomato (Marlatt et al. 1992). GBVs comprise a sugar moiety (a glycone moiety) and a non-sugar moiety (an aglycone moiety), and VOCs are liberated by β-glycosidases that catalyze the hydroxylation of glycosidic bonds between glycone and aglycone (Mizutani et al. 2002) (Fig. 10.1). Aroma compounds in foods with rich flavors, such as wine (grape),

Research Institute of Green Science and Technology, Shizuoka University, 836 Ohya, Shizuoka, Shizuoka 422-8529, Japan.
Email: ohnishi.toshiyuki@shizuoka.ac.jp

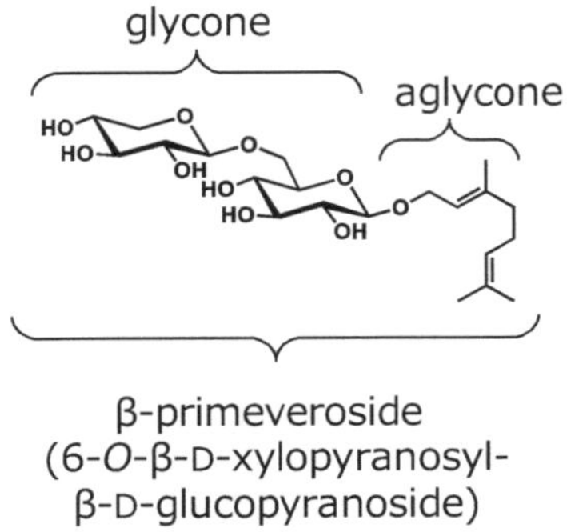

Fig. 10.1: Chemical structure of glycosidically bounds volatiles (GBVs).

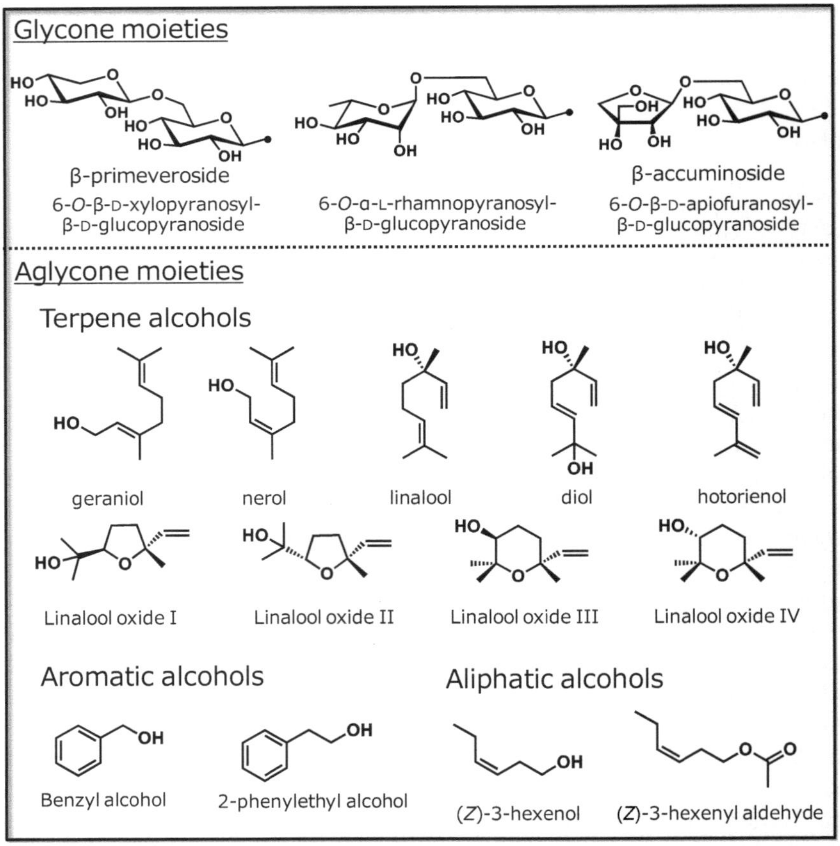

Fig. 10.2: Chemical structure of aglycone moieties and glycone moieties in plant glycosidically bounds volatiles (GBVs).

tea (tea plant), and shochu (a Japanese liquor; sweet potato), predominantly comprise VOCs that are derived from aglycone moieties of GBVs during manufacture. Thus, GBVs are considered as aroma precursors (Fig. 10.2). This chapter describes the plant metabolic processes behind VOCs that contribute to chemical defense systems and food quality.

2. Plant Specialized Metabolites

Because plants are stationary, they cannot escape from biological and non-biological stresses, such as drought, heat, cold and herbivorous feeding damage. Consequently, plants employ adaptive mechanisms to cope with environmental stresses and these include various defense systems that are tailored to the habitat (Pichersky and Gershenzon 2002, Sugimoto et al. 2014). Hydrophytes and land plants reportedly produce over 200,000 compounds with various physiological activities that can be used to respond to external factors, such as attack from herbivore or pathogenic fungi. VOCs usually have molecular weights of < 300 g/mol and readily volatilize into the atmosphere at ambient temperatures and pressures. Multiple VOCs are known to repel herbivores and attract natural predators of herbivores, reflecting their anti-microbial activities against pathogenic fungi, their attractiveness to pollinators and their activities as signaling molecules for communication between plants and other organisms. Due to their low molecular weight and hydrophobicity, VOCs easily evaporate in the air. Therefore, land plants have developed several intracellular mechanisms for stably storing VOCs. There are at least two primary systems that plants employ for this, including morphological developments that allow accumulation of hydrophobic or highly volatile compounds in specialized storage cells and chemical conversion of VOCs to more soluble GBVs with limited volatility. It is well known that various seed plants accumulate GBVs.

Fresh tea leaves (*Camellia sinensis*) emit slight green notes before manufacture and processing and these reflect the presence of monoterpene alcohols (linalool and geraniol), phenylpropanoids (2-phenylethanol and benzyl alcohol), and aliphatic alcohols (*cis*-3-hexen-1-ol) as odorless GVBs, predominantly diglycoside and β-primeveroside (6-*O*-β-ᴅ-xylopyranosyl-β-ᴅ-glucopyranoside) (Guo et al. 1993, 1994, Kawakami et al. 1995, Kobayashi et al. 1994, Kumazawa and Masuda 2002, Moon et al. 1994, 1996, Yano et al. 1991, Wang et al. 2000) (Fig. 10.2). During the tea manufacturing process, these volatiles are released by endogenous diglycoside-specific β-glycosidases. Similarly, vanillin, which is one of the most commercially valuable flavors from vanilla (*Vanilla planifolia*), is also biosynthesized and stored as vanillin glucoside. Moreover, isothiocyanates are bound to glucose via sulfur atoms and are stably stored as a glucosinolates, such as sinigrin (Koroleva et al. 2010). GBVs have also been found in plant species, such as grapes, *Prunus armeniaca, Prunus persica, Prunus domestica*, kiwifruit (*Actinidia chinensis*), strawberries, raspberries (*Rubus species*), hops (*Humulus lupulus*), sweet potatoes (*Ipomoea batatas*) and tomatoes. These findings indicate that the glycosylation of volatiles is common to a broad range of land plants and regulates the physiological activities of specialized metabolites. Glycosylation of volatile compounds also allows stable retention of volatile compounds inside cells. Thus, GBVs are the predominant chemical regulator of volatile metabolism in plants (Fig. 10.2).

3. Tea Aroma

Tea is manufactured from fresh tea leaves and is the most popular beverage in the world. Black tea, green tea and oolong tea are representative tea products that are manufactured by withering, rolling and fermenting. Desirable qualities of tea are taste, scent and color. Following visual evaluation of food qualities, flavor wheels offer convenient methods for assessing the qualities of scent and taste. Accordingly, flavor wheels are widely used to assess the quality of whiskey, coffee, chocolate and honey, and have been developed specifically for each of these consumables and many others. A new character wheel was recently developed for tea, and includes assessments of taste and scent and color. Scent is classified into nine groups ("Green," "Woody," "Dry," "Flowery," "Fruity," "Sweet," "Roast," "Spicy," and "Cool") and these groups are further divided into several sub groups using a total 50 words relating to scent (Ohno 2014). For example, flower scent is classified using the terms "Flowery," "Fruity," and "Sweet" in the tea character wheel, and reflects the presence of compounds, such as benzyl alcohol, 2-phenylethanol, linalool, and geraniol. Fresh scents, such as fresh green, are caused by *cis*-3-hexen-1-ol and *cis*-3-hexenyl acetate, which are classified in "Green."

More than 630 volatile compounds have been identified in fresh tea leaves and tea products (Chaturvedula and Prakash 2011). In tea, aromas are stored as GBVs comprising aroma compounds bound to sugar moieties with glycosyl groups. It is thought that tea aroma follows enzymatic hydrolysis of GBVs during the tea manufacturing process. Major GBVs in fresh tea leaves include various disaccharide glycosides. Among these, β-primeveroside is a non-volatile substance that lacks a scent, and when endogenous diglycosidases in tea leaves perform hydrolysis of β-primeveroside into its constituent aroma compounds and sugar moieties (β-primeverose), aroma components are released into the air.

3.1 The UGT 85 Enzyme Family Produces Aroma Monoglycosides

Qualitative and quantitative analyses of aroma glycosides are predominantly performed using Gas Chromatography-Mass Spectrometry (GC-MS) followed by TMS/TFA derivatization. The benefits of GC-MS include high resolution and high sensitivity, but sample preparation is very complicated and derivatization leads to heat-degradation and loss of aroma glycosides. During the past decade, various instruments for chemical analysis have been developed. Liquid Chromatography-Mass Spectrometry (LC-MS) is a powerful chemical analysis tool for quantitative analysis of GBVs at the picomolar level, and does not require derivatization. In recent years, analytical methods for GBVs have also been developed using ultra performance LC (UPLC) and MS/MS. These convenient GBV analytical tools have improved tea manufacturing processes by informing selections of tea varieties. It is known that benzyl alcohol, 2-phenylethanol, *cis*-3-hexen-1-ol, geraniol, and linalool are primary aroma compounds in tea and are stored as aroma glucosides (monoglycosides) and β-primeverosides. We developed analytical conditions for determining the quantities of 10 highly accumulated GBVs in tea leaves using LC-MS. Although analytical times were significantly shortened by this method, the resolution of isomers was sacrificed. For example, chemical isomers, such as β-primeveroside and β-visianoside, which are primary diglycosides in tea leaves, have the same molecular weight and carry hexose and pentose moieties. Moreover, β-primeveroside and β-visianoside differ only in the stereochemistry of the hydroxy group at C-4 of the pentose moiety. Because MS/MS techniques fail to distinguish between stereoisomers, technical methods for quantitative/qualitative analysis using LC-MS/MS were required to improve resolution. These techniques required the use of a solid phase extraction cartridge to eliminate interfering substances, but were performed within 85 minutes and had detection limits of 2–30 pmol for 10 kinds of GBV. These detection limits were comparable to those of GC-MS analyses of TMS/TFA derivatives of GBVs and show that the LC-MS conditions are sufficient for qualitative and quantitative analyses of the GBVs involved in tea leaf odors. Diglycosides such as geranyl β-primeveroside (geranyl-pri) and linalyl β-primeveroside (linalyl-pri) were detected mainly in young organs, whereas

2-phenylethyl β-primeveroside (2PE-pri), benzyl β-primeveroside (benzyl-pri), and *cis*-3-hexenyl β-primeveroside (hexenyl-pri) were detected in mature leaves (Cui et al. 2016, Ohgami et al. 2015). These results indicate that the enzymatic activities of terpene synthases (geranyl synthase and linalool synthase) and glycosyltransferases that perform aroma glycoside biosynthesis are higher in younger tissues than in mature organs.

3.2 UGT85 Family Responsible for Aroma Monoglycosides

To investigate the aroma glycoside biosynthesis pathway, we identified the glycosyltransferases responsible for producing aroma monoglycosides and diglycosides in young tea leaves. In a study published in 2001 (Lücker et al. 2001), a *Clarkia breweri* linalool synthase was overexpressed in *Petunia hybrid*, and the resulting strain did not produce free linalool, but markedly accumulated non-volatile linalyl glucoside (linalyl-glc) and 8-hydroxy linalyl glucoside (8-OH linalyl-glc) through the actions of endogenous glucosyltransferase and transgenic Arabidopsis plants expressing a strawberry terpene synthase (*FaNES1*) produced (*S*)-linalool, nerolidol, and the glycosylated derivatives (Aharoni et al. 2003). This report supported our hypothesis that linalool synthase, oxygenase, and glucosyltransferase are synchronously expressed in plants. Coexpression analyses are performed with powerful tools for surveying genes that encode aroma glycosyltransferases (Ginglinger et al. 2013, Usadel et al. 2009). Similar coexpression patterns have been described in the representative model dicot and monocot plants *Arabidopsis thaliana* and rice, respectively. In addition, ATTED II (http://atted.jp) is a web based analytical tool for determining coexpression of genes in Arabidopsis and rice plants, but cannot be applied to coexpression analysis in tea plants (Ohbayashi et al. 2011). Thus, to identify glucosyltransferases that perform the first glycosylation step in the biosynthesis of the aroma β-primeveroside, we surveyed target UGTs (UDP-glycosyltransferases) that are expressed during monoterpene biosynthesis in Arabidopsis plants. It is known that Arabidopsis plants accumulate linalyl-glc. Thus, we used linalool synthase genes as probes to find UGTs that produce monoterpenyl monoglucoside (Ginglinger et al. 2013, Obayashi et al. 2011). In these experiments, expression profiles of the Arabidopsis UGT85A3 gene were highly correlated with the Arabidopsis linalool synthase gene. To confirm that UGT85A3 converts linalool to linalyl-glc, we heterologously expressed recombinant UGT85A3 in *E. coli*. Subsequent enzyme assays of UGT85A3 indicated that UGT85A3 produces linalyl-glc from linalool as the sugar acceptor and UDP-Glu as the sugar donor (Ohgami et al. 2015) (Fig. 10.3). Moreover, UGT85A3 converted *cis*-3-hexen-1-ol and geraniol to hexenyl-glc and geranyl-glc, respectively. Glycosyltransferase homologues have also been described in other plants, including CsGT1 (UGT85K11) in tea; UGT85A33, UGT85A28, and UGT85A30 in grapes; UGT85A32 in sweet potato; and AdGT4 in kiwi (Bönisch et al. 2014a, Bönisch et al. 2014b, Ohgami et al. 2015, Yauk et al. 2014). All recombinant UGTs (UGT85A28, 30, 32, 33, and UGT85K11) showed positive enzymatic activities in the presence of geraniol or *cis*-3-hexen-1-ol, indicating that UGT85s are widely conserved in various plants that can catalyze the first glycosylation step for aroma diglycosides. The glycon moiety of tea GBVs is predominantly primeverose, whereas aglycon moieties are chemically diversified into compounds such as terpene alcohols, aromatic alcohols, and aliphatic alcohols. Relative enzyme activities of CsGT1 (UGT85K11) with various aroma alcohols revealed that geraniol (100%) was the best substrate, followed by eugenol (84%), *cis*-3-hexen-1-ol (62%), and benzyl alcohol (48%) (Fig. 10.4). CsGT1 also glycosylated monoterpene, aromatic and aliphatic alcohols, and is hence considered a critical enzyme for the production of aroma monoglucosides in tea (Ohgami et al. 2015).

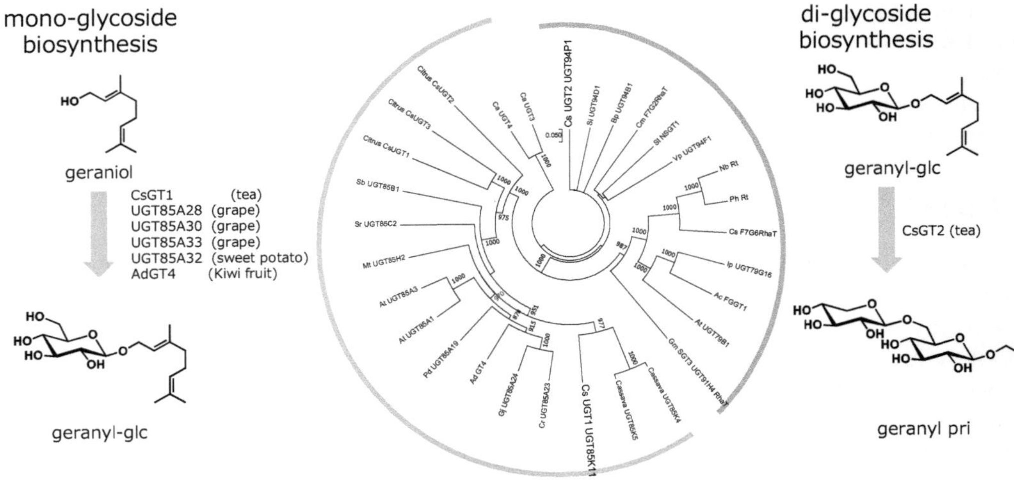

Fig. 10.3: Phylogenetic tree analysis of aroma mono- and di- glycosyltransferases.

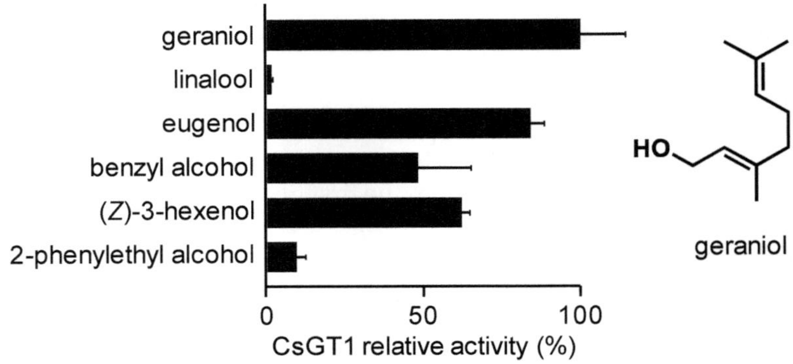

Fig. 10.4: Relative activity of CsGT1 against various volatiles.

3.3 UGT94P1 (CsGT2) Catalyzes the Second Glycosylation Reaction to Produce Aroma β-Primeverosides

UGTs that catalyze the second glycosylation in the conversion of aroma monoglucosides to aroma β-primeverosides (6-*O*-xylosylation of the glucose moiety of aroma monoglucoside) were identified following purification of xylosyltransferases from fresh tea leaves. Xylosidation activities were increased by 13.0-fold through a seven-step purification procedure, and the purified fraction included a dominant protein band with a molecular weight of approximately 50 kDa on sodium dodecyl sulfate-polyacrylamide gel electrophoresis (SDS-PAGE). Moreover, LC-MS/MS analyses of partial amino acid sequences identified three peptide sequences, and subsequent tBLASTn searches of these sequences gave significant identity with an in-house contig EST database that was constructed using a new generation sequencer. The isolated full-length cDNA includes a 1362-bp open reading frame (453 amino acid residues), and the identified UGT was named UGT94P1 (CsGT2) (Mackenzie et al. 1997, Ohgami et al. 2015). To determine whether CsGT2 catalyzes the second glycosylation (xylosylation) of aroma monoglucosides, we performed heterologous expression studies in *E. coli* followed by CsGT2 enzyme activity assays using UDP-xylose as a sugar donor and geranyl-glc as a sugar acceptor. These assays gave a new peak on LC-MS chromatograms that corresponded with authentic geranyl-pri. This result indicates that CsGT2 catalyzes the second glycosylation (xylosylation) of the C-6′ position of geranyl-glc to produce geranyl-pri. The substrate specificity of CsGT2 was also indicated by relative activities toward the following substrates: geranyl-glc (100%), 2-phenylethyl β-D-glucopyranoside (2PE-glc; 16%), linalyl-glc (12%), and eudinyl-glc (2%) (Ohgami et al. 2015). However, unlike CsGT1, CsGT2 did not catalyze the first glycosidation of monoterpene alcohols and aromatic alcohols to produce aroma monoglycosides. In further studies, the sugar donor specificity of CsGT2 was examined using various sugar donors and geranyl-glc as a sugar acceptor. CsGT2 preferentially metabolized UDP-xylose (100%) as a sugar donor, and had only weak activity with UDP-glucose (30%) (Ohgami et al. 2015). These results demonstrate that CsGT2 is a dominant UGT that catalyzes the second glycosylation (xylosylation) of aroma glucosides to produce aroma β-primeveroside. It is shown that CsGT1 and CsGT2 belong to different clades based on phylogenetic analysis (Yonekura-Sakakibara and Hanada 2011). The clade belongs to CsGT1 included cassava UGT85K4 and 5 involved in cyanogenic glucosides biosynthesis (Kannangara et al. 2011). While CsGT2 belongs to a new member of "sugar–sugar UGT" catalyzes glycosylation at the sugar moiety of various phytochemical glycosides. This clade included the morning glory Dusky (UGT79G16), and tomato NSGT1 involved in the glucosylation of anthocyanin and volatile glycosides, respectively (Morita et al. 2005, Tikunov et al. 2013).

4. Two Leaves and a Bud of Young Tea Leaves

High-grade tea products are manufactured from young tea leaves (two leaves and a bud) at the top of the tea plant (Fig. 10.5). These buds are tiny and their leaves are only partially

Fig. 10.5: Two Leaves and a Bud of young tea leaves in *Camellia sinensis*.

formed. Yet, tea products that are produced in this way are prized for their rich flower- and muscat-like aromas. To investigate why these high-quality tea products release high quantities of aroma compounds, we analyzed endogenous aroma glycosides and transcript levels of *CsGT1*, *CsGT2*, and *β-primimerosidase* (*β-PRD*) (Mizutani et al. 2002). We found that quantities of endogenous aroma glycosides were remarkably higher in young tissues than in mature leaves and roots, and transcripts of CsGT1 and CsGT2 were correspondingly slightly more abundant in young leaves. These results are explanatory of the qualities of teas that are produced from buds and their associated two leaves.

4.1 Quantitative Analysis of GBVs in the Black Tea Manufacturing Process

The dominant aroma compounds in black tea are β-ionone, geraniol, linalool, 2-phenylethanol, benzyl alcohol, linalool oxide, methyl salicylate and *cis*-3-hexen-1-ol. Among these, volatile alcohols are predominantly glycosylated and stored in tea leaves, and in black tea or oolong tea, are likely produced by enzymatic or chemical hydrolysis of GBV during the tea manufacturing process (Fig. 10.6). The related molecular mechanisms may be as follows. The aroma β-primeveroside is stored in vacuoles in fresh tea leaves, and the disaccharide hydrolase β-PRD, which specifically hydrolyzes β-primeveroside, is located in intercellular spaces or in plant cell walls. During tea manufacturing, disruption of plant cells leads to an encounter of the substrate β-primeveroside with β-PRD, leading to the release of volatile aglycons into the air. To validate this putative molecular mechanism, various reported analytical results show that quantities of aroma compounds increase

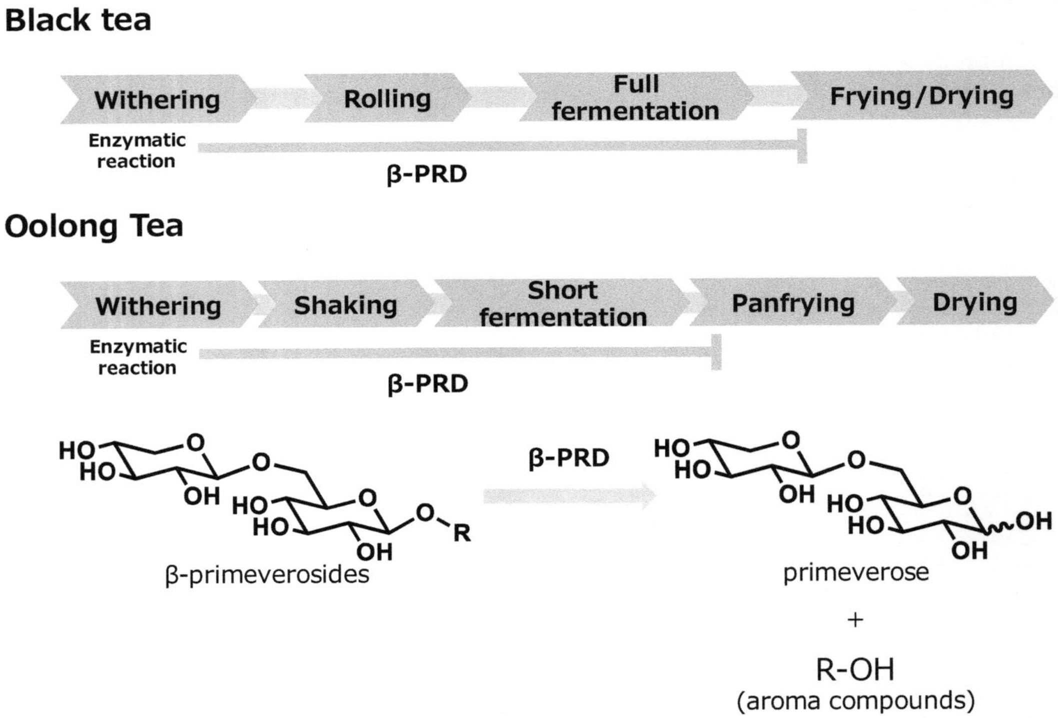

Fig. 10.6: Tea manufacturing process in Oolong tea and black tea.

during wilting, rolling and fermentation processes in black tea and oolong tea. However, it remains unknown whether aroma compounds are derived from aglycones of GBVs in fresh tea leaves. Quantitative changes in GBVs are, however, indicative of the molecular mechanisms behind aroma production. Hence, we analyzed GBVs (geraniol, linalool, benzyl alcohol, 2-phenylethanol, and *cis*-3-hexen-1-ol glycosides) contents at the black tea manufacturing steps of plucking, withering, rubbing, fermenting and drying. Whereas β-primeveroside contents did not change during plucking and wilting processes, all aroma β-primeverosides were dramatically decreased by rolling, to 0.66–0.81% of that in freshly plucked tea leaves.

Endogenous aroma monoglycoside contents were not substantially changed during withering, rolling or plucking processes. Previously, Mizutani et al. (2002) reported that β-PRD, the primary sugar hydrolase, has high activity against aroma diglycosides (especially β-primeveroside), but low activity against aroma monoglycosides. In line with these results, the aroma compounds in black tea were likely derived from hydrolysates of aroma β-primeverosides that were hydrolyzed by β-PRD during the rolling process, and the source of "scent" of black tea primarily reflects aroma diglycosides (Mizutani et al. 2002).

4.2 Quantitative Analysis of GBVs in the Oolong Tea Manufacturing Process

Oolong tea manufacturing processes are more complicated than those of black tea, with plucking, solar withering, indoor withering, rotation fermentation and roasting/drying steps. Endogenous enzymes in flesh tea leaves are likely inactivated during the roasting/drying procedure. In contrast, it is accepted that the aroma compounds in oolong tea are produced by hydrolysis of aroma glycosides during the withering process, as in black tea (Fig. 10.6). Recent electron microscopic analysis of tea leaves during the oolong tea manufacturing process show that plant cells are not damaged and their organelles are maintained. These findings suggest that β-PRD in the cell walls is not made available for the hydrolysis of aroma β-primeverosides in vacuoles, because the enzyme and the substrate are spatially isolated from each other. To investigate the above inconsistencies, Cui et al. quantified GBVs at each step of the oolong tea manufacturing process (Cui et al. 2016). In their study, aroma glycoside contents gradually increased with oolong tea production processes, but aroma diglycoside contents were remarkably increased during the roasting/drying process. These findings suggest that aroma compounds that are generated during withering are independent of GBVs in oolong tea. In addition, Cui et al. (2016) demonstrated that transcripts of linalool synthase gene accumulate during solar and indoor withering processes and are further induced by drying and mechanical stresses. Accordingly, aroma compounds that are released from oolong tea are synthesized *de novo* during the withering process.

4.3 Aroma Malonyl Glycosides in Tea Leaves

Aroma mono- and di-glycoside contents increased markedly during the drying step in the oolong tea manufacturing process. Some aroma compounds, such as benzyl alcohol and 2-phenylethyl alcohol, are stored as malonyl β-glucoside in strawberries and tomatoes. Hence, tea plants may similarly accumulate aroma glycosides, such as malonyl β-glycosides, although these are very chemically unstable (Fig. 10.7). Hence, to determine whether aroma

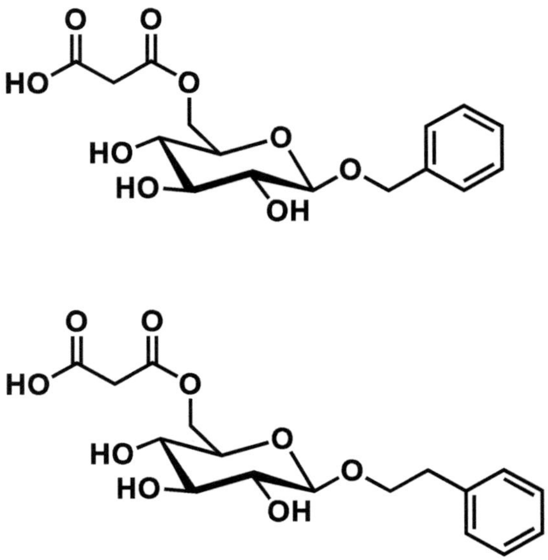

Fig. 10.7: Chemical structure of aroma malonyl glycosides.

malonyl β-glycosides degrade and increase the production of aroma glycosides during the dry process, we performed stability tests at 230°C. In these experiments, tea leaves are heated on a frying pan at 230°C, and degradation rates of the synthetic compound phenyl malonyl β-glucoside were monitored at 25°C, 40°C and 90°C. As expected, malonyl groups were eliminated under these conditions, and phenyl β-glucoside was produced at 90°C within 6 hours. In addition, phenyl malonyl β-glucoside was stable at 25°C and 40°C for at least 6 hours. These findings suggest that GBV contents increase due to hydrolysis of aroma malonyl β-glycosides. Thus, analyses of endogenous aroma malonyl β-glycosides were conducted and compared with those in fresh tea leaves. In these experiments, monomaleate esters of benzyl β-glucoside and 2-phenylethyl β-glucoside were detected at plucking, solar withering, and indoor wilting processes, suggesting that aroma malonyl β-glycosides accumulate stably. Tikunov and colleagues (2013) suggested that disaccharide glycosides and di-malonyl esters of trisaccharide glycosides are present in tomatoes. However, further studies are needed to determine the roles of aroma malonyl β-glycoside in food chemistry and plant physiology (Fig. 10.8).

5. Conclusions

The aroma glycoside biosynthetic enzymes CsGT1 and CsGT2 have recently been characterized for their contributions to aroma compounds. Humans actively utilize aroma compounds from plants for religious ceremonies and to preserve food. As a consequence, research into VOCs and GBVs is expanding both from viewpoints of phytochemical and plant physiology and perspectives of food chemistry and medical science. In the future studies, new analyses of aromas, which are important quality factors of teas, will identify aroma glycosides and aroma malonyl glycosides and will define the *de novo* synthetic processes for the production of aroma components during the tea manufacturing process.

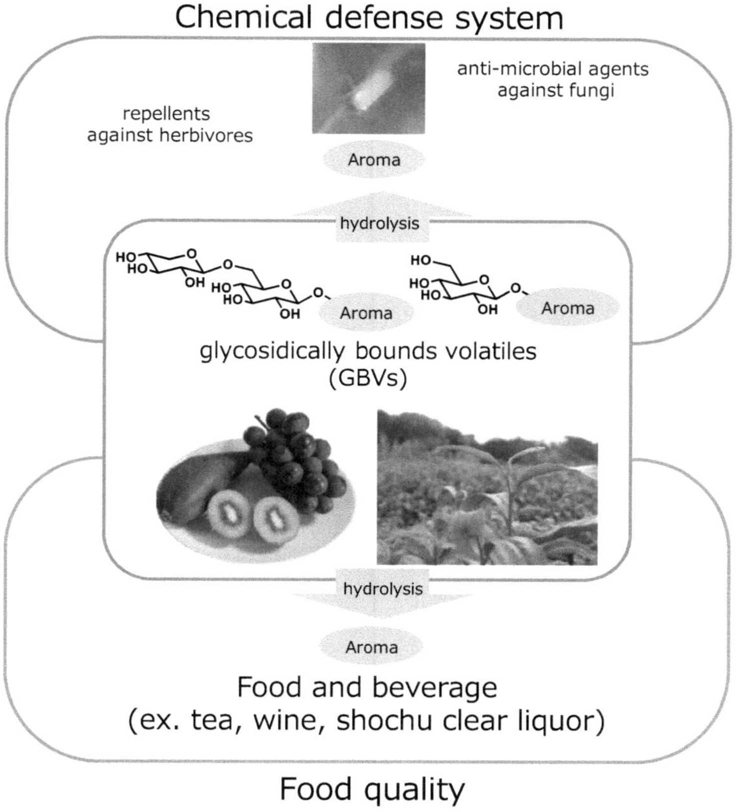

Fig. 10.8: The role of aroma β-glycoside in food chemistry and plant physiology.

Acknowledgments

The major part of this work was conducted at the Research Institute of Green Science and Technology, Shizuoka University.

References

Aharoni, A., A.P. Giri, S. Deuerlein, F. Griepink, W.J. de Kogel, F.W. Verstappen et al. 2003. Terpenoid metabolism in wild-type and transgenic Arabidopsis plants. Plant Cell 15: 2866–2884.

Arimura, G., K. Matsui and J. Takabayashi. 2009. Chemical and molecular ecology of herbivore-induced plant volatiles: proximate factors and their ultimate functions. Plant Cell Physiol. 50: 911–923.

Bönisch, F., J. Frotscher, S. Stanitzek, E. Rühl, M. Wüst, O. Bitz et al. 2014a. Activity based profiling of a physiologic aglycone library reveals sugar acceptor promiscuity of family 1 UDP-glucosyltransferases from Vitis vinifera. Plant Physiol. 166: 23–39.

Bönisch, F., J. Frotscher, S. Stanitzek, E. Rühl, M. Wüst, O. Bitz et al. 2014b. A UDP-glucose: Monoterpenol glucosyltransferase adds to the chemical diversity of the grapevine metabolome. Plant Physiol. 165: 561–581.

Chaturvedula, V.S.P. and I. Prakash. 2011. The aroma, taste, color and bioactive constituents of tea. J. Med. Plants Res. 5: 2110–2124.

Cui, J., T. Katsuno, K. Totsuka, T. Ohnishi, H. Takemoto, N. Mase et al. 2016. Characteristic fluctuations in glycosidically bound volatiles during tea processing and identification of their unstable derivatives. J. Agric. Food Chem. 64: 1151–1157.

Ginglinger, J.F., B. Boachon, R. Höfer, C. Paetz, T.G. Köllner, L. Miesch et al. 2013. Gene coexpression analysis reveals complex metabolism of the monoterpene alcohol linalool in Arabidopsis flowers. Plant Cell 25: 4640–4657.

Günata, Y.Z., C.L. Bayonove, R.L. Baumes and R.E. Cordonnier. 1985. The aroma of grapes: I. Extraction and determination of free and glycosidically bound fractions of some grape aroma components. J. Chromatogr. A 331: 83–90.

Guo, W., K. Sakata, N. Watanabe, R. Nakajima, A. Yagi, K. Ina et al. 1993. Geranyl 6-*O*-β-D-xylopyranosyl-β-D-glucopyranoside isolated as an aroma precursor from tea leaves for oolong tea. Phytochemistry 33: 1373–1375.

Guo, W., R. Hosoi, K. Sakata, N. Watanabe, A. Yagi, K. Ina et al. 1994. (*S*)-linalyl, 2-phenylethyl, and benzyl disaccharide glycosides isolated as aroma precursors from oolong tea leaves. Biosci. Biotechnol. Biochem. 58: 1532–1534.

Izumi, S., O. Takashima and T. Hirata. 1999. Geraniol is a potent inducer of apoptosis-like cell death in the cultured shoot primordia of Matricaria chamomilla. Biochem. Biophys. Res. Commun. 259: 519–522.

Kannangara, R., M.S. Motawia, N.K. Hansen, S.M. Paquette, C.E. Olsen, B.L. Møller et al. 2011. Characterization and expression profile of two UDP-glucosyltransferases, UGT85K4 and UGT85K5, catalyzing the last step in cyanogenic glucoside biosynthesis in cassava. Plant J. 68: 287–301.

Kawakami, M., S.N. Ganguly, J. Banerjee and A. Kobayashi. 1995. Aroma composition of oolong tea and black tea by brewed extraction method and characterizing compounds of darjeeling tea aroma. J. Agric. Food Chem. 43: 200–207.

Kobayashi, A., K. Kubota, Y. Joki, E. Wada and M. Wakabayashi. 1994. (Z)-3-Hexenyl β-D-glucopyranoside in fresh tea leaves as a precursor of green odor. Biosci. Biotechnol. Biochem. 58: 592–593.

Koroleva, O.A., T.M. Gibson, R. Cramer and C. Stain. 2010. Glucosinolate-accumulating S-cells in Arabidopsis leaves and flower stalks undergo programmed cell death at early stages of differentiation. Plant J. 64: 456–469.

Krammer, G., P. Winterhalter, M. Schwab and P. Schreier. 1991. Glycosidically bound aroma compounds in the fruits of Prunus species: apricot (*P. armeniaca*, L.), peach (*P. persica*, L.), yellow plum (*P. domestica*, L. ssp. syriaca). J. Agric. Food Chem. 39: 778–781.

Kumazawa, K. and H. Masuda. 2002. Identification of potent odorants in different green tea varieties using flavor dilution technique. J. Agric. Food Chem. 50: 5660–5663.

Lücker, J., H.J. Bouwmeester, W. Schwab, J. Blaas, L.H. van der Plas and H.A. Verhoeven. 2001. Expression of Clarkia S-linalool synthase in transgenic petunia plants results in the accumulation of *S*-linalyl-β-D-glucopyranoside. Plant J. 27: 315–324.

Mackenzie, P.I., I.S. Owens, B. Burchell, K.W. Bock, A. Bairoch, A. Bélanger et al. 1997. The UDP glycosyltransferase gene superfamily: Recommended nomenclature update based on evolutionary divergence. Pharmacogenetics 7: 255–269.

Marlatt, C., C. Ho and M.J. Chien. 1992. Tomato: Studies of aroma constituents bound as glycosides in tomato. J. Agric. Food Chem. 40: 249–252.

Mizutani, M., H. Nakanishi, J. Ema, S.J. Ma, E. Noguchi, M. Inohara-Ochiai et al. 2002. Cloning of β-primeverosidase from tea leaves, a key enzyme in tea aroma formation. Plant Physiol. 130: 2164–2176.

Moon, J.H., N. Watanabe, K. Sakata, A. Yagi, K. Ina and S. Luo. 1994. *Trans*- and *cis*-linalool 3,6-oxide 6-*O*-β-D-xylopyranosyl-β-D-glucopyranosides isolated as aroma precursors from leaves for oolong tea. Biosci. Biotechnol. Biochem. 58: 1742–1744.

Moon, J.H., N. Watanabe, Y. Ijima, A. Yagi and K. Sakata. 1996. *Cis*- and *trans*-linalool 3,7-oxides and methyl salicylate glycosides and (Z)-3-hexenyl β-D-glucopyranoside as aroma precursors from tea leaves for oolong tea. Biosci. Biotechnol. Biochem. 60: 1815–1819.

Morita, Y., A. Hoshino, Y. Kikuchi, H. Okuhara, E. Ono, Y. Tanaka et al. 2005. Japanese morning glory dusky mutants displaying reddish-brown or purplish-gray flowers are deficient in a novel glycosylation enzyme for anthocyanin biosynthesis, UDP-glucose: anthocyanidin 3-*O*-glucoside-2″-*O*-glucosyltransferase, due to 4-bp insertions in the gene. Plant J. 42: 353–363.

Obayashi, T., K. Nishida, K. Kasahara and K. Kinoshita. 2011. ATTED-II updates: condition-specific gene coexpression to extend coexpression analyses and applications to a broad range of flowering plants. Plant Cell Physiol. 52: 213–219.

Ohgami, S., E. Ono, M. Horikawa, J. Murata, K. Totsuka, H. Toyonaga et al. 2015. Volatile glycosylation in tea plants: Sequential glycosylations for the biosynthesis of aroma β-primeverosides are catalyzed by two *Camellia sinensis* glycosyltransferases. Plant Physiol. 168: 464–477.

Ohno, A. 2014. Black tea character wheel and pictorial charts define tea characteristics. J. Japan Association on Odor Environment 45: 344–350.

Pabst, A., A. Barron, E.P. Eti and P. Schreier. 1991. Studies on the enzymatic hydrolysis of bound aroma constituents from raspberry fruit pulp. J. Agric. Food Chem. 39: 173–175.

Pattnaik, S., V.R. Subramanyam, M. Bapaji and C.R. Kole. 1997. Antibacterial and antifungal activity of aromatic constituents of essential oils. Microbios. 89: 39–46.

Pichersky, E. and J. Gershenzon. 2002. The formation and function of plant volatiles: perfumes for pollinator attraction and defense. Curr. Opin. Plant Biol. 5: 237–243.

Roscher, R., M. Herderich, J.P. Steffen, P. Schreier and W. Schwab. 1997. 2,5-dimethyl-4-hydroxy-3(2H)-furanone 6′-*O*-malonyl-β-D-glucopyranoside in strawberry fruits. Phytochemistry 43: 155–159.

Sugimoto, K., K. Matsui, Y. Iijima, Y. Akakabe, S. Muramoto, R. Ozawa et al. 2014. Intake and transformation to a glycoside of (*Z*)-3-hexenol from infested neighbors reveals a mode of plant odor reception and defense. Proc. Natl. Acad. Sci. USA 111: 7144–7149.

Tikunov, Y.M., J. Molthoff, R.C. de Vos, J. Beekwilder, A. van Houwelingen, J.J. van der Hooft et al. 2013. Non-smoky glycosyltransferase1 prevents the release of smoky aroma from tomato fruit. Plant Cell 25: 3067–3078.

Usadel, B., T. Obayashi, M. Mutwil, F.M. Giorgi, G.W. Bassel, M. Tanimoto et al. 2009. Co-expression tools for plant biology: Opportunities for hypothesis generation and caveats. Plant Cell Environ. 32: 1633–1651.

Wang, D., T. Yoshimura, K. Kubota and A. Kobayashi. 2000. Analysis of glycosidically bound aroma precursors in tea leaves: I. Qualitative and quantitative analyses of glycosides with aglycons as aroma compounds. J. Agric. Food Chem. 48: 5411–5418.

Yano, M., Y. Joki, H. Mutoh, K. Kubota and A. Kobayashi. 1991. Benzyl glucoside from tea leaves. Agric. Biol. Chem. 55: 1205–1206.

Yauk, Y.K., C. Ged, M.Y. Wang, A.J. Matich, L. Tessarotto, J.M. Cooney et al. 2014. Manipulation of flavour and aroma compound sequestration and release using a glycosyltransferase with specificity for terpene alcohols. Plant J. 80: 317–330.

Yazaki, K., G.I. Arimura and T. Ohnishi. 2017. 'Hidden' terpenoids in plants: their biosynthesis, localization and ecological roles. Plant Cell Physiol. 58: 1615–1621.

Yonekura-Sakakibara, K. and K. Hanada. 2011. An evolutionary view of functional diversity in family 1 glycosyltransferases. Plant J. 66: 182–193.

Young, H. and V.J. Paterson. 1995. Characterization of bound flavour components in kiwi fruit. J. Sci. Food Agric. 68: 257–260.

Chapter 11

Heat Tolerance Enhancers of Plants for Agricultural Use

Masakazu Hara

1. Introduction

Heat waves are climate phenomena in which hotter weather than usual continues for a few weeks. The extremely high temperatures during a heat wave often negatively influence plant production in various regions of the world (Ainsworth and Ort 2010, Hall 2001, Teskey et al. 2015). Angiosperms were generated in the Mesozoic era and evolved during long warm climate periods. The mean temperature of the earth during the Triassic period ranged from 32°C to 40°C, which was far higher than that of the equatorial sea surface temperature in modern times (i.e., 25°C to 30°C) (Sun et al. 2012). It is likely that mankind started to practice systematic agriculture and conducted primitive breeding during the Atlantic period (from 5500 to 3000 BC). The climate for this period was 2–3 degrees warmer than at present, suggesting that most agricultural crops had been originally developed under a high-temperature environment. However, the earth at present is in a cool environment. Modern agricultural technology was developed over the course of the Little Ice Age, which suggests that today's crops have adapted to the cooler environment. This is one of the reasons that the recent warming climate, and especially the occurrence of sudden heat waves, can severely impact crop production.

The seaboard areas of the Far East such as Japan have a stable climate. In 2010, however, Japan had an unusual summer which the government certified as "the hottest summer ever recorded." The quality and yield of rice, vegetables and fruits declined. Since then, the highest temperatures have increased further, with the record high reaching 41.1°C in 2018. Although many technologies to reduce the agricultural damage caused by heat are under development, their practical application has not yet been well established because it would be costly and time-consuming. Here we propose a new methodology involving chemicals that enhance the heat tolerance of plants.

Research Institute of Green Science and Technology, Shizuoka University, 836 Ohya, Shizuoka, Shizuoka 422-8529, Japan.
Email: hara.masakazu@shizuoka.ac.jp

2. Enhancement of Heat Tolerance

2.1 Reduction of Thermal Damage in Crop Production

2.1.1 Plant Response to Extreme Heat

Since the ancestors of today's crops originated and have prospered during the warm periods, they are thought to have the potential for heat tolerance. Generally, plants in cold regions show apparent tolerance for high temperatures after heat acclimation (Larcher 2003). The maximum heat tolerance potential of cold-region plants is similar to that of plants grown in high-temperature regions. However, crops in the field are easily damaged by sudden heat stress (i.e., heat waves) because their heat tolerance has not been established. If we can efficiently activate the heat-tolerance system in plants before a heat wave arrives, the risk of severe damage due to heat can be reduced. Practical approaches to suppressing the loss of crop productivity due to heat are needed.

It has been well documented that plants exposed to heat show various symptoms of injury, such as protein denaturation and degradation, the fluidization of membrane lipids, the generation of Reactive Oxygen Species (ROS), and the inhibition of photosynthesis (Allakhverdiev et al. 2008, Ruelland and Zachowski 2010, Wahid et al. 2007) (Fig. 11.1). These symptoms of damage in plants can be reduced by the accumulation of compatible solutes, the alteration of hormone signaling, the enhancement of antioxidative activities, and the expression of heat shock proteins (HSPs) (Iba 2002, Kotak et al. 2007). HSPs prevent the heat denaturation of proteins by facilitating protein folding (Fig. 11.2). HSPs are classified into HSP60s, HSP70s, HSP90s, HSP100s, and small HSPs on the basis of molecular weight. The

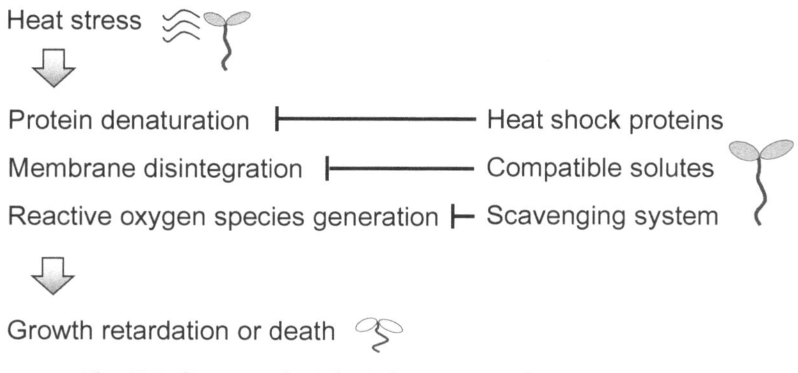

Fig. 11.1: Common physiological responses to heat stress in plants.

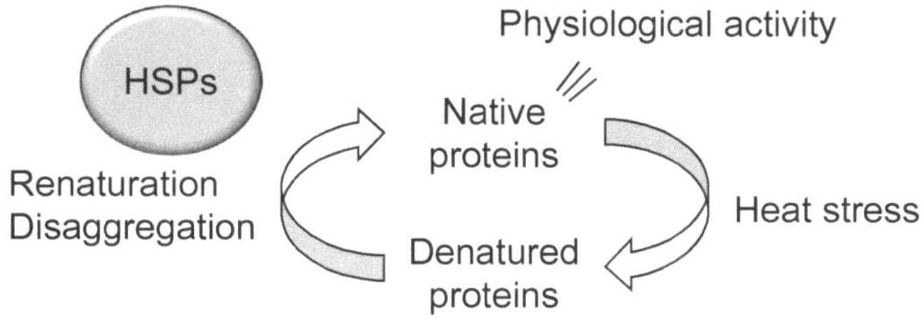

Fig. 11.2: Renaturation and disaggregation of denatured proteins by heat shock proteins (HSPs). Native proteins are denatured by heat stress.

expression of HSP genes is a crucial response in establishing the heat tolerance of plants (Huang and Xu 2008), because there are positive correlations between the expression levels of HSP genes and the adaptability of plants to high temperatures (Wang et al. 2004, Wahid et al. 2007, Waters 2013, Zhang et al. 2014). For instance, the overexpression of HSP genes elevated the heat tolerance of plants (Masand and Yadav 2016, Sung and Guy 2003, Wang et al. 2016), and conversely, mutations in HSP genes led to hypersensitivity to heat (Leng et al. 2017, Su and Li 2008). This means that HSPs are strongly related to the heat-tolerance system of plants; therefore efficient activation of the heat-tolerance system before heat waves arrive is a practical approach to reducing the loss of crop productivity due to heat.

2.1.2 Countermeasures to Crop Damage Due to Heat

Diverse approaches to enhancing the heat tolerance of crops have been described in academic papers and patents. They can be divided into three categories: control of environmental conditions by agricultural equipment, breeding of thermo-tolerant cultivars, and Heat Tolerance Eenhancers (HTLEs) (Table 11.1). The first category includes indoor farming systems in which the growing temperature, moisture and illumination conditions are highly controlled. Although the indoor environment of the system is not influenced even by extremely high outdoor temperatures, the primary and operation costs of the system are expensive.

The second category, i.e., cross-breeding and selective breeding, includes proven ways to establish thermo-tolerant crops. However, such traditional breeding tends to be time consuming. If the breeding targets are major crops whose genomes were sequenced, and if the genes and loci involved in heat tolerance have been determined, genomic breeding can be used. But this technique is difficult to apply to the majority of cultivated crops. Transgenic breeding including the application of genome editing is a powerful technique. However, if thermo-tolerant crops can be bred by this method, the crops must be cultivated under precise control.

The third category involves the use of agrochemicals, so-called HTLEs, which prevent heat damage in crops. Because some HTLEs stimulate the common heat-responsive mechanism in plants, they can be applied to various plant species. Therefore, agricultural producers can continue to cultivate traditional varieties which are sensitive to high temperatures without changing to thermo-tolerant cultivars that have not been

Table 11.1: Countermeasures to high temperatures in practical agriculture.

1) Control of environmental conditions	
	e.g., Indoor farming system
	AD, No influence of outdoor environment
	DA, Costly
2) Breeding	
	e.g., Thermotolerant rice cultivars
	AD, Genetically acquired tolerance
	DA, Depends on breeding efficiency
3) Heat tolerance enhancers (HTLEs)	
	e.g., Sanguinarine agent
	AD, Simple application (with watering)
	DA, Short history of study

well established. Moreover, HTLEs can be supplied to crops simply with daily watering. Recently, we developed a sanguinarine agent as an HTLE which is now available on the market in Japan. Below the sanguinarine agent and the candidates for HTLEs are described.

3. HTLEs as Chemical Stimulators for Plants

3.1 Chemical Induction of HSP Expression

The controlled induction of HSP expression is a promising methodology to improve the heat tolerance of crops. Although the most effective cue for the induction of HSPs is heat, it is difficult to supply equivalent thermal energy to crops in the agricultural field. It has been reported that some chemicals can induce HSP expression in plants. Salicylic acid (Clarke et al. 2004), benzyl alcohol (Saidi et al. 2005), celastrol (Saidi et al. 2007), isothiocyanates (Hara et al. 2013), sanguinarine (Hara and Kurita 2014), monoterpenes (Hara et al. 2018b), and (*E*)-2-hexenal (Yamauchi et al. 2015) are heat shock response inducers in plants. In addition, HSP90 inhibitors such as macrocyclic types (e.g., geldanamycin, radicicol and monocillin I), benzamide types (e.g., PF-04929113 and SNX-2112), purine types (e.g., BIIB021 and PU-H71), and resorcinol types (e.g., AT13387, AUY922, and Ganetespib) induced the heat shock response in *Arabidopsis* (McLellan et al. 2007, Murano et al. 2017, Yamada et al. 2007, Yoshida et al. 2011). Although it is theoretically possible for such chemicals to be applied to enhance the heat tolerance of crops, most of the compounds are difficult to use as commercial HTLEs due to the following problems. Salicylic acid and benzyl alcohol show weak HSP-inducing activities. Quantitative applications of monoterpenes and (*E*)-2-hexenal in the field are likely to be difficult because these compounds are volatile. Celastrol and HSP90 inhibitors are so expensive that it is not practical to apply them in the field. However, sanguinarine has various properties that make it suitable for practical use. A low concentration (at the micromolar level) of sanguinarine is sufficient to express the activity. Sanguinarine can be applied as an aqueous solution. The production cost of the sanguinarine agent is low, because the agent can be prepared from the extract of *Macleaya cordata,* which contains a large amount of sanguinarine. Sanguinarine has been approved to be used as a food additive for livestock (trade name: Sangrovit), indicating that the safety of sanguinarine is acceptable for agricultural use. The characteristics of sanguinarine that make it a practical choice as an HTLE are first described and then information is provided about isothiocyanates and miscellaneous chemicals.

3.2 Sanguinarine

Sanguinarine is a benzophenanthridine alkaloid found in Papaveraceae plants such as *Sanguinaria canadensis, Papaver somniferum, Eschscholzia californica, Chelidonium majus,* and *M. cordata.* Since sanguinarine shows strong antimicrobial activity (Facchini and St-Pierre 2005), this alkaloid is believed to contribute to the defense system in plant-microbe interaction. Sanguinarine has been demonstrated to have allelopathic effects in plants (Wink and Twardowsky 1992). Sanguinarine has been reported to be medicinally active compounds showing antihypertensive, antiplatelet, and anticarcinogenic activities in animals (Mackraj et al. 2008). Sanguinarine is used as a feed additive for livestock such as pigs and broiler chickens (Kantas et al. 2015, Kosina et al. 2004, Lee et al. 2015, Vieira et al. 2008). Sangrovit is a commercial product containing sanguinarine as a major ingredient which is prepared from *M. cordata*. Supplementation of a livestock diet with Sangrovit increased the body weight and improved the growth performance of the livestock. A high dose of Sangrovit did not result in toxic effects (Zhao et al. 2017).

Recently, we revealed that sanguinarine enhanced heat tolerance in plants. This effect was originally found during a research program for industrial development conducted by Shizuoka University, Shizuoka City, and the Shizuoka Chamber of Commerce and Industry. Pot-culture experiments using leaf vegetables indicated that the extract from *M. cordata* (Japanese name: Takenigusa), which is native to East Asia, ameliorated the heat damage suffered by the vegetables. Subsequently, the active component of the extract was determined to be sanguinarine. Field tests indicated that the administration of sanguinarine showed a tendency to inhibit the reduction of the grain quality of rice that occurs due to high temperatures. A functional fertilizer containing sanguinarine is commercially available through the Japan Agricultural Cooperatives. The fertilizer is supplied to crops grown in the field as an aqueous solution.

3.3 Isothiocyanates

Isothiocyanates are sulfur-containing volatile compounds which are generated mainly from Brassicaceae plants. The isothiocyanate group (–N=C=S) is a consensus structure. More than 100 kinds of isothiocyanates which have different R moieties have been found in nature. They are produced through the hydrolysis of glucosinolates by a specific enzyme, myrosinase (glucosinolate-myrosinase system). The strong pungency of mustard (*Brassica juncea*) and wasabi (*Wasabia japonica*) is attributed to isothiocyanates. The biosynthetic pathways of glucosinolates, the regulation of the biosynthesis and the tissue and cellular localization of the glucosinolate-myrosinase system have been well characterized (Halkier and Gershenzon 2006, Kissen et al. 2009). Because isothiocyanates have anti-microorganism, animal-repellent, allelopathic and herbicidal activities (Halkier and Gershenzon 2006, Hara et al. 2010, Hopkins et al. 2009), isothiocyanates are thought to be related to the defense system in plants. We found that high concentrations of isothiocyanates showed herbicidal activity, whereas low concentrations enhanced the heat tolerance of *Arabidopsis* (Hara et al. 2010). Regarding the latter effect, we found that phenethyl isothiocyanate enhanced the heat tolerance of spinach and alfalfa in the pot cultivation test. Isothiocyanates have practical merits because they are approved to be used as food additives and can be obtained in large quantities. Appropriate emulsifying agents and pungency reducers should be developed for the commercialization of isothiocyanates as HTLEs.

3.4 Monoterpenes

Monoterpenes are C10-isoprenoid derivatives which are produced by diverse plant species. The biosynthesis of monoterpenes proceeds via the cyclization of geranyl diphosphate and neryl diphosphate by monoterpene synthases (Dudareva et al. 2004). Many monoterpenes show volatility and toxicity to insects, fungi, bacteria and animals (Mithöfer and Boland 2012). In plants, monoterpenes show allelopathic activities (Macías et al. 2007) and heat-tolerance-enhancing activities (Delfine et al. 2000, Loreto et al. 1998, Peñuelas and Llusià 2002). Since some monoterpenes scavenge ROS, monoterpenes have been believed to protect leaves by reducing ROS that are generated due to heat stress (Peñuelas and Llusià 2002).

3.5 Reactive Short-Chain Leaf Volatiles

Plants produce a series of short-chain volatiles which possess α, β-unsaturated carbonyl bonds from peroxidized polyunsaturated fatty acids. These volatiles, including (*E*)-2-hexenal and (*E*)-2-butenal, are called Reactive Short-chain Leaf Volatiles (RSLVs). The addition of (*E*)-2-hexenal to *Arabidopsis* induced heat-shock transcription factor (HSF) gene expression, enhanced the accumulation of HSPs and then increased the heat tolerance of the plant (Yamauchi et al. 2015).

3.6 Others

A heat-shock promoter system using moss *Physcomitrella patens* was studied to find chemicals that induced the heat shock response (Saidi et al. 2007). Benzene derivatives (a series of chlorophenols, benzyl alcohol, and salicylate), heavy metals (cadmium ion and zinc ion), and triterpene celastrol activated the heat shock promoter.

4. Action Mechanism of HTLEs

4.1 Common Responses to HTLEs in Plants

How do HTLEs enhance the heat tolerance of plants? The common action mechanisms of HTLEs by focusing on sanguinarine and isothiocyanates are described here. When these compounds have been administered to various crops, the reduction of heat damage has been observed on one level or another. This result suggests that sanguinarine and isothiocyanates do not act via a specific mechanism in limited plant species, but rather that they stimulate a universal machinery to establish the heat tolerance of plants. Microarray analyses indicated that sanguinarine and isothiocyanates up-regulated a similar series of HSP genes of *Arabidopsis* such as small HSPs, HSP70s, and HSP90s. Accordingly, these HTLEs are the inducers of the heat shock response.

As mentioned above, HSPs are related to the establishment of heat tolerance in plants. The transcription of HSP genes are regulated by HSF. Under non-stressed conditions, HSF is inactivated by HSPs, e.g., HSP90 and HSP70, by forming HSF-HSP complexes (Jacob et al. 2017). When proteins are damaged by heat stress, HSP70 and HSP90 release HSF from the complexes and are recruited to restore the denatured proteins. As a result, HSF is activated and promotes the expression of HSP genes. According to this theory, sanguinarine and isothiocyanates may activate HSF by facilitating the release of HSF from the HSF-HSP complexes, which are the universal machineries of the heat shock response in plants. This idea suggests that these agents protect diverse plant species against heat stress.

However, the question of why sanguinarine and isothiocyanates induced the expression of HSP genes in spite of the large structural difference remains unresolved. Since sanguinarine is an alkaloid, we tested the heat-tolerance-enhancing activity in *Arabidopsis* of 12 alkaloids belonging to different structural groups, i.e., the isoquinoline, indole, purine, quinoline and pyridine types, and found that only sanguinarine showed the activity (Hara and Kurita 2014). Intriguingly, berberine, which has a similar structure to sanguinarine, did not enhance the heat tolerance of the plant. In the case of isothiocyanates, when the isothiocyanate group was changed to the amino or carboxyl group, the heat-tolerance-enhancing activity was diminished (Hara et al. 2013). This means that sanguinarine and isothiocyanates may individually have specific modes of action to induce the HSP expression and enhance the heat tolerance in *Arabidopsis*.

4.2 Specific Modes of Action

4.2.1 Sanguinarine

At the beginning of our investigation, it was difficult to speculate on how sanguinarine induced the HSP expression in plants, because the physiological effect of sanguinarine on plants has been little studied. However, we obtained some evidence regarding the mode of action of sanguinarine through comparative studies by using sanguinarine and geldanamycin. Geldanamycin, which is an HSP90 inhibitor, has been used to induce the heat shock response in *Arabidopsis*. The induction mechanism was proposed to be as follows (Yamada et al. 2007). Under non-stressed conditions, HSP90 inactivates HSF by forming a complex with HSF. Geldanamycin inhibits the HSP90 activity and then releases HSF from the complex. The activated HSF induces the transcription of HSP genes. Our studies demonstrated that sanguinarine and geldanamycin induced the expression of HSPs (HSP90.1, HSP70, and HSP17.6C-CI) in *Arabidopsis* at the transcript and protein levels, and enhanced the heat tolerance of the plant (Hara and Kurita 2014, Matsuoka et al. 2016). The dose and time-course responses of sanguinarine were similar to those of geldanamycin, suggesting that the mode of action of sanguinarine may be close to that of geldanamycin. HSP90, which is a target of geldanamycin, and is a major component of chaperones in many organisms. It was proved that wheat germ lysate contained HSP90 (Reddy et al. 1998) and efficiently refolded heat-denatured luciferase (McLellan et al. 2007). Indeed, geldanamycin was shown to significantly inhibit the refolding of the heat-denatured luciferase (i.e., the chaperone activity) of wheat germ lysate at the concentration of 1 μM (Matsuoka et al. 2016). Surprisingly, sanguinarine also significantly inhibited the chaperone activity of wheat germ lysate at 1 μM. This strongly suggests that sanguinarine induces the expression of HSPs by reducing the stability of HSF-HSP complexes like geldanamycin. However, it has not been determined which HSPs are inhibited by sanguinarine.

Another benzophenanthridine alkaloid, chelerythrine, whose structure is close to sanguinarine's, also induced the heat shock response in *Arabidopsis*, although the activity was weaker than that of sanguinarine (Fig. 11.3). Berberine and papaverine, which are benzylisoquinoline alkaloids, did not show the heat-shock-response-inducing activity. With regard to the chaperone inhibition activity determined by using wheat germ lysate, the magnitudes of inhibition have been listed in the following order: sanguinarine (strong inhibition), chelerythrine (slight inhibition), and berberine and papaverine (no inhibition) (Matsuoka et al. 2016). Taken together, these findings suggest that benzophenanthridine alkaloids, especially sanguinarine, likely have a structure optimal for producing the heat-shock-response-inducing activity and the chaperone-inhibiting activity. Moreover, the structural difference between sanguinarine and chelerythrine is very small because sanguinarine has two methylenedioxy bridges and chelerythrine has one methylenedioxy bridge and one dimethoxy group (Matsushima et al. 2012). This suggests that the methylenedioxy bridge attached to the isoquinoline moiety was related to the expression of the heat-shock-response-inducing activity of the benzophenanthridine alkaloids in plants.

Sanguinarine biosynthesis has been studied using Papaveraceae plants, especially *E. californica* (Yamada et al. 2017) and *P. somniferum* (Desgagné-Penix et al. 2010). A major intermediate of sanguinarine biosynthesis is norcoclaurine, which is derived from a

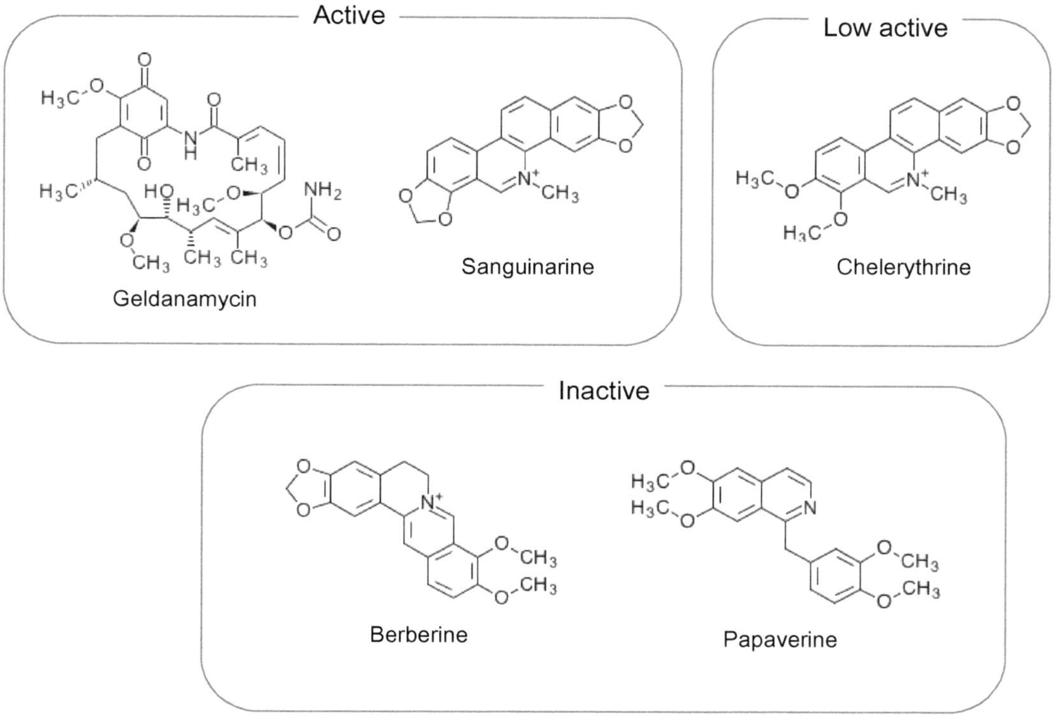

Fig. 11.3: Structures of geldanamycin, sanguinarine, and sanguinarine-related alkaloids. Geldanamycin and sanguinarine activated the induction of heat shock protein genes in *Arabidopsis*. Chelerythrine showed low activity. Berberine and papaverine were inactive.

precursor tyrosine. Recently, it was reported that the expression of the HSP70 gene was synchronized with that of the norcoclaurine-*O*-methyltransferase gene, which is involved in the rate-limiting step of sanguinarine biosynthesis in *E. californica* (Angelova et al. 2010). Although Papaveraceae plants produce sanguinarine to defend themselves against microbial attacks, this sanguinarine may simultaneously disturb the chaperone system of the plants. It is likely that co-expression of sanguinarine biosynthetic genes and HSP genes is crucial for Papaveraceae plants to express their defense system in a safe way.

4.2.2 Isothiocyanates

Unlike in the case of sanguinarine, the mode of action of isothiocyanates on the heat-shock-response-inducing activities has not been well understood. Although the isothiocyanate group (–N=C=S) was necessary for the heat-tolerance-enhancing activity (Hara et al. 2013), the role of moieties other than the isothiocyanate group in the activity was unclear. In order to clarify it, we applied the reporter gene assay using *Arabidopsis* seedlings to evaluate the degree of heat-shock-response-inducing activities of 16 naturally occurring isothiocyanates, i.e., five aromatic isothiocyanates, seven aliphatic isothiocyanates, and four sulfur-containing aliphatic isothiocyanates (Hara et al. 2018a). On the whole, aliphatic isothiocyanates tended to show potent activities for inducing the heat shock response (Fig. 11.4). The activities of aliphatic isothiocyanates were several times stronger than

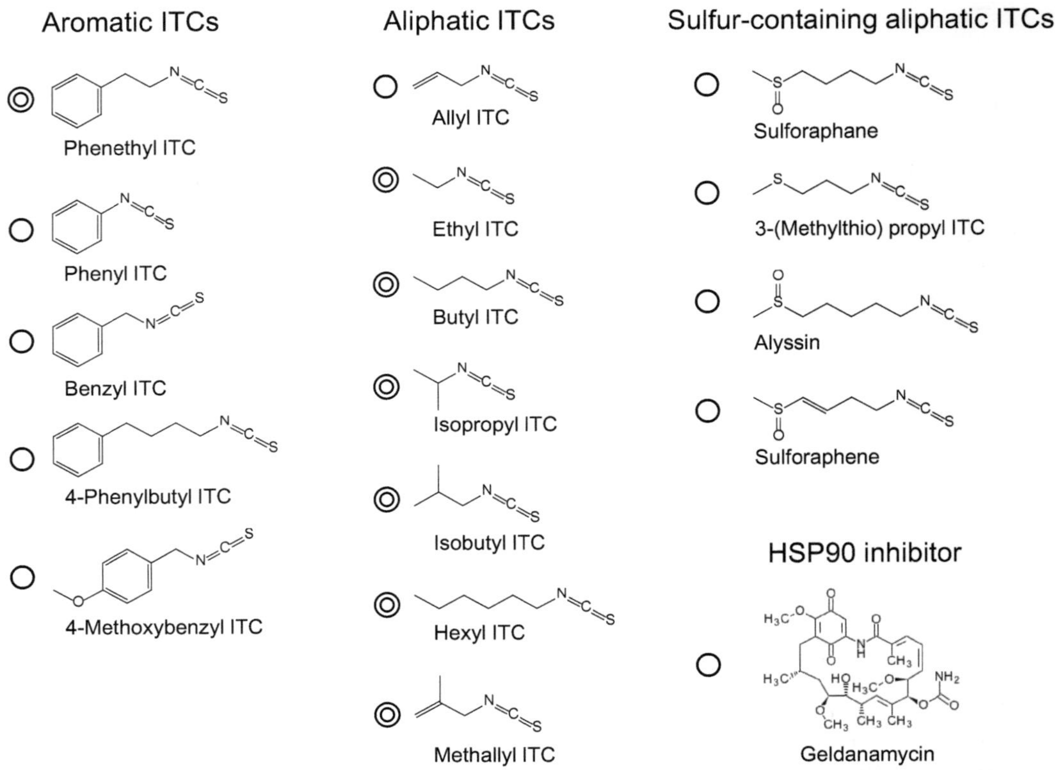

Fig. 11.4: Structures of isothiocyanates. Single and double circles indicate the levels of heat-shock-response-inducing activity in *Arabidopsis*. Single circles: activities similar to that of geldanamycin. Double circles: activities remarkably higher than that of geldanamycin. ITC represents isothiocyanate.

those of geldanamycin. Among the aliphatic isothiocyanates, short-branched aliphatic isothiocyanates, i.e., isopropyl isothiocyanate and isobutyl isothiocyanate, showed the highest activities (Hara et al. 2018a). On the other hand, aromatic isothiocyanates and sulfur-containing aliphatic isothiocyanates showed lower activities. Aromatic isothiocyanates and sulfur-containing aliphatic isothiocyanates have greater molecular weights than short-branched aliphatic isothiocyanates, suggesting that their low molecular weight allows easy access to the active sites of the plant cell. Moreover, sulfur-containing aliphatic isothiocyanates are likely more unstable under the physiological environment than normal aliphatic isothiocyanates because the thioether bond is weaker than the C-C bond. In a similar way, aliphatic isothiocyanates possessing C-C double bonds, i.e., allyl isothiocyanate and methallyl isothiocyanate, showed lower activities than isothiocyanates with C-C single bonds, probably due to their stability. Taken together, these findings suggest that the heat-shock-response-inducing activities of isothiocyanates can be determined by their molecular sizes and stability.

Preliminary experiments showed that very high concentrations of isothiocyanates were needed to inhibit the chaperone activities of wheat germ lysate. If isothiocyanates do not greatly affect the HSF-HSP complexes in the physiological condition, isothiocyanates must induce the expression of HSPs via mechanisms different from the HSF-HSP system. In general, isothiocyanates can react with the thiol and amino groups of proteins as strong electrophiles. This may induce the denaturation of proteins. Since the denatured proteins are targets of HSPs, isothiocyanates may release HSF from the HSF-HSP complexes and activate it. Further studies are needed to elucidate the mechanism of the isothiocyanates' action in plants.

4.2.3 Monoterpenes

Although several monoterpenes showed heat-shock-response-inducing activities in plants, the magnitude of the activities was different according to their structures. We tested 13 monoterpenes (i.e., three acyclic, five monocyclic, and five bicyclic monoterpenes) as to whether they induce the heat shock response in *Arabidopsis* using the reporter gene assay (Hara et al. 2018b). Although six monoterpenes induced the heat shock response, (–)-perillaldehyde, a monocyclic monoterpene, exhibited the highest inducing activity (Fig. 11.5). One monocyclic monoterpene, i.e., (+)-limonene, and four bicyclic monoterpenes, i.e., (+)-camphor, (-)-borneol, (+)-1,8-cineol, and α-pinene, also induced the heat shock response, whereas their activities were weaker than that of (–)-perillaldehyde. (–)-Perillaldehyde inhibited the chaperone activity of wheat germ lysate, suggesting that (–)-perillaldehyde promoted the expression of HSP genes by destabilizing the HSF-HSP complexes. However, a high concentration (> 100 μM) of (–)-perillaldehyde was needed to significantly inhibit the wheat chaperone. Thus, it is unclear whether plant chaperones are inhibited by exogenously applied (–)-perillaldehyde under physiological conditions.

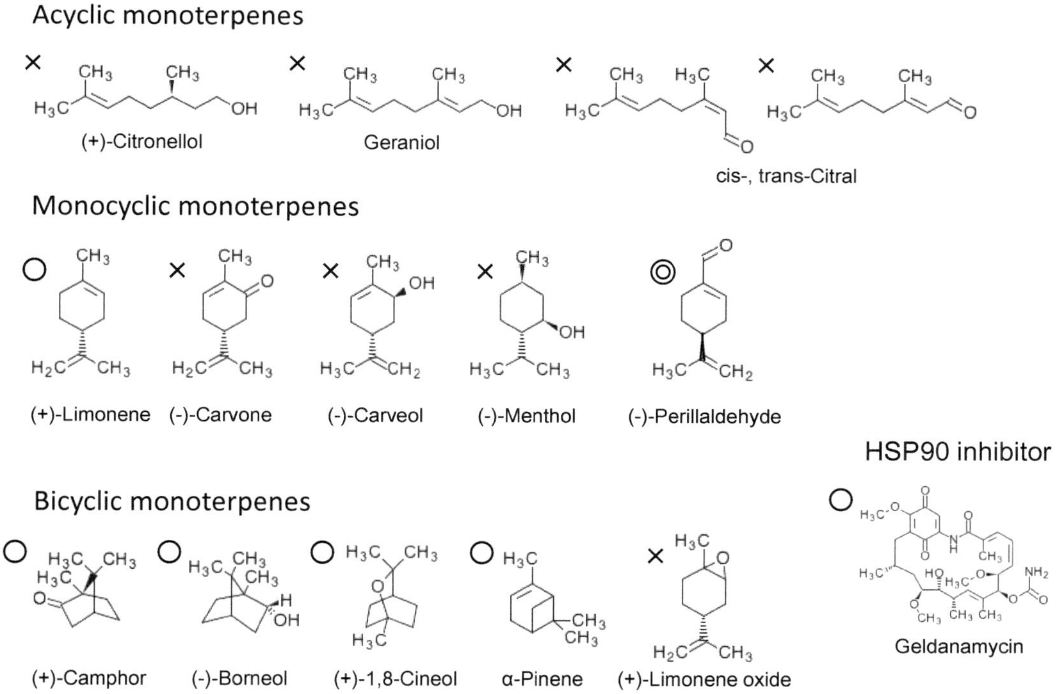

Fig. 11.5: Structures of monoterpenes. Single and double circles indicate the levels of heat-shock-response-inducing activity in *Arabidopsis*. Single circles: activities similar to that of geldanamycin. Double circles: activities remarkably higher than that of geldanamycin. X-marks signify that the compounds were inactive.

It has been reported that monoterpenes were emitted from heat-treated plants such as *Quercus ilex* (Loreto et al. 1998), *Solanum lycopersicum* (Copolovici et al. 2012), and *Musa* species (Jardine et al. 2017). Since monoterpene mixtures containing (+)-limonene and α-pinene were used to enhance the heat tolerance of *Q. ilex* (Loreto et al. 1998, Peñuelas and Llusià 2002), these monoterpenes may enhance the heat tolerance of the plant by up-regulating the level of HSP expression in the plants. (–)-Perillaldehyde, which is the potent inducer of the heat shock response in *Arabidopsis*, is a unique monoterpene that is the major component of the essential oil of *Perilla frutescens*. Since the plant is known as a traditional Chinese medicine, (–)-perillaldehyde itself and *P. frutescens* oil may be safely used in agriculture.

4.3 Categories of Action Mechanisms

In summary, there are different action mechanisms in HTLEs. The HSP90 inhibitor geldanamycin was used to induce the heat shock response in the *Arabidopsis* system. Geldanamycin significantly inhibited the chaperone activity of wheat germ lysate at 1 μM. Similarly, sanguinarine clearly inhibited the activity at 1 μM. This suggests that geldanamycin and sanguinarine are potent inhibitors of plant chaperones.

(–)-Perillaldehyde inhibited the wheat chaperones, whereas a concentration of at least 100 μM was needed to significantly inhibit the chaperone activity (Hara et al. 2018b). Our unpublished results indicated that the inhibitory concentration of isopropyl isothiocyanate for the wheat chaperone activity was at the millimolar level. It is supposed that (–)-perillaldehyde and isopropyl isothiocyanate may not efficiently inhibit the chaperones in plants, because their inhibition concentrations were high.

Based on the results of studies on geldanamycin, sanguinarine, (–)-perillaldehyde, and isopropyl isothiocyanate, it is proposed that HTLEs may be categorized into two types: chaperone-inhibition-type HTLEs (currently, geldanamycin and sanguinarine) and non-chaperone-inhibition-type HTLEs [probably, for example, (–)-perillaldehyde and isopropyl isothiocyanate] (Fig. 11.6).

The action mechanisms of HTLEs besides chaperone inhibition have not been elucidated. It is noteworthy, however, that some previously found HTLEs such as isothiocyanates and RSLVs were typical electrophiles. Since the center carbon of the isothiocyanate group (–N=C=S) is electron deficient, isothiocyanates show an electrophilic character. The α, β-unsaturated carbonyl moiety of (*E*)-2-hexenal is the site of nucleophilic attack. According to this point of view, (–)-perillaldehyde and celastrol also have α, β-unsaturated carbonyl moieties in their structures; therefore (–)-perillaldehyde and celastrol can be targets of nucleophilic attack. Generally, electrophiles non-specifically bind proteins that form adducts with thiols and primary amines (Grimsrud et al. 2008, Hinman et al. 2006) and affect the structures and functions of the proteins (Murakami 2013). Simultaneous and multiple protein modifications may occur in cells treated with electrophilic HTLEs. This can induce the dissociation of the HSF-HSP complex because HSPs may be released from the complex to bind the modified proteins. HSF is then activated to promote the heat-shock response in plants.

The role of ROS in the action mechanism of HTLEs should be considered. It was demonstrated that isothiocyanates enhanced hydrogen peroxide generation in *Arabidopsis* (Hara et al. 2010). Treatment with (–)-perillaldehyde elevated the ROS production in *Ceratocystis fimbriata,* which is a fungal pathogen of the sweet potato (Tian et al. 2018). This suggests that these HTLEs can stimulate ROS generation in plants. Intriguingly, the

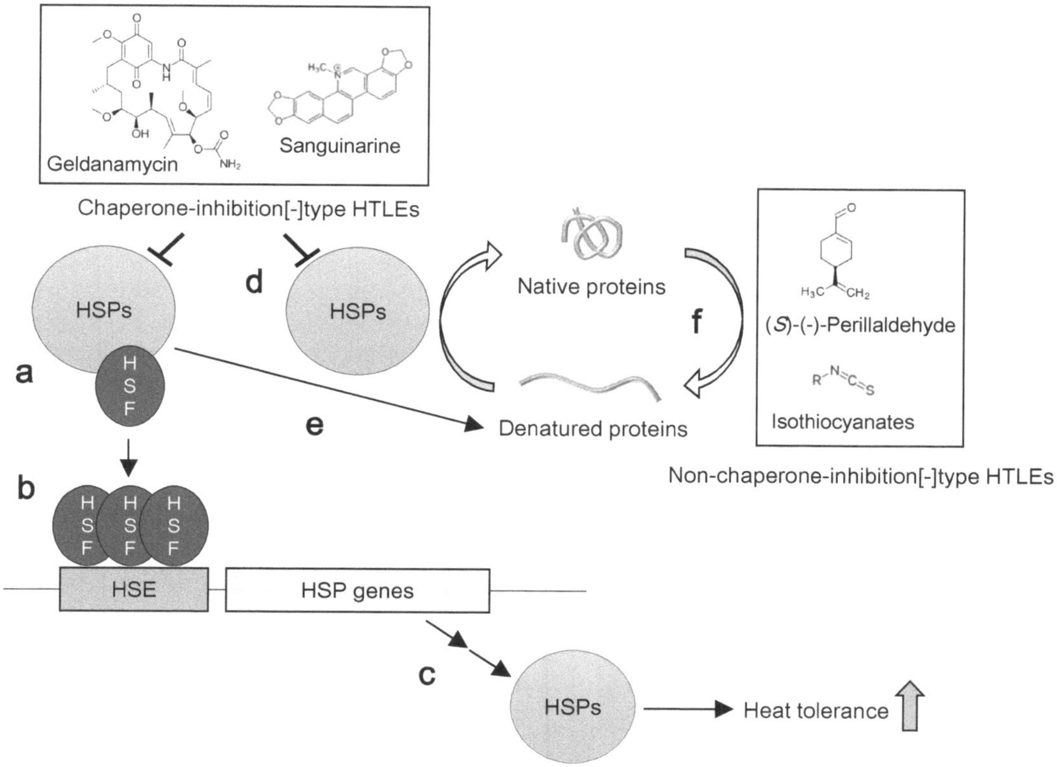

Fig. 11.6: Hypothetical action mechanisms of heat tolerance enhancers (HTLEs) in plants. Heat shock factors (HSFs) are held by HSPs under non-stressed conditions (Jacob et al. 2017) (a). Chaperone-inhibition-type HTLEs such as geldanamycin and sanguinarine may release and activate HSFs (b) by inhibiting HSPs. This enhances the expression of HSP genes (c) and subsequently heat tolerance. In addition, HSPs refold denatured proteins. Chaperone-inhibition-type HTLEs may accumulate denatured proteins by inhibiting HSPs (d). The accumulation of denatured proteins can trigger the release of active HSFs from the HSF-HSP complexes because HSPs are associated with the denatured proteins (e). Non-chaperone-inhibition-type HTLEs may accumulate denatured proteins through direct action on native proteins (f). This also activates HSFs via the same process described above (e).

HSP17.6C-CI gene of *Arabidopsis* was not only the most up-regulated in a catalase-deficient mutant (Vandenabeele et al. 2004), but it was also highly induced by excess light in the rosettes (Bechtold et al. 2008), indicating that the *HSP17.6C-CI* gene, whose promoter was used to analyze the heat shock response of HTLEs, is related to the ROS response of *Arabidopsis*. Since ROS are known to cause the oxidative modification of proteins (Beal 2002, Romero-Puertas et al. 2002), ROS can induce the denaturation of proteins as well as electrophiles. Accordingly, HTLEs like isothiocyanates may activate HSF by denaturing diverse proteins directly and via ROS generation. Further studies are required to determine whether HTLEs induce ROS generation in various plants.

4.4 Unfavorable Side Effects of HTLEs

Inappropriately high concentrations of HTLEs may affect the growth and yield of crops. HSP90 and HSP70 were related to signals of plant hormones such as auxin (Wang et al.

2016), jasmonate (Zhang et al. 2015), brassinosteroid (Samakovli et al. 2014), and abscisic acid (Clément et al. 2011). Moreover, HSP90 is required to regulate the plant immune system (Huang et al. 2014, Kadota and Shirasu 2012). This suggests that chaperone-inhibition-type HTLEs may influence the growth and development of crops. In the case of the sanguinarine agents on sale, however, we have little information on the damage to the growth and yield of crops under typical use of the agent.

The use of non-chaperone-inhibition-type HTLEs at high concentrations also requires attention. As mentioned above, some non-chaperone-inhibition-type HTLEs may denature proteins. The great magnitude of protein denaturation promoted by the application of non-chaperone-inhibition-type HTLEs may result in a decrease in the yield of crops. Even chaperone-inhibition-type HTLEs may accumulate denatured proteins, because the inactivated chaperone cannot restore the denatured proteins. Overproduction of ROS also causes severe oxidative stress in plants. Indeed, high doses of isothiocyanates showed herbicidal effects by inducing an oxidative burst-like response in *Arabidopsis* (Hara et al. 2010). Any type of HTLEs should be used while securing the appropriate concentration. On the other hand, there is still not enough information on the timing of the application, the optimal dosage, the application methods, the parts or organs which should be applied, which HTLEs are effective in individual crops, etc. Therefore, practical field research is necessary to maximize the availability of HTLEs.

5. Requirements for Practical Use

The research and development of HTLEs has just begun. Considering the practical application of HTLEs, the following items are important. First, it is preferable that practical HTLEs are water soluble. Although water-insoluble HTLEs can be applied as emulsions, liquid delivery systems and spraying apparatuses in the field are sometimes clogged by the emulsified solution. Evaporation systems are applicable for the volatile HTLEs, but it is not easy to control the supplemental dose. Second, considerable stability in air, water, under illumination, and under extreme temperatures is needed for HTLEs, because the storage and application conditions are harsh in the farm environment. Third, the candidates for HTLEs should be safe. In particular, ingredients which have been used in diets for a long time or compounds which have been proven to be safe in animal tests are desirable. HTLEs need to show low ecotoxicity. Safety assessments must be conducted not only for the pure active ingredients but also for the final products, which may contain inorganic fertilizers and biomaterials. Even if the candidates for HTLEs fulfill the above requirements, costly HTLEs are difficult to commercialize.

6. Advantages and Disadvantages of HTLEs

6.1 Advantages

Whether HTLEs are chaperone-inhibition types or non-chaperone-inhibition types, their core action mechanisms are the same regardless of the plant species. This means that HTLEs can enhance the heat tolerance of diverse plants, such as angiosperms, gymnosperms, ferns and mosses. They have great potential for wide application. Although breeding is a powerful technique for producing thermotolerant crops, it can take a long time, especially in the case of crops whose breeding methods have not been established. HTLEs may help increase the heat tolerance in crops that are difficult to breed.

Even when thermotolerant varieties have been bred, quite a few agricultural producers prefer not to immediately use new varieties because they are accustomed to cultivating the previously established varieties, especially when the former varieties have high marketability. HTLEs will provide a chance for agricultural producers to choose crops that they want to produce. Moreover, HTLEs can be useful in protected horticulture. The use of HTLEs can decrease the air conditioning expense, which is a major running cost in protected horticulture, by increasing the heat tolerance of crops.

6.2 Disadvantages

The timing of the application of HTLEs is very important, because crops should be treated with HTLEs before the arrival of a heat wave. However, sufficient field tests on the optimal timing have not been performed. Although HTLEs enhance the heat tolerance of plants, the tolerance cannot surpass a plant's heat tolerance capability, which has been genetically acquired in the plants. Even if the HTLEs show low toxicity to organisms in acute and chronic tests, their long-term influence on the ecosystem cannot be predicted. Continuous ecological assessments of the use of HTLEs are needed.

7. Conclusion

HTLEs are promising agents for reducing the plant damage that occurs due to heat stress. Examples of HTLEs are HSP90 inhibitors (e.g., geldanamycin), sanguinarine, isothiocyanates, monoterpenes, RSLVs, etc. HSP90 inhibitors and sanguinarine induced the heat shock response in plants by inhibiting their chaperones (chaperone-inhibition-type HTLEs). In addition, some HTLEs potently induced the heat shock response by weakly or scarcely inhibiting the chaperones (non-chaperone-inhibition-type HTLEs). Since HTLEs stimulate common physiological responses in plants, the plant species to which they are applicable are not logically restricted. In practical agriculture, however, the methods and timing of the application of HTLEs to the target crops should be well investigated. HTLEs will not only be administered through a single application but also through collaboration applications using other methods such as thermotolerance breeding or protected horticulture.

Acknowledgments

The major part of this work was conducted at the Research Institute of Green Science and Technology, Shizuoka University. Some of the studies were supported by Shizuoka City, the Shizuoka Chamber of Commerce & Industry, and Menicon Co., Ltd. The commercialization of the sanguinarine agent was achieved by Fujimikogyo Co., Ltd. by utilizing our study.

References

Ainsworth, E.A. and D.R. Ort. 2010. How do we improve crop production in a warming world? Plant Physiol. 154: 526–530.

Allakhverdiev, S.I., V.D. Kreslavski, V.V. Klimov, D.A. Los, R. Carpentier and P. Mohanty. 2008. Heat stress: an overview of molecular responses in photosynthesis. Photosynth. Res. 98: 541–550.

Angelova, S., M. Buchheim, D. Frowitter, A. Schierhorn and W. Roos. 2010. Overproduction of alkaloid phytoalexins in California poppy cells is associated with the co-expression of biosynthetic and stress-protective enzymes. Mol. Plant 3: 927–939.

Beal, M.F. 2002. Oxidatively modified proteins in aging and disease. Free Radic. Biol. Med. 32: 797–803.
Bechtold, U., O. Richard, A. Zamboni, C. Gapper, M. Geisler, B. Pogson et al. 2008. Impact of chloroplastic- and extracellular-sourced ROS on high light-responsive gene expression in *Arabidopsis*. J. Exp. Bot. 59: 121–133.
Clarke, S.M., L.A. Mur, J.E. Wood and I.M. Scott. 2004. Salicylic acid dependent signaling promotes basal thermotolerance but is not essential for acquired thermotolerance in *Arabidopsis thaliana*. Plant J. 38: 432–447.
Clément, M., N. Leonhardt, M.J. Droillard, I. Reiter, J.L. Montillet, B. Genty et al. 2011. The cytosolic/nuclear HSC70 and HSP90 molecular chaperones are important for stomatal closure and modulate abscisic acid-dependent physiological responses in *Arabidopsis*. Plant Physiol. 156: 1481–1492.
Copolovici, L., A. Kannaste, L. Pazouki and U. Niinemets. 2012. Emissions of green leaf volatiles and terpenoids from *Solanum lycopersicum* are quantitatively related to the severity of cold and heat shock treatments. J. Plant Physiol. 169: 664–672.
Delfine, S., O. Csiky, G. Seufert and F. Loreto. 2000. Fumigation with exogenous monoterpenes of a non-isoprenoid-emitting oak (*Quercus suber*): monoterpene acquisition, translocation, and effect on the photosynthetic properties at high temperatures. New Phytol. 146: 27–36.
Desgagné-Penix, I., M.F. Khan, D.C. Schriemer, D. Cram, J. Nowak and P.J. Facchini. 2010. Integration of deep transcriptome and proteome analyses reveals the components of alkaloid metabolism in opium poppy cell cultures. BMC Plant Biol. 10: 252.
Dudareva, N., E. Pichersky and J. Gershenzon. 2004. Biochemistry of plant volatiles. Plant Physiol. 135: 1893–1902.
Facchini, P.J. and B. St-Pierre. 2005. Synthesis and trafficking of alkaloid biosynthetic enzymes. Curr. Opin. Plant Biol. 8: 657–666.
Grimsrud, P.A., H. Xie, T.J. Griffin and D.A. Bernlohr. 2008. Oxidative stress and covalent modification of protein with bioactive aldehydes. J. Biol. Chem. 283: 21837–21841.
Halkier, B.A. and J. Gershenzon. 2006. Biology and biochemistry of glucosinolates. Annu. Rev. Plant Biol. 57: 303–333.
Hall, A.E. 2001. Crop Responses to Environment. CRC Press LLC, Boca Raton, Florida, USA.
Hara, M., Y. Yatsuzuka, K. Tabata and T. Kuboi. 2010. Exogenously applied isothiocyanates enhance glutathione *S*-transferase expression in *Arabidopsis* but act as herbicides at higher concentrations. J. Plant Physiol. 167: 643–649.
Hara, M., A. Harazaki and K. Tabata. 2013. Administration of isothiocyanates enhances heat tolerance in *Arabidopsis thaliana*. Plant Growth Regul. 69: 71–77.
Hara, M. and I. Kurita. 2014. The natural alkaloid sanguinarine promotes the expression of heat shock protein genes in *Arabidopsis*. Acta Physiol. Plant 36: 3337–3343.
Hara, M., T. Matsubara, I. Takahashi and H. Murano. 2018a. Isobutyl isothiocyanate is a potent heat tolerance enhancer for *Arabidopsis*. Environ. Cont. Biol. 56: 121–126.
Hara, M., N. Yamauchi and Y. Sumita. 2018b. Monoterpenes induce the heat shock response in *Arabidopsis*. Z. Naturforsch. C 73: 177–184.
Hinman, A., H.H. Chuang, D.M. Bautista and D. Julius. 2006. TRP channel activation by reversible covalent modification. Proc. Natl. Acad. Sci. USA 103: 19564–19568.
Hopkins, R.J., N.M. van Dam and J.J. van Loon. 2009. Role of glucosinolates in insect-plant relationships and multitrophic interactions. Annu. Rev. Entomol. 54: 57–83.
Huang, S., J. Monaghan, X. Zhong, L. Lin, T. Sun, O.X. Dong et al. 2014. HSP 90s are required for NLR immune receptor accumulation in *Arabidopsis*. Plant J. 79: 427–439.
Huang, B. and C. Xu. 2008. Identification and characterization of proteins associated with plant tolerance to heat stress. J. Integr. Plant Biol. 50: 1230–1237.
Iba, K. 2002. Acclimative response to temperature stress in higher plants: approaches of gene engineering for temperature tolerance. Annu. Rev. Plant Biol. 53: 225–245.
Jacob, P., H. Hirt and A. Bendahmane. 2017. The heat-shock protein/chaperone network and multiple stress resistance. Plant Biotechnol. J. 15: 405–414.
Jardine, K.J., A.B. Jardine, J.A. Holm, D.L. Lombardozzi, R.I. Negron-Juarez, S.T. Martin et al. 2017. Monoterpene 'thermometer' of tropical forest-atmosphere response to climate warming. Plant Cell Environ. 40: 441–452.
Kadota, Y. and K. Shirasu. 2012. The HSP90 complex of plants. Biochim. Biophys. Acta (BBA)-Molecular Cell Research 1823: 689–697.

Kantas, D., V.G. Papatsiros, P.D. Tassis, L.V. Athanasiou and E.D. Tzika. 2015. Effect of a natural feed additive (*Macleaya cordata*), containing sanguinarine, on the performance and health status of weaning pigs. Anim. Sci. J. 86: 92–98.

Kissen, R., J.T. Rossiter and A.M. Bones. 2009. The 'mustard oil bomb': not so easy to assemble?! Localization, expression and distribution of the components of the myrosinase enzyme system. Phytochem. Rev. 8: 69–86.

Kosina, P., D. Walterová, J. Ulrichová, V. Lichnovský, M. Stiborová, H. Rýdlová et al. 2004. Sanguinarine and chelerythrine: assessment of safety on pigs in ninety days feeding experiment. Food Chem. Toxicol. 42: 85–91.

Kotak, S., J. Larkindale, U. Lee, P. von Koskull-Döring, E. Vierling and K.D. Scharf. 2007. Complexity of the heat stress response in plants. Curr. Opin. Plant Biol. 10: 310–316.

Larcher, W. 2003. Physiological Plant Ecology 4th edn. Springer, Berlin, Heidelberg.

Lee, K.W., J.S. Kim, S.T. Oh, C.W. Kang and B.K. An. 2015. Effects of dietary sanguinarine on growth performance, relative organ weight, cecal microflora, serum cholesterol level and meat quality in broiler chickens. J. Poult. Sci. 52: 15–22.

Leng, L., Q. Liang, J. Jiang, C. Zhang, Y. Hao, X. Wang et al. 2017. A subclass of HSP70s regulate development and abiotic stress responses in *Arabidopsis thaliana*. J. Plant Res. 130: 349–363.

Loreto, F., A. Förster, M. Dürr, O. Csiky and G. Seufert. 1998. On the monoterpene emission under heat stress and on the increased thermotolerance of leaves of *Quercus ilex* L. fumigated with selected monoterpenes. Plant Cell Environ. 21: 101–107.

Macías, F.A., J.M. Molinillo, R.M. Varela and J.C. Galindo. 2007. Allelopathy—a natural alternative for weed control. Pest Manag. Sci. 63: 327–348.

Mackraj, I., T. Govender and P. Gathiram. 2008. Sanguinarine. Cardiovasc. Ther. 26: 75–83.

Masand, S. and S.K. Yadav. 2016. Overexpression of MuHSP70 gene from *Macrotyloma uniflorum* confers multiple abiotic stress tolerance in transgenic *Arabidopsis thaliana*. Mol. Biol. Rep. 43: 53–64.

Matsuoka, E., T. Matsubara, I. Takahashi, H. Murano and M. Hara. 2016. The isoquinoline alkaloid sanguinarine which inhibits chaperone activity enhances the production of heat shock proteins in *Arabidopsis*. Plant Biotechnol. 33: 409–413.

Matsushima, Y., H. Minami, K. Hori and F. Sato. 2012. Pathway engineering of benzylisoquinoline alkaloid biosynthesis in transgenic California poppy cells with ectopic expression of tetrahydroberberine oxidase from *C. japonica*. Plant Biotechnol. 29: 473–481.

McLellan, C.A., T.J. Turbyville, E.M. Wijeratne, A. Kerschen, E. Vierling, C. Queitsch et al. 2007. A rhizosphere fungus enhances *Arabidopsis* thermotolerance through production of an HSP90 inhibitor. Plant Physiol. 145: 174–182.

Mithöfer, A. and W. Boland. 2012. Plant defense against herbivores: chemical aspects. Annu. Rev. Plant Biol. 63: 431–450.

Murakami, A. 2013. Modulation of protein quality control systems by food phytochemicals. J. Clin. Biochem. Nutr. 52: 215–227.

Murano, H., T. Matsubara, I. Takahashi and M. Hara. 2017. A purine-type heat shock protein 90 inhibitor promotes the heat shock response in *Arabidopsis*. Plant Biotechnol. Rep. 11: 107–113.

Peñuelas, J. and J. Llusià. 2002. Linking photorespiration, monoterpenes and thermotolerance in *Quercus*. New Phytol. 155: 227–237.

Reddy, R.K., I. Kurek, A.M. Silverstein, M. Chinkers, A. Breiman and P. Krishna. 1998. High-molecular-weight FK506-binding proteins are components of heat-shock protein 90 heterocomplexes in wheat germ lysate. Plant Physiol. 118: 1395–1401.

Romero-Puertas, M.C., J.M. Palma, M. Gómez, L.A. Del Rio and L.M. Sandalio. 2002. Cadmium causes the oxidative modification of proteins in pea plants. Plant Cell Environ. 25: 677–686.

Ruelland, E. and A. Zachowski. 2010. How plants sense temperature. Environ. Exp. Bot. 69: 225–232.

Saidi, Y., A. Finka, M. Chakhporanian, J.P. Zrÿd, D.G. Schaefer and P. Goloubinoff. 2005. Controlled expression of recombinant proteins in *Physcomitrella patens* by a conditional heat-shock promoter: a tool for plant research and biotechnology. Plant Mol. Biol. 59: 697–711.

Saidi, Y., M. Domini, F. Choy, J.P. Zryd, J.P. Schwitzguebel and P. Goloubinoff. 2007. Activation of the heat shock response in plants by chlorophenols: transgenic *Physcomitrella patens* as a sensitive biosensor for organic pollutants. Plant Cell Environ. 30: 753–763.

Samakovli, D., T. Margaritopoulou, C. Prassinos, D. Milioni and P. Hatzopoulos. 2014. Brassinosteroid nuclear signaling recruits HSP90 activity. New Phytol. 203: 743–757.

Su, P.H. and H.M. Li. 2008. *Arabidopsis* stromal 70-kD heat shock proteins are essential for plant development and important for thermotolerance of germinating seeds. Plant Physiol. 146: 1231–1241.

Sun, Y., M.M. Joachimski, P.B. Wignall, C. Yan, Y. Chen, H. Jiang et al. 2012. Lethally hot temperatures during the Early Triassic greenhouse. Science 338: 366–370.

Sung, D.Y. and C.L. Guy. 2003. Physiological and molecular assessment of altered expression of Hsc70-1 in *Arabidopsis*. Evidence for pleiotropic consequences. Plant Physiol. 132: 979–987.

Teskey, R., T. Wertin, I. Bauweraerts, M. Ameye, M.A. McGuire and K. Steppe. 2015. Responses of tree species to heat waves and extreme heat events. Plant Cell Environ. 38: 1699–1712.

Tian, J., C. Pan, M. Zhang, Y.Y. Gan, S.Y. Pan, M. Liu et al. 2018. Induced cell death in *Ceratocystis fimbriata* by pro-apoptotic activity of a natural organic compound, perillaldehyde, through Ca^{2+} overload and accumulation of reactive oxygen species. Plant Pathol. Published online.

Vandenabeele, S., S. Vanderauwera, M. Vuylsteke, S. Rombauts, C. Langebartels, H.K. Seidlitz et al. 2004. Catalase deficiency drastically affects gene expression induced by high light in *Arabidopsis thaliana*. Plant J. 39: 45–58.

Vieira, S.L., O.A. Oyarzabal, D.M. Freitas, J. Berres, J.E.M. Pena, C.A. Torres et al. 2008. Performance of broilers fed diets supplemented with sanguinarine-like alkaloids and organic acids. J. Appl. Poult. Res. 17: 128–133.

Wahid, A., S. Gelani, M. Ashraf and M.R. Foolad. 2007. Heat tolerance in plants: An overview. Environ. Exp. Bot. 61: 199–223.

Wang, W., B. Vinocur, O. Shoseyov and A. Altman. 2004. Role of plant heat-shock proteins and molecular chaperones in the abiotic stress response. Trends Plant Sci. 9: 244–252.

Wang, R., Y. Zhang, M. Kieffer, H. Yu, S. Kepinski and M. Estelle. 2016. HSP90 regulates temperature-dependent seedling growth in *Arabidopsis* by stabilizing the auxin co-receptor F-box protein TIR1. Nat. Commun. 7: 10269.

Waters, E.R. 2013. The evolution, function, structure, and expression of the plant sHSPs. J. Exp. Bot. 64: 391–403.

Wink, M. and T. Twardowski. 1992. Allelochemical properties of alkaloids. pp. 129–150. *In*: S.J.H. Rizvi and V. Rizvi (eds.). Effects on Plants, Bacteria and Protein Biosynthesis. Allelopathy. Springer, Dordrecht, Netherlands.

Yamada, K., Y. Fukao, M. Hayashi, M. Fukazawa, I. Suzuki and M. Nishimura. 2007. Cytosolic HSP90 regulates the heat shock response that is responsible for heat acclimation in *Arabidopsis thaliana*. J. Biol. Chem. 282: 37794–37804.

Yamada, Y., T. Shimada, Y. Motomura and F. Sato. 2017. Modulation of benzylisoquinoline alkaloid biosynthesis by heterologous expression of CjWRKY1 in *Eschscholzia californica* cells. PLoS One 12: e0186953.

Yamauchi, Y., M. Kunishima, M. Mizutani and Y. Sugimoto. 2015. Reactive short-chain leaf volatiles act as powerful inducers of abiotic stress-related gene expression. Sci. Rep. 26: 8030.

Yoshida, T., N. Ohama, J. Nakajima, S. Kidokoro, J. Mizoi, K. Nakashima et al. 2011. *Arabidopsis* HsfA1 transcription factors function as the main positive regulators in heat shock-responsive gene expression. Mol. Gen. Genomics 286: 321–332.

Zhang, X.C., Y.A. Millet, Z. Cheng, J. Bush and F.M. Ausubel. 2015. Jasmonate signalling in *Arabidopsis* involves SGT1b–HSP70–HSP90 chaperone complexes. Nat. Plants 1: 15049.

Zhang, L., Q. Zhang, Y. Gao, H. Pan, S. Shi and Y. Wang. 2014. Overexpression of heat shock protein gene PfHSP21.4 in *Arabidopsis thaliana* enhances heat tolerance. Acta. Physiol. Plant. 36: 1555–1564.

Zhao, L., S. von Alvensleben, G. Fusconi and M. Morlacchini. 2017. Safety evaluation of a standardized *Macleaya cordata* extract in a ninety day feeding study in weaned piglets. Open J. Anim. Sci. 7: 213–231.

Chapter 12

Small Molecules that Regulate Plant Hormone Signaling

Yasushi Todoroki

1. Introduction

Some phytohormones trigger the signal transduction by modulating Protein-Protein Interactions (PPIs) (Kumari et al. 2011, Lumba et al. 2010) (Fig. 12.1). This type of plant hormone is classified into two types, molecular glues and allosteric modulators, according to the difference in the role in induction of PPI. Molecular glues interact with two target proteins without inducing major conformational changes of both proteins to mediate complex formation. Brassinosteroids, indole-3-acetic acid and jasmonate-isoleucine function as molecular glues. On the other hand, gibberellin (GA) and abscisic acid (ABA) act as allosteric modulators, that is, introduce conformational changes in their receptors to stabilize the complex with the partner proteins. Although PPI is being actively researched in the field of drug discovery, it is still challenging to develop small molecules that control PPI. Because proteins interact with each other through the relatively flat and large surfaces, it is difficult for small molecules to mimic or disrupt these interactions (Wells et al. 2007). On the other hand, plant hormones, such as ABA and GA, bind within a cavity to induce conformational changes. Thus in the case of these hormones the challenge is different. This feature may make easier manipulation of these types of PPIs.

ABA is a phytohormone that protects plants from adverse environments and temporarily suppresses their growth until they are suitable for growth (Zhang 2014). Chemical tools for modulating ABA signaling, alone or in combination with genetic analysis, may be useful to explore the function of ABA and its signaling molecules in the physiological processes of plants. In addition, a modulator of ABA signaling would be able to control seed dormancy, fruit ripening and water use. Because this is crucial for efficacious and reliable agricultural production, it would be exploited in agriculture.

Research Institute of Green and Technology, Shizuoka University, 836 Ohya, Shizuoka 422-8529, Japan.
Email: todoroki.yasushi@shizuoka.ac.jp

ABA signaling is elicited by ABA-dependent activation of the ABA receptors family (PYR/PYL/RCAR, PYL), which was first identified in *Arabidopsis* thaliana (Ma et al. 2009, Park et al. 2009) (Fig. 12.2). Binding with ABA induces a conformational change (gate closure) in the flexible loop of PYL, which allows it to bind to negative regulators of ABA signaling, group-A protein phosphatases 2C (PP2C) such as HAB1, ABI1 and ABI2 (Fujii et al. 2009, Miyazono et al. 2009, Soon et al. 2012, Vlad et al. 2009). Fortunately, many crystal structures of PYL–ABA and PYL–ABA–PP2C complexes were revealed soon after PYL identification (Santiago et al. 2012, Sun et al. 2012, Zhang et al. 2013). This structural information enabled us to logically design PYL antagonists. ABA closes the gate loop of PYL by interacting hydrophobically with the conserved gate loop and α helix, in the ABA-recognition site. Although ABA is fully embedded into the pocket, two small tunnels (referred to as the 3′- and 4′-tunnels) are opened at the PP2C binding interface (Fig. 12.3a–c). The 3′-tunnel has no function, whereas the 4′-tunnel functions critically to form and stabilize PYL–PP2C complex; it accommodates an indole moiety of a conserved Trp residue of PP2C which forms a π-π interaction with the side chain benzene moiety in the

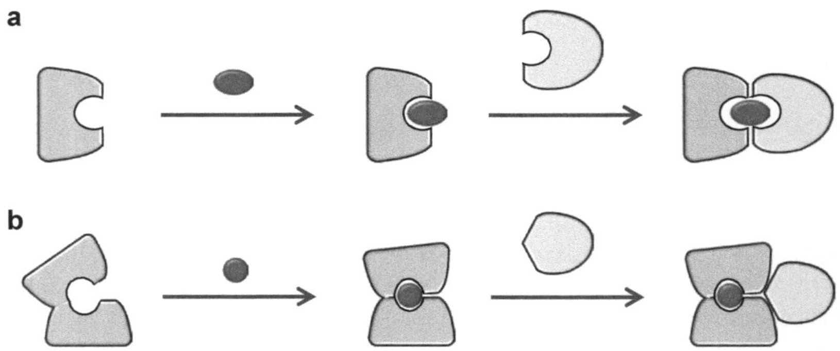

Fig. 12.1: Plant hormones as modulators of protein-protein interactions: (a) molecular glues (auxin, jasmonic acid, and brassinosteroid); (b) allosteric modulators (ABA, GA, and strigolactones).

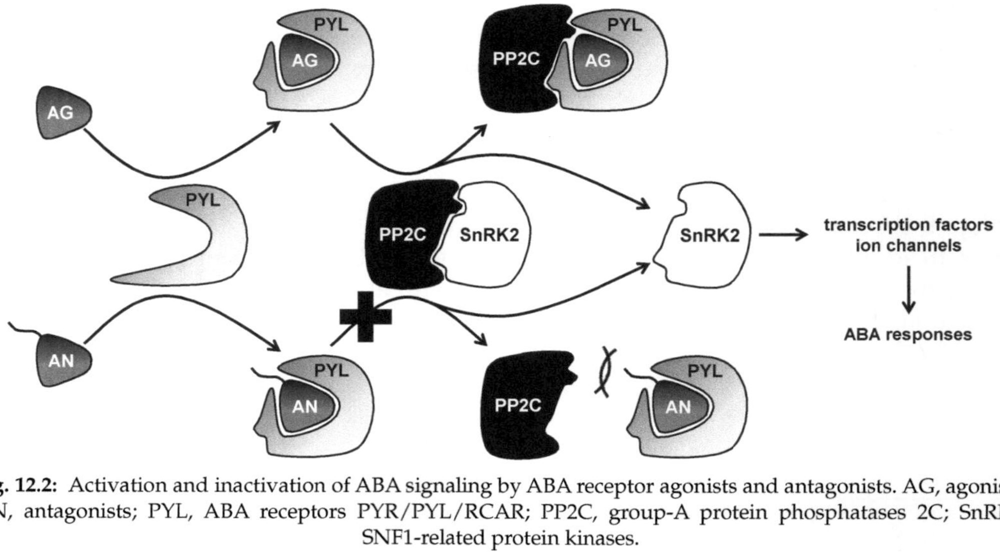

Fig. 12.2: Activation and inactivation of ABA signaling by ABA receptor agonists and antagonists. AG, agonists; AN, antagonists; PYL, ABA receptors PYR/PYL/RCAR; PP2C, group-A protein phosphatases 2C; SnRK2, SNF1-related protein kinases.

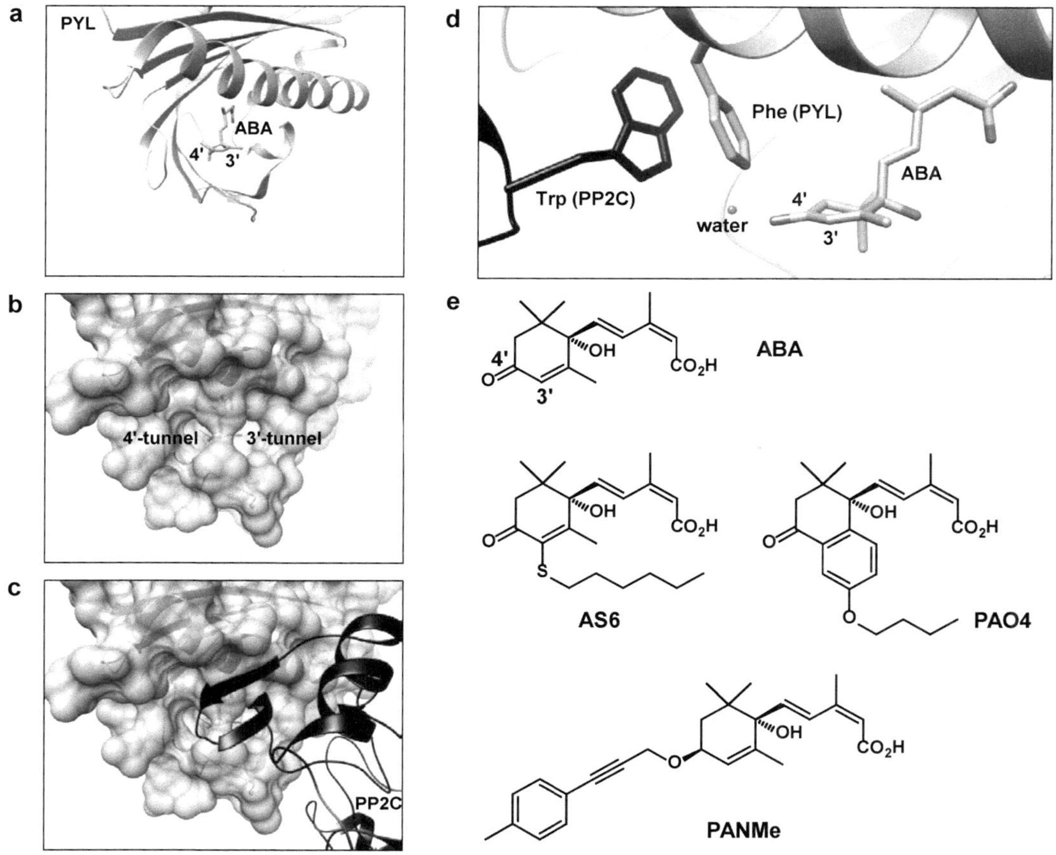

Fig. 12.3: Crystal structures of PYL–ABA and PYL–ABA–PP2C complexes: (a) the ribbon drawing of the PYL1–ABA complex (3JRS); (b) the solvent excluded surface of the complex; (c) the PYL1–ABA–ABI1 complex (3JRQ); and (d) the detailed view of the 4′-tunnel, and chemical structures of ABA and synthetic PYL antagonists (e).

Phe of PYL and a hydrogen bond via a water molecule with the ABA C4′ carbonyl oxygen (Jiang et al. 2014, Weiner et al. 2010) (Fig. 12.3d). The binding interface is close to the active center of PP2C; the side chain hydroxy group of Ser on the PYL gate loop is plugged into the region where aspartic acid residues are located to chelate magnesium ions in PP2C, physically constraining PP2C and releasing SNF1-related protein kinases (SnRK2) from PP2C. The released SnRK2, which is activated by self or mutual phosphorylation, activates ion channels and transcription factors to induce ABA responses (Weiner et al. 2010).

We created an ABA analog, AS6, based on this structural feature of the ABA-induced PYL–PP2C interaction, (Takeuchi et al. 2014), PAO4 (Takeuchi et al. 2015), and PANMe (Takeuchi et al. 2018), as antagonists of PYL (Fig. 12.3e). This chapter focuses mainly on the design, synthesis and characterization of AS6 and PANMe.

2. Molecular Design and Synthesis

The 3′-tunnel is constructed simply by conserved hydrophobic amino acid residues. Because some 3′-modified ABA analogs acted as ABA agonists (Todoroki and Ueno 2010,

Todoroki and Hirai 2002), the C3′ alkyl substituents would be accepted by the 3′-tunnel to form hydrophobic interactions with the conserved residues. That is, 3′-modified ABA analogs might have been expected to act as a potential stabilizer or an instability factor of the gate-closed conformer. Even if a 3′-modified ABA analog stabilize the gate-closed form, it may disturb PYL–PP2C binding if it has a 3′-substituent long enough to protrude out of the tunnel. Based on this consideration, we designed 3′-alkylsulfanyl ABAs (AS*n* compound series), where *n* is the number of carbons in the alkyl chain. We constructed a molecular model of a PYR1–AS6 complex with reference to crystal structures of PYR1–ABA complexes (Fig. 12.4). On the basis of this model, a hexyl chain was long enough to protrude out of the tunnel and overlap with the highly conserved hydrophobic residue of PP2C (Val393, HAB1 numbering). This suggested that AS6 binds to the ligand-binding site of PYR1 and prevents binding with PP2C. To investigate whether these predictions are valid, we synthesized 11 AS*n* compounds (n = 2–12).

The 4′-tunnel is inserted by the Trp of PP2C to stabilize a PYL–ABA–PP2C ternary complex (Fig. 12.3d). Thus, we first designed ABA analogs with a simple alkyl chain at the ABA C4′, 4′-*O*-butyl/pentyl-ABAs. The C4′ chains were expected to disturb the binding of the Trp-indole of PP2C with the 4′-tunnel to inhibit forming the PYL-PP2C complex. These compounds exhibited slight agonist and antagonist effects on *Arabidopsis* seed germination. Because the length of the *O*-butyl and pentyl chains (4.9 and 6.2 Å) seemed long enough to inhibit the binding of the Trp-indole, the weak biological effects suggested that the affinity for PYL was reduced by the C4′ linear alkyl chains, probably because of the zigzag conformations to induce the steric hindrance. Based on this result, five ABA analogs with a triple bond and a phenyl moiety in the 4′-*O*-substituent were designed as PYL antagonists (PAN) (Fig. 12.4). The triple bond was inserted to give rigid linearity without bending associated with free rotation around single bonds, whereas the phenyl moiety was introduced to imitate the PP2C Trp-indole.

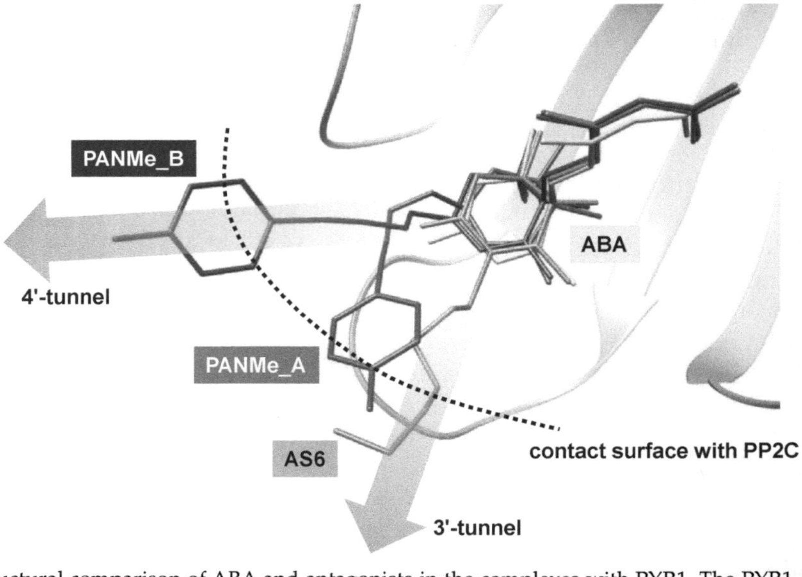

Fig. 12.4: Structural comparison of ABA and antagonists in the complexes with PYR1. The PYR1–AS6 complex (3WG8) and the PYR1–PANMe complexes A and B (5YGV) were superimposed on the PYR1–ABA complex (3K3K). The dotted line represents the contact surface with PP2C. The 3′- or 4′-side chains protrude sufficiently to interfere with binding to PP2C.

3. Physiological Effects of AS*n* and PAN on *Arabidopsis*

We examined the *in vivo* effects of AS*n* and PAN in the *Arabidopsis* seed germination assay. ABA agonists delay seed germination, whereas ABA antagonists alleviate the delay of seed germination by coapplied ABA. For AS*n*, AS2 and AS3 acted as ABA agonists. AS4 acted as a weaker ABA agonist than AS2 and AS3, and AS*n* with $n > 4$ did not act as ABA agonists. When coapplied with ABA, AS*n* with $n > 4$ acted as ABA antagonists, suggesting that these compounds were ABA receptor antagonists. However, these ABA antagonists were not able to fully bring back the delayed germination time course of ABA treated seeds to that of control even at 100 times the concentration of ABA. On the other hand, all PAN acted only as ABA antagonists and completely recovered the germination time course at concentrations of more than three-fold that of ABA. PANMe was more effective than AS6 as an ABA antagonist on *Arabidopsis* seedling growth (Fig. 12.5).

AS6 and PANMe functioned as strong ABA antagonists also at the gene expression level. AS6 and PANMe never promoted the ABA-responsive genes expression when they were not co-coapplied with ABA. On the other hand, when co-applied with ABA, PANMe attenuated ABA-responsive gene expression by ABA treatment more strongly than AS6. Not only when ABA treatment is performed, had AS6 and PANMe strongly suppressed the ABA-responsive genes expression, but also when treated by mannitol, which promoted ABA biosynthesis by mimicking desiccation caused by drought. This indicates that AS6 and PANMe acted as antagonists not only to administered ABA but also to ABA biosynthesized

Fig. 12.5: Effects of AS6 and PANMe on *Arabidopsis* seedling growth. Ten-day-old plants were sprayed with 1 mL of ABA and antagonists daily for 3 weeks. Reprinted with permission from (Takeuchi et al. 2018). Copyright (2018) American Chemical Society.

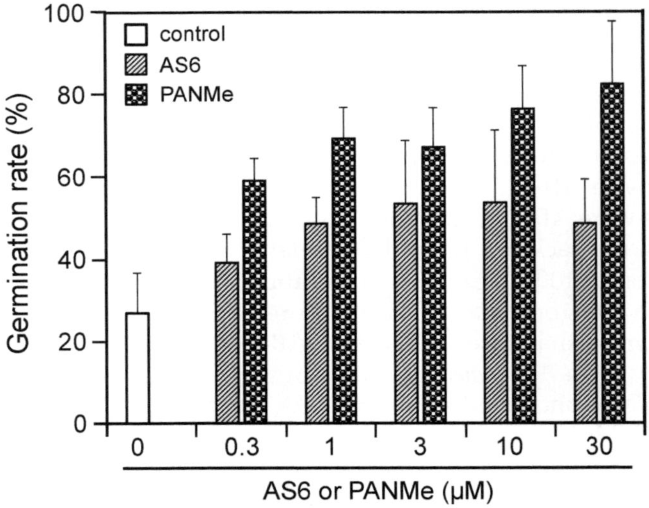

Fig. 12.6: Effect of AS6 and PANMe on thermoinhibition of *Arabidopsis* seed germination. *Arabidopsis* seeds were treated with AS6 (gray) or PANMe (black) at 33°C, and the germination rate was determined at 72 h after stratification ($n = 3$, error bars represent s.d.). Reprinted with permission from (Takeuchi et al. 2018). Copyright (2018) American Chemical Society.

in plants. Nevertheless, because the endogenous ABA level or ABA sensitivity is low under unstressed conditions, AS6 and PANMe does not significantly affect plant phenotype when ABA was not administered. At high temperature (33°C), which causes thermoinhibition of *Arabidopsis* seed germination by promoting ABA biosynthesis and raising endogenous ABA content (Toh et al. 2008), AS6 and PANMe reduced the thermoinhibition of *Arabidopsis* seeds germination dose-dependently (Fig. 12.6).

4. Biochemical Characterization of AS*n* and PANs

Because the 3′- and 4′-tunnels are constructed by highly conserved residues, AS*n* and PANs should act as a broad-spectrum antagonist that inhibits all the PYL subtypes. We examined the effects of these compounds on *Arabidopsis* PYL using the PP2C assays (Fig. 12.7), in which effective PYL agonists inhibit near completely PP2C activity at saturating concentrations, whereas partial PYL agonists fail to inhibit completely it. All the tested AS*n*, AS2, AS4 and AS6, acted as the agonist of some receptors (PYR1 and PYL1–3) although varying in degree. The agonistic ability of AS2 was comparable to that of ABA, whereas AS4 and AS6 worked as the partial agonist. On the other hand, all the tested PANs at 50 μM did not significantly activate all the tested PYL.

The *in vitro* PYL antagonist activities were evaluated by examining their ability to relieve ABA-dependent inhibition of PP2C activity. AS6 and PAN acted as the antagonist for all of the receptors tested, indicating that these compounds were broad-spectrum PYL antagonists. PANMe retrieved PP2C activity more greatly than did AS6 and the other PAN. For PYR1 and PYL5, 100 μM AS6 retrieved PP2C activity to about 60% of its original activity, whereas PANMe at the same concentration almost completely recovered. This indicates that PANMe is a full antagonist of PYL, whereas AS6 is a partial antagonist of PYL.

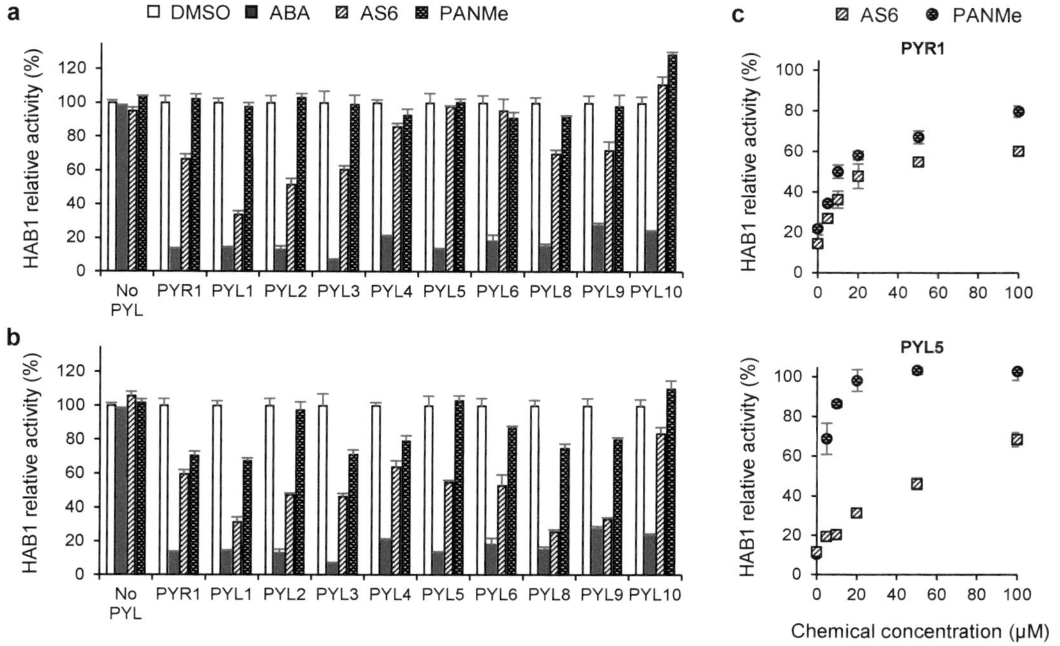

Fig. 12.7: Effects of AS6 and PANMe on HAB1 inhibition by ABA receptors: (a) chemical inhibition of HAB1 by various ABA receptors in the presence of 50 μM of each test compound; (b) Antagonistic effect of each test compound against various ABA receptors; and (c) effect of AS6 and PANMe on PYR1 and PYL5. Assays were performed in the presence of 5 μM ABA and (b) 50 μM of each compound or (c) various concentrations (0, 5, 10, 20, 50 and 100 μM) of AS6 and PANMe. The concentrations of each PYL were set with molar rations to HAB1 as 1:1 (PYR1, PYL1–6 and PYL10) or 2:1 (PYL8 and PYL9). HAB1 phosphatase activity of each reaction was expressed as a relative activity normalized to a control (DMSO-treated) value of 100% ($n = 3$, error bars represent s.d.). Reprinted with permission from (Takeuchi et al. 2018). Copyright (2018) American Chemical Society.

5. Thermodynamic and Structural Analysis of Complexes with PYL

The thermodynamic behavior of AS6 and PANMe in binding to PYL was investigated using Isothermal Titration Calorimetry (ITC) (Table 12.1). The dissociation constant (K_d) of AS6 and PANMe was lower than that of ABA, indicating that AS6 and PANMe interacted with PYL more greatly compared with ABA. The 4′-*O*-chain of PAN has a phenyl moiety to mimic the interaction between the PP2C Trp-indole and the PYL Phe-phenyl. Therefore, PAN is expected to interact more strongly with PYL than ABA. However, the major reason for the high affinity was different from what was expected in the design of PAN. The formation of PYL-PANMe complex was driven not more enthalpically but more entropically compared to that of PYL-ABA/AS6 complex. The PYL5–AS6 complex formation was not favored entropically, probably because the C–C rotations of 3′-*S*-chain was limited by interacting with the 3′-tunnel of PYL5. In fact, an AS6 analog with a more restricted 3′-chain, PAO4, has no entropy loss in binding to PYL5 (Takeuchi et al. 2015). The PANMe 4′-*O*-chain is less flexible than the AS6 3′-*S*-hexyl chain. The single bond rotation in the alkyne (–C≡C–) has no effect on the chain linearity. However, rotation of the terminal phenyl ring attached to an alkyne with the low energy barrier generates several conformations. In addition, because the rotational barrier of the ether bond is lower than that of the alkyl bond, the conformational variation of the 4′-*O*-chain of PANMe is much larger than expected from

Table 12.1: Apparent AS6 and PANMe binding affinity for PYL5.

Compounds	K_d (nM)	ΔH (kcal/mol)	$-T\Delta S$ (kcal/mol)	ΔG (kcal/mol)	ΔC_p (kcal/mol/K)
ABA	890 ± 100	−7.8 ± 0.1	−0.3	−8.1 ± 0.1	−0.56
AS6	480 ± 100	−9.7 ± 0.1	1.2	−8.5 ± 0.1	not measured
PANMe	47 ± 22	−6.4 ± 0.1	−3.5	−9.8 ± 0.3	−0.38

the apparent linear structure. Thus, the entropically driven formation of PYL5–PANMe complex could not depend on the conformational inflexibility.

To structurally explain the activity of AS6 and PANMe and the thermodynamic behavior in complex formation with PYL, the crystal structures of PYR1–AS6 and PYR1–PANMe were elucidated at 2.3 and 2.5 Å resolutions, respectively. Interestingly, the structure of PYR1–PANMe was obtained as a dimer composed of two different complexes. Three complexes resembled that of the PYR1–ABA complex; the ABA-skeletons of AS6 and PANMe were found in the similar position normally occupied by ABA, and PYR1 adopted a gate-closed conformation (Fig. 12.4). In the PYR1–AS6 complex, the 3′-tunnel accepted the AS6 *S*-hexyl chain, which also protruded out onto the PP2C-interaction surface of PYR1. These observations revealed the PYL antagonistic mechanism of AS6; it interferes with the interaction with PP2C through steric effects of the *S*-hexyl chain, even though it induces PYL to the gate-closed, that is, active form. On the other hand, the electron density of the PANMe 4′-*O*-chain and the phenyl moiety of PYR1-Phe in the 4′-tunnel was not clearly determined. The B-factor of the 4′-tunnel region of PYR1–PANMe was higher than that of PYR1–ABA, suggesting that the PANMe 4′-*O*-chain increased the flexibility of this region. These suggested that the flexibility of the PANMe 4′-*O*-chain remained in the complex with PYL. The 4′-*O*-chain may fluctuate between the two tunnels due to this residual mobility. In any case, the PYL-PP2C complex formation is obstructed by the terminal tolyl group.

This structural analysis gave us a key to understanding the thermodynamic behavior of the PYL5–PANMe complex formation. We considered that the entropically driven formation of the PYL5–PANMe complex might be induced by the residual mobility of 4′-*O*-chain and the partial fluctuation of PYL5 caused thereby. The heat capacity change (ΔC_p) is involved in folding stability of proteins; protein denaturation shows positive ΔC_p, whereas the protein–ligand complexes formation shows negative ΔC_p (Cooper 2005), which may be caused by changes in solvation, hydrophobic effects, or other weak interactions (Cooper 2005, Olsson et al. 2008, Snyder et al. 2011). Although the ΔC_p was negative for the formation of both PYL5–ABA and PYL5–PANMe complexes, the latter was less negative than the former (Table 12.1). Less restrained water molecules in a protein–ligand complex induce a less negative ΔC_p (Snyder et al. 2011). The ordered water molecules in PYL5–PANMe may have been less than those in PYL5–ABA, because of fluctuation of the 4′-*O*-chain.

6. Physiological Effects of PANMe on other Plants

PYL antagonists are beneficial for elucidating the ABA functions in plants, especially in non-model plants without genetic resources or not enough. If PYL antagonists are applicable to many crops, it is of great agricultural value. PANMe acted as an ABA antagonist in many plants, but in some plants including rice it did not act as an ABA antagonist. Of particular interest to us was that rice was insensitive to this compound co-applied with ABA. *Arabidopsis* PYL (AtPYL)–like proteins in other plants including rice function as ABA receptors (Hauser et al. 2011). There are no significant differences in the primary and higher order structures of the ABA-binding site and 4′-tunnel between rice PYL (OsPYL) and AtPYL (He et al. 2014). Additionally, PANMe reactivated OsPP2C06 in the presence of OsPYL2 and ABA in our *in vitro* assay system, that is, PANMe acted as an effective OsPYL at least *in vitro*. AS6 also did not act as ABA antagonist for rice, although it inhibited OsPYL2 *in vitro*. A large number of factors including absorption, distribution, metabolism, and excretion (so-called ADME in the field of drug discovery) affect the activity of chemicals also in plants. We have not yet examined PANMe and AS6's ADME. Therefore, it is not clear why these chemicals do not act as ABA antagonists for intact rice plants although they act as OsPYL antagonists. Some reports suggest that not all ABA responses in plants can be elicited by ABA-dependent PYL activation mechanism (Yin et al. 2013). In this context, our chemicals that act as PYL antagonists but not ABA antagonists could be useful to unravel the complicated regulatory mechanism of ABA responses.

7. Summary and Discussion

We created ABA receptor PYL antagonists AS6 and PANMe based on structure-guided design. The structural, thermodynamic, biochemical and physiological analysis revealed that these compounds conformationally activate PYL but inhibited PYL–PP2C complexes formation by direct steric hindrance of the elongated substituents at the 3′ or 4′ position of ABA's ring that is close to two tunnels on the interface with PP2C. Although AS6 and PANMe never exhibited the agonistic activity in the *in vivo* assays using *Arabidopsis*, the biochemical data *in vitro* suggested that AS6 is a partial agonist of PYL, whereas PANMe not. This may be one of the reasons that PANMe exhibited the greater antagonistic activity than AS6 both *in vitro* and *in vivo*, in addition to the greater affinity of PANMe to PYL compared to that of AS6. Currently, we believe that the partial agonist activity of AS6 *in vitro* is due to lack of rigidity of the 3′-*S*-hexyl chain. An isolated linear alkyl compound will take a rounded shape rather than straight in order to minimize contact with water molecules in aqueous solution. For this, AS6 in the complex with PYL should be difficult to protrude straight the terminal portion of the 3′-*S*-hexyl chain. In fact, it does not protrude straight in the crystal structures of PYL5–AS6 complex. This incomplete protrusion may have allowed weak PYL–PP2C interaction to induce the partial agonist activity. It also explains that PANMe with rigid side chains do not function as a partial agonist.

The phenyl moiety of PANMe was expected to have the π-π interaction with the Phe-phenyl in the 4′-tunnel. Although PANMe stabilized the complex with PYL more greatly than did ABA and AS6, the complex formation was not enthalpically but entropically driven by increasing the conformational fluctuations in the complex. There are several recent findings that increased protein conformational flexibility contributes significantly to stable complex formation (Wand 2013, Akke 2012, Diehl et al. 2010, Baron and McCammon 2008, MacRaild et al. 2007, Frederick et al. 2007). An increase in the restriction of the ligand in binding to the target protein negatively contributes to the affinity. This is known as entropic penalty. Conformational restriction of the flexible ligand is one of the effective strategies in lead optimization to reduce this entropic penalty. We also adopted this strategy in the design of PAO4. On the other hand, we had no idea of improving the affinity by ligand design that increases protein flexibility. Although we did not design PANMe for this purpose, as a result, PANMe bound with high affinity to the target protein by partially increasing the flexibility of the target protein. Our findings suggest that molecular design to increase the local flexibility of the target protein may be a new approach to develop high affinity ligands.

Acknowledgments

I appreciate the following collaborators: J. Takeuchi, T. Ohnishi, T. Muto and N. Mimura (Shizuoka Univ.); M. Okamoto (Utsunomiya Univ.); T. Akiyama, S. Yajima and M. Sue (Tokyo Univ. of Agriculture); M. Seo and Y. Kanno (RIKEN); T. Kamo and A. Endo (NARO); E. Nambara (Univ. of Toronto); N. Hirai (Kyoto Univ.); S. R. Cutler (UC Riverside); K. Monda and K. Iba (Kyushu Univ.).

References

Akke, M. 2012. Conformational dynamics and thermodynamics of protein–ligand binding studied by NMR relaxation. Biochem. Soc. Trans. 40: 419–423.

Baron, R. and J.A. McCammon. 2008. (Thermo)dynamic role of receptor flexibility, entropy, and motional correlation in protein-ligand binding. ChemPhysChem. 9: 983–988.

Cooper, A. 2005. Heat capacity effects in protein folding and ligand binding: A re-evaluation of the role of water in biomolecular thermodynamics. Biophys. Chem. 115: 89–97.

Diehl, C., O. Engström, T. Delaine, M. Håkansson, S. Genheden, K. Modig et al. 2010. Protein flexibility and conformational entropy in ligand design targeting the carbohydrate recognition domain of galectin-3. J. Am. Chem. Soc. 132: 14577–14589.

Frederick, K.K., M.S. Marlow, K.G. Valentine and A.J. Wand. 2007. Conformational entropy in molecular recognition by proteins. Nature 448: 325–329.

Fujii, H., V. Chinnusamy, A. Rodrigues, S. Rubio, R. Antoni, S.-Y. Park et al. 2009. *In vitro* reconstitution of an abscisic acid signalling pathway. Nature 462: 660–664.

Hauser, F., R. Waadt and J.I. Schroeder. 2011. Evolution of abscisic acid synthesis and signaling mechanisms. Curr. Biol. 21: R346–R355.

He, Y., Q. Hao, W. Li, C. Yan, N. Yan and P. Yin. 2014. Identification and characterization of ABA receptors in *Oryza sativa*. PLoS One 9: DOI: 10.1371/journal.pone.0095246.

Jiang, L., X. Zhang and Z. Chen. 2014. Structural basis of ABA perception by PYR/PYL/RCAR receptors. pp. 117–135. *In*: D.-P. Zhang (ed.). Abscisic Acid: Metabolism, Transport and Signaling. Springer, Netherlands.

Kumari, S. and R.A.L. van der Hoorn. 2011. A structural biology perspective on bioactive small molecules and their plant targets. Curr. Opin. Plant Biol. 14: 480–488.

Lumba, S., S. Cutler and P. McCourt. 2010. Plant nuclear hormone receptors: a role for small molecules in protein-protein interactions. Annu. Rev. Cell Dev. Biol. 26: 445–469.

Ma, Y., I. Szostkiewicz, A. Korte, D. Moes, Y. Yang, A. Christmann et al. 2009. Regulators of PP2C Phosphatase activity function as abscisic acid sensors. Science 324: 1064–1068.

MacRaild, C.A., A.H. Daranas, A. Bronowska and S.W. Homans. 2007. Global changes in local protein dynamics reduce the entropic cost of carbohydrate binding in the arabinose-binding protein. J. Mol. Biol. 368: 822–832.

Miyazono, K.-I., T. Miyakawa, Y. Sawano, K. Kubota, H.-J. Kang, A. Asano et al. 2009. Structural basis of abscisic acid signalling. Nature 462: 609–614.

Olsson, T.S.G., M.A. Williams, W.R. Pitt and J.E. Ladbury. 2008. The Thermodynamics of protein-ligand interaction and solvation: Insights for ligand design. J. Mol. Biol. 384: 1002–1017.

Park, S.Y., P. Fung, N. Nishimura, D.R. Jensen, H. Fujii, Y. Zhao et al. 2009. Abscisic acid inhibits type 2C protein phosphatases via the PYR/PYL family of START proteins. Science 324: 1068–1071.

Santiago, J., F. Dupeux, K. Betz, R. Antoni, M. Gonzalez-Guzman, L. Rodriguez et al. 2012. Structural insights into PYR/PYL/RCAR ABA receptors and PP2Cs. Plant Sci. 182: 3–11.

Snyder, P.W., J. Mecinovic, D.T. Moustakas, S.W. Thomas, M. Harder, E.T. Mack et al. 2011. Mechanism of the hydrophobic effect in the biomolecular recognition of arylsulfonamides by carbonic anhydrase. Proc. Natl. Acad. Sci. 108: 17889–17894.

Soon, F.-F., L.-M. Ng, X.E. Zhou, G.M. West, A. Kovach, M.H.E. Tan et al. 2012. Molecular mimicry regulates ABA signaling by SnRK2 kinases and PP2C phosphatases. Science 335: 85–88.

Sun, D., H. Wang, M. Wu, J. Zang, F. Wu and C. Tian. 2012. Crystal structures of the *Arabidopsis thaliana* abscisic acid receptor PYL10 and its complex with abscisic acid. Biochem. Biophys. Res. Commun. 418: 122–127.

Takeuchi, J., M. Okamoto, T. Akiyama, T. Muto, S. Yajima, M. Sue et al. 2014. Designed abscisic acid analogs as antagonists of PYL–PP2C receptor interactions. Nat. Chem. Biol. 10: 477–482.

Takeuchi, J., T. Ohnishi, M. Okamoto and Y. Todoroki. 2015. Conformationally restricted 3′-modified ABA analogs for controlling ABA receptors. Org. Biomol. Chem. 13: 4278–4288.

Takeuchi, J., N. Mimura, M. Okamoto, S. Yajima, M. Sue, T. Akiyama et al. 2018. Structure-based chemical design of abscisic acid antagonists that block PYL–PP2C receptor interactions. ACS Chem. Biol. 13: 1313–1321.

Todoroki, Y. and N. Hirai. 2002. Abscisic acid analogs for probing the mechanism of abscisic acid reception and inactivation. Stud. Nat. Prod. Chem. 27: 321–360.

Todoroki, Y. and K. Ueno. 2010. Development of specific inhibitors of CYP707A, a key enzyme in the catabolism of abscisic acid. Curr. Med. Chem. 17: 3230–3244.

Toh, S., A. Imamura, A. Watanabe, K. Nakabayashi, M. Okamoto, Y. Jikumaru et al. 2008. High temperature-induced abscisic acid biosynthesis and its role in the inhibition of gibberellin action in *Arabidopsis* seeds. Plant Physiol. 146: 1368–1385.

Vlad, F., S. Rubio, A. Rodrigues, C. Sirichandra, C. Belin, N. Robert et al. 2009. Protein phosphatases 2C regulate the activation of the Snf1-related kinase OST1 by abscisic acid in *Arabidopsis*. Plant Cell 21: 3170–3184.

Wand, A.J. 2013. The dark energy of proteins comes to light: Conformational entropy and its role in protein function revealed by NMR relaxation. Curr. Opin. Struct. Biol. 23: 75–81.

Weiner, J.J., F.C. Peterson, B.F. Volkman and S.R. Cutler. 2010. Structural and functional insights into core ABA signaling. Curr. Opin. Plant Biol. 13: 495–502.

Wells, J.A. and C.L. McClendon. 2007. Reaching for high-hanging fruit in drug discovery at protein-protein interfaces. Nature 450: 1001–1009.

Yin, Y., Y. Adachi, W. Ye, M. Hayashi, Y. Nakamura, T. Kinoshita et al. 2013. Difference in abscisic acid perception mechanisms between closure induction and opening inhibition of stomata. Plant Physiol. 163: 600–610.

Zhang, D.-P. 2014. Abscisic Acid: Metabolism, Transport and Signaling. Springer, Netherlands.

Zhang, X., L. Jiang, G. Wang, L. Yu, Q. Zhang, Q. Xin et al. 2013. Structural insights into the abscisic acid stereospecificity by the ABA receptors PYR/PYL/RCAR. PLoS One 8: e67477.

Chapter 13

Wilt Detection Soft Sensor for Plant Water Stress Management

Hiroshi Mineno

1. Introduction

Agricultural artificial intelligence (AI), which quantifies and formalizes tacit knowledge—the "craftsmanship" held by expert farmers (so-called master farmers)—by applying knowledge of information sciences, including artificial intelligence to the agricultural field, has been gathering attention (Shinjo 2017). Based on knowledge of crops and farmland, expert farmers judge what kind of work they ought to do and carry it out. Here, when considering how the environment relates to crops, the judgment of expert farmers must be analyzed from the perspective that the response to the surrounding environment differs depending on the variety of the plant, and the condition of its growth. The main environmental elements in horticulture relate largely to the part of the plant above ground and the part below. If, however, appropriate agricultural work can be carried out at the appropriate time, in response to gradual changes in the condition of plants' growth and environmental elements, as with expert farmers, it should be possible to raise plants quicker, by promoting photosynthesis and thereby increasing yield, producing functional plants with increased fruit sugar content and heightened specific nutritional value. It should also be possible to suppress the occurrence of disease and physiological disorders, while striking a balance between vegetative and reproductive growth.

Many of the advanced industrialized countries of Europe and the US are also advanced agricultural countries, and not only are they introducing automated labor, based on the latest engineering technology, they are also introducing the selective breeding technology necessary to rapidly develop new varieties and cultivation management systems, using various sensors. On the other hand, in addition to its high cost land, labor, and agricultural production materials, along with its rapidly ageing society, Japan is facing the loss of its

Research Institute of Green Science and Technology, Shizuoka University, 3-5-1 Johoku, Naka-ku, Hamamatsu, Shizuoka 432-8011, Japan.
Email: mineno@inf.shizuoka.ac.jp

collective farming "craftsmanship," due to the retirement of its expert farmers, who boast outstanding productivity. Moreover, in recent years, profitability has also decreased, due to unstable annual yields caused by climate fluctuations, and, due to the harsh nature of the work, the labor shortage has deepened. However, if, through AI, we can formalize the advanced cultivation technology being lost, not only can the know-how that expert farmers have acquired based on years of experience and intuition be efficiently passed onto the next generation, there is the related possibility of establishing new, innovative methods of cultivating agricultural products.

Based on knowledge from the combined agricultural sciences fields of plant physiology and plant ecology, here we apply an information sciences approach of the Internet of Things (IoT) and AI (machine learning, deep-learning) to plant growth. We consider the physiological state and environmental response, and introduce a wilt-detection soft sensor, researched and developed with software, which can detect wilting plants with high accuracy.

2. A Soft Sensor for Wilt Detection

Among the advanced cultivation techniques known to master farmers, stress cultivation, which subjects plants to moderate water stress (thirst), can produce fruit with high sugar content, and because this leads to the increased market value of the fruit, it would be useful for these techniques to be formalized.

Various methods for estimating water stress have been proposed, in order to reproduce stress cultivation using Information and Communications Technology (ICT), and these methods are necessary to estimate the water stress plants are subject to with high accuracy, and to control irrigation through the appropriate timing. Among these, an image processing method, which uses images of plant posture created using general consumer cameras, and which can estimate water stress without coming into contact with or damaging the plant, is expected to be easily used, even by new farmers with relatively limited agricultural knowledge. However, the existing method of estimating water stress using image processing requires a clear separation of the plant body from the background; this is challenging to apply in situations with complex backdrops, and its use at real cultivation sites is difficult.

Therefore, by measuring the factors that cause plants to wilt, one of the determining indicators of water stress sensed by master farmers, we believe it may be possible to reproduce the water stress sensing ability of master farmers, by creating a suitable model for plant wilting.

As shown in Fig. 13.1(a), in plants, water evaporates from pores on the surface of leaves during good weather to promote the uptake of water and nutrients from the roots. When there is little water available to be drawn up, however, some of the water inside the body of the plant is lost, through transpiration, and it is thought that wilting is a phenomena where, through excessive transpiration, the stem becomes unable to bear the weight of its withering leaves, and this is manifested in plant posture. The amount by which the pores open and close, which is one factor controlling transpiration amount, changes with light intensity, temperature and humidity, and is also related to the surface area of the leaf, and the extent to which the plant colony is overgrown. Therefore, as shown in Fig. 13.1(b), in addition to measurement of temperature and humidity, it is thought possible to understand the number of pores and the extent of their opening using leaf surface area. This can be expressed as a ratio of the light scattered on the upper and lower

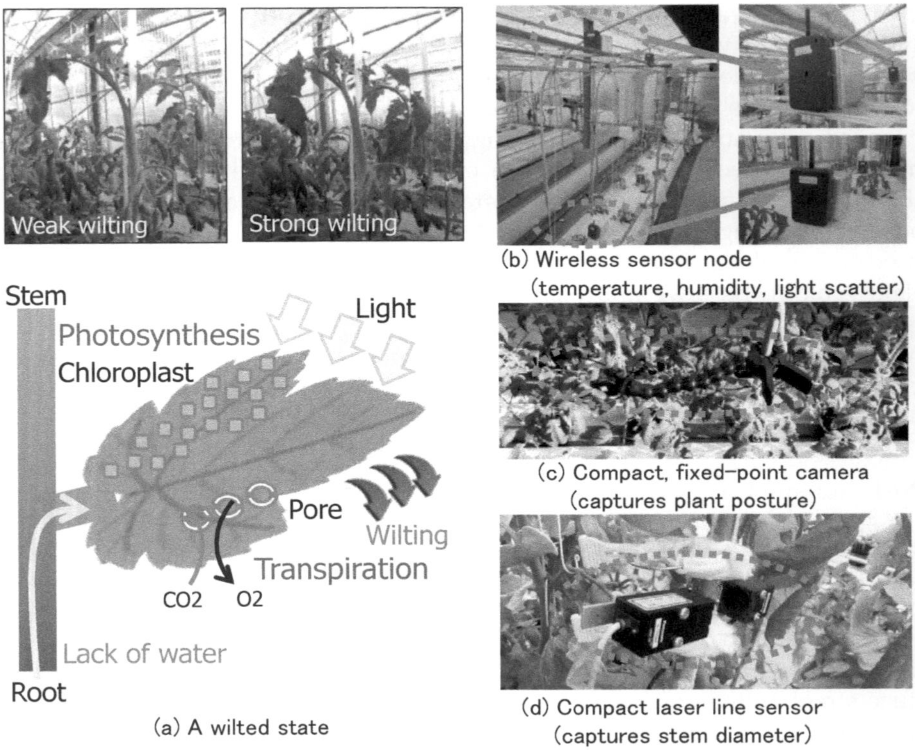

Fig. 13.1: Sensors for modelling plant wilting.

Color version at the end of the book

parts of the plant colony, and the extent of overgrowth of the plants. Moreover, as a result of transpiration, it was thought that the visible extent of wilting could be captured through images of plant posture, taken periodically using a small, fixed-point camera, as shown in Fig. 13.1(c).

Using non-destructive data, which is comparatively easy to measure, and by using machine learning tied to stem diameter, which is thought to directly represent the main cause of wilting, we trialed the creation of a model for the plant wilting understood by master farmers through experience and intuition. Here, stem diameter linked to machine learning was measured using a compact laser line sensor, which was able to measure even minute changes in position, as shown in Fig. 13.1(d). This compact laser line sensor was used only when applying machine learning to the plant wilting model, and once the plant wilting model was constructed—that is, once it was applied in actual operations—it reported the amount of relative change in stem diameter, using only temperature, humidity, scattering of light and an image of the plant's posture, as inputs. It becomes, in effect, a soft sensor for the detection of plant wilting.

An overview of the wilt detection soft sensor is shown in Fig. 13.2. Specifically, visual characteristics relating to plant wilting are extracted, using a Convolutional Neural Network (CNN), which is one type of deep learning that has been effective in its application to the field of image recognition. Meanwhile, environmental data that is considered difficult to obtain visually, temperature, humidity and the amount of light, were collected using

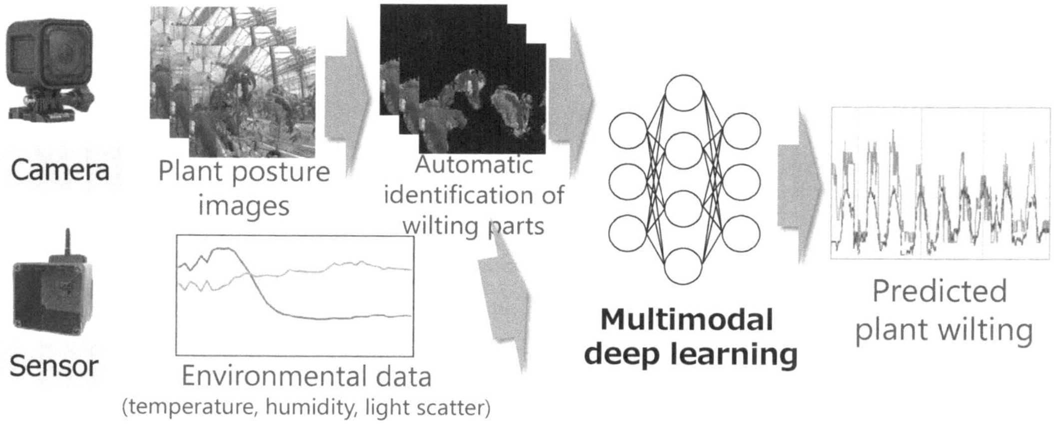

Fig. 13.2: Overview of the plant wilt detection soft sensor.

hardware sensors. Finally, it was possible to estimate and predict relative changes in stem diameter with high accuracy, through multimodal deep learning, which was performed by superimposing these multidimensional characteristic values (Kaneda et al. 2017).

In general, training datasets with thousands to tens of thousands of data points are necessary, in order to improve the accuracy of machine learning and deep learning. In fields such as general object recognition, where enhanced labeled image databases, such as ImageNet, are available on the internet, these very large published datasets can be appropriated to facilitate training. There were, however, hardly any image databases with labeled data relating to plant wilting, and independently preparing a large-scale dataset requires great effort. Moreover, when the size of training datasets relating to the number of parameters of neural networks using deep learning is small, overfitting is likely to occur, so the structure of the neural network should be as simple as possible, and it is important to limit the number of parameters that need to be tuned. Therefore, in order to limit the effort required to create a very large training dataset independently, we added pre-processing, to enable high accuracy estimation with even a small amount of training data. We achieved this in such a way that it was easy to extract motion characteristics relating to plant wilting, by processing current images of plant posture. In other words, by processing the input dataset to make it easy to find the characteristics one wished to learn, compared to the conventional method of training high-accuracy AI using spartan education with huge training datasets, we applied a method whereby regular AI can learn well, even with a small training dataset.

Figure 13.3 shows an example of processing an original image. First, by applying Optical Flow (OF) used for moving object detection against two original images separated by n points in time, an image extracting the movement of the plant is formed. Next, by applying a pooling process to the image obtained using OF, we created a Pooled Optical Flow (POF). Since OF is differential displacement between two certain points in time, it was not possible to extract wilting arising outside of the two points in time using OF alone. Therefore, by applying maximum pooling to each pixel of the OF for the set time period, the image in which wilting occurred outside the two points in time was also extracted, as a POF. Finally, by masking the original image with the POF, a masked image ROAF (Remarkable moving Object detected by Adjacent optical Flow), which identifies the periphery of the wilted part of the plant, is generated. The POF contains the position information of parts in

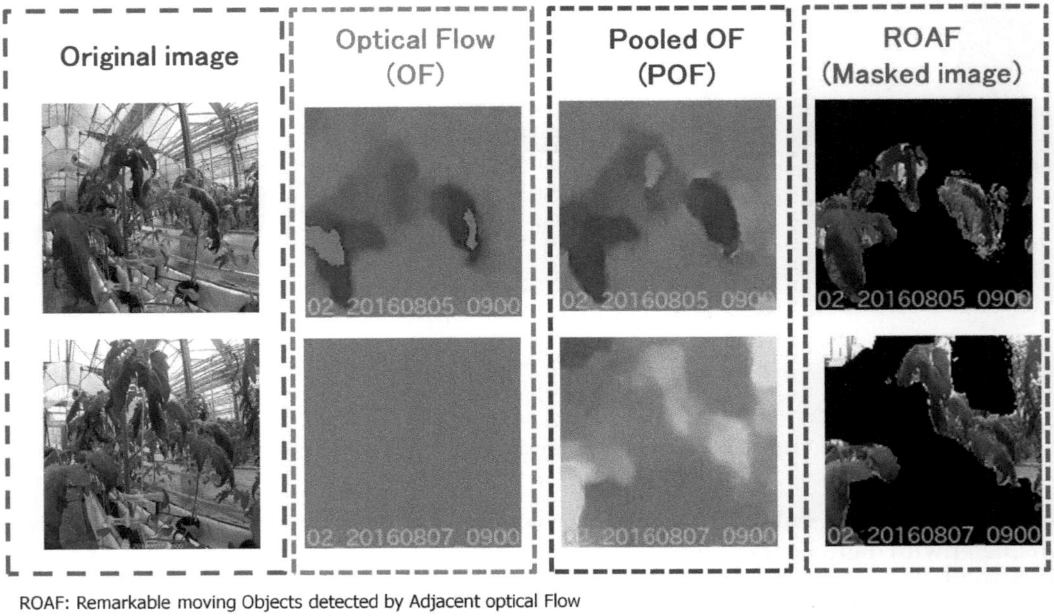

Fig. 13.3: An example of processing the original image.

which wilting occurs, due to the applications of water stress, however, the original image contains information not found in the POF, such as the color and shape of the plant. By masking the POF containing different information with the original image, an image can be created in which the parts to which attention must be paid, in regard to plant wilting, are emphasized. Moreover, when masking at this time, by filling in areas where wilting has not occurred in black, feature extraction—focusing on the parts where wilting has occurred in the convolution process of the CNN—becomes easier.

In applying this processing to the original image, in addition to the ROAF introduced, it is also desirable to apply a method of superimposing information concerning movement over time, size and direction, to the two-dimensional image. This method is referred to as MDCD (Motion-specialized Deep Convolutional Descriptor) (Shibata et al. 2017), and in MDCD, by superimposing the characteristics of the gentle movements over several minutes to several tens of minutes as a plant wilts, we found that it was also possible to train the AI effectively, using a training dataset with just a few temporal features.

Multimodal data, quantified multidimensionally through characteristics obtained from images of plant posture and environmental data, included environmental conditions and growth states affecting the growth and development periods of plants, is expected to gradually change. We proceeded to cluster the multi-dimensional, multimodal data for each point in time, and extracted a large number of sub-datasets with various characteristics in multi-dimensional space. We then applied an adaptive machine learning device known as SW-SVR (Sliding Window-based Support Vector Regression) (Kaneda and Mineno 2017), which performs optimal estimation and forecasting at each point in time, through weighted ensemble learning from a number of specialized constructed sub-models. The SW-SVR allowed development of a model that used a wide variety of complex plant physiological information, realistic measurement data and machine learning with time granularity, to create a soft sensor for wilt detection in plants.

3. The Amount and Quality of Training Data

In general image recognition, the recognition accuracy of machine learning models is said to be influenced by the quantity and quality of training data used when constructing the model, the number of features, and the actual machine learning device itself, in that order. For this reason, fields and organizations that can collect large volumes of training data can construct high accuracy machine learning models with deeper deep learning, but this is not the case for fields and organizations for which obtaining very large datasets is difficult. Therefore, during construction of the model for wilt detection in plants, we analyzed the influence of the amount and quality of training data on the accuracy of stem diameter change prediction outputs by the soft sensor, using the multimodal deep learning described above, and by processing original images.

Figure 13.4 shows the results of analysis of the quantity and quality of training data used when constructing the machine learning model for the wilt detection soft sensor. For

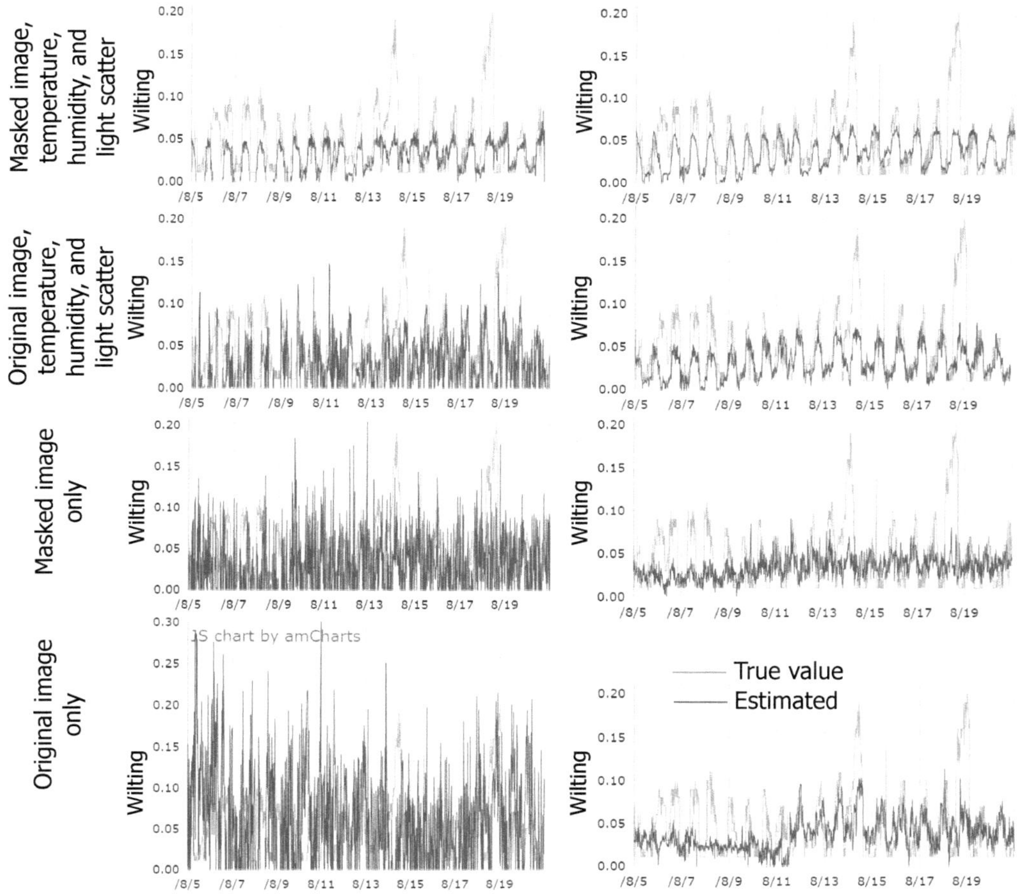

(a) 1,400 points of multidimensional data (b) 14,000 points of multidimensional data

Fig. 13.4: Results from analyzing the quantity and quality of training data.

Color version at the end of the book

multidimensional, multimodal datasets with original images of plant posture, and masked images processed to make it easier to extract characteristics and environmental data such as temperature, humidity and light intensity, we verified the accuracy of stem diameter change predictions. Temporal datasets with 1,400 points in time and 14,000 points in time were used. For 1,400 points in time, for original images only, the true value of the change in stem diameter over time was difficult to estimate. It became possible to predict the change in stem diameter over time, to some extent however, even with a dataset of only 1,400 points in time, using multimodal deep learning, superimposing processed masks, and temperature, humidity and light intensity data. Increasing the training dataset to 14,000 points in time, it became possible to predict stem diameter change more sharply, and it was seen that the issue of the size of the dataset can be at least partly resolved by focusing on quality.

Based on the balance between transpiration and water absorption, which determines the physical phenomenon of wilting, we constructed an AI irrigation control system incorporating the wilt detection soft sensor, that is, we developed a system able to control irrigation by using the plant's physical appearance. Following some basic evaluations and improvements over a period of 12 months, we carried out comparative cultivation experiments, using medium size tomato, low-density hydroponics, and the same greenhouse. In these experiments, irrigation was programmed to occur according to the growth and state of stalk diameter displacement (A. Stem diameter (dynamic)), with a fixed threshold for stem diameter displacement (B. stem diameter (static)), based on the proportion of sunlight (C. Sunlight promotion), and using the agricultural AI's stem diameter predicted values. As a result, we confirmed it was possible to mass produce high sugar content tomatoes (Average Brix, 8.87, standard deviation 0.845, maximum 16.9, saleable proportion 0.917) mechanically, with a low labor cost, and without lowering the fruit yield during the harvest period, using AI irrigation control (Table 13.1).

Table 13.1: Results of a real cultivation experiment.

Process area	**Labor burden**	**Sweetness [brix]**				**Num. of samples**	**Avg. fruit weight [g]**	**10a yield (t/10a)***	**Saleable rate**
		Avg.	**Max.**	**Min.**	**SD**				
A: Stem diameter (dynamic)	Moderate (setup)	8.75	16.9	3.00	0.900	14,089 (728 items)	22.5	8.7	0.965
B: Stem diameter (fixed)	Moderate (setup)	8.10	16.9	4.80	0.717	17,885 (823 items)	27.5	11.7	0.955
C: Proportion of sunlight	High (control)	8.73	15.7	5.10	0.941	10,591 (604 items)	22.8	9.5	0.826
D: AI	Low	8.87	16.9	4.30	0.845	15,901 (822 items)	20.8	8.6	0.917

*4.5 items/10a, 4 tiers/item, 8 weeks/tier, 4 cropping/year.

4. Conclusion

By using our wilt detection soft sensor, inputting only images of plant posture and easy to measure, non-destructive environmental data (temperature, humidity, sunlight intensity), and based on transpiration excess which does not depend on the volume of cultivation medium or the plant variety, we have shown that it is possible to construct a system capable of correctly-timed, automatic irrigation control. For protected horticulture environments, home gardens, etc., where cameras and sensors can be powered, not only can this technology be easily introduced, but if the AI component is developed as a cloud service, automatic irrigation control systems may spread worldwide, via the internet. Moreover, by adjusting the hydroponic concentration and the extent of wilting, not only are quality controls, such as functional improvements and high sugar content, expected, but improvements to production planning technology are also anticipated. In addition, regarding the promotion of photosynthesis—which is as fundamental for cultivation as irrigation—by controlling the use of CO2, as well as saturation deficits, it will be possible to maximize effective use of sunlight, in response to the state of plant growth and development. Based on this additional concept, we aim to achieve an agricultural AI with the flexibility to manage cultivation in sometimes harsh, sometimes gentle situations. This agricultural AI will be able to support the transfer of agricultural success across time to new farmers and generations.

If an IoT and AI-based Information Sciences approach, based on knowledge from cross-disciplinary cooperation between the agricultural fields of plant physiology and plant ecology, can be used, it will be possible to understand plant growth, physiological states and environmental responses in detail, and the formalization of tacit knowledge will advance dramatically. If field phenomics (Fritsche-Neto and Borem 2015), which can perform environmental measurement at cultivation sites and phenotyping simultaneously, and the agricultural information that lies behind tacit knowledge can be linked together and formalized, then we believe that establishing an innovative cultivation model that exceeds the extent of human knowledge thus far, may be feasible.

References

Fritsche-Neto, R. and A. Borem. 2015. Phenomics: How Next-Generation Phenotyping is Revolutionizing Plant Breeding.

Kaneda, Y. and H. Mineno. 2016. Sliding window-based support vector regression for predicting micrometeorological data. Expert. Syst. Appl. 59: 217–225.

Kaneda, Y., S. Shibata and H. Mineno. 2017. Multi-modal sliding window-based support vector regression for predicting plant water stress. Knowl. Based Syst. 134: 135–148.

Shibata, S., Y. Kaneda and H. Mineno. 2017. Motion-specialized Deep Convolutional Descriptor for Plant Water Stress Estimation. Int'l Conf. on Engineering Applications of Neural Networks (EANN).

Shinjo, A. 2017. Achieving an Agricultural AI through IT and the Skills of Expert Farmers. Nikkei BP.

Chapter 14

Removal of Toxic Anions from Environmental Water by using Cage Compounds

Mitsuru Kondo

1. Introduction

The perchlorate ion, ClO_4^-, is a well-known inorganic oxyanion that has a tetrahedral shape with a Cl atom at the center. Perchlorate salts are produced in industry for many purposes, such as fireworks, road flares, matches, commercial explosives, airbag inflators, etc. (Cole-Dai et al. 2018, Furdui et al. 2018, Logan 2001). As a byproduct, ClO_4^- is also produced when Cl bleach is produced electrically from an aqueous solution that includes chloride, Cl^- (Greliner et al. 2008, Jasper et al. 2017, Poghosyan et al. 2014, Srinivasan and Viraraghavan 2009). This anion is also contained in natural minerals considered to be produced naturally by oxidation of Cl^- by ozone in air (Dasgupta et al. 2005).

Iodine is an essential component of the thyroid hormones that are formed in the thyroid gland (Chung 2014, Wolff and Halmi 1963). The concentration of Cl^- around the thyroid tissue is about 0.01 to 0.1 M, whereas that of I^- is only about 10^{-8}–10^{-7} M. Although the concentration of I^- is remarkably lower than that of Cl^-, the thyroid gland shows selective uptake for I^- over the much more abundant Cl^-. However, in addition to I^-, the thyroid gland shows uptake activities for other anions with the following potency series, $TcO_4^- \geqq ClO_4^- > ReO_4^- > SCN^- > BF_4^- > I^- > NO_3^- > Br^- > Cl^-$ (Wolff 1998). The order means that even at the ppb-level concentration, ClO_4^- inhibits I^- uptake in the thyroid gland, causing a decrease in thyroid hormone production (Greer et al. 2002, Lawrence et al. 2000, Rubin et al. 2017), for ClO_4^-, about 10^{-8} M is nearly equal to 1 ppb with 99.5 of the formula weight. Because the thyroid hormones are essential for neural growth, which is particularly imperative in the early stages of human development, deficiency in these hormones would

Research Institute of Green Science and Technology, Shizuoka University, 836 Ohya, Suruga-ku, Shizuoka 422-8529, Japan.
Email: kondo.mitsuru@shizuoka.ac.jp

seriously damage the nervous system of developing fetuses, newborns, and infants (Furin et al. 2015, Srinivasan and Viraraghavan 2009, Yu et al. 2002).

In the last two decades, ClO_4^- has been detected around the world not only in groundwater and surface water, but also in daily milk, vegetables such as lettuce, wine, and other agricultural products (Asami et al. 2009, Dyke et al. 2007, Kosaka et al. 2007, Liu et al. 2016). Detection of the anion in human breast milk was also reported at a significant concentration, avg. 10.5 μg L^{-1} (10.6 ppb) in 82 samples (Kirk et al. 2005). The U.S. Environmental Protection Agency (EPA) proposed the acceptable exposure level of this anion to be 15 μg L^{-1} (15 ppb) in drinking water (Cole-Dai et al. 2018, Jackson 2011). The states of California and Massachusetts regulated the concentrations for the anion more strictly at 6 and 2 μg L^{-1} (6 and 2 ppb), respectively (California Department of Health Services 2007, Hutchison et al. 2017, Massachusetts Department of Environmental Protection 2016).

Because of its high solubility and noncomplexing property, traditional wastewater treatment technologies for ClO_4^- removal from aqueous solutions were ineffective (Urbansky and Schock 1999). Even $KClO_4$, which has the lowest solubility in water among the general ClO_4^- salts, has a solubility of about 15 g in 1 L of water (USEPA 2008). Various materials have been examined toward the development of ClO_4^- removal methods from aqueous solutions, but sufficient removal levels were still not attained. For example, activated carbon is a low-cost material with a large surface area that is known to adsorb many small molecules in the pores, but it showed too little adsorption capacity and weak ion-selectivity for ClO_4^- (Kim and Choi 2014, Parette and Cannon 2005, Yoon et al. 2009). Although reverse osmosis has been shown to be effective in removing the anion from aqueous solutions, this method was excluded from the practical treatment methods for drinking water because of its high costs (Fritzmann et al. 2007, Hutchison et al. 2017).

Anion exchange and biological reduction of ClO_4^- have been considered to be available technologies for ClO_4^- removal from aqueous solutions (Gu et al. 2000, Hutchison et al. 2013, Hutchison and Zilles 2015). However, there still remain some problems in the removal activity of aqueous ClO_4^- solutions with low concentration, namely severe disturbing effects by other anions that generally coexist at higher concentrations in the environmental water (Hutchison et al. 2017). Environmental effects during the regeneration processes of the resins and/or disposal after treatments, as well as the overall costs are also unresolved subjects for these technologies (Kim and Choi 2014). Based on this background, developments of new materials and/or new ClO_4^- removal methods are important themes in environmental chemistry.

2. Removal of ClO_4^- from Aqueous Solutions by Inclusion in Self-assembled Cationic M_2L_4 Cages

2.1 Artificial Anion Hosts

The host materials that exhibit selective recognition and encapsulation of the target anions are known as "anion hosts," often also called "anion receptors." Their host frameworks usually recognize the target anions by forming H- and/or coordination-bonds with the target anions, as if the shapes and sizes are accommodated in the host frameworks. Because water molecules significantly disturb their chemical bond formations, recognition of anions in aqueous solutions is difficult for artificial anion hosts (Kubik 2010, Langton et al. 2016, Załubiniak et al. 2017). In particular, the recognition of ClO_4^- by anion hosts in aqueous solutions is quite difficult because the anion hardly forms chemical bonds with other molecules, i.e., anion hosts, and also because of the high solubility of the anion in water. Moreover, coexisting anions often disturb the ClO_4^- recognition by the anion hosts because of their higher affinity to the anion hosts.

We have developed some ClO_4^- removal methods by using metal complex compounds that are constructed by metal ions and organic bridging ligands. The recognition and removal of the anion function even in aqueous solutions, and the removal proceeds unaffected by the disturbing coexisting anions. The removal strategies established in this work are classified in the following four patterns: (i) trap in the self-assembled M_2L_4 cages, (ii) trap in the self-assembled Metal-Organic Frameworks (MOFs), (iii) removal by anion exchange reactions by the cationic M_2L_4 cage, and (iv) removal by anion exchange reactions by MOFs. The syntheses of new materials and the new methods for ClO_4^- removal from aqueous solutions are described in this chapter.

2.2 Cationic M_2L_4 Cages for Anions

Flexible bis-imidazole- and bis-benzimidazole-type ligands have yielded many cage-type coordination compounds (Han et al. 2014). Structures of the cage compounds are expressed as M_nL_m cages depending on the number of the metal ions (*n*) and ligands (*m*) that construct the cage frameworks. For example, M_2L_2 (Liu et al. 2009, Liu et al. 2001a, Nikolayenko et al. 2017), M_2L_4 (Amouri et al. 2007, Liu et al. 2009, Su et al. 2003, Su et al. 2001), M_2L_3 (Su et al. 2003, Su et al. 2001), M_3L_4 (Koike et al. 2014, Liu et al. 2001b), and $M_4(\mu\text{-}OR)_4L_4$ (R = Me, Et) cages (Inoue et al. 2013) have been reported to date. The bridging ligands, 1,3-bis(benzimidazol-1-ylmethyl)-2,4,6-trimethylbenzene (*m*-bbitrb), 1,4-bis(benzimidazol-1-ylmethyl)-2,3,5,6-tetramethylbenzene (*p*-bbiteb), 1,3-bis(imidazol-1-ylmethyl)-2,4,6-tetramethylbenzene (*m*-bitrb), 1,4-bis(imidazol-1-ylmethyl)-2,3,5,6-tetramethylbenzene (*p*-biteb), and 1,3-(benzimidazol-1-ylmethyl)-2,5-dimethoxy-4,6-dimethylbenzene were used for the syntheses. The structures of each ligand are summarized in Scheme 14.1.

Among the various M_nL_m cages, M_2L_4 cages have trapped many kinds of anions. The kinds of anions trapped in the cages were affected by the positions of the two methylimidazole or methylbenzimidazole groups. Only a few limited anions such as ClO_4^- and BF_4^- are trapped in the M_2L_4 cages constructed by ligands with *p*-substituted ligands, whereas many kinds of anions, ClO_4^-, BF_4^-, $CF_3SO_3^-$, NO_3^-, SO_4^{2-}, Cl^-, Br^-, etc., are trapped in M_2L_4 cages with *m*-substituted ligands. The M_2L_4 cages create a hydrophobic space. Among

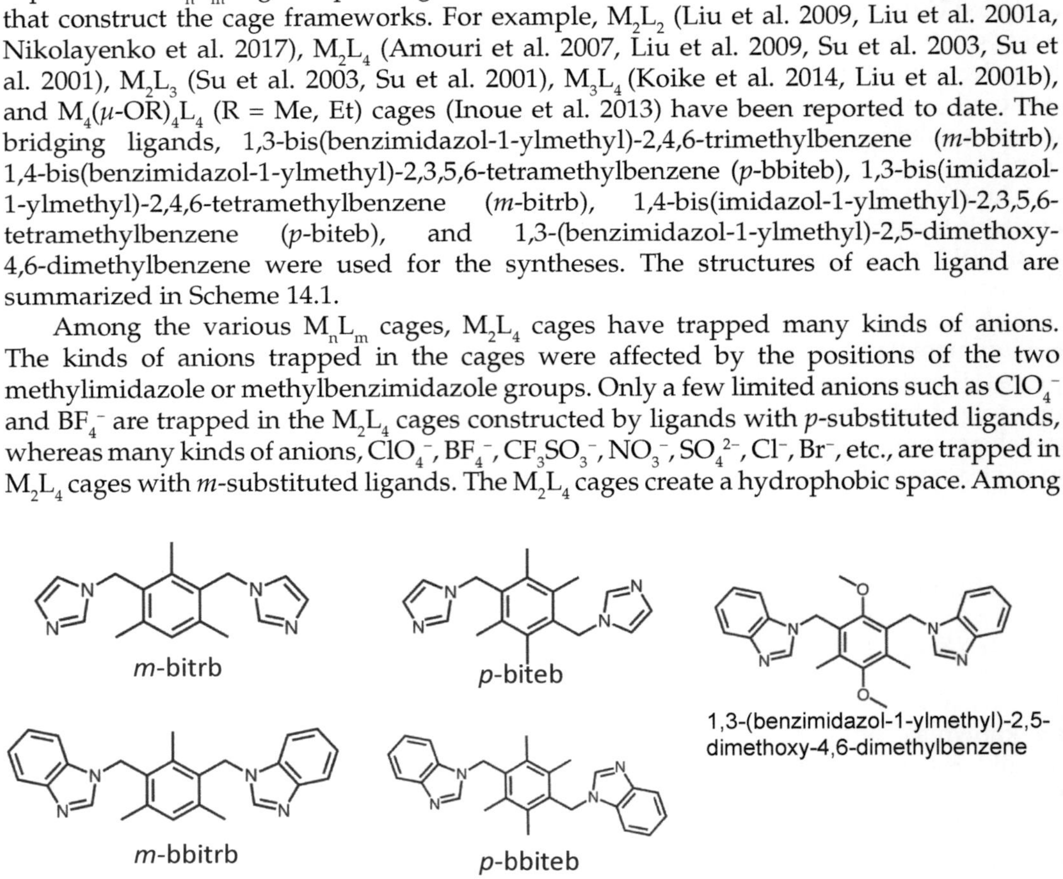

Scheme 14.1: Bis-imidazole and bis-benzimidazole ligands that have been used for M_nL_m cages.

the anions listed above, three (ClO_4^-, BF_4^-, and $CF_3SO_3^-$) have more hydrophobic surfaces, and other anions have hydrophilic surfaces. That is, M_2L_4 cages with *p*-substituted ligands might accommodate anions with a hydrophilic surface preferentially. The mechanism of the inclusion selectivity is discussed in detail below.

The M_2L_4 cages that trap Cl^- give a unique example. The reaction of $CuCl_2$ with *p*-bbiteb produces an MOF constructed by a 2D framework and M_2L_4 cages (vide infra). Although these cages trap Cl^-, they have been unsuccessful in isolating the M_2L_4 compound without connections to the 2D framework. The M_2L_4 cages constructed by the *p*-substituted ligands would be more unstable than those constructed by *m*-substituted ligands.

3. Removal of ClO_4^- from Aqueous Solution by Inclusion in Self-assembled M_2L_4 Cages

3.1 Synthesis and Structure of M_2L_4 Cages Constructed by Cu^{II} Ions and p-Biteb

An M_2L_4 compound $[ClO_4 \subset Cu_2(p\text{-biteb})_4(ClO_4)_2]ClO_4$ (**1**) has been isolated by the reaction of *p*-biteb with $Cu(ClO_4)_2{\cdot}6H_2O$ (Hirakawa et al. 2009). A single-crystal X-ray analysis of Compound **1** showed that the ClO_4^- trapped in the cage was weakly associated with two Cu^{II} sites with a Cu–O distance of 2.87(1) Å, but other significant interactions between the included anion and the M_2L_4 cage framework were not observed. The M_2L_4 cage unit of Compound **1** is shown in Fig. 14.1a later. The size of the cage was about 6.5 × 6.5 × 5.0 Å^3. The two other perchlorate anions located outside the cage were weakly associated with each Cu^{II} site with a Cu-O distance of 2.642(8) Å. The other ClO_4^- was trapped in the space created between the cationic M_2L_4 cages without significant interactions with the cages.

The electrospray ionization time-of-flight (ESI-TOF) Mass (MS) Spectrum showed two peaks ascribed to $[ClO_4 \subset Cu_2(p\text{-biteb})_4]^{3+}$ and $[ClO_4 \subset Cu_2(p\text{-biteb})_4(ClO_4)]^{2+}$. From the spectrum, it has been suggested that the M_2L_4 cage structure of Compound **1** is retained in the MeCN solution.

When an aqueous solution of $NaClO_4$ was added to a mixed $MeCN/H_2O$ (1:1) solution, which contained $CuSO_4{\cdot}5H_2O$ and 2 equiv of *p*-biteb, a shift of the absorption spectrum due to the d-d transition was observed; the maximum at 610 nm gradually shifts to shorter wavelength. The shift essentially stops when 0.5 equiv of $NaClO_4$ (aq) is added to 1.0 equiv of $CuSO_4{\cdot}5H_2O$. The reaction ratio of ClO_4^- to Cu^{II} ion, as estimated from the

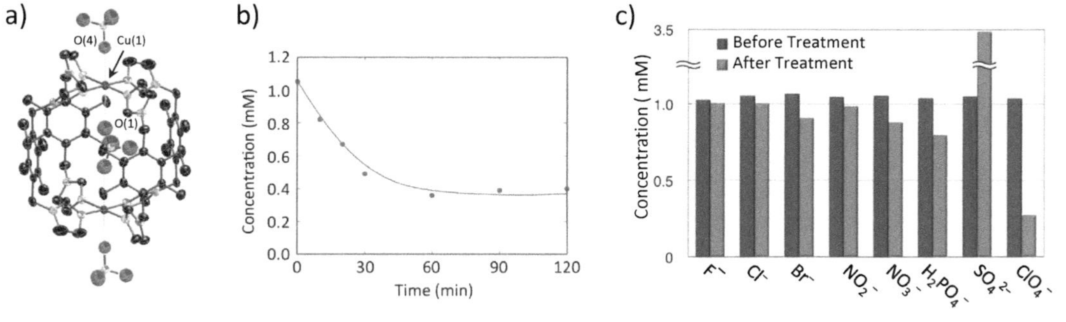

Fig. 14.1: Thermal ellipsoids of the M_2L_4 cage structure of **1** (a). The changes of ClO_4^- concentration in an aqueous solution of $NaClO_4$ after addition of $\{CuSO_4{\cdot}5H_2O + 2\ p\text{-biteb}\}$ (b), and the changes in concentration of each anion before and after 2 h (c).

resulting absorption spectrum, was 1 to 2, which is consistent with that of the composition of Compound **1**. This result indicated that the $[ClO_4 \subset Cu_2(p\text{-biteb})_4]^{3+}$ cage was formed by self-assembly even in the presence of SO_4^{2-} in the aqueous solution.

3.2 Removal of ClO_4^- from Aqueous Solutions by Inclusion in Self-assembled M_2L_4 Cages by Additions of $CuSO_4{\cdot}5H_2O$ and p-Biteb

It was found that the formation of Compound **1** from $CuSO_4{\cdot}5H_2O$ and *p*-biteb proceeds in aqueous solutions as well as in organic solutions. Because Compound **1** was insoluble in water, the reaction performed in the aqueous solution led to the removal of ClO_4^- from aqueous solutions. Figure 14.1b shows the changes in ClO_4^- concentration in an aqueous solution (100 mL) that contained 1.0 mM $NaClO_4$ after addition of $CuSO_4{\cdot}5H_2O$ (49.9 mg, 0.20 mmol) and *p*-biteb (117.8 mg, 0.40 mmol). The changes in concentration were monitored by ion chromatography. The experiment showed that the ClO_4^- concentration reduced from 1.0 mM to 0.4 mM within 1 hour.

The preferential removal of ClO_4^- was also studied. Figure 14.1c summarizes the changes in concentration of each anion from additions of $CuSO_4{\cdot}5H_2O$ (63 mg, 0.25 mmol) and *p*-biteb (148 mg, 0.5 mmol) in an aqueous solution (100 mL) that contained 1.0 mM each of NaF, NaCl, NaBr, $NaNO_2$, $NaNO_3$, NaH_2PO_4, Na_2SO_4, and $NaClO_4$, before and after 2 hours. The changes in the concentration of each anion in the supernatant were monitored by ion chromatography. A white powder of *p*-biteb, which is insoluble in water, changed to a pale purple powder because of the formation of Compound **1** as the reaction proceeded. The result shows the preferential removal of ClO_4^- from among the various anions in the aqueous solution by addition of $\{CuSO_4{\cdot}5H_2O + 2\ p\text{-biteb}\}$. This is the first report of ion removal by inclusion in an M_2L_4 cage.

3.3 Removal of ClO_4^- from Aqueous Solutions by Inclusion in Self-assembled M_2L_4 Cages by Additions of $Cu(CO_3)_{0.5}(OH){\cdot}5H_2O$ and p-Biteb

Although the formation of Compound **1** from $CuSO_4{\cdot}5H_2O$ and *p*-biteb is useful for preferential ClO_4^- removal from aqueous solutions, the solutions obtained after the treatment were severely contaminated by the Cu^{II} ion. To develop a new ClO_4^- removal method without Cu^{II} contamination in the resulting solutions, the use of $Cu(CO_3)_{0.5}(OH)$ as a Cu^{II} source was explored (Yamaguchi et al. 2010).

The copper salts, $Cu(CO_3)(OH)$, i.e., basic copper (II) carbonate, and $Cu(OH)_2$, are known as Cu^{II} compounds that are insoluble in water. For the two materials, it has been shown that $Cu(CO_3)(OH)$ is useful as the source of the M_2L_4 cages that trap ClO_4^- in aqueous solutions. The changes in ClO_4^- concentration after addition of $\{Cu(CO_3)_{0.5}(OH) + 2\ p\text{-biteb}\}$ or $\{Cu(OH)_2 + p\text{-biteb}\}$ in a $NaClO_4$ aqueous solution (1.0 mM) were determined by using ion chromatography. The changes in ClO_4^- concentration for the two Cu^{II} sources as a function of time revealed that although combining $\{Cu(OH)_2 + 2\ p\text{-biteb}\}$, it was difficult to remove ClO_4^- from the aqueous solution; the other combination of $\{Cu(CO_3)_{0.5}(OH) + 2\ p\text{-biteb}\}$ reduced the concentration from 1.0 mM to about 0.3 within 2 hours.

The solution was colorless during the ClO_4^- removal experiment by using $Cu(CO_3)_{0.5}(OH)$ and *p*-biteb, and the concentration of the Cu^{II} ion in the resulting solution was only 0.05 mg/L, as determined by Inductively Coupled Plasma–Optical Emission Spectrometry (ICP–OES). This value was lower than that required by the Japanese regulations for wastewater (3 mg/L) and even for drinking water (1 mg/L). This value is also lower than the concentration in drinking water mandated by the World Health Organization (WHO) (2 mg L^{-1}). These results showed that addition of {$Cu(CO_3)_{0.5}(OH)$ + 2 *p*-biteb} offers a new method for ClO_4^- removal from aqueous solutions.

Tetrafluoroborate (BF_4^-) is a tetrahedral anion that has similar shape and properties as ClO_4^-. This anion is also difficult to remove from aqueous solutions for similar reasons to ClO_4^-. This anion was also trapped in the M_2L_4 cage constructed by two Cu^{II} ions and four *p*-biteb. The affinity of the cationic cage to the anions was studied. When {$CuSO_4{\cdot}5H_2O$ + 2 *p*-biteb} was added to an aqueous solution containing equimolar $NaClO_4$ and $NaBF_4$, ClO_4^- was preferentially removed, indicating that the $[Cu_2(p\text{-biteb})_4]^{4+}$ cage had a higher encapsulation selectivity for ClO_4^- than for BF_4^-. It was found that the combination {$Cu(CO_3)_{0.5}(OH)$ + 2 *p*-biteb} also reduced the BF_4^- concentration from 1.0 mM to about 0.4 within 2 hours. Similarly to the case of ClO_4^- removal experiment, the combination of {$Cu(OH)_2$ + 2 *p*-biteb} did not show BF_4^- removal activity in an aqueous solution.

4. Removal of ClO_4^- by Incorporation in the Self-assembled Formation of Co–2,2′-Bis(1*H*-Imidazolyl-1-Yl)ether (Bie) MOF

4.1 Synthesis and Structure of an MOF Constructed by Co^{II} Ions and Bie

In addition to ClO_4^- removal by trapping in self-assembled M_2L_4 cages, it was found that trapping ClO_4^- in a self-assembled polymeric framework, i.e., MOF, is also useful for selective removal of ClO_4^- from aqueous solutions. The reaction of $Co(ClO_4)_2{\cdot}6H_2O$ and bie produces $[Co(bie)_3](ClO_4)_2$ (**2-ClO$_4$**) in the presence of ClO_4^- in aqueous solutions (Murakami et al. 2015). Because Compound **2-ClO$_4$** is insoluble in water, this reaction removed ClO_4^- from aqueous solutions. The crystal structure of Compound **2-ClO$_4$** was determined by using single-crystal X-Ray Diffraction (XRD) analysis. Figure 14.2a shows the crystal structure of Compound **2-ClO$_4$**. The octahedral Co^{II} centers are connected by bie, yielding a three-dimensional (3D) network. Although bie was remarkably disordered, the disorder is not shown in the figure for the sake of clarity. The framework creates channel-like cavities whose channel window is about 2.9 Å × 1.0 Å along the *a* and *b* axes. The cavities are filled with ClO_4^-. Treatment of $Co(NO_3)_2{\cdot}6H_2O$ with bie gives $[Co(bie)_3](NO_3)_2$ (**2-NO$_3$**), which has a 3D network structure similar to Compound **2-ClO$_4$**, consisting of an octahedral Co^{II} ion and bie.

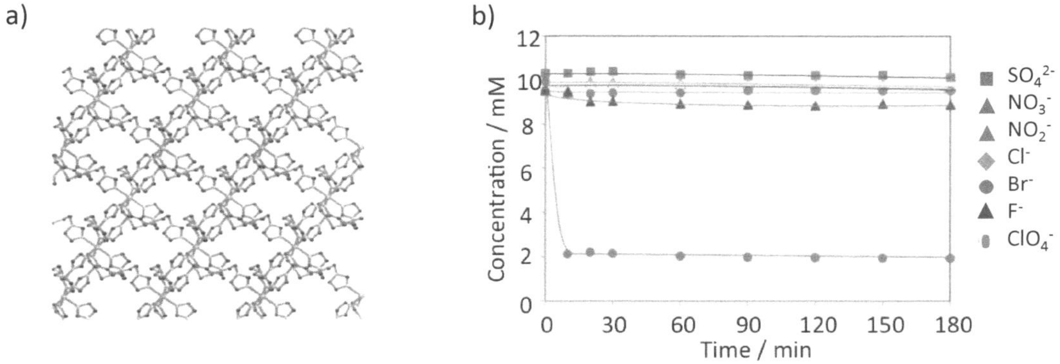

Fig. 14.2: Crystal structure of **2-ClO$_4$** (a), and the changes in anion concentrations in the aqueous solution after additions of bie as a function of time (b). The figures are reproduced from (Murakami et al. 2015) with permission from the Chemical Society of Japan.

The changes in anion concentrations vs. time in an aqueous solution (50.0 mL) that contained 5.00 mM of Co^{II} ion and 10 mM each of F^-, Cl^-, Br^-, NO_2^-, NO_3^-, SO_4^{2-}, and ClO_4^-, after addition of bie (154 mg, 0.75 mmol) at room temperature are shown in Fig. 14.2b. This chart demonstrates that the ClO_4^- concentration is rapidly reduced from 10 mM to 2.1 within 10 minutes, whereas the concentrations of other anions are not much reduced. The pink solid obtained after the treatment was confirmed to be Compound **2-ClO$_4$** by measurements of the Powder XRD pattern (PXRD), which was well consistent with the simulated PXRD pattern calculated from the crystal structure of Compound **2-ClO$_4$**. These results indicated that formation of Compound **2-ClO$_4$** proceeded even in aqueous solutions, and removed ClO_4^- from the aqueous solutions selectively by including in the cationic 3D network. This is the first unique example of ClO_4^- removal from aqueous solutions by using self-assembled constructions of MOFs, i.e., coordination polymers.

Single crystal X-ray structures confirmed that the anions were not bound to the Co^{II} sites, but were located at the space created in the cationic 3D networks for Compound **2-ClO$_4$** and **2-NO$_3$**. The 3D network structure had an ability to accommodate NO_3^- as well as ClO_4^-. Nevertheless, the anion removal experiment showed that the treatment of Co^{2+} ion with bie trapped the ClO_4^- selectively. This result indicated that the creation of hydrophobic space was the key for the preferential trapping of ClO_4^-, which has a hydrophobic surface. The mechanism for this selectivity is discussed below.

5. Removal of ClO_4^- by an Anion Exchange Reaction of a Water-insoluble M_2L_4 Cage

5.1 Syntheses of Cationic M_2L_4 Cages that have ClO_4^- Removal Functions

Earlier it was shown that in aqueous solutions ClO_4^- can be removed by encapsulating it in a water-insoluble M_2L_4 cage as well as in a space created in the cationic 3D network structures in MOFs. The ClO_4^- removal by anion exchange reactions by water-insoluble M_2L_4 cage compounds was studied in continuing works. In this chapter, ClO_4^- removal functions of two kinds of M_2L_4 cages are described. One has a cationic M_2L_4 cage that traps SO_4^{2-} in the cage, whereas the other involves cationic M_2L_4 cages that trap Cl^- and Br^-. For the cationic M_2L_4 cage compounds that showed ClO_4^- removal, four compounds were structurally characterized, namely [$SO_4 \subset Cu_2$(*m*-bbitrb)$_4$]SO_4 (**3-SO$_4$**), [$SO_4 \subset Cu_2$(*m*-bbitrb)$_4$]Cl_2·**H_2O** (**3-Cl**), [$SO_4 \subset Cu_2$(*m*-bbitrb)$_4$]Br_2·**H_2O** (**3-Br**), [$Cl \subset Cu_2$(*m*-bbitrb)$_4$(H_2O)$_2$]Cl_3 (**4-Cl**), and [$Br \subset Cu_2$(*m*-bbitrb)$_4$(H_2O)$_2$]Br_3 (**4-Br**) (Mochizuki et al. 2014, Sone et al. 2016a). In addition to these compounds, [$SO_4 \subset Cu_2$(*m*-bbitrb)$_4$](ClO_4)$_2$·**4MeCN** (**3-ClO$_4$**) were characterized as a reaction product of ClO_4^- removal by **3-SO$_4$**.

The compound **3-SO$_4$** was obtained by the reaction of $CuSO_4{\cdot}5H_2O$ with *m*-bbitrb in mixed EtOH/H_2O solution, while the compounds **3-Cl** and **3-Br** were prepared by treatments of $CuSO_4{\cdot}5H_2O$ with *m*-bbitrb and NaCl (**3-Cl**) or NaBr (**3-Br**). The compounds **4-Cl** and **4-Br** were obtained conveniently by a reaction of $CuCl_2{\cdot}2H_2O$ or $CuBr_2$ with *m*-bbitrb. Compound **3-ClO$_4$** was isolated from the treatment of **3-SO$_4$** with excess $NaClO_4$ in an aqueous solution.

5.2 Structures of Cationic M_2L_4 Cages that include SO_4^{2-} in the Cage

The cationic M_2L_4 cage structures of the compounds **3-SO$_4$**, **3-Br**, and **3-ClO$_4$**, which were determined by single crystal XRD analyses, are shown in Fig. 14.3. The compound **3-SO$_4$**

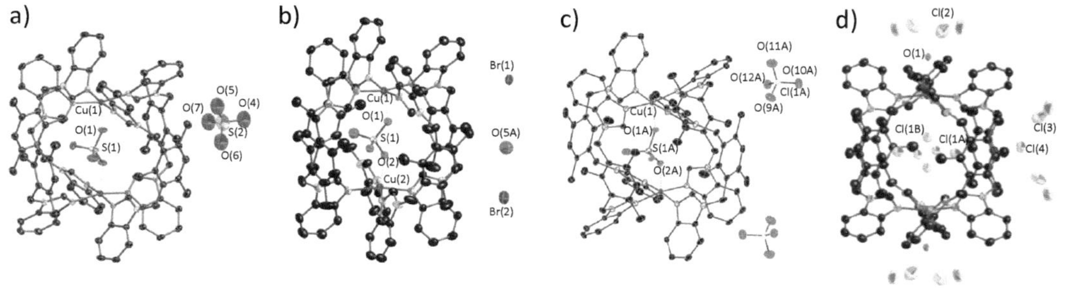

Fig. 14.3: Thermal ellipsoids of the cage structures of **3-SO$_4$** (a), **3-Br** (b), **3-ClO$_4$** (c), and **4-Cl** (d) at 30% probability. Disorders of SO_4^{2-} and H atoms are omitted for clarity.

has SO_4^{2-} both inside and outside the cage (Fig. 14.3a). The SO_4^{2-} inside the cage coordinates with the two Cu^{II} with Cu–O distances of 2.129 (7) and 2.126 (7) Å. These distances are significantly shorter than the corresponding Cu–O distances (2.388 (4) Å) between ClO_4^- and Cu^{II} observed in [$ClO_4 \subset Cu_2$(*m*-bbitrb)$_4$(ClO_4)$_2$](ClO_4) (**5**), reported by Su et al. (Su et al. 2001) Compound **5** is a M_2L_4-cage compound that has a cage framework similar to Compound **3-SO$_4$**. The coordination of SO_4^{2-} pulls the Cu^{II} ions toward the center of the cage in Compound **3-SO$_4$**; the Cu^{II} ion of Compound **3-SO$_4$** is withdrawn by about 0.21 Å to the center of the cage from the N_4 plane that is defined by four imidazole N atoms in the Cu^{II} environment, while the Cu^{II} ion of Compound **5** remains in the N_4 plane. As a result, the Cu–Cu distance of the M_2L_4 cage is shorter for Compound **3-SO$_4$** (~ 6.6 Å) compared with that of Compound **5** (7.36 Å). This is the first M_2L_4-type cage capturing divalent anions or hydrophilic anions in M_2L_4 cationic cages constructed by bis(imidazole)- or bis(benzimidazole)-type ligands.

Compounds **3-Cl** and **3-Br** are composed of a cationic M_2L_4 part [$SO_4 \subset Cu_2$(*m*-bbitrb)$_4$]$^{2+}$ and counter anions, two Cl^- or two Br^- ions. These compounds have different anions inside and outside the cage. The preferential formations of these structures indicated that the space outside the cationic cage had a higher affinity to Cl^- or Br^- than to SO_4^{2-}, and the space inside the cage had a higher affinity to SO_4^{2-} than to Cl^- and Br^-. As Compounds **3-Cl** and **3-Br** were crystallographically isostructural and had essentially the same structure, the structure of Compound **3-Br**, which was solved at a higher quality, is described for their structure.

The cage structure of Compound **3-Br**, which crystallized in the tetragonal space group *P4/ncc* (# 130) is shown in Fig. 14.3b. The two Cu^{II} ions are bridged by four *m*-bbitrb. The crystallographic *C*4 axis runs through the two Cu^{II} ions in the M_2L_4 cage. The obtained cationic cage traps the SO_4^{2-} ions, which are disordered at the four positions by the crystallographic *C*4 axis. The cationic part, including a SO_4^{2-} of this structure, is essentially the same to those of Compounds **3·SO$_4$** and **3·ClO$_4$**. Synthesis of Compound **3·ClO$_4$** is mentioned below. The SO_4^{2-} in the cage is bound to the two Cu^{II} ions in the cage framework with Cu–O distances of 2.173(5) and 2.164(5) Å. Two Br^- ions that are included as counter anions per the cationic cage are located between the cationic cages.

5.3 Structures of M_2L_4 Cages that include Cl^- or Br^-

The M_2L_4 cage structure of Compound **4-Cl**, which was determined by single-crystal X-ray analysis is shown in Fig. 14.3d earlier. The crystallographic *C*4 axis runs through the two Cu^{II} centers, and a crystallographic mirror exists in the equatorial plane of the cage. The two Cu^{II} ions are bridged by four *m*-bbitrb to give the cationic M_2L_4 cage. This compound incorporates a Cl^- inside the cage and three Cl^- ions outside the cage. The Cl^- in the cage is disordered at the six positions, which are four crystallographically equivalent positions for Cl(1A) and two crystallographically equivalent positions for Cl(1B). Their occupancies were refined as 0.2 and 0.1, respectively. The Cl(1A) associates with the imidazole H atom with a Cl(1A)–H(1) distance of about 2.77 Å. Although Cl(1B) occupies the axial site of the Cu(1) environment, the interaction with Cu(1) is weak as shown by the long Cu(1)–Cl(1B) distance (2.97 (3) Å) (Sone et al. 2016a).

For the coordination circumstance around the Cu^{II} center, a water molecule outside the cage associates with the metal center with a Cu–O distance of 2.637 (5) Å, which is significantly shorter than the Cu(1)–Cl(1B) distance, and longer than the Cu–N bond distances, which are 2.002 (4) Å. The elongated octahedral geometry and square pyramidal geometries are formed around the Cu^{II} center. The geometries are dependent on whether Cl^- occupies the axial site of the Cu^{II} center or not. The water molecule (O(1)) located in the axial position of the two Cu^{II} centers bind to a Cl^- (Cl(2)) via a hydrogen bond with Cl(2)–O(1) distance of 3.113 (12) Å. The occupancy of Cl(2), which was disordered at the four positions by the crystallographic C4 axis, was refined as 0.25. The other Cl^- was located between the cationic M_2L_4 cages without significant interactions with the cage.

Compound **4-Br** crystallized with a similar cell parameter and the same tetragonal space group *I*4/*m* (# 87) to Compound **4-Cl**. Although Compound **4-Br** has a quite similar structure to that of Compound **4-Cl**, the disordered aspect of the anion inside the cage is significantly different. This compound also has the crystallographic *C*4 axis that runs through the two Cu^{II} atoms in the cage framework, and a crystallographic mirror on the equatorial plane. The cationic cage of Compound **4-Br** contains a Br^-, which was disordered at the eight positions (four crystallographically equivalent positions for Br(1A) and four crystallographically equivalent positions for Cl(1B)). Their occupancies were refined as 0.14 and 0.11, respectively. The Br^- in the eight disordered positions in the cage interacted with an imidazole H atom (Br(1A)–H(1) = ca. 2.80 Å, Br(1B)–H(1) = ca. 3.06 Å). In contrast to Compound **4-Br**, Br^- was not observed in the axial position in the coordination environment of the Cu^{II} center.

As observed in Compound **4-Cl**, a water molecule (O(1)) outside the cage weakly associates to the Cu^{II} center in the M_2L_4 cage of Compound **4-Br**. The Cu–O bond distance (2.658 (10) Å) is significantly longer than that of Cu–N bonds (1.992 (4) Å) in the coordination environment. Coordination of four imidazole N atoms at the basal positions and a water molecule in the apical position form the square pyramidal geometry around the Cu^{II} center. Similar to Compound **4-Cl**, a water molecule associated with the Cu^{II} center interacted with a Br^- ion (Br(2A)–O(1) = 3.143 (9) Å), which was disordered at the four positions by the C4 axis. The occupancy of the water molecule (O(1)) was refined as 0.25. The other Br^- was located between the M_2L_4 cages without significant interactions with the cationic cage.

To estimate the stability of the cage frameworks of $[Cl \subset Cu_2(m\text{-bbitrb})_4]^{3+}$ and $[SO_4 \subset Cu_2(m\text{-bbitrb})_4]^{2+}$, ESI-TOF MS spectra were measured for Compounds **4-Cl** and **3-SO$_4$** in MeCN solution. For Compound **3-SO$_4$**, intense peaks ascribed to $[SO_4 \subset Cu_2(m\text{-bbitrb})_3]^{2+}$ and $[SO_4 \subset Cu_2(m\text{-bbitrb})_4]^{2+}$ were observed at m/z 682.2 (calcd. 682.2) and 872.3 (calcd. 872.3). In the case of **4-Cl**, intense peaks ascribed to $[Cl \subset Cu_2(m\text{-bbitrb})_3]Cl^{2+}$, $[Cl \subset Cu_2(m\text{-bbitrb})_2]^{3+}$, $[Cl \subset Cu_2(m\text{-bbitrb})_2]Cl_2^+$, $[Cl \subset Cu_2(m\text{-bbitrb})_3]Cl_2^+$, $[Cl \subset Cu_2(m\text{-bbitrb})_4]Cl_2^+$ were observed at 669.3 (calcd. 669.2), 858.4 (858.3), 993.3 (993.2), 1373.5 (1373.4), and 1755.8 (1755.6), respectively. Compound **3-SO$_4$** showed M_2L_3 and M_2L_4 cages, whereas **4-Cl** showed M_2L_2, M_2L_3, and M_2L_4 cages by the measurements, meaning that [Cl ⊂ Cu$_2$(*m*-

bbitrb)$_4$]$^{3+}$ cage was less stable than the [$SO_4 \subset Cu_2$(*m*-bbitrb)$_4$]$^{2+}$ cage in the solution state. The higher stability of [$SO_4 \subset Cu_2$(*m*-bbitrb)$_4$]$^{2+}$ would be due to the stronger associations of SO_4^{2-} with the Cu^{II} centers (Sone et al. 2016a).

5.4 Removal of ClO_4^- by 3-SO$_4$

Compound **3-SO$_4$** had a high ClO_4^- removal activity from aqueous solutions despite the insolubility of the compound in water. Figure 14.4 exhibits changes in the anion concentration in the aqueous solution as a function of time when 0.20 g (0.10 mmol) of Compound **3-SO$_4$** was added to a solution (100 mL) containing NaF, NaCl, NaBr, $NaNO_3$, $NaClO_4$, and Na_2SO_4 at the same concentration (1.0 mM). Ion chromatography measurements monitored the changes in the concentration of each anion. An amount of 1.0 mM of ClO_4^- was reduced to about 0.070 mM at 5 minutes, and then 0.013 mM within 30 minutes (Fig. 14.4). Other anions, NO_3^- and Br^-, were also reduced from 1.0 mM to about 0.55 and 0.80 mM, respectively, within 30 minutes. Although the high activity was obtained when the powder of Compound **3-SO$_4$** was well ground, the activity remarkably decreased when the microcrystalline sample was used for the experiment. This is because the SO_4^{2-} embedded in the particle cannot be exchanged with ClO_4^-.

To estimate the ClO_4^- removal ability of Compound **3-SO$_4$**, the ClO_4^- removal abilities were examined for an anion-exchange resin (Purolite A530E), which has been widely used for ClO_4^- removal under similar reaction conditions. The resin reduced 1.0 mM of ClO_4^- to about 0.42 mM at 30 minutes and 0.15 mM at 180 minutes. Concentrations of three other anions, NO_3^-, Br^-, and SO_4^{2-}, were also reduced from 1.0 mM to 0.71, 0.79, and 0.91, respectively, at 30 minutes. This study indicated that Compound **3-SO$_4$** had a high ClO_4^- selective removal ability from aqueous solution.

The structure of the reaction product was determined by single crystal XRD analysis for the study of the reaction mechanism. The reaction product obtained after ClO_4^- removal was prepared by immersion of Compound **3-SO$_4$** in an aqueous solution that contained excess $NaClO_4$ for a few days. After careful washing of the blue product, it was

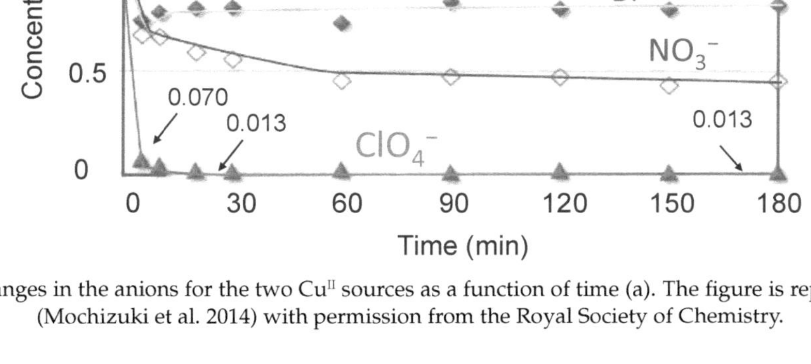

Fig. 14.4: Changes in the anions for the two Cu^{II} sources as a function of time (a). The figure is reproduced from (Mochizuki et al. 2014) with permission from the Royal Society of Chemistry.

recrystallized from MeCN. The single-crystal XRD study confirmed that the product was $[SO_4 \subset Cu_2(m\text{-bbitrb})_4](ClO_4)_2$·**4MeCN (3-ClO$_4$)** (Fig. 14.3c earlier). The M_2L_4 cage structure that trapped SO_4^{2-} of Compound **3-ClO$_4$** was essentially same as that of Compound **3-SO$_4$**, and the two ClO_4^- were located outside the cages. That is, the only SO_4^{2-} outside the cage was exchanged with two ClO_4^- by the ClO_4^- removal reactions, clearly demonstrating that the high ClO_4^- removal activity of Compound **3-SO$_4$** is the result of the counter anion exchange reaction of Compound **3-SO$_4$**. It was also confirmed that Compound **3-ClO$_4$** was insoluble in water.

The concentration of the Cu^{II} ion dissolved in the aqueous solution obtained after the ClO_4^- removal experiment, which was determined using ICP–OES, was only 0.13 mg L^{-1}. This value is lower than the values regulated by Japanese law for waste water (3 mg/L), drinking water (1 mg/L), and mandated by the WHO (2 mg L^{-1}) for drinking water. These results demonstrate that Compound **3-SO$_4$** has a high potential as a new material for ClO_4^- removal from aqueous solutions.

5.5 Perchlorate Removals by 3-Cl, 3-Br, 4-Cl, and 4-Br

The ClO_4^- removal activities from aqueous solutions were characterized for four Compounds, **3-Cl**, **3-Br**, **4-Cl**, and **4-Br**. Figure 14.5a shows the changes of concentrations of ClO_4^- (1.0 mM) depending on time for each aqueous solution (50 mL) of $NaClO_4$ (0.05 mmol) after addition of the powder sample of each compound (0.05 mmol). The changes of the concentrations of anions of each solution were monitored by ion chromatography. As shown in the chart, concentrations of ClO_4^- are reduced significantly in the aqueous solution with similar activities. For the treatments, release of Cl^- (**3-Cl**, **4-Cl**) or Br^- (**3-Br**, **4-Br**) was observed responding to the removal of ClO_4^- from aqueous solutions. For example, in the case of **3-Cl**, 1.0 mM of ClO_4^- was reduced to 0.44 mM at 30 minutes, and then 0.27 mM at 180 minutes, while Cl^-, which was not included in the initial solution, was observed after addition of **3-Cl**. The concentration was 0.37 mM at 10 minutes, and 0.83 mM at 180 minutes. These results exhibit that the four cage compounds removed ClO_4^- by anion exchange reactions.

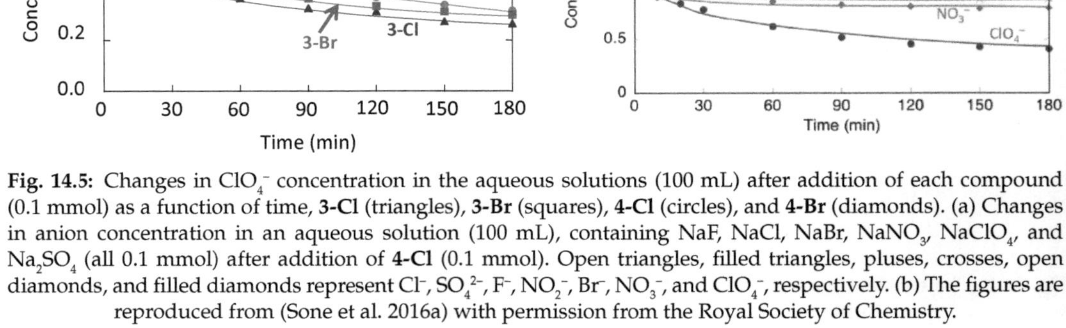

Fig. 14.5: Changes in ClO_4^- concentration in the aqueous solutions (100 mL) after addition of each compound (0.1 mmol) as a function of time, **3-Cl** (triangles), **3-Br** (squares), **4-Cl** (circles), and **4-Br** (diamonds). (a) Changes in anion concentration in an aqueous solution (100 mL), containing NaF, NaCl, NaBr, $NaNO_3$, $NaClO_4$, and Na_2SO_4 (all 0.1 mmol) after addition of **4-Cl** (0.1 mmol). Open triangles, filled triangles, pluses, crosses, open diamonds, and filled diamonds represent Cl^-, SO_4^{2-}, F^-, NO_2^-, Br^-, NO_3^-, and ClO_4^-, respectively. (b) The figures are reproduced from (Sone et al. 2016a) with permission from the Royal Society of Chemistry.

Among the four compounds, **3-Cl**, **3-Br**, **4-Cl**, and **4-Br**, Compound **3-Cl** showed a somewhat high activity for ClO_4^- removal. We have studied the selectivity of the anions removed from aqueous solutions by **3-Cl** (Fig. 14.5b). Changes in anion concentration were noticed after the addition of Compound **3-Cl** (0.10 mmol) in an aqueous solution (100 mL) containing NaF, NaCl, NaBr, $NaNO_3$, $NaClO_4$, and Na_2SO_4 at the same concentration (1.0 mM) as a function of time. ClO_4^- was preferentially reduced from 1.0 mM to 0.63 at 30 minutes, and then 0.41 at 180 minutes. Other anions, NO_3^-, NO_2^-, and Br^- were slightly reduced from 1.0 mM to 0.79, 0.88, and 0.93, respectively, at 180 minutes. The reductions in the F^- and SO_4^{2-} concentrations were even smaller. The Cl^- concentration increased from 1.0 mM to 2.1 by the anion exchange reaction of Compound **3-Cl** with the anions removed. The total quantity of anions removed from the aqueous solution was about 0.10 mmol, close to the Cl^- released in the aqueous solution (0.11 mmol).

A M_2L_4 cage of $[X \subset Cu_2(m\text{-bbitrb})_4]^{3+}$ (X = Cl^-, Br^-) would accommodate three ClO_4^- outside the cage as the counter anions, whose number is larger than the capacity of $[SO_4 \subset Cu_2(m\text{-bbitrb})_4]^{2+}$. That is, Compounds **4-Cl** and **4-Br** would have a higher ClO_4^- removal capacity compared to Compound **3-SO$_4$**. However, the ClO_4^- removal abilities of Compounds **4-Cl** and **4-Br** are remarkably lower than that of Compound **3-SO$_4$**, and are comparable to those of Compounds **3-Cl** and **3-Br**. Despite the same cationic units, i.e., $[SO_4 \subset Cu_2(m\text{-bbitrb})_4]^{2+}$, for Compounds **4-Cl**, **4-Br**, and **3-SO$_4$**, the ClO_4^- removal abilities are remarkably high for Compound **3-SO4**. These results indicate that the kind of counter anions that are exchanged with ClO_4^- largely affects the ClO_4^- removal activity. The larger size of counter anions such as SO_4^{2-} would give a larger space, sufficient for ClO_4^- incorporation by the anion exchange reactions. This is useful for inclusion of the ClO_4^- ions that are removed from the solution. In contrast, the narrower space that remained after release of the smaller anions such as Cl^- would make incorporation of ClO_4^- with larger size more difficult.

6. Removal of ClO_4^- and NO_3^- by Anion Exchange Reactions by MOF

6.1 Synthesis and Structure of 2D MOF with Fixed M_2L_4 Cages

The fixation of functional discrete molecules on an infinite polymer framework is a promising method for the preparation of new polymer compounds that exhibit functionality due to discrete molecular units. For example, many molecular catalysts are bound to 1D organic polymers as pendant units, yielding new polymer compounds that exhibit catalytic reactivity due to the discrete units. Furthermore, sensors and molecular devices have been prepared in addition to various heterogeneous catalysts by attaching discrete functional molecules to the surface of 2D organic films or inorganic layer compounds.

The treatment of metal ions with organic bridging ligands using self-assembly processes has provided numerous MOFs with well-ordered structures. Although some coordination polymers incorporating polyoxometalate units in the frameworks have been reported, the preparation of MOFs bearing discrete molecules as pendant units is not developed. The one-step construction of such functional and outstanding structures by the simple treatment of one type of metal source and one type of organic ligand remains an unexplored topic.

Flexible bis-imidazole- and bis-benzimidazole-type ligands, as illustrated in Scheme 14.1, have been used for the synthesis of MOFs and for various discrete cage molecules, as mentioned above. The structures of the compounds obtained from these ligands are largely dependent on the type of metal ions, counter anions, and reaction solvents used in the synthesis. As shown above, Compound **3-SO$_4$** efficiently removes ClO_4^- from aqueous solutions by exchange with a counter anion SO_4^{2-} located outside the cationic cage, despite the water insolubility of this compound. This shows that the space created among the M_2L_4 cages is effective for selective ClO_4^- traps from aqueous solutions.

For ClO_4^- removal by Compound **3-SO$_4$**, it was necessary to grind the solid sample well to achieve sufficient anion removal activity. It is well-known that MOFs often show unique functions that are independent of particle size due to their porous structures. On continuing the study, we synthesized a new MOF with M_2L_4 cages that were fixed in the 2D framework. The compound obtained showed ClO_4^- removal activity from aqueous solutions irrespective of particle size. The new water-insoluble MOF [Cu(*p*-bbiteb)$_2$ (-μ-Cl-[Cl·(H$_2$O)$_2$ ⊂ Cu$_2$(*p*-bbiteb)$_4$])]Cl$_4$·MeCN·(H$_2$O)$_8$ (**6**) was conveniently obtained by the treatment of $CuCl_2$ with *p*-bbiteb in a mixed MeCN/H_2O solution (Sone et al. 2016b). This compound consisted of a sheet framework expressed as Cu(*p*-bbiteb)$_2$ and cationic Cu$_2$(*p*-bbiteb)$_4$ cages that were bound to the sheet. The regular arrangement of the cationic cages on the sheet created a channel-like space surrounded by the cationic cages. The new self-assembly Compound **6** showed NO_3^- and ClO_4^- removal activities from aqueous solutions. The activities were independent of the particle size due to their porous structure, and the reaction proceeded by anion exchange reactions with the Cl^- ions located in the channels.

Compound **6** crystallized in the tetragonal centric space group ($P4/n$), the structure of which was clarified by single-crystal XRD (Fig. 14.6). The framework consisted of a 2D sheet formulated as [Cu(*p*-bbiteb)$_2$]$^{2+}$ (Fig. 14.6a) and cationic cages (Fig. 14.6b) that trap one Cl^- ion and two H_2O molecules, formulated as [Cl·(H$_2$O)$_2$ ⊂ Cu$_2$(*p*-bbiteb)$_4$]$^{3+}$. Each cage was bound to the Cu^{II} site in the sheet by a μ-Cl^- bridge (Fig. 14.6c). The two Cu^{II} centers were connected by four *syn*-formed *p*-bbiteb in the M_2L_4 cage. The two Cu^{II} ions in the M_2L_4 cage were essentially in the N_4 plane, defined by the four imidazole N atoms in the coordination circumstance. The cage space was well isolated with a gap of about 1.8 × 1.3 Å^2 between the two *p*-bbiteb units in the M_2L_4 framework.

The guest anion Cl^- and the two H_2O molecules trapped in the cage do not show strong interactions with the cage framework. The Cl^- associates with the Cu^{II} site of the M_2L_4 cage with a relatively long Cu(1)–Cl(1) distance of 2.531 (3) Å. The distance is remarkably longer

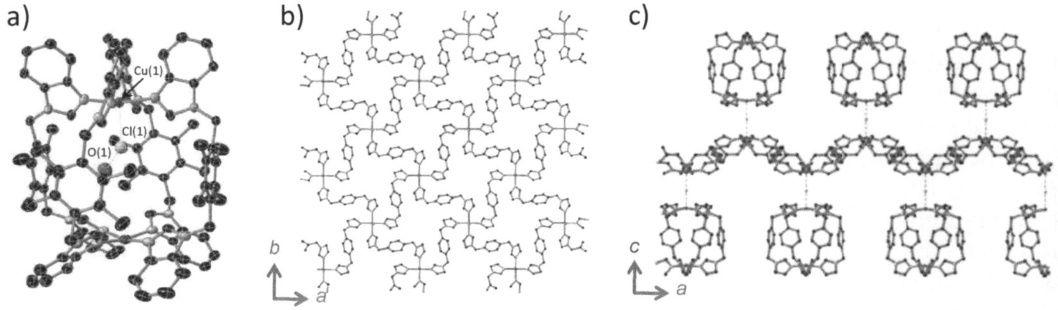

Fig. 14.6: Structures of the cationic cage unit (a), the sheet framework (b), and the 2D structure that bears the cationic cage units (c) of **6**. The following were omitted for clarity: H atoms and cocrystallized MeCN molecules in (a)–(c), disorder of water molecules in the cage in (a) and (c), and phenyl rings of benzimidazole and methyl groups of *p*-bbiteb in (b) and (c). The figures are reproduced from (Sone et al. 2016b) with permission from the Royal Society of Chemistry.

than the Cu–N distances (avg. 2.014 Å) in the coordination environment because of the Jahn–Teller effect. The Cl^- also weakly associates with H atoms of the imidazole ring of the *p*-bbiteb (Cl(1)–H(1) = 2.882 Å) and water molecules (Cl(1)–O(1) = 3.026 (8) Å). The water molecules are disordered at the four crystallographic equivalent positions.

The 2D framework of Compound **6** is shown in Fig. 14.6b. Each Cu^{II} ion is bridged by the *anti*-formed *p*-bbiteb to give a 2D framework in the *ab* plane. The Cu–N distances (2.015 (4) Å) are similar to those observed in the M_2L_4 cages. The sheet is not planar but has waves (Fig. 14.6c). The Cu^{II} ions located at the top of the waves are bound to the cationic M_2L_4 cages via the μ-Cl^- bridge on both sides of the sheet. The Cu–Cl distance is similar for the two sets of Cu and Cl atoms, (2.504 (2) Å and 2.553 (2) Å). The formed sheet bearing cationic M_2L_4 cages was stacked along the *c* axis. Compound **6** created a channel-like space surrounded by cationic M_2L_4 cages that were bound to the sheet. The channel space, with a size of about 8.3 × 3.4 $Å^2$ in the channel window, was filled with four Cl^- ions, one MeCN unit, and eight H_2O molecules per three Cu^{II} ions. Among the guest molecules, MeCN was easily removed under air or reduced pressure, giving [Cu(*p*-bbiteb)$_2$(-μ-Cl-[Cl·$(H_2O)_2$ ⊂ Cu_2(*p*-bbiteb)$_4$])]$Cl_4(H_2O)_8$ (**6b**), which was characterized by elemental analysis and PXRD measurements. The PXRD measurement demonstrated that Compound **6b** retained the overall structure of Compound **6**. The ClO_4^- removal properties of the MOF were characterized by using Compound **6b** because of the property of convenient storage.

6.2 ClO_4^- Removal Activities of 6b

Characterization of the anion removal property of Compound **6b** revealed that this compound is useful for removing NO_3^-, as well as ClO_4^-, from aqueous solutions. NO_3^- is also an anion that is difficult to remove from aqueous solutions because of its high solubility. The inorganic anion NO_3^- is also toxic to humans, and is particularly harmful to infants and children. The anion disturbs O transport by hemoglobin, a phenomenon known as blue-baby syndrome. Recent reports also suggest that NO_3^- causes various cancers. In addition to ClO_4^-, NO_3^- has also been detected in the environment and in drinking water; therefore, the development of new materials that remove these anions is very important.

To determine whether Compound **6** has useful toxic anion removal properties based on the cationic cages arranged on the 2D sheet, the removal properties for the ClO_4^- and NO_3^- were characterized for Compound **6b**. The changes in the concentration of ClO_4^- and NO_3^- (1.0 mM) in 100 mL of an aqueous solution are shown in Fig. 14.7 as a function of time after the addition of Compound **6b** (147.5 mg, 0.05 mmol) with different particle sizes. The concentration of ClO_4^- in an aqueous solution of $NaClO_4$ (0.1 mmol in 100 mL) was reduced from 1.0 mM to 0.43 and to 0.26 after 10 and 180 minutes, respectively, after addition of a powdered sample of Compound **6b** (147.5 mg, 0.05 mmol) (Fig. 14.7). The concentration of NO_3^- was reduced from 1.0 mM to 0.58 and then to 0.28 after 10 and 180 minutes, respectively, by a similar treatment. The anion removal activity increased significantly when the microcrystalline form of Compound **6b** was used. The powdered sample of Compound **6b** became aggregated during the treatment, but the microcrystalline sample did not, causing the intense activity of the latter. In contrast to Compound **3-SO$_4$**, Compound **6b** showed less dependency of the particle size on the anion removal activity. This is the effect of the channel structure, as generally observed in porous materials such as zeolites, graphite and many MOFs, with channel frameworks.

In the ClO_4^- removal experiment by Compound **6b**, Cl^- appeared, and the concentration in the solution increased, in response to the reduction of the ClO_4^- concentration. This result

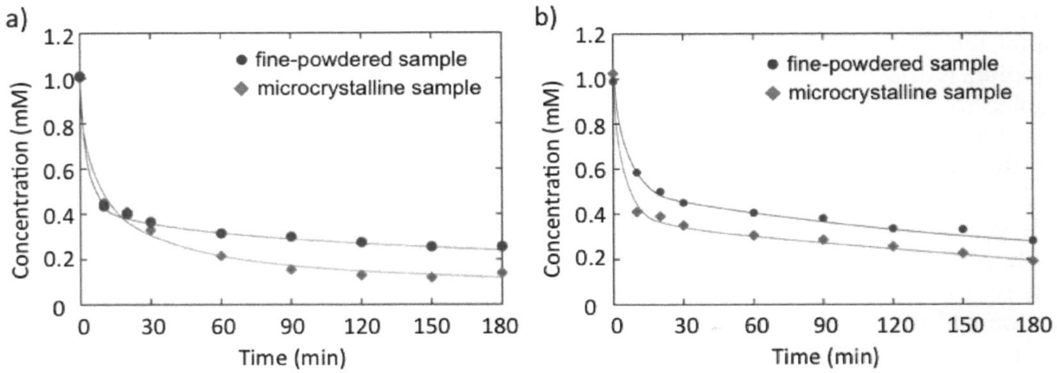

Fig. 14.7: Changes in concentration of ClO_4^- (1.0 mM) (a) and NO_3^- (1.0 mM) (b) in 100 mL aqueous solution as a function of time after the addition of a fine-powdered (circles) and microcrystalline sample (diamonds) of **6b** (147.5 mg, 0.05 mmol). The figures are reproduced from (Sone et al. 2016a) with permission from the Royal Society of Chemistry.

indicated that the anions were removed in anion-exchange reactions. To study the ClO_4^- and NO_3^- removal process by Compound **6b**, the reaction products obtained by the treatment of Compound **6b** with 10 equiv ClO_4^- or NO_3^- in aqueous solution were characterized. The PXRD patterns of the reaction products were essentially identical to those of Compound **6** and **6b**, indicating that the framework of **6** was retained after the anion-exchange reactions. The inclusion of ClO_4^- and NO_3^- in the products was also confirmed by IR spectroscopy, which showed new intense peaks at around 1100 and 1400 cm^{-1}; these peaks were ascribed to ClO_4^- and NO_3^-.

The d–d absorption of the M_2L_4 cage shifts depending on the kinds of anions trapped in the cage because the anion affects the d orbital levels, particularly the d_{x2-y2} orbital level, of the Cu^{II} ions in the cage. For example, the M_2L_4 cage that traps ClO_4^- shows the d–d absorption at a higher energy region due to an increased d_{x2-y2} orbital caused by the weak crystal field of ClO_4^-. The reflectance spectra of the reaction products show similar d–d absorptions compared to those of **6** and **6b**, indicating that Cl^- in the cationic Cu_2(*p*-bbiteb)$_4$ cage is not exchanged with ClO_4^- or NO_3^- during the anion removal experiments. That is, only Cl^- in the channel-like space is exchanged with ClO_4^- and NO_3^- by the anion removal process.

It was investigated whether **6b** could remove ClO_4^- and NO_3^- from an aqueous solution that mimicked groundwater, which contained $NaHCO_3$ (3.0 mM), $CaCl_2$ (1.0 mM), $MgCl_2$ (0.5 mM), Na_2SO_4 (0.5 mM), and KNO_3 (0.5 mM). For the removal experiments, $NaClO_4$ or $NaNO_3$ (0.10 mmol) was dissolved in this solution (100 mL). Changes in concentrations of ClO_4^- and NO_3^- as a function of time after the addition of **6b** (299 mg, 0.10 mmol) are shown in Fig. 14.8. The concentration of ClO_4^- was reduced from 1.0 mM to 0.32 and 0.13 after 10 and 180 minutes, respectively (Fig. 14.8a), and the concentration of NO_3^- was reduced from 1.0 mM to 0.42 and 0.28 after 10 and 180 minutes, respectively (Fig. 14.8b). The removal activity was higher for ClO_4^- than for NO_3^-. This tendency was consistent with those of other systems described in this chapter. The preferential ClO_4^- removal activity would be the result of higher affinity of the anion to the hydrophobic space surrounded by the cationic cages. Although many MOFs showed the anion exchange property, this is the first example that shows selective removals for ClO_4^- and NO_3^- from aqueous solutions.

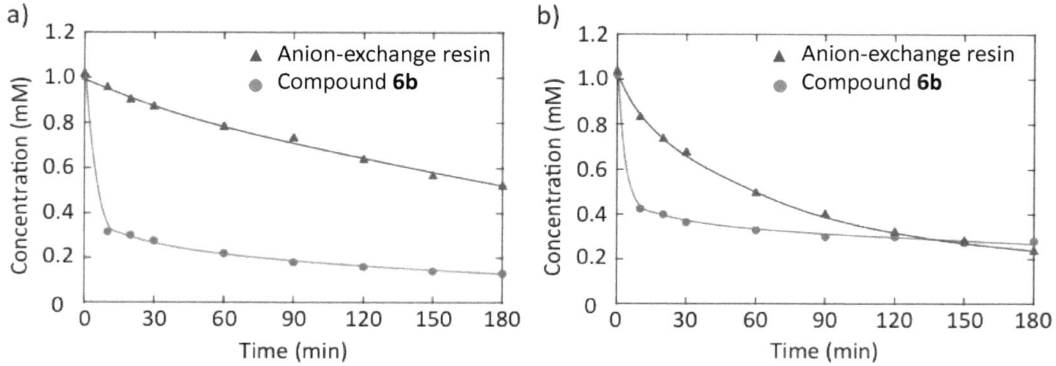

Fig. 14.8: Changes in ClO_4^- and NO_3^- (1.0 mM each) concentration (a) in aqueous solutions (100 mL each) that mimic groundwater as a function of time after the addition of **6b** (299 mg, 0.10 mmol) (circles) or anion-exchange resins (triangles): Purolite A530E (299 mg) (a) and Purolite A520E (299 mg) (b). The figures are reproduced from (Sone et al. 2016a) with permission from the Royal Society of Chemistry.

To estimate the anion removal activities of **6b**, similar anion removal experiments were carried out for the anion-exchange resins Purolite A530E and A520E (299 mg of each), which are widely used for the removal of ClO_4^- and NO_3^- from aqueous solutions, respectively. The changes in the ClO_4^- or NO_3^- (1.0 mM each) concentration dissolved in the mimicked groundwater after addition of **6b** or the resin are shown in Fig. 14.8. As shown, with Compound **6b** the anion removals occur within a shorter period than with the anion-exchange resins. That is, the ClO_4^- and NO_3^- removals were remarkably faster for Compound **6b** than for the resins, indicating that Compound **6b** is superior in terms of ClO_4^- and NO_3^- removals.

7. Regeneration of M_2L_4 Cage for ClO_4^- Removal after the Treatment

Regeneration of the resins after ClO_4^- adsorption is an important subject for anion-exchange resins in practical use. However, with a higher ClO_4^- adsorption activity, the regeneration, i.e., desorption of ClO_4^-, would be difficult. As Compounds **3-SO$_4$** and **6** (**6b**) showed remarkably high ClO_4^- or NO_3^- removal activities, the regeneration process after ClO_4^- removal was examined for these compounds.

It was found that Compound **3-ClO$_4$** is soluble in an organic solvent such as MeCN. Treatment of the extracted MeCN solution with a strong base, such as NaOH, decomposed the cage compound into $Cu(OH)_2$, *m*-bbitrb, and $NaClO_4$. Analytically pure *m*-bbitrb was conveniently recovered from the decomposed products by a general procedure with about 53% yield. The recovered *m*-bbitrb was useful for the repreparation of Compound **3-SO$_4$**.

For Compound **6b**, the simple regeneration process from the anion-exchanged products by simple reverse anion-exchange reactions, and the chemical degradation process were examined. When ClO_4^-- and NO_3^-- absorbed products were treated with excess NaCl in aqueous solutions, the anions trapped in the channels were re-exchanged to Cl^- at 12 and 40% yield, respectively. A lower yield for the ClO_4^--containing product is due to the high affinity of ClO_4^- in the channel space. In contrast, the general chemical degradation procedure of the anion-exchanged products gives analytically pure *p*-bbiteb at 47% yield. The ligand can be used for repreparation of Compound **6b**.

8. Mechanism of Selective Inclusion of ClO_4^- in the M_2L_4 Cage

8.1 Anion Exchanges Inside and Outside the Cage

It has been shown that cationic M_2L_4 cage compounds remove ClO_4^- preferentially from aqueous solutions by anion exchange reactions. To study the affinity of the anions to the inside and outside space created by the cationic cage $[Cu_2(m\text{-bbitrb})_4]^{4+}$, the exchanges of the anions between the inside and outside of the cage for Compound **3-ClO$_4$** were studied. The reflectance spectra of solid state $[ClO_4 \subset Cu_2(m\text{-bbitrb})_4(ClO_4)_2](ClO_4)$, which was prepared by the reaction of $Cu(ClO_4)_2{\cdot}6H_2O$ with *m*-bbitrb and Compound **3-ClO$_4$**, reveal the d–d absorptions at λ_{max} = 568 and 628 nm, respectively. The Cu(II) center of the former compound is based on the elongated octahedral geometry with four imidazole N atoms at the equatorial positions and two O atoms of ClO_4^- at the axial positions. The higher d–d absorption energy for $[ClO_4 \subset Cu_2(m\text{-bbitrb})_4(ClO_4)_2](ClO_4)$ is due to the d_{x2-y2} orbital that increased due to the weak crystal field of ClO_4^- as discussed above.

The absorption spectra of $[ClO_4 \subset Cu_2(m\text{-bbitrb})_4(ClO_4)_2](ClO_4)$ and Compound **3-ClO$_4$** dissolved in MeCN, which show d–d absorptions at λ_{max} = 573 and 626 nm, respectively. Although each compound shows the slight shift of their λ_{max}, their d–d absorption energies are essentially similar to those of the solid-state reflectance spectra, indicating that their Cu^{II} coordination environments, i.e., their molecular structures, found in the solid-state are retained in the MeCN solution.

The absorption spectra of $[ClO_4 \subset Cu_2(m\text{-bbitrb})_4(ClO_4)_2](ClO_4)$ and Compound **3-ClO$_4$** measured in a MeCN/H_2O (v/v; 1/1) solution exhibit the λ_{max} of the d–d absorption at 573 and 574 nm, respectively. The absorption spectrum of Compound **3-ClO$_4$** significantly changed when it was measured in the MeCN/H_2O mixed solution. The absorption spectrum of Compound **3-ClO$_4$** became quite similar to that of $[ClO_4 \subset Cu_2(m\text{-bbitrb})_4(ClO_4)_2](ClO_4)$ in the MeCN/H_2O mixed solution. This result indicates that SO_4^{2-} inside the cage was exchanged with ClO_4^- located outside the cage.

To confirm this anion exchange reaction, the ESI-TOF mass spectrum of Compound **3-ClO$_4$** was measured in the MeCN and MeCN/H_2O mixed solution. Compound **3-ClO$_4$** showed two intense peaks ascribed to $[SO_4 \subset Cu_2(m\text{-bbitrb})_3]^{2+}$ and $[SO_4 \subset Cu_2(m\text{-bbitrb})_4]^{2+}$ at *m/z* 682.2 (calcd. 682.2) and *m/z* 872.3 (calcd. 872.3), respectively. The ESI-TOF mass spectrum of Compound **3-ClO$_4$** measured in the MeCN/H_2O (v/v; 1/1) solution did not exhibit the intense peaks at (*m/z* 682.2 and 872.3), which is observed in the MeCN solution, but showed other peaks at *m/z* 582.5 and 923.2, which were assigned to $[ClO_4 \subset Cu_2(m\text{-bbitrb})_4]^{2+}$ (calcd. 582.5) and $[ClO_4 \subset Cu_2(m\text{-bbitrb})_4(ClO_4)]^+$ (calcd. 923.2), respectively. This result demonstrated that the SO_4^{2-} ion that was trapped in the cage in the MeCN solution was replaced with the counter anion ClO_4^- located outside the cage when **3-ClO$_4$** was dissolved in the MeCN/H_2O mixed solution. The results observed in ESI-TOF mass spectra were consistent with the results of the absorption spectra change mentioned above.

The changes in the absorption spectra of the MeCN solution of **3-ClO$_4$** by additions of H_2O aliquots are also shown in Fig. 14.9. The intensity of absorption at λ_{max} = 629 nm of **3-ClO$_4$** gradually reduces and shifts to the higher energy region by the addition of water. The changes in the absorption spectra stop essentially when 0.8 mL of water is added in 2.0 mL of MeCN solution. The absorption spectrum finally obtained is consistent with that of $[ClO_4 \subset Cu_2(m\text{-bbitrb})_4(ClO_4)_2](ClO_4)$, indicating that their anions are completely exchanged when about 30% of water is present in the MeCN solution.

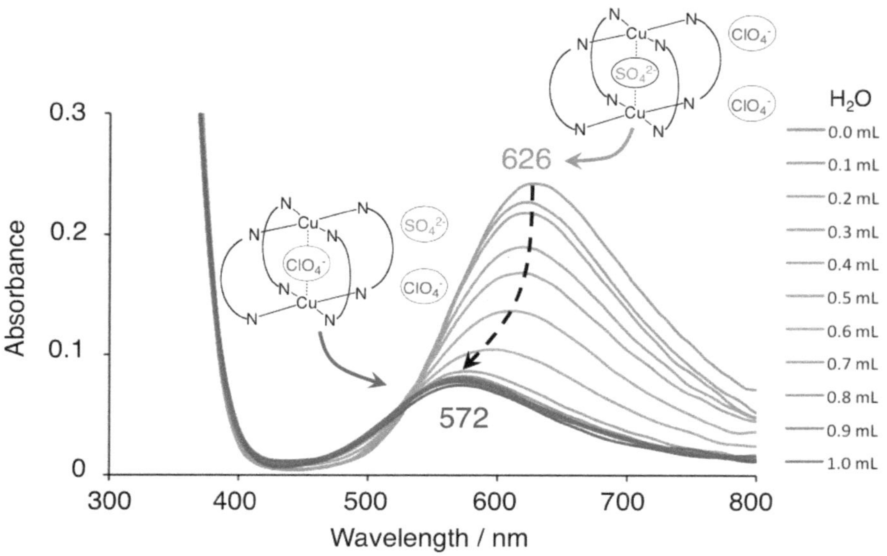

Fig. 14.9: Absorption titration spectra of **3-ClO$_4$** in MeCN (2.0 ml) upon addition of sequential 0.1 mL aliquots of water. The figure is reproduced from (Sone et al. 2016a) with permission from the Royal Society of Chemistry.

The space in the cationic cage [Cu_2(*m*-bbitrb)$_4$] is hydrophobic because it is surrounded by the hydrophobic ligands *m*-bbitrb. ClO_4^- has a high hydrophobic surface, whereas SO_4^{2-} has a hydrophilic surface. The space inside and outside the [Cu_2(*m*-bbitrb)$_4$]$^{4+}$ cage is hydrophobic both in the solid state or in the MeCN solutions. Because Compound **3-SO$_4$** is insoluble in water, when the powder sample is immersed in water, the space created between the [Cu_2(*m*-bbitrb)$_4$]$^{4+}$ cages is hydrophobic. Therefore, Compound **3-SO$_4$** changes to Compound **3-ClO$_4$** by the treatment with ClO_4^- in aqueous solutions. The solid-state structure of **3-ClO$_4$** is retained in MeCN because MeCN was sufficiently hydrophobic. In contrast, when the hydrophilicity of the solution increased by the addition of H_2O, the space outside the cage increased the hydrophilic circumstance, inducing an exchange of hydrophilic ClO_4^- outside the cage and hydrophilic SO_4^{2-} inside the cage. The ability for selective removal of ClO_4^- observed in some compounds increases based on the hydrophobic space of the compounds.

9. Preparation of ClO_4^- Removal Filter

The ClO_4^- removal speed of **3-SO$_4$** was very fast, prompting us to prepare a **3-SO$_4$**-fixed filter to remove ClO_4^- from aqueous solutions by using a commercially available syringe filter (ϕ 40 mm, 0.2 μm pores). The fixation of **3-SO$_4$** on the membrane was carried out by filtration of a powder sample of **3-SO$_4$** suspended in water using the syringe filter. After the aqueous solution of $NaClO_4$ (ca. 1,000 ppb for ClO_4^-) was passed through the filter under hand-pressure, the ClO_4^- concentration in the filtrate was determined by ion chromatography. The filter reduced the 1000 ppb ClO_4^- to ca. 12 ppb, a value that is lower than the value (15 ppb) declared by EPA as the acceptable concentration level for ClO_4^- in drinking water. This result indicated that **3-SO$_4$** was useful for convenient ClO_4^- removal from aqueous solutions within a short time.

10. Summary

New coordination compounds that show ClO_4^- removal activities from aqueous solutions are described in this chapter. Some compounds trap ClO_4^- during the self-assembled constructions of the structure, while other compounds trap the anion by anion-exchange reactions. For the M_2L_4 cage compounds, not only do the cationic cages preferentially trap the anion inside the cage, but the compounds also trap the anion outside the cage. The functions of the selective ClO_4^- removal from aqueous solutions of the various compounds are the results of the high affinity of the hydrophobic ClO_4^- to the hydrophobic space created in the host compounds.

References

Agency, U.S.E.P. 2011. Drinking water: Regulatory determination on perchlorate. Federal Register 76 FR 7762. 76: 7762–7767.

Amouri, H., C. Desmarets, A. Bettoschi, M.N. Rager, K. Boubekeur, P. Rabu et al. 2007. Supramolecular cobalt cages and coordination polymers templated by anion guests: Self-assembly, structures, and magnetic properties. Chem. Eur. J. 13: 5401–5407.

Asami, M., K. Kosaka and N. Yoshida. 2009. Occurrence of chlorate and perchlorate in bottled beverages in Japan. J. Health Sci. 55(4): 549–553.

California Department of Health Services. 2007. Perchlorate in Drinking Water. R-16-04.

Chung, H.R. 2014. Iodine and thyroid function. Ann. Pediatr. Endocrinol Metab. 19: 8–12.

Cole-Dai, J., K.M. Peterson, J.A. Kennedy, T.S. Cox and D.G. Ferris. 2018. Evidence of influence of human activities and volcanic eruptions on environmental perchlorate from a 300-year greenland ice core record. Environ. Sci. Technol. 52: 8373–8380.

Dasgupta, P.K., P.K. Martinelango, W.A. Jackson, T.A. Anderson, K. Tian, R.W. Tock et al. 2005. The origin of naturally occurring perchlorate: the role of atmospheric processes. Environ. Sci. Technol. 39: 1569–1575.

Dyke, J.V., K. Ito, T. Obitsu, Y. Hisamatsu, P.K. Dasgupta and B.C. Blount. 2007. Perchlorate in dairy milk. Comparison of Japan versus the United States. Environ. Sci. Technol. 41: 88–92.

Fritzmann, C., J. Löwenberg, T. Wintgens and T. Melin. 2007. State-of-the-art of reverse osmosis desalination. Desalination 216: 1–76.

Furdui, V.I., J. Zheng and A. Furdui. 2018. Anthropogenic perchlorate increases since 1980 in the Canadian high arctic. Environ. Sci. Technol. 52: 972–981.

Furin, C.G., F.A. von Hippel, J.H. Postlethwait, C.L. Buck, W.A. Cresko and T.M. O'Hara. 2015. Developmental timing of sodium perchlorate exposure alters angiogenesis, thyroid follicle proliferation and sexual maturation in stickleback. Gen. Comp. Endocrinol. 219: 24–35.

Greer, M.A., G. Goodman, R.C. Pleus and S.E. Greer. 2002. Health effects assessment for environmental perchlorate contamination: The dose response for inhibition of thyroidal radioiodine uptake in humans. Environ. Health Perspect. 110: 927–937.

Greliner, P., C. McLellan, D. Bennett and A. Ewing. 2008. Occurrence of perchlorate in sodium hypochlorite. AWWA 100(11): 68–74.

Gu, B., G.M. Brown, P.V. Bonnesen, L. Liang, B.A. Moyer, R. Ober et al. 2000. Development of novel bifunctional anion-exchange resins with improved selectivity for pertechnetate sorption from contaminated groundwater. Environ. Sci. Technol. 34: 1075–1080.

Han, M., D.M. Engelhard and G.H. Clever. 2014. Self-assembled coordination cages based on banana-shaped ligands. Chem. Soc. Rev. 4: 1848–1860.

Hirakawa, T., M. Yamaguchi, N. Ito, M. Miyazawa, N. Nishina, M. Kondo et al. 2009. Removal of perchlorate anion from an aqueous solution by encapsulation in an anion-templated self-assembled molecular capsule. Chem. Lett. 38: 290–291.

Hutchison, J.M., S.K. Poust, M. Kumar, D.M. Cropek, I.E. MacAllister, C.M. Amett et al. 2013. Perchlorate reduction using free and encapsulated *Azospira oryzae* enzymes. Environ. Sci. Technol. 47: 9934–9941.

Hutchison, J.M. and J.L. Zilles. 2015. Biocatalytic perchlorate reduction: kinetics and effects of groundwater characteristics. Environ. Sci. Water Res. Technol. 1: 913–921.

Hutchison, J.M., J.S. Guest and J.L. Zilles. 2017. Evaluating the development of biocatalytic technology for the targeted removal of perchlorate from drinking water. Environ. Sci. Technol. 51: 7178–7186.
Inoue, T., K. Yamanishi and M. Kondo. 2013. Solvent-soluble coordination polymer that reconstructs cyclic frameworks that trap a kinetically labile $[Cu(CO_3)_2]^{2-}$ unit. Inorg. Chem. 52: 4765–4767.
Jackson, L.P. 2011. Drinking water. Regulatory determination on perchlorate. EPA-HQ-OW-2008-0692, EPA-HQ-OW-2009-0297; FRL-9262-8, Fed. Reg. 76: 7762–7767. Washington. www.gpo.gov/fdsys/pkg/FR-2011-02-11/pdf/2011-2603.pdf.
Jasper, J.T., Y. Yang and M.R. Hoffmann. 2017. Toxic byproduct formation during electrochemical treatment of latrine wastewater. Environ. Sci. Technol. 51: 7111–7119.
Kim, Y.-N. and M. Choi. 2014. Synergistic integration of ion-exchange and catalytic reduction for complete decomposition of perchlorate in waste water. Environ. Sci. Technol. 48: 7503–7510.
Kirk, A.B., P.K. Martinelango, K. Tian, A. Dutta, E.E. Smith and P.K. Dasgupta. 2005. Perchlorate and iodide in daily and breast milk. Environ. Sci. Technol. 39: 2011–2017.
Koike, S., T. Hirakawa, K. Yamanishi and M. Kondo. 2014. Two new coordination polymers, a trinuclear metal complex and their interconversion depending on the solvent. Dalton Trans. 43: 12832–12835.
Kosaka, K., M. Asami, Y. Matsuoka, M. Kamoshita and S. Kunikane. 2007. Occurrence of perchlorate in drinking watersources of metropolitan are in Japan. Water Res. 41: 3474–3482.
Kubik, S. 2010. Anion recognition in water. Chem. Soc. Rev. 39: 3648–3663.
Langton, M.J., C.J. Serpell and P.D. Beer. 2016. Anion recognition in water: Recent advances from supramolecular and macromolecular perspective. Angew. Chem. Int. Ed. 55: 1974–1987.
Lawrence, J.E., S.H. Lamm, S. Pino, K. Richman and L.E. Braverman. 2000. The effect of short-term low-dose perchlorate on various aspects of thyroid function. Thyroid 10: 659–663.
Liu, H.-K., J. Hu, T.-W. Wang, X.-L. Yu, J. Liu and B. Kang. 2001a. Synthesis and crystal structures of silver(I), palladium(II), zinc(II) and cobalt(II) complexes of the diptotic ligand 1,3-bis(imidazole-1-ylmethyl)2,4-6-trimethylbenzene. J. Chem. Soc. Dalton Trans. 3534–3540.
Liu, H.-K., C.-Y. Su, C.-M. Qian, J. Liu, H.-Y. Tan and B.-S. Kang. 2001b. Assembly of a twin-cage complex containing a linear array of Pd_3. J. Chem. Soc., Dalton Trans. 1167–1168.
Liu, H.-K., Y. Cai, W. Luo, F. Tong, C. You, S. Lü et al. 2009. Discrete M_2L_2 metallacycle and M_2L_4 cage frameworks and anion competitive reactions of Cu_2L_4 type receptor. Inorg. Chem. Commun. 12: 457–460.
Liu, J., M. Han, D. Wu, X. Chen, J.K. Choe, C.J. Werth et al. 2016. A new bioinspired perchlorate reduction catalyst with significantly enhanced stability via rational tuning of rhenium coordination chemistry and heterogeneous reaction pathway. Environ. Sci. Technol. 50: 5874–5881.
Logan, B.E. 2001. Assessing the outlook for perchlorate remediation. Environ. Sci. Technol. 35: 482A–487A.
Massachusetts Department of Environmental Protection. 2016. Massachusetts Drinking Water Regulations. 310 CMR 22, 2016.
Mochizuki, M., T. Inoue, K. Yamanishi, S. Koike, M. Kondo, L. Zhang et al. 2014. Efficient removal of perchlorate ion from water by a water-insoluble M2L4 type compound. Dalton Trans. 43: 17924–17927.
Murakami, A., K. Yamanishi, E. Sone and M. Kondo. 2015. Preferential removal of perchlorate ion from water using self-assembled constructions of cationic 3D coordination frameworks with methylene units. Chem. Lett. 44: 1007–1009.
Nikolayenko, V.I., A. Heyns and L.J. Barbour. 2017. Threading the needle: guest transport in a versatile 0D porous molecular crystal. Chem. Commun. 53: 11306–11309.
Parette, R. and F.S. Cannon. 2005. The removal of perchlorate from groundwater by activated carbon tailored with cationic surfactants. Water Res. 39(16): 4020–4028.
Poghosyan, A., N.C. Sturchio, C.G. Morrison, A.D.J. Beloso, Y. Guam, J.M. Eiler et al. 2014. Perchlorate in the great lakes: Isotopic composition and origin. Environ. Sci. Technol. 48: 11146–11153.
Rubin, R., M. Pearl, M. Kharrazi, B.C. Blount, M.D. Mille, E.N. Pearce et al. 2017. Maternal perchlorate exposure in pregnancy and altered birth outcomes. Environ. Res. 158: 72–81.
Sone, E., M. Sato, M. Mochizuki, C. Kamio, K. Yamanishi and M. Kondo. 2016a. Cationic M2L4 cages for perchlorate removal from aqueous solutions and preferential perchlorate incorporation in hydrophilic solutions. CrystEngComm. 18: 5004–5011.
Sone, E., M. Sato, K. Yamanishi, C. Kamio, H. Takemoto and M. Kondo. 2016b. Self-assembled construction of a sheet-type coordination polymer bearing cationic M2L4 cages: Creation of channel-like space for removal of ClO_4^- and NO_3^- from aqueous solutions. Dalton Trans. 45: 894–898.

Srinivasan, A. and T. Viraraghavan. 2009. Perchlorate: Health effects and technologies for its removal from water resources. Int. J. Environ. Res. Public Health 6: 1418–1442.

Su, C.-Y., Y.-P. Cai, C.-L. Chen, H.-X. Zhang and B.-S. Kang. 2001. Coordination-directed assembly of trigonal and tetragonal molecular boxes encapsulating anionic guests. J. Chem. Soc. Dalton Trans.: 359–361.

Su, C.-Y., Y.-P. Cai, C.-L. Chen, M.D. Smith, W. Kaim and H.-C. zur Loye. 2003. Ligand-directed molecular architectures: self-assembly of two-dimensional rectangular metallacycles and three-dimensional trigonal or tetragonal prisms. J. Am. Chem. Soc. 125: 8595–8613.

Urbansky, E.T. and M.R. Schock. 1999. Issues in managing the risks associated with perchlorate in drinking water. J. Environ. Manage. 56: 79–95.

USEPA. 2008. Interim drinking water health advisory for perchlorate. Health and Ecological Criteria Division, Office of Science and Technology, Office of Water, US Environmental Protection Agency, Washington, DC 20460, www.epa.gov/waterscience. EPA 822-R-08-025. www.epa.gov/safewater/contaminants/unregulated/perchlorate.html.

Wolff, J. and N.S. Halmi. 1963. Thyroidal iodide transport. J. Biol. Chem. 238: 847–851.

Wolff, J. 1998. Perchlorate and the thyroid gland. Pharmacol. Rev. 50: 89–105.

Yamaguchi, M., T. Hirakawa, N. Nishina, M. Kondo, H. Aoki, E. Okuda et al. 2010. Removal of perchlorate and tetrafluoroborate anions from aqueous solutions by a self-assembled Cu(II) molecular capsule system using water-insoluble Cu(II) source. Chem. Lett. 39(11): 1192–1193.

Yoon, I.-H., X. Meng, C. Wang, K.-W. Kim, S. Bang, E. Choe et al. 2009. Perchlorate adsorption and desorption on activated carbon and anion exchange resin. J. Hazard. Mater. 164: 87–94.

Yu, K.O., L. Narayanan, D.R. Mattie, R.J. Godfrey, P.N. Todd, T.R. Sterner et al. 2002. The pharmacokinetics of perchlorate and its effect on the Hypothalamus–Pituitary–Thyroid axis in the male rat. Toxicol. Appl. Pharmacol. Rev. 182(2): 148–159.

Załubiniak, D., J. Kos and P. Piatek. 2017. Exploiting cooperative binding of ion-pair to boost anion recognition in water/acetonitrile mixtures. Tetrahedron 73: 7190–7194.

Chapter 15

Zirconocene Derivatives having α-Keggin-type Mono-Aluminum-coordinated Polyoxometalates

Synthesis, Molecular Structures, and Application to Heterogeneous Catalysts

Chika Nozaki Kato[1,2]

1. Introduction

Alumina (Al_2O_3) is widely used as a support for industrial catalysts due to its mechanical strength, thermal stability, surface property and strong interaction with metals and metal oxides (Cotton and Wilkinson 1988, Ono and Hattori 2011). Generally, alumina is prepared from aluminum hydroxide ($Al(OH)_3$) and aluminum oxyhydroxide (AlO(OH)) by dehydration at elevated temperature, and it exists in different crystalline phases depending on the precursors and the conditions of heat treatment (Ono and Hattori 2011). Although the property and activity of alumina strongly depend on surface aluminum sites, construction of well-defined aluminum structures is insufficient even when conditions are controlled during preparations.

Polyoxometalates are metal-oxo clusters that are of great interest as catalysts (Pope 1983, Pope and Muller 1991, 1994, Wang and Yang 2015). Particularly, the introduction of metal ions as heteroatoms to the vacant sites of lacunary polyoxometalates is one of the unprecedented effective technologies for constructing well-defined, active metal sites. Various types of aluminum-coordinated polyoxometalates have also been synthesized and

[1] Department of Chemistry, Shizuoka University, Ohya, Suruga-ku, Shizuoka 422-8529, Japan.
[2] Research Institute of Green Science and Technology, Shizuoka University, 836 Ohya, Suruga-ku, Shizuoka 422-8529, Japan.
Email: kato.chika@shizuoka.ac.jp

structurally characterized, as follows: $[(n\text{-}C_4H_9)_4N]_4[H]ClAlW_{11}PO_{39}$ (Knoth et al. 1983), $[(n\text{-}C_4H_9)_4N]_4[\alpha\text{-}PW_{11}\{Al(OH_2)\}O_{39}]$ (Kato et al. 2012), $[(CH_3)_4N]_4[\alpha\text{-}PW_{11}\{Al(OH_2)\}O_{39}]\cdot 8H_2O$ (Kato et al. 2013a), $Cs_4[\alpha\text{-}PW_{11}\{Al(OH_2)\}O_{39}]\cdot 8H_2O$ (Kato et al. 2013a), $K_5[\alpha\text{-}SiW_{11}\{Al(OH_2)\}O_{39}]\cdot 7H_2O$ (Ma et al. 2011), $[(n\text{-}C_4H_9)_4N]_4K_{0.5}H_{0.5}[\alpha\text{-}SiW_{11}\{Al(OH_2)\}O_{39}]\cdot H_2O$ (Kato et al. 2016), $K_7[\alpha_2\text{-}P_2W_{17}\{Al(OH_2)\}O_{61}]\cdot 14H_2O$ (Kato et al. 2013a, Zonnevijlle et al. 1982), $[(n\text{-}C_4H_9)_4N]_3H[\gamma\text{-}SiW_{10}O_{36}\{Al(OH_2)\}_2(\mu\text{-}OH)_2]\cdot 4H_2O$ (Kikukawa et al. 2008a), $K_6H_3[ZnW_{11}O_{40}Al]\cdot 9.5H_2O$ (Yang et al. 1997), $K_6Na[(A\text{-}PW_9O_{34})_2\{W(OH)(OH_2)\}\{Al(OH)(OH_2)\}\{Al(\mu\text{-}OH)(OH_2)_2\}_2]\cdot 19H_2O$ (Kato et al. 2010), $K_6[B\text{-}\alpha\text{-}H_3P_2W_{15}\{Al(OH_2)\}_3O_{59}]\cdot 14H_2O$ (Kato et al. 2014a), $Cs_6[B\text{-}\alpha\text{-}H_3P_2W_{15}\{Al(OH_2)\}_3O_{59}]\cdot 13H_2O$ (Kato et al. 2014a), $K_{14}Na_2[B\text{-}\alpha\text{-}H_3P_2W_{15}\{Al(OH)\}_2\{Al(OH_2)\}O_{59}]_2\cdot 30H_2O$ (Kato et al. 2014a), $[(CH_3)_4N]_{14}Na_2[B\text{-}\alpha\text{-}H_3P_2W_{15}\{Al(OH)\}_2\{Al(OH_2)\}O_{59}]_2\cdot 39H_2O$ (Kato et al. 2014a), $Rb_2Na_2[Al_4(H_2O)_{10}(\beta\text{-}AsW_9O_{33}H)_2]\cdot 20H_2O$ (Carraro et al. 2013), and $[NH_4]_2Na_2[Al_4(H_2O)_{10}(\beta\text{-}AsW_9O_{33}H)_2]\cdot 20H_2O$ (Carraro et al. 2013). Some of these exhibited high catalytic activities for Lewis acid catalysis (Kikukawa et al. 2008a), oxidation catalysis (Carraro et al. 2013, Kato et al. 2013a, 2014b), and photocatalysis (Hattori et al. 2015, Liu et al. 2012). However, there are no reported examples, so far, of aluminum-substituted polyoxometalates being used as molecular supports for catalyst preparation.

To date, complete and intact polyoxometalates that can be used as molecular supports have been limited to compounds containing d^0 transition metals (Day and Klemperer 1985a, Gouzerh and Proust 1998). The examples are: $Nb_6O_{19}{}^{8-}$ (Flynn and Stucky 1969a, 1969b, 1969c), $Nb_2W_4O_{19}{}^{4-}$ (Besecker et al. 1982, 1984, 1985, Besecker and Klemperer 1980, Day et al. 1990, 1993, Klemperer and Main 1990), $A\text{-}\beta\text{-}SiW_9M_3O_{40}{}^{7-}$ (M = Nb^V, V^V) (Finke et al. 1986a, 1986b, 1990a, 1992, Finke and Droege 1984, Lin et al. 1993, Rapko et al. 1994), $B\text{-}\alpha\text{-}P_2W_{15}M_3O_{62}{}^{9-}$ (M = Nb^V, V^V) (Edlund et al. 1988, Finke et al. 1990b, 1990c, Lyon and Finke 1990, Nomiya et al. 1994, 1995, 1997, 2003, Pohl et al. 1995a, 1995b, Pohl and Finke 1993, Trovarelli and Finke 1993), $\alpha\text{-}PW_{11}NbO_{40}{}^{4-}$ (Radkov et al. 1999, Radkov and Beer 2000), $ClTiW_5O_{18}{}^{3-}$ (Day et al. 1985b, Klemperer and Zhong 1993), $ClTiMo_5O_{18}{}^{3-}$ (Che et al. 1985, Day et al. 1981, Klemperer and Shum 1979), and $MW_5O_{19}{}^{3-}$ (M = Nb^V, Ta^V) (Day et al. 1985c) and some of them have been applied to oxidation catalysts (Edlund et al. 1988, Mizuno et al. 1996, Weiner et al. 2003), precursors to prepare nanoclusters (Hornstein and Finke 2004, Lin and Finke 1994a, 1994b, Ozkar and Finke 2002, Watzky et al. 2008), and transfer reagents (Radkov et al. 1999).

Polyoxometalate-supported complexes still have great potential to develop polyoxometalate chemistry. However, the available metal-substituted sites to graft organometallic fragments still are limited. Therefore, we would like to propose using α-Keggin-type mono-aluminum-substituted polyoxometalates, $[(n\text{-}C_4H_9)_4N]_4[\alpha\text{-}PW_{11}\{Al(OH_2)\}O_{39}]$ (Kato et al. 2012) and $[(n\text{-}C_4H_9)_4N]_4K_{0.5}H_{0.5}[\alpha\text{-}SiW_{11}\{Al(OH_2)\}O_{39}]\cdot H_2O$ (Kato et al. 2016), as molecular supports, and also report on how the grafting reactions of organometallic species onto the aluminum sites affect that improving the catalytic activities.

2. Synthesis and Molecular Structures of Zirconocene Derivatives having α-Keggin-type Mono-aluminum-substituted Polyoxotungstates

$[(n\text{-}C_4H_9)_4N]_4[\alpha\text{-}PW_{11}\{Al(OH_2)\}O_{39}]$ (**TBA-3**) was formed by a reaction of aluminum nitrate with γ-Keggin-type di-lacunary polyoxometalate $[\gamma\text{-}PW_{10}O_{36}]^{7-}$ (the molar ratio of Al^{3+}:$[\gamma\text{-}PW_{10}O_{36}]^{7-}$ was ~ 1.0) in an aqueous solution at approximately 40°C, followed by adding excess tetra-*n*-butylammonium bromide (Kato et al. 2012). The white precipitate was crystallized by a slow-evaporation from acetonitrile at approximately 25°C, as shown in Scheme 15.1. For the synthesis of $[(n\text{-}C_4H_9)_4N]_4K_{0.5}H_{0.5}[\alpha\text{-}SiW_{11}\{Al(OH_2)\}O_{39}]\cdot H_2O$ (**TBA-4**), a stoichiometric amount of aluminum nitrate was added to the aqueous solution of α-Keggin-type mono-lacunary polyoxometalate $[\alpha\text{-}SiW_{11}O_{39}]^{8-}$ at approximately 25°C, and then excess tetra-*n*-butylammonium bromide was added to the solution (Kato et al. 2016). The obtained product was crystallized via vapor diffusion from acetonitrile/methanol, as shown in Scheme 15.2. **TBA-3** and **TBA-4** were characterized by elemental analysis, X-ray crystallography, thermogravimetric/differential thermal analysis (TG/DTA), potentiometric titration, Fourier Transform Infrared spectroscopy (FTIR) and solution {^{31}P, ^{29}Si, ^{27}Al, and ^{183}W} Nuclear Magnetic Resonance (NMR) spectroscopies, and density-

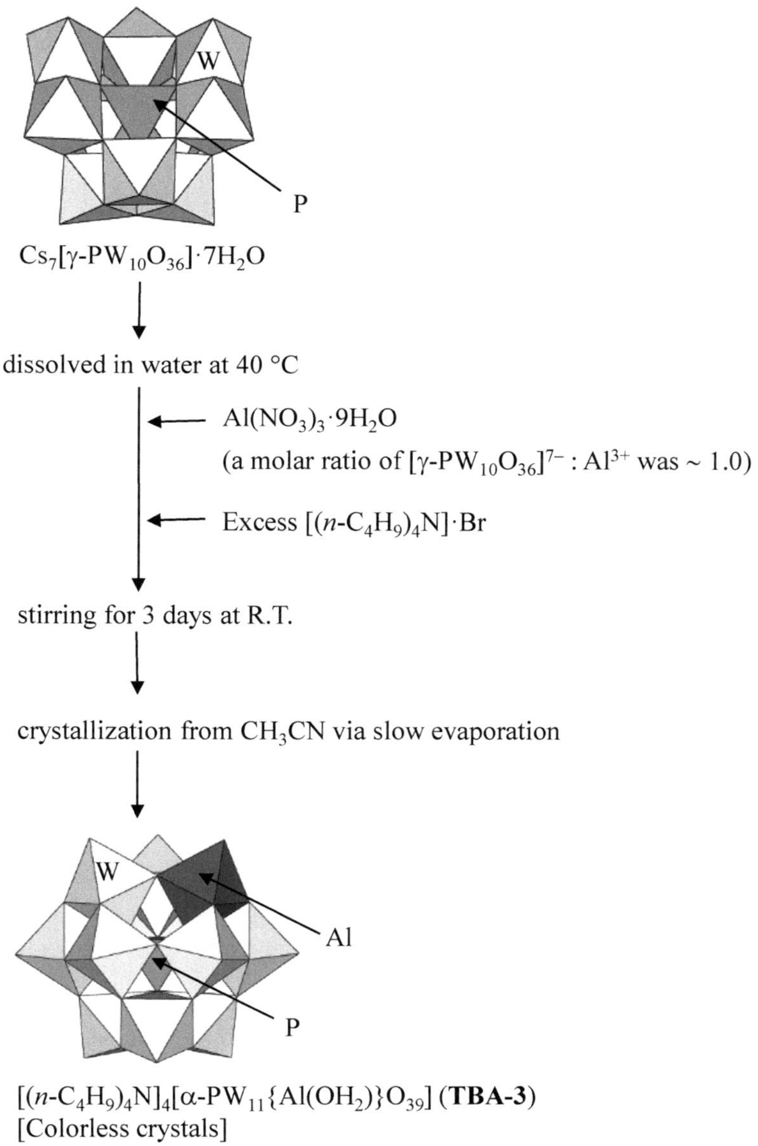

Scheme 15.1: Synthetic diagram of **TBA-3**.

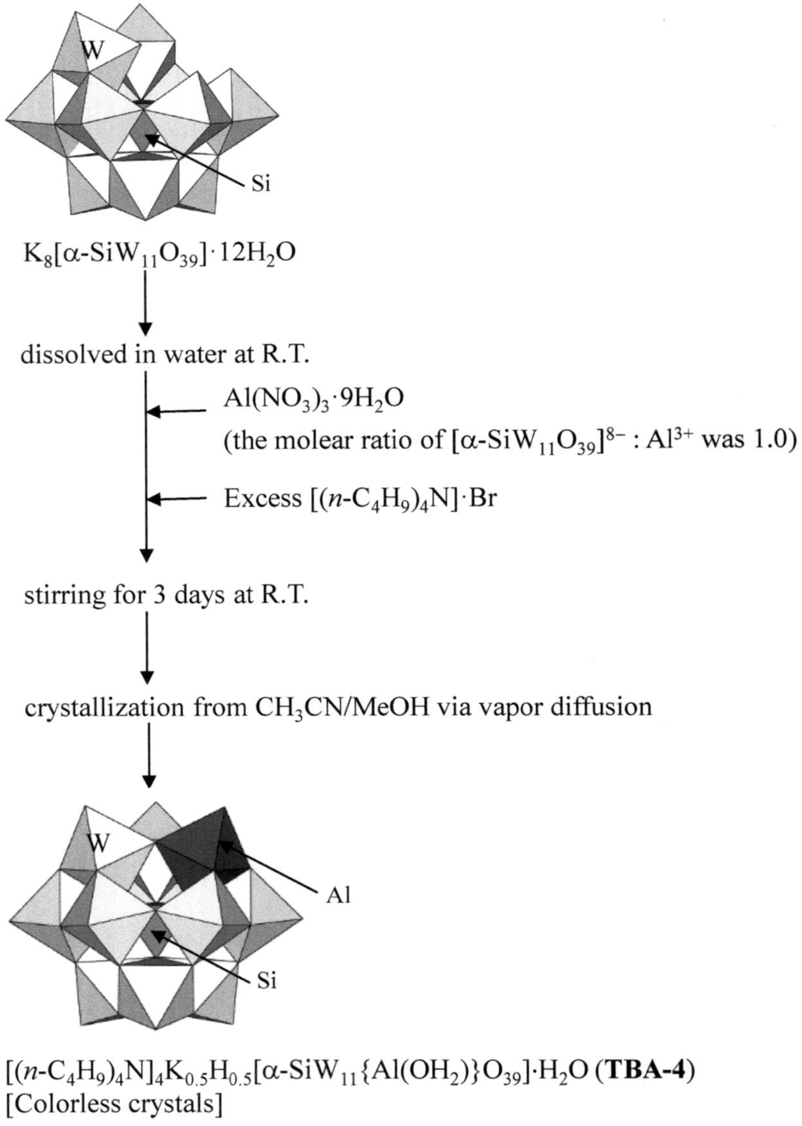

Scheme 15.2: Synthetic diagram of **TBA-4**.

functional-theory calculations. The characterization results revealed that both compounds had a monomeric α-Keggin-type structure, and a water molecule was coordinated to the aluminum site.

Here, the selection of organometallic species was extremely important for manifestation of excellent catalytic activities. We focused on the chemistry of organometallic moieties with η^5-cyclopentadienyl ligands because they have been widely used in various fields including catalytic chemistry and materials science (Glockner and Tamm 2013, Krut'ko 2009, Raith et al. 2010). Only a few examples (e.g., $\{[(\eta^5\text{-}C_5H_5)_2U]_2(\mu\text{-}k^2O\text{-}TiW_5O_{19})_2\}^{4-}$ (Day et al. 1985b, Klemperer and Zhong 1993), $[(PW_{11}O_{39}NbO)_2Zr(\eta^5\text{-}C_5H_5)_2]^{6-}$ (Radkov et al.

1999, Radkov and Beer 2000), $[(\eta^5\text{-}C_5H_5)_3U(MW_5O_{19})_2]^{5-}$, and $[(\eta^5\text{-}C_5H_5)_3Th(MW_5O_{19})_2]^{5-}$ (M = Nb, Ta) (Day et al. 1985c)) have been reported, however, the synthetic chemistry regarding the grafting reactions of organometallic moieties with η^5-cyclopentadienyl ligands onto the metal-substituted sites in polyoxometalates has scarcely been advanced. Here, we demonstrated the synthetic strategies of bis(η^5-cyclopentadienyl)zirconium(IV) compounds having α-Keggin-type mono-aluminum-substituted phosphotungstates $[(n\text{-}C_4H_9)_4N]_6[\alpha\text{-}PW_{11}Al(OH)O_{39}Zr(\eta^5\text{-}C_5H_5)_2]_2$ (**TBA-1**) (Kato et al. 2013b) and $[(n\text{-}C_4H_9)_4N]_6[\alpha\text{-}SiW_{11}Al(OH)_2O_{38}Zr(\eta^5\text{-}C_5H_5)_2]_2 \cdot 2H_2O$ (**TBA-2**) (Kato et al. 2016).

TBA-1 was synthesized by the reaction of **TBA-3** with $(\eta^5\text{-}C_5H_5)_2Zr(O_3SCF_3)_2 \cdot THF$ (approximately 1:1 ratio) in acetonitrile under argon atmosphere (Kato et al. 2013b). Then, a yellow precipitate was obtained by the addition of water to the acetonitrile solution in air. Crystallization was performed by slow evaporation from acetonitrile under argon atmosphere at approximately 25°C (Scheme 15.3). **TBA-1** was characterized based on X-ray structure analysis, elemental analysis, TG/DTA, FTIR and solution {^{1}H, ^{13}C, ^{31}P, ^{27}Al, and ^{19}F} NMR spectroscopy. Single-crystal X-ray structure analysis of **TBA-1** showed that the two {$PW_{11}AlO_{40}$} units were bridged by two bis(η^5-cyclopentadienyl) zirconium fragments. Each zirconium fragment was linked to the terminal oxygen atom coordinated to the Al and W sites and the edge-sharing oxygen atom at the Al-O-W bond.

TBA-2 can also be obtained by a stoichiometric reaction of **TBA-4** with $(\eta^5\text{-}C_5H_5)_2Zr(O_3SCF_3)_2 \cdot THF$ in acetonitrile under an argon atmosphere, followed by precipitation from water in air (Kato et al. 2016). However, **TBA-2** can be easily synthesized in air using $(\eta^5\text{-}C_5H_5)_2ZrCl_2$ instead of $(\eta^5\text{-}C_5H_5)_2Zr(O_3SCF_3)_2 \cdot THF$ as a starting complex (Scheme 15.3). The crystallization of **TBA-2** was carried out by vapor diffusion from

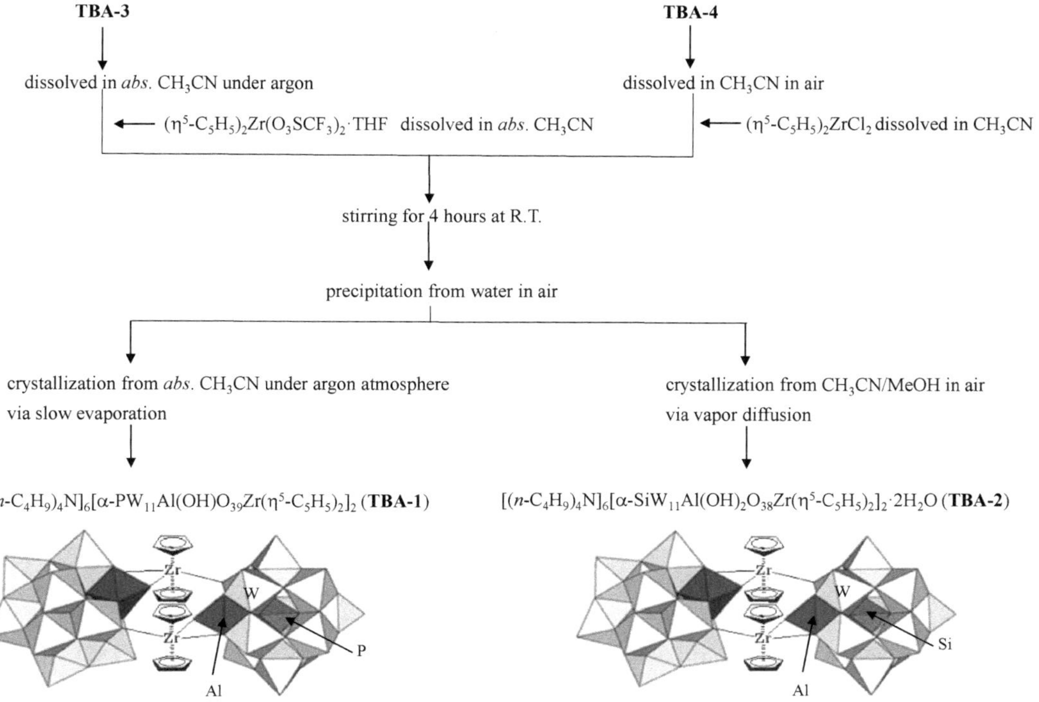

Scheme 15.3: Synthetic diagram of **TBA-1** and **TBA-2**.

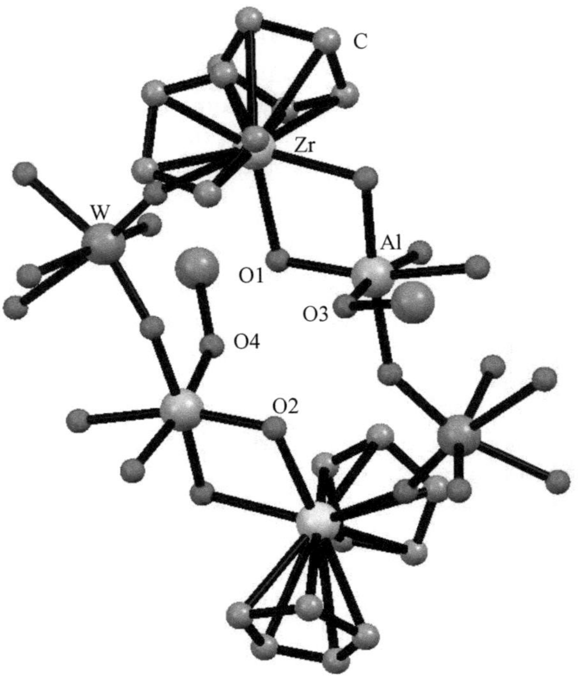

Fig. 15.1: The partial structure around the zirconium and aluminum sites of **TBA-2**. Copyright (2017) Elsevier.

acetonitrile/methanol in air. For the characterization of **TBA-2**, elemental analysis, X-ray crystallography, TG/DTA, cold-spray ionization mass (CSI-MS), potentiometric titration, FTIR and solution {^{1}H, ^{13}C, and ^{27}Al} and solid-state {^{13}C, ^{27}Al, and ^{29}Si} NMR spectroscopy were performed.

X-ray structural analysis also showed that the **TBA-2** was dimeric, containing two {$SiW_{11}AlO_{40}$} units bridged by two $(\eta^5\text{-}C_5H_5)_2Zr^{2+}$ fragments. Each zirconium center was five-coordination geometry, same as those of **TBA-1**. The values of the Bond Valence Sums (BVSs) (Brown 1992, 1996, Brown and Altermatt 1985, Brown and Shannon 1973) for **TBA-2** showed that **TBA-2** has a pair of protons on the oxygen atoms (O1 and O2) at Al-O-Zr linkage and another pair of protons on the oxygen atoms (O3 and O4) at Al-O-W linkage, as shown in Fig. 15.1. While, **TBA-1** has only a pair of protons on the oxygen atoms at Al-O-Zr linkage (O1 and O2). Thus, the number and position of protons in **TBA-2** were different from those in **TBA-1**.

The catalytic activities of the two organozirconium compounds for esterification and Meerwein-Ponndorf-Verley (MPV) reduction are described below.

3. Heterogeneous Esterification of Fatty Acids with Methanol Catalyzed by TBA-1 and TBA-2

From the viewpoint of environmental and economic consideration, the production of renewable resource fuels has become a very important task in maintaining our way of life. Among various approaches to renewable energy manufacturing technology, biodiesel production has attracted a great deal of attention because it can be prepared from a variety of vegetable oils and animal fats and it is readily used in existing diesel engines (Demirbas 2009, Helwani et al. 2009, Lopez et al. 2007).

Biodiesel is manufactured industrially by transesterification of triglycerides with methanol or ethanol catalyzed by bases (e.g., NaOH, KOH, or methoxides) in homogeneous systems (Corma et al. 2007, Huber et al. 2006, Su and Guo 2014a). However, the process is inhibited by the contamination of Free Fatty Acids (FFAs) which cause saponification side reactions. To overcome the problem, the synthetic methods of various solid acid catalysts suitable for the esterification of FFAs with methanol or ethanol have been investigated. The examples are sulfated zirconia (Banerjee et al. 2015, Patel et al. 2013), sulfonated carbon (Liu et al. 2008, Mo et al. 2008), Nafion-based composites (Ni and Meunier 2007), sulfonated and incompletely carbonized sugar, starch, or cellulose (Takagaki et al. 2006, Toda et al. 2005), zirconium-containing metal organic frameworks (Cirujano et al. 2015), layered zinc laurate and zinc stearate (de Paiva et al. 2015), and pyrene-based porous organic polymers (Kundu and Bhaumik 2015).

Among various types of polyoxometalates, heteropolyacids (HPAs), e.g., $H_3[PW_{12}O_{40}]$ and/or $H_3[PMo_{12}O_{40}]$, are well-known Bronsted acid catalyst (Okuhara et al. 1996), and supported HPAs on activated carbon fibers, zirconia, silica, and zeolites showed high catalytic activities for esterification of FFAs with alcohols (Alcaniz-Monge et al. 2013, Baroi and Dalai 2014, Narkhede et al. 2015, Su et al. 2014b). However, since HPAs are generally soluble in alcohol solutions, leaching of HPAs into the reaction solutions has become a problem. Here, we focused on the fact that **TBA-1** and **TBA-2** were insoluble in methanol, and attempted to use them as solid acid catalysts for the esterification of FFAs with methanol (Kato et al. 2017).

Figure 15.2 showed the catalytic activities of **TBA-1** and **TBA-2** for esterification of linoleic acid with methanol at 80 ± 2°C. The main product was methyl linoleate with > 99% selectivity. It was also confirmed that the reaction hardly progresses without the catalyst. For **TBA-1**, the conversion was 83% after 6 hours. After 12 hours, it reached to 99%. In contrast, **TBA-2** showed 15% of conversion after 6 hours, which was lower than that of **TBA-1**. Although **TBA-1** and **TBA-2** were different only in their internal elements and retained the same dimeric structure (see Scheme 15.3), the catalytic activities of **TBA-1** were about six times higher than those of **TBA-2**. Even in a homogeneous system (where acetonitrile was used as a solvent), **TBA-1** showed the higher conversion (83% after 96 hours) for esterification of benzoic acid than that of **TBA-2** (22%).

In order to confirm the leaching of these compounds during the catalytic reactions, we carried out the experiments as follows: after stirring at 80 ± 2°C for 6 hours, the solid of **TBA-1** and **TBA-2** were immediately separated from the reaction solutions by filtration. The ^1H NMR spectra in acetonitrile-d_3 of the filtrates did not show the signals of η^5-C_5H_5 molecules, $(\eta^5\text{-}C_5H_5)_2Zr^{2+}$, and tetra-*n*-butylammonium ions. No signal was also observed in the ^{31}P NMR spectrum in acetonitrile-d_3 of the filtrate recovered from the reaction solution containing **TBA-1**. These results showed that (1) **TBA-1** and **TBA-2** did not leach into the mixture of linoleic acid and methanol, (2) the η^5-C_5H_5 ligands were coordinated to the zirconium atoms, and (3) $(\eta^5\text{-}C_5H_5)_2Zr^{2+}$ fragments were present on the surface of **TBA-3** and **TBA-4**. It was also confirmed in the following experiment that **TBA-1** and **TBA-2** are heterogeneous catalysts: after stirring the methanol suspension containing **TBA-1** or **TBA-2** for 6 hours at 80 ± 2°C, **TBA-1** and **TBA-2** were separated from the solution by filtration. Then, linoleic acid was added to the filtrates, and stirred the reaction solutions at 80 ± 2°C again. No reaction was observed after 6 hours.

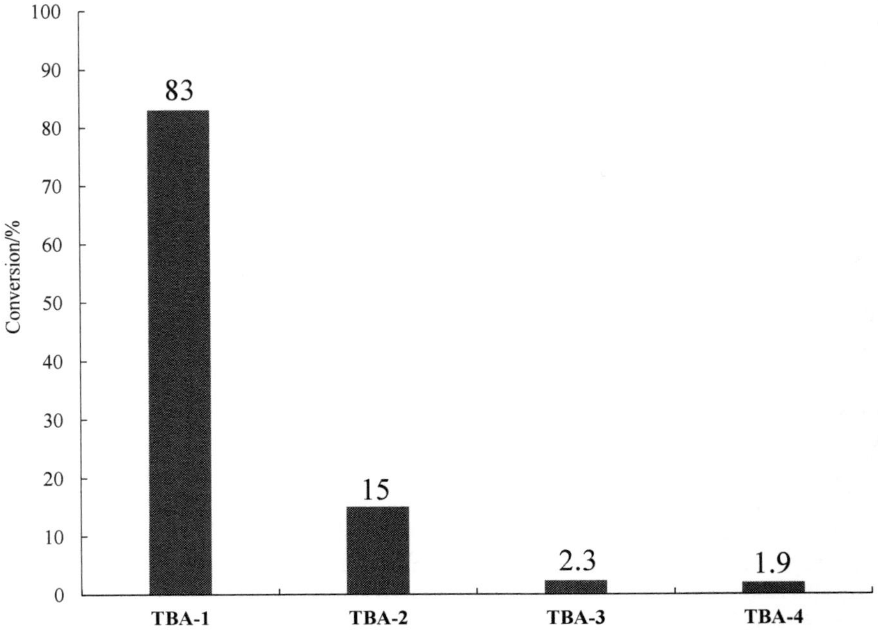

Fig. 15.2: Esterification of linoleic acid with methanol catalyzed by **TBA-1**, **-2**, **-3**, and **-4**. Reaction conditions: catalyst (19 μmol), linoleic acid (0.28 mmol), methanol (49.3 mmol), 80 ± 2°C, and reaction time 6 h. Conversion (%) = {[mol of linoleic acid]$_0$ − [mol of linoleic acid]$_t$}/[mol of linoleic acid]$_0$ × 100.

Under the reported reaction conditions, **TBA-3** and **TBA-4** showed a conversion of less than 3%. Although $(\eta^5\text{-}C_5H_5)_2Zr(O_3SCF_3)_2\cdot THF$ and $(\eta^5\text{-}C_5H_5)_2ZrCl_2$ exhibited > 98% conversion, they acted as homogeneous catalysts. These results suggested that the grafting reactions of $(\eta^5\text{-}C_5H_5)_2Zr^{2+}$ onto the surfaces of α-Keggin-type mono-aluminum-substituted polyoxometalaes were effective in generating the active sites in a heterogeneous system. $H_3[PW_{12}O_{40}]\cdot 23H_2O$ also showed > 99% conversion, however, it also acted as a homogeneous catalyst. Even if it is supported on the zirconium oxide surface, $H_3[PW_{12}O_{40}]\cdot 23H_2O$ was leached into the solution. Zirconium oxide showed lower conversion (15% after 6 hours) in a heterogeneous system, which was lower than that of **TBA-1**. Some Lewis acidic polyoxometalates $[(n\text{-}C_4H_9)_4N]_3H[\gamma\text{-}SiW_{10}O_{36}\{Al(OH_2)\}_2(\mu\text{-}OH)_2]\cdot 4H_2O$ (Kikukawa et al. 2008a), $K_{15}H[Zr(a_2\text{-}P_2W_{17}O_{61})_2]\cdot 25H_2O$ (Absillis and Parac-Vogt 2012, Kato et al. 2006, Stroobants et al. 2014), $Na_8[\{Zr_4(\mu\text{-}OH)_2(m_3\text{-}O)_2(H_2O)_4\}(\alpha\text{-}1,4\text{-}PW_{10}O_{37})_2]\cdot 17H_2O$ (Nomiya et al. 2013), and $Cs_8[(\gamma\text{-}SiW_{10}O_{36})_2\{Zr(H_2O)\}_4(\mu_4\text{-}O)(\mu\text{-}OH)_6]\cdot 26H_2O$ (Kikukawa et al. 2008b) showed no reaction under the reported reaction conditions. **TBA-1** was also active in the heterogeneous esterification of oleic acid (the conversion after 6 hours was 90%), palmitic acid (86%), myristic acid (75%), and lauric acid (69%) with methanol, as shown in Fig. 15.3. The main products were the corresponding ester molecules with > 99% selectivity. Thus, **TBA-1** exhibited excellent catalytic activities for esterification of various FFAs with methanol even in heterogeneous system.

As shown in X-ray crystal data, **TBA-1** and **TBA-2** have two and four protons located at the oxygen atoms ((O1 and O2) and (O1, O2, O3, and O4), respectively), as shown in Fig. 15.1. To investigate the effect of protons for the esterification, we observed the Bronsted acidity of **TBA-1** and **TBA-2** by using dicinnamalacetone, which is a kind of

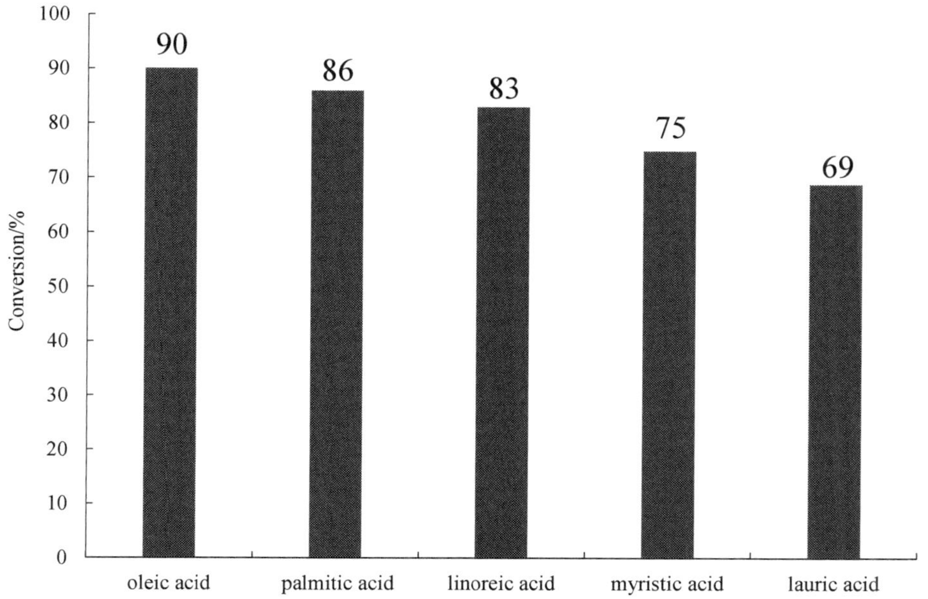

Fig. 15.3: Esterification of various FFAs with methanol catalyzed by **TBA-1**.
Reaction conditions: catalyst (19 micromol), substrates (0.28–0.34 mmol), methanol (49.3 mmol), 80 ± 2°C, and reaction time 6 h.
Conversion (%) = {[mol of substrate]$_0$ – [mol of substrate]$_t$}/[mol of substrate]$_0$ × 100.

Hammett indicators (Okuhara et al. 1994). When a solution dissolving **TBA-1** or **TBA-2** in acetonitrile was added to the dicinnamalacetone solution, the color of solution did not change. In contrast, the color of solution dissolving $H_3[PW_{12}O_{40}]\cdot 23H_2O$ changed from colorless to red. The coloration to red indicated that $H_3[PW_{12}O_{40}]\cdot 23H_2O$ had Bronsted acidity. It was also confirmed that deprotonation of **TBA-1** and **TBA-2** had little effect on the catalytic activities. These results suggested that the protons in **TBA-1** and **TBA-2** did not act as Bronsted acid sites, and the Lewis acidic centers in **TBA-1** and **TBA-2** caused the catalysis under the reported reaction conditions.

In addition to the Lewis acidic polyoxometalates used as comparisons above, several compounds have been reported to show Lewis acid catalysis. Examples are $Na_{10}K_{22}[Zr_{24}O_{22}(OH)_{10}(H_2O)_2(W_2O_{10}H)_2(GeW_9O_{34})_4(GeW_8O_{31})_2]\cdot 85H_2O$ (Huang et al. 2014), $[(n\text{-}C_4H_9)_4N]_6[\{W_5O_{18}Zr(\mu\text{-}OH)\}_2]\cdot 2H_2O$ (Absillis and Parac-Vogt 2012, Stroobants et al. 2014), $[(C_2H_5)_2NH_2]_{10}[Zr(PW_{11}O_{39})_2]\cdot 7H_2O$ (Absillis and Parac-Vogt 2012, Stroobants et al. 2014), $Na_{14}[Zr_4(P_2W_{16}O_{59})_2(m_3\text{-}O)_2(OH)_2(H_2O)_4]\cdot 10H_2O$ (Absillis and Parac-Vogt 2012, Stroobants et al. 2014), $Rb_2Na_2[Al^{III}_4(H_2O)_{10}(\beta\text{-}As^{III}W_9O_{33}H)_2]\cdot 20H_2O$ (Carraro et al. 2013), and $[NH_4]_2Na_2[Al^{III}_4(H_2O)_{10}(\beta\text{-}Sb^{III}W_9O_{33}H)_2]\cdot 20H_2O$ (Carraro et al. 2013). However, there has been no report that these polyoxometalates showed catalytic activities for esterification of FFAs. As the only example of polyoxometalate-based, heterogeneous catalysts, $[C_{16}TA]H_4TiPW_{11}O_{40}$ ($C_{16}TA$ = cetyltrimethylammonium) exhibited 94.7% conversion after 6 hours for the esterification of palmitic acid with methanol at 65°C (Zhao et al. 2012). However, it has both Bronsted and Lewis acid sites. Thus, **TBA-1** would be the first example of a polyoxometalate-based, solid Lewis acidic catalyst for the esterification of FFAs with methanol.

4. Heterogeneous Meerwein-Ponndorf-Verley (MPV) Reduction of Carbonyl Compounds with 2-propanol Catalyzed by TBA-1 and TBA-2

Meerwein-Ponndorf-Verley (MPV) reduction is an effective reaction for the selective reduction of carbonyl groups in the presence of other reducible functions (Chuah et al. 2006, Cohen et al. 2004, de Graauw et al. 1994, Klomp et al. 2004, Li et al. 2015, Ruiz and Jimenez-Sanchidrian 2007, Shylesh et al. 2009, Song et al. 2015). In particular, an attractive possibility is to use secondary alcohols as the hydrogen (hydride) resource. For the classical MPV reduction catalyzed by metal alkoxide, like aluminum and zirconium complexes in the presence of secondary alcohols, the reaction mechanism involved a reversible hydride transfer from a secondary alcohol to a carbonyl substrate via a six-membered transition state, initiated by the activation of the carbonyl groups by coordination to the Lewis acidic metal center (Shylesh et al. 2009). A zirconium complex with η^5-C_5H_5 ligand, e.g., $(\eta^5\text{-}C_5H_5)_2ZrH_2$, has also been used as a homogeneous Lewis acidic catalyst for MPV reduction, and it exhibited high catalytic activities (Ishii et al. 1986, Nakano et al. 1988).

Therefore, we would like to introduce the catalysis of **TBA-1** and **TBA-2** for the MPV reduction of carbonyl compounds with 2-propanol (Kato et al. 2016). Even in this reaction system, **TBA-1** and **TBA-2** were insoluble in 2-propanol, thus, they also were used as heterogeneous catalysts. Figure 15.4 showed the catalytic activities of **TBA-1** and **TBA-2** for the MPV reduction of cyclohexanone with 2-propanol at 80 ± 2°C under argon. The

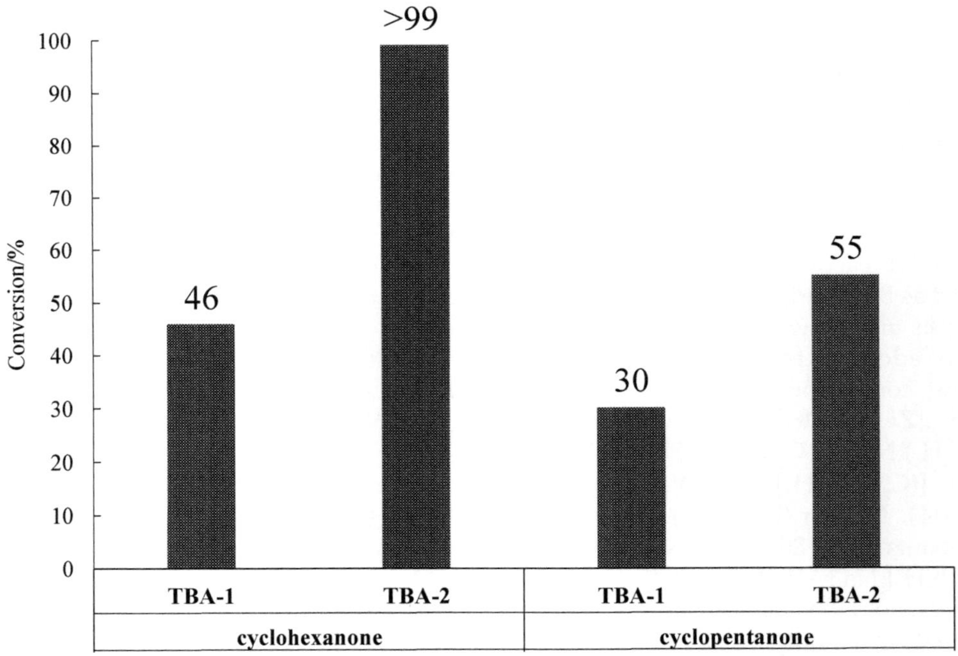

Fig. 15.4: Meerwein-Ponndorf-Verley reduction of ketones with 2-propanol catalyzed by **TBA-1** and **TBA-2** in a heterogeneous system.

Reaction conditions: catalyst (27 μmol), substrates (1.9 mmol for cyclohexanone, 2.2 mmol for cyclopentanone), 2-propanol (78 mmol), reaction temperature 80 ± 2°C, reaction time 24 h (for cyclohexanone) and 96 h (for cyclopentanone), under argon.

Conversion (%) = $\{[\text{mol of substrate}]_0 - [\text{mol of substrate}]_t\}/[\text{mol of substrate}]_0 \times 100$.

products were cyclohexanol and acetone with > 99% selectivity. It has also been confirmed that the reaction does not proceed without a catalyst. **TBA-2** exhibited > 99% of conversion after 24 hours, which was higher than that of **TBA-1** (46%). As control experiments, it was confirmed that **TBA-3** and **TBA-4** showed no reaction during 24 hours. $(\eta^5\text{-}C_5H_5)_2ZrCl_2$ had 45% conversion, suggesting that the grafting reaction of $(\eta^5\text{-}C_5H_5)_2Zr^{2+}$ onto the surface of **TBA-4** caused the most effective active centers. In contrast, $H_3[PW_{12}O_{40}]\cdot 23H_2O$ showed no reaction, thus, Bronsted acid sites did not act as the active centers, as reported for various metal oxides (Ivanov et al. 1994). When ZrO_2 (Komanoya et al. 2015) was used as a heterogeneous catalyst, the conversion was 29% under the reported reaction conditions. When cyclopentanone was used as a substrate, cyclopentanol and acetone were products with > 99% selectivity. The conversion after 96 hours of **TBA-2** (55%) was higher than that of **TBA-1** (30%).

Here, it should be noted that the order of catalytic activity of **TBA-1** and **TBA-2** was reversed by esterification and MPV reduction under the reported reaction conditions. For this reason, although it is still necessary to obtain further experimental results, it is presumed that not only Lewis acid sites but also Lewis basic sites affect the reaction mechanism in MPV reduction, as reported for MgO (Okamoto et al. 1972, Ono and Hattori 2011) and ZrO_2 (Komanoya et al. 2015). In the future, we will elucidate the expression mechanism of the Lewis acid and base sites, and will also investigate the effects of the parts constituting the aluminum-containing polyoxometalates (internal elements, aluminum-substituted structures, polyoxometalate skeleton structures, etc.), on the properties and catalytic activities in various organic reactions.

5. Summary

We introduced the syntheses and molecular structures of zirconocene derivatives, $[(n\text{-}C_4H_9)_4N]_6[\alpha\text{-}PW_{11}Al(OH)O_{39}Zr(\eta^5\text{-}C_5H_5)_2]_2$ (**TBA-1**) and $[(n\text{-}C_4H_9)_4N]_6[\alpha\text{-}SiW_{11}Al(OH)_2O_{38}Zr(\eta^5\text{-}C_5H_5)_2]_2\cdot 2H_2O$ (**TBA-2**), which were formed by the reaction of α-Keggin-type mono-aluminum-substituted polyoxotungstates, $[(n\text{-}C_4H_9)_4N]_4[\alpha\text{-}PW_{11}\{Al(OH_2)\}O_{39}]$ (**TBA-3**) and $[(n\text{-}C_4H_9)_4N]_4K_{0.5}H_{0.5}[\alpha\text{-}SiW_{11}\{Al(OH_2)\}O_{39}]\cdot H_2O$ (**TBA-4**), with a bis(η^5-cyclopentadienyl)zirconium(IV) complex in an acetonitrile solution. **TBA-1** and **TBA-2** were characterized based on elemental analysis, X-ray structure analysis, TG/DTA, CSI-MS, potentiometric titration, FTIR, and NMR spectroscopy. Single-crystal X-ray structure analyses of **TBA-1** and **TBA-2** revealed that the two $\{XW_{11}AlO_{40}\}$ (X = P or Si) units were linked by two sandwiched $(\eta^5\text{-}C_5H_5)_2Zr^{2+}$ fragments with C_2 symmetry. Each zirconium fragments was linked to the terminal oxygen atom coordinated to the Al and W sites and the edge-sharing oxygen atom at Al-O-W bond.

We also introduced the catalytic activities of **TBA-1** and **TBA-2** in the esterification of FFAs with methanol and MPV reduction of carbonyl compounds with 2-propanol. During the reactions, the two organozirconium compounds acted as heterogeneous catalysts, showing that an alcohol-soluble organozirconium fragment can be utilized in a heterogeneous system. For the esterification of FFAs with methanol, **TBA-1** had greater catalytic activities than those of **TBA-2**. In contrast, **TBA-2** had greater catalytic activities than those of **TBA-1** in the MPV reduction of ketones with 2-propanol, demonstrating that the internal elements in α-Keggin mono-aluminum-substituted polyoxometalates significantly affected catalytic activities.

Acknowledgments

This work was supported by the Ministry of Education, Culture, Sports, Science and Technology of Japan. CNK acknowledges Miss Akari Kondo (Shizuoka University) for providing the catalytic data of ZrO_2 on MPV reduction.

References

Absillis, G. and T.N. Parac-Vogt. 2012. Peptide bond hydrolysis catalyzed by the Wells-Dawson $Zr(\alpha_2\text{-}P_2W_{17}O_{61})_2$ polyoxometalate. Inorg. Chem. 51: 9902–9910.

Alcaniz-Monge, J., G. Trautwein and J.P. Marco-Loza. 2013. Biodiesel production by acid catalysis with heteropolyacids supported on activated carbon fibers. Appl. Catal. A 468: 432–441.

Banerjee, B., S. Bhunia and A. Bhaumik. 2015. Self-assembled sulfated zirconia nanocrystals with mesoscopic void space synthesized via ionic liquid as a porogen and its catalytic activity for the synthesis of biodiesels. Appl. Catal. A 502: 380–387.

Baroi, C. and A.K. Dalai. 2014. Esterification of free fatty acids (FFA) of green seed canola (GSC) oil using H-Y zeolite supported 12-tungstophosphoric acid (TPA). Appl. Catal. A 485: 99–107.

Besecker, C.J. and W.G. Klemperer. 1980. Polyoxoanion-supported metal carbonyls: syntheses of the $[(OC)_3M(Nb_2W_4O_{19})]^{3-}$ anions (M = rhenium and manganese). J. Am. Chem. Soc. 102: 7598–7600.

Besecker, C.J., W.G. Klemperer and V.W. Day. 1982. Synthesis and structure of the $\{[(C_7H_8)Rh]_5(cis\text{-}Nb_2W_4O_{19})_2\}^{3-}$ anion. J. Am. Chem. Soc. 104: 6158–6159.

Besecker, C.J., V.W. Day, W.G. Klemperer and M.R. Thompson. 1984. $[(Me_5C_5)Rh(cis\text{-}Nb_2W_4O_{19})]^{2-}$ isomers: synthesis, structure, and dynamics. J. Am. Chem. Soc. 106: 4125–4136.

Besecker, C.J., V.W. Day, W.G. Klemperer and M.R. Thompson. 1985. Synthesis and structure of the $[(OC)_3Mn(cis\text{-}Nb_2W_4O_{19})]^{3-}$ and $[(OC)_3Re(cis\text{-}Nb_2W_4O_{19})]^{3-}$ anions. Inorg. Chem. 24: 44–50.

Brown, I.D. and R.D. Shannon. 1973. Empirical bond-strength-bond-length curves for oxides. Acta Crystallogr. A 29: 266–282.

Brown, I.D. and D. Altermatt. 1985. Bond-valence parameters obtained from a systematic analysis of the inorganic crystal structure database. Acta Crystallogr. B 41: 244–247.

Brown, I.D. 1992. Chemical and steric constraints in inorganic solid. Acta Crystallogr. B 48: 553–572.

Brown, I.D. 1996. VALENCE: a program for calculating bond valences. J. Appl. Crystallogr. 29: 479–480.

Carraro, M., B.S. Bassil, A. Soraru, S. Berardi, A. Suchopar, U. Kortz et al. 2013. A Lewis acid catalytic core sandwiched by inorganic polyoxoanion caps: selective H_2O_2-based oxidations with $[Al^{III}_4(H_2O)_{10}(\beta\text{-}XW_9O_{33}H)_2]^{6-}$ (X = As^{III}, Sb^{III}). Chem. Commun. 49: 7914–7916.

Che, T.M., V.W. Day, L.C. Francesconi, M.F. Fredrich, W.G. Klemperer and W. Shum. 1985. Synthesis and structure of the $[(\eta^5\text{-}C_5H_5)Ti(Mo_5O_{18})]^{3-}$ and $[(\eta^5\text{-}C_5H_5)Ti(W_5O_{18})]^{3-}$ anions. Inorg. Chem. 24: 4055–4062.

Chuah, G.K., S. Jaenicke, Y.Z. Zhu and S.H. Liu. 2006. Meerwein-Ponndorf-Verley reduction over heterogeneous catalysts. Curr. Org. Chem. 10: 1639–1654.

Cirujano, F.G., A. Corma and F.X. Llabrés i Xamena. 2015. Zirconium containing metal organic frameworks as solid acid catalysts for the esterification of free fatty acids: Synthesis of biodiesel and other compounds of interest. Catal. Today 257: 213–220.

Cohen, R., C.R. Graves, S.B.T. Nguyen, J.M.L. Martin and M.A. Ratner. 2004. The mechanism of aluminum-catalyzed Meerwein-Schmidt-Ponndorf-Verley reduction of carbonyls to alcohols. J. Am. Chem. Soc. 126: 14796–14803.

Corma, A., S. Iborra and A. Velty. 2007. Chemical routes for the transformation of biomass into chemicals. Chem. Rev. 107: 2411–2502.

Cotton, F.A. and G. Wilkinson. 1988. Advanced Inorganic Chemistry, Fifth Edition. John Wiley & Sons, New York.

Day, V.W., M.F. Fredrich, M.R. Thompson, W.G. Klemperer, R.S. Liu and W. Shum. 1981. Reactivity of the $[(\eta^5\text{-}C_5H_5)Ti(Mo_5O_{18})]^{3-}$ anion: synthesis and structure of MoO_2Cl^+ and $Mn(CO)_3^+$ adducts. J. Am. Chem. Soc. 103: 3597–3599.

Day, V.W. and W.G. Klemperer. 1985a. Metal oxide chemistry in solution: the early transition metal polyoxoanions. Science 228: 533–541.

Day, V.W., C.W. Earley, W.G. Klemperer and D.J. Maltbie. 1985b. Synthesis, structure, and reactivity of the $\{[(\eta^5\text{-}C_5H_5)_2U]_2(\mu\text{-}TiW_5O_{19}\text{-}\kappa_2O)_2\}^{4-}$ anion. J. Am. Chem. Soc. 107: 8261–8262.

Day, V.W., W.G. Klemperer and D.J. Maltbie. 1985c. Polyoxo anion-supported organoactinides: the $[(C_5H_5)_3U(MW_5O_{19})_2]^{5-}$ anions (M = Nb, Ta) and their thorium analogs $[(C_5H_5)_3Th(MW_5O_{19})_2]^{5-}$. Organometallics 4: 104–111.

Day, V.W., W.G. Klemperer and D.J. Main. 1990. Niobotungstate, $Nb_2W_4O_{19}^{4-}$ and triphosphate, $P_3O_9^{3-}$, complexes of (cyclooctadiene)iridium(I): Synthesis, structure, and stability of tetra-*n*-butylammonium salts of $\{[(C_8H_{12})Ir]_5(Nb_2W_4O_{19})_2\}^{3-}$, $\{[(C_8H_{12})Ir]_2H(Nb_2W_4O_{19})_2\}^{5-}$, and $[(C_8H_{12})Ir(P_3O_9)]^{2-}$. Inorg. Chem. 29: 2345–2355.

Day, V.W., T.A. Eberspacher, W.G. Klemperer, R.P. Planalp, P.W. Schiller, A. Yagasaki et al. 1993. Synthesis and characterization of ruthenium complexes containing tridentate $P_3O_9^{3-}$, $Nb_2W_4O_{19}^{4-}$, and $Cp^*TiW_5O_{18}^{3-}$ ligands. Inorg. Chem. 32: 1629–1637.

de Graauw, C.F., J.A. Peters, H. van Bekkum and J. Huskens. 1994. Meerwein-Ponndorf-Verley reductions and Oppenauer oxidations: an integrated approach. Synthesis 10: 1007–1017.

Demirbas, A. 2009. Biodiesel from waste cooking oil via base-catalytic and supercritical methanol transesterification. Energy Convers. Manage. 50: 923–927.

de Paiva, E.J.M., S. Sterchele, M.L. Corazza, D.Y. Murzin, F. Wypych and T. Salmi. 2015. Esterification of fatty acids with ethanol over layered zinc laurate and zinc stearate—Kinetic modeling. Fuel 153: 445–454.

Edlund, D.J., R.J. Saxton, D.K. Lyon and R.G. Finke. 1988. Trisubstituted heteropolytungstates as soluble metal oxide analogues. 4. The synthesis and characterization of organic solvent-soluble $(Bu_4N)_{12}H_4P_4W_{30}Nb_6O_{123}$ and $(Bu_4N)_9P_2W_{15}Nb_3O_{62}$ and solution spectroscopic and other evidence for the supported organometallic derivatives $(Bu_4N)_7[(C_5Me_5)Rh{\cdot}P_2W_{15}Nb_3O_{62}]$ and $(Bu_4N)_7[(C_6H_6)Ru{\cdot}P_2W_{15}Nb_3O_{62}]$. Organometallics 7: 1692–1704.

Finke, R.G. and M.W. Droege. 1984. Trisubstituted heteropolytungstates as soluble metal oxide analogs. 1. The preparation, characterization, and reactions of organic solvent soluble forms of the silicon-niobium heteropolytungstates $Si_2W_{18}Nb_6O_{77}^{8-}$, $SiW_9Nb_3O_{40}^{7-}$, and the $SiW_9Nb_3O_{40}^{7-}$ supported organometallic complex $[(C_5Me_5)Rh{\cdot}SiW_9Nb_3O_{40}]^{5-}$. J. Am. Chem. Soc. 106: 7274–7277.

Finke, R.G., B. Rapko and P.J. Domaille. 1986a. Trisubstituted heteropolytungstates as soluble metal oxide analog. 2. 1,2,3-β-$SiW_9V_3O_{40}^{7-}$ supported $CpTi^{3-}$, $(Bu_4N)_4[CpTi{\cdot}SiW_9V_3O_{40}]$. Organometallics 5: 175–178.

Finke, R.G., B. Rapko, R.J. Saxton and P.J. Domaille. 1986b. Trisubstituted heteropolytungstates as soluble metal oxide analogs. III. Synthesis, characterization, phosphorus-31, silicon-29, vanadium-51, and 1- and 2-D tungsten-183 NMR, deprotonation, and proton mobility studies of organic solvent solute forms of $H_xSiW_9V_3O_{40}^{x-7}$ and $H_xP_2W_{15}V_3O_{62}^{x-9}$. J. Am. Chem. Soc. 108: 2947–2960.

Finke, R.G., C.A. Green and B. Rapko. 1990a. Trisubstituted heteropolytungstates as soluble metal oxide analog. 2. 1,2,3-β-$SiW_9V_3O_{40}^{7-}$ supported $CpTi^{3-}$, $(Bu_4N)_4[CpTi{\cdot}SiW_9V_3O_{40}]$. Inorg. Synth. 27: 128–135.

Finke, R.G., D.K. Lyon, K. Nomiya, S. Sur and N. Mizuno. 1990b. Polyoxoanions as soluble metal oxide analogs. 5. Synthesis and characterization of polyoxoanion-supported, atomically dispersed iridium(I), $(1,5\text{-COD})Ir{\cdot}P_2W_{15}Nb_3O_{62}^{8-}$. Inorg. Chem. 29: 1784–1787.

Finke, R.G., D.K. Lyon, K. Nomiya and T.J.R. Weakley. 1990c. Structure of nonasodium α-triniobatopen tadecatungstatodiphosphate-acetonitrile-water (1/2/23), $Na_9[P_2W_{15}Nb_3O_{62}]{\cdot}2CH_3CN{\cdot}23H_2O$. Acta Crystallogr. C 46: 1592–1596.

Finke, R.G., K. Nomiya, C.A. Green and M.W. Droege. 1992. Disodium tris(tetrabutylammonium) [β-hexatricontaoxo(m12-tetraoxosilicato)(10,11,12-triniobiumnonatungsten)ato(7−)](η^5-pentamethylcyclopentadienyl)rhodate(5−), $Na_2(Bu_4N)_3[Rh[\beta\text{-}Nb_3SiW_9O_{40}]\{\eta^5\text{-}C_5(CH_3)_5\}]$. Inorg. Synth. 29: 239–247.

Flynn, C.M. Jr. and G.D. Stucky. 1969a. Sodium 6-niobo(ethylenediamine)cobaltate(III) and its chromate(III) analog. Inorg. Chem. 8: 178–180.

Flynn, C.M. Jr. and G.D. Stucky. 1969b. Heteropolyniobate complexes of manganese(IV) and nickel(IV). Inorg. Chem. 8: 332–334.

Flynn, C.M. Jr. and G.D. Stucky. 1969c. Crystal structure of sodium 12-niobomanganate(IV), $Na_{12}MnNb_{12}O_{38.50}H_2O$. Inorg. Chem. 8: 335–344.

Glockner, A. and M. Tamm. 2013. The organometallic chemistry of cycloheptatrienyl zirconium complexes. Chem. Soc. Rev. 42: 128–142.

Gouzerh, P. and A. Proust. 1998. Main-group element, organic, and organometallic derivatives of polyoxometalates. Chem. Rev. 9: 77–111.

Hattori, S., Y. Ihara and C.N. Kato. 2015. A novel photocatalytic system constructed using eosin Y, titanium dioxide, and Keggin-type platinum(II)-and aluminum(III)-coordinated polyoxotungstates for hydrogen production from water under visible light irradiation. Catal. Lett. 145: 1703–1709.

Helwani, Z., M.R. Othman, N. Aziz, W.J.N. Fernando and J. Kim. 2009. Technologies for production of biodiesel focusing on green catalytic techniques: A review. Fuel Process. Technol. 90: 1502–1514.

Hornstein, B.J. and R.G. Finke. 2004. Transition-metal nanocluster kinetic and mechanistic studies emphasizing nanocluster agglomeration: demonstration of a kinetic method that allows monitoring of all three phases of nanocluster formation and aging. Chem. Mater. 16: 139–150.

Huber, G.W., S. Iborra and A. Corma. 2006. Synthesis of transportation fuels from biomass: chemistry, catalysts, and engineering. Chem. Rev. 106: 4044–4098.

Huang, L., S.S. Wang, J.W. Zhao, L. Cheng and G.Y. Yang. 2014. Synergistic combination of multi-Zr^{IV} cations and lacunary Keggin germanotungstates leading to a gigantic Zr_{24}-cluster-substituted polyoxometalate. J. Am. Chem. Soc. 136: 7637–7642.

Ishii, Y., T. Nakano, A. Inada, Y. Kishigami, K. Sakurai and M. Ogawa. 1986. Meerwein-Ponndorf-Verley type reduction of ketones and Oppenauer type oxidation of alcohols under the influence of bis(cyclopentadienyl)zirconium dihydride. J. Org. Chem. 51: 240–242.

Ivanov, V.A., J. Baschelier, F. Audry and J.C. Lavalley. 1994. Study of the Meerwein-Pondorff-Verley reaction between ethanol and acetone on various metal oxides. J. Mol. Catal. 91: 45–59.

Kato, C.N., A. Shinohara, K. Hayashi and K. Nomiya. 2006. Syntheses and X-ray crystal structures of zirconium(IV) and hafnium(IV) complexes containing monovacant Wells-Dawson and Keggin polyoxotungstates. Inorg. Chem. 45: 8108–8119.

Kato, C.N., Y. Katayama, M. Nagami, M. Kato and M. Yamasaki. 2010. A sandwich-type aluminium complex composed of tri-lacunary Keggin-type polyoxotungstate: synthesis and X-ray crystal structure of $[(A\text{-}PW_9O_{34})_2\{W(OH)(OH_2)\}\text{-}\{Al(OH)(OH_2)\}\{Al(\mu\text{-}OH)(OH_2)_2\}_2]^{7-}$. Dalton Trans. 39: 11469–11474.

Kato, C.N., Y. Makino, M. Yamasaki, Y. Kataoka, Y. Kitagawa and M. Okumura. 2012. Synthesis and X-ray crystal structure of α-Keggin-type aluminum-substituted polyoxotungstate. pp. 601–618. *In*: Y. Mastai (ed.). Advances in Crystallization Processes, InTech, Croatia.

Kato, C.N., M. Nagami and N. Ukai. 2013a. Strong influence of structures around aluminum centers constructed in polyoxotungstates for catalytic oxidation of alcohols with dioxygen in water. Appl. Catal. A: General 452: 69–74.

Kato, C.N., Y. Makino, W. Unno and H. Uno. 2013b. Synthesis, molecular structure, and stability of a zirconocene derivative with α-Keggin mono-aluminumsubstituted polyoxotungstate. Dalton Trans. 42: 1129–1135.

Kato, C.N., T. Kashiwagi, W. Unno, M. Nakagawa and H. Uno. 2014a. Syntheses and molecular structures of monomeric and hydrogen-bonded dimeric Dawson-type tri-aluminum-substituted polyoxotungstates derived under acidic and basic conditions. Inorg. Chem. 53: 4824–4832.

Kato, C.N. 2014b. Mono- and multi-aluminum-coordinated polyoxotungstates: syntheses, molecular structures, and application to oxidation catalysts and molecular supports. Catalysts and Catalysis 56: 347–352.

Kato, C.N., W. Unno, S. Kato, T. Ogasawara, T. Kashiwagi and H. Uno. 2016. Organozirconium complex with Keggin-Type mono-aluminum-substituted silicotungstate: synthesis, molecular structure, and catalytic performance for Meerwein–Ponndorf–Verley Reduction. Catal. Lett. 146: 2119–2128.

Kato, C.N., T. Ogasawara, A. Kondo and D. Kato. 2017. Heterogeneous esterification of fatty acids with methanol catalyzed by Lewis acidic organozirconium complexes with Keggin-type mono-aluminum-substituted polyoxotungstates. Catal. Commun. 96: 41–45.

Kikukawa, Y., S. Yamaguchi, Y. Nakagawa, K. Uehara, S. Uchida, K. Yamaguchi et al. 2008a. Synthesis of a dialuminum-substituted silicotungstate and the diasteroselective cyclization of citronellal derivatives. J. Am. Chem. Soc. 130: 15872–15878.

Kikukawa, Y., S. Yamaguchi, K. Tsuchida, Y. Nakagawa, K. Uehara, K. Yamaguchi et al. 2008b. Synthesis and catalysis of di- and tetranuclear metal sandwich-type silicotungstates $[(\gamma\text{-}SiW_{10}O_{36})_2M_2(\mu\text{-}OH)_2]^{10-}$ and $[(\gamma\text{-}SiW_{10}O_{36})_2M_4(\mu_4\text{-}O)(\mu\text{-}OH)_6]^{8-}$ (M = Zr or Hf). J. Am. Chem. Soc. 130: 5472–5478.

Klemperer, W.G. and W. Shum. 1979. Synthesis and oxygen-17 NMR structure determination of a η-cyclopentadienyl titanomolybdate, $[(\eta\text{-}C_5H_5)Ti(Mo_5O_{18})]^{3-}$. J. Chem. Soc. Chem. Commun. 60–61.

Klemperer, W.G. and D.J. Main. 1990. Polyoxoanion-supported metal carbonyls: synthesis and characterization of $\{[(OC)_2Rh]_5(Nb_2W_4O_{19})_2\}^{3-}$, $\{[(OC)_2Rh]_3(Nb_2W_4O_{19})_2\}^{5-}$, $\{[(OC)_2Ir]_2H(Nb_2W_4O_{19})_2\}^{5-}$, and $[(OC)_2Ir(P_3O_9)]^{2-}$. Inorg. Chem. 29: 2355–2360.
Klemperer, W.G. and B. Zhong. 1993. Synthesis and characterization of the polyoxoanion-supported ruthenium(I)-ruthenium(I) tetracarbonyls $[(P_3O_9)_2Ru_2(CO)_4]^{4-}$ and $[(Cp^*TiW_5O_{18})_2Ru_2(CO)_4]^{4-}$. Inorg. Chem. 32: 5821–5826.
Klomp, D., T. Maschmeyer, U. Hanefeld and J.A. Peters. 2004. Mechanism of homogeneously and heterogeneously catalyzed Meerwein-Ponndorf-Verley-Oppenauer reactions for the racemization of secondary alcohols. Chem. Eur. J. 10: 2088–2093.
Knoth, W.H., P.J. Domaille and D.C. Roe. 1983. Halometal derivatives of $W_{12}PO_{40}{}^{3-}$ and related ^{183}W NMR studies. Inorg. Chem. 22: 198–201.
Komanoya, T., K. Nakajima, M. Kitano and M. Hara. 2015. Synergistic catalysis by lewis acid and base sites on ZrO_2 for Meerwein-Ponndorf-Verley reduction. J. Phys. Chem. 119: 26540–26546.
Krut'ko, D.P. 2009. Zirconium and titanium complexes with side-chain functionalized cyclopentadienyl ligands. Russ. Chem. Bull. 58: 1745–1771.
Kundu, S.K. and A. Bhaumik. 2015. Pyrene-based porous organic polymers as efficient catalytic support for the synthesis of biodiesels at room temperature. ACS Sustainable Chem. Eng. 3: 1715–1723.
Li, G., W.H. Fu and Y.M. Wang. 2015. Meerwein-Ponndorf-Verley reduction of cyclohexanone catalyzed by partially crystalline zirconosilicate. Catal. Commun. 62: 10–13.
Lin, Y., K. Nomiya and R.G. Finke. 1993. Polyoxoanions as soluble metal oxide analogs. Synthesis and characterization of the polyoxoanion-supported iridium(I) complex $(1,5\text{-COD})Ir\cdot SiW_9Nb_3O_{40}{}^{6-}$. Inorg. Chem. 32: 6040–6045.
Lin, Y. and R.G. Finke. 1994a. A more general approach to distinguishing "Homogeneous" from "Heterogeneous" catalysis: discovery of polyoxoanion- and Bu_4N^+-stabilized, isolable and redissolvable, high-reactivity Ir~190-450 nanocluster catalysts. Inorg. Chem. 33: 4891–4910.
Lin, Y. and R.G. Finke. 1994b. Novel polyoxoanion- and Bu_4N^+-stabilized, isolable, and redissolvable, 20-30-Å Ir 300-900 nanoclusters: the kinetically controlled synthesis, characterization, and mechanism of formation of organic solvent-soluble, reproducible size, and reproducible catalytic activity metal nanoclusters. J. Am. Chem. Soc. 116: 8335–8353.
Liu, R., X. Wang, X. Zhao and P. Feng. 2008. Sulfonated ordered mesoporous carbon for catalytic preparation of biodiesel. Carbon 46: 1664–1669.
Liu, X., Y. Li, S. Peng, G. Lu and S. Li. 2012. Photocatalytic hydrogen evolution under visible light irradiation by the polyoxometalate $\alpha\text{-}[AlSiW_{11}(H_2O)O_{39}]^{5-}$-Eosin Y system. Int. J. Hydrogen Energy 37: 12150–12157.
Lopez, D.E., K. Suwannakarn, D.A. Bruce and J.G. Goodwin Jr. 2007. Esterification and transesterification on tungstated zirconia: Effect of calcination temperature. J. Catal. 247: 43–50.
Lyon, D.K. and R.G. Finke. 1990. Polyoxoanions as soluble metal oxide analogs. 6. Catalytic activity and initial kinetic and mechanistic studies of polyoxoanion-supported, atomically dispersed iridium(I), $(1,5\text{-COD})Ir\cdot P_2W_{15}Nb_3O_{62}{}^{8-}$. Inorg. Chem. 29: 1787–1789.
Ma, R., T. Wei and C. Zhao. 2011. Synthesis and thermal decomposition reaction kinetics of $\alpha\text{-}SiW_{11}Al$ and tetrabutylammonium charge-transfer complex. Huaxue Shiji 33: 307–310.
Mizuno, N., H. Weiner and R.G. Finke. 1996. Co-oxidative epoxidation of cyclohexene with molecular oxygen, isobutyraldehyde reductant, and the polyoxoanion-supported catalyst precursor $[(n\text{-}C_4H_9)_4N]_5Na_3[(1,5\text{-COD})Ir\cdot P_2W_{15}Nb_3O_{62}]$. The importance of key control experiments including omitting the catalyst and adding radical-chain initiators. J. Mol. Catal. A: Chemical 114: 15–28.
Mo, X., E. Lotero, C. Lu, Y. Liu and J.G. Goodwin. 2008. A novel sulfonated carbon composite solid acid catalyst for biodiesel synthesis. Catal. Lett. 123: 1–6.
Nakano, T., S. Umano, Y. Kino, Y. Ishii and M. Ogawa. 1988. Meerwein-Ponndorf-Verley-type reduction of dicarbonyl compounds to hydroxy carbonyl compounds and α, β-unsaturated carbonyl compounds to allylic alcohols catalyzed by zirconocene and hafnocene complexes. J. Org. Chem. 53: 3752–3757.
Narkhede, N., S. Singh and A. Patel. 2015. Recent progress on supported polyoxometalates for biodiesel synthesis via esterification and transesterification. Green Chem. 17: 89–107.
Ni, J. and F.C. Meunier. 2007. Esterification of free fatty acids in sunflower oil over solid acid catalysts using batch and fixed bed-reactors. Appl. Catal. A 333: 122–130.

Nomiya, K., M. Kaneko, N. Kasuga, R.G. Finke and M. Pohl. 1994. The nonasodium salt of the tri-niobium-substituted polyoxoanion $P_2W_{15}Nb_3O_{62}^{9-}$: A water soluble, readily crystallized form of this Dawson-based soluble metal-oxide organometallics-support system. Inorg. Chem. 33: 1469–1472.

Nomiya, K., C. Nozaki, M. Kaneko, R.G. Finke and M. Pohl. 1995. The all-sodium salt of a polyoxoanion-supported organometallic complex: synthesis and characterization of $Na_7[(\eta^5\text{-}C_5Me_5)Rh\cdot P_2W_{15}Nb_3O_{62}]\cdot 7DMSO\cdot 5H_2O$. J. Organometal. Chem. 505: 23–28.

Nomiya, K., M. Pohl, N. Mizuno, D.K. Lyon and R.G. Finke. 1997. Polyoxoanion-supported, atomically dispersed iridium(I) and rhodium(I): $Na_3[(C_4H_9)_4N]_5[Ir[\alpha\text{-}Nb_3P_2W_{15}O_{62}]\{\eta^4\text{-}C_8H_{42}\}]$ and $Na_3[(C_4H_9)_4N]_5[Rh[\alpha\text{-}Nb_3P_2W_{15}O_{62}]\{\eta^4\text{-}C_8H_{12}\}]$. Inorg. Synth. 31: 186–201.

Nomiya, K., H. Torii, C.N. Kato and Y. Sado. 2003. Synthesis, characterization, and oxidation catalysis of a novel Dawson polyoxometalate-supported platinum(II) complex, $[\{Pt(cod)\}(P_2W_{15}V_3O_{62})]^{7-}$ (cod = 1,5-cyclooctadiene). Chem. Lett. 32: 664–665.

Nomiya, K., K. Ohta, Y. Sakai, T. Hosoya, A. Ohtake, A. Takakura et al. 2013. Tetranuclear hafnium(IV) and zirconium(IV) cationic complexes sandwiched between two di-lacunary species of α-Keggin polyoxometalates: Lewis acid catalysis of the Mukaiyama–Aldol reaction. Bull. Chem. Soc. Jpn. 86: 800–812.

Okamoto, Y., T. Imanaka and S. Terasaki. 1972. Hydrogen transfer reaction between ketones and alcohols over metal oxide catalysts. Bull. Chem. Soc. Jpn. 45: 3207–3208.

Okuhara, T., C. Hu, M. Hashimoto and M. Misono. 1994. Acid strength of heteropolyacids and its correlation with catalytic activity. Bull. Chem. Soc. Jpn. 67: 1186–1188.

Okuhara, T., N. Mizuno and M. Misono. 1996. Catalytic chemistry of heteropoly compounds. Adv. Catal. 41: 113–252.

Ono, Y. and H. Hattori. 2011. Solid Base Catalysis. Springer-Verlag, Berlin and Tokyo Institute of Technology Press, Tokyo.

Ozkar, S. and R.G. Finke. 2002. Nanocluster formation and stabilization fundamental studies: ranking commonly employed anionic stabilizers via the development, then application, of five comparative criteria. J. Am. Chem. Soc. 124: 5796–5810.

Patel, A., V. Brahmkhatri and N. Singh. 2013. Biodiesel production by esterification of free fatty acid over sulfated zirconia. Renew. Energy 51: 227–233.

Pohl, M. and R.G. Finke. 1993. Polyoxoanion-supported, atomically dispersed transition-metal precatalyst $[(1,5\text{-COD})Ir\cdot P_2W_{15}Nb_3O_{62}]^{8-}$: direct oxygen-17 NMR evidence for $Ir\text{-}ONb_2$ bonding and for a C_{3v} average symmetry, iridium-to-polyoxoanion support interaction. Organometallics 12: 1453–1457.

Pohl, M., Y. Lin, T.J.R. Weakley, K. Nomiya, M. Kaneko, H. Weiner et al. 1995a. Trisubstituted heteropolytungstates as soluble metal-oxide analogs. Isolation and characterization of $[(C_5Me_5)Rh\cdot P_2W_{15}Nb_3O_{62}]^{7-}$ and $[(C_6H_6)Ru\cdot P_2W_{15}Nb_3O_{62}]^{7-}$, including the first crystal structure of a Dawson-type polyoxoanion-supported organometallic complex. Inorg. Chem. 34: 767–777.

Pohl, M., D.K. Lyon, N. Mizuno, K. Nomiya and R.G. Finke. 1995b. Polyoxoanion-supported catalyst precursors. Synthesis and characterization of the iridium(I) and rhodium(I) precatalysts $[(n\text{-}C_4H_9)_4N]_5Na_3[(1,5\text{-COD})M\cdot P_2W_{15}Nb_3O_{62}]$ (M = Ir, Rh). Inorg. Chem. 34: 1413–1429.

Pope, M.T. 1983. Heteropoly and Isopoly Oxometalates. Springer-Verlag. Berlin.

Pope, M.T. and A. Müller. 1991. Chemistry of polyoxometallates. Actual variation on an old theme with interdisciplinary references. Angew. Chem. Int. Ed. Engl. 30: 34–48.

Pope, M.T. and A. Müller. 1994. Polyoxometalates: From Platonic Solids to Anti-Retroviral Activity. Kluwer Academic Publishers, Dordrecht, The Netherlands.

Radkov, E.V., V.G. Young, Jr. and R.H. Beer. 1999. A polyoxometalate transfer reagent: Synthesis, structure, and reactivity of zirconocene polyoxometalate $[(PNbW_{11}O_{40})_2ZrCp_2]^{6-}$. J. Am. Chem. Soc. 121: 8953–8954.

Radkov, E.V. and R.H. Beer. 2000. A comprehensive series of Group 14 derivatives of the plenary mixed-metal polyoxometalate $[PNbW_{11}O_{40}]^{4-}$. Inorg. Chim. Acta 297: 191–198.

Raith, A., P. Altmann, M. Cokoja, W.A. Herrmann and F.E. Kühn. 2010. η^5,η^1-Coordinated cyclopentadienyl transition metal complexes featuring s-metal-carbon ansa bridges. Coord. Chem. Rev. 254: 608–634.

Rapko, B.M., M. Pohl and R.G. Finke. 1994. Trivanadium polyoxoanion chemistry. Part 4. Synthesis, isolation, and spectroscopic characterization of trivanadium polyoxoanion-supported $(C_5H_5)Ti^{3+}$: $(Bu_4N)_4[CpTi\cdot SiW_9V_3O_{40}]$ and $(Bu_4N)_6[CpTi\cdot P_2W_{15}V_3O_{62}]$. Inorg. Chem. 33: 3625–3634.

Ruiz, J.R. and C. Jimenez-Sanchidrian. 2007. Heterogeneous catalysis in the Meerwein-Ponndorf-Verley reduction of carbonyl compounds. Curr. Org. Chem. 11: 1113–1125.

Shylesh, S., M.P. Kapoor, L.R. Juneja, P.P. Samuel, C. Srilakshmi and A.P. Singh. 2009. Catalytic Meerwein-Ponndorf-Verley reductions over mesoporous silica supports: Rational design of hydrophobic mesoporous silica for enhanced stability of aluminum doped mesoporous catalysts. J. Mol. Catal. A: Chem. 301: 118–126.

Song, J., B. Zhou, H. Zhou, L. Wu, Q. Meng, Z. Liu et al. 2015. Porous Zirconium-phytic acid hybrid: A highly efficient catalyst for Meerwein-Ponndorf-Verley reductions. Angew. Chem. Int. Ed. 54: 9399–9403.

Stroobants, K., G. Absillis, E. Moelants, P. Proost and T.N. Parac-Vogt. 2014. Regioselective hydrolysis of human serum albumin by Zr^{IV}-substituted polyoxotungstates at the interface of positively charged protein surface patches and negatively charged amino acid residues. Chem. Eur. J. 20: 3894–3897.

Su, F. and Y. Guo. 2014a. Advancements in solid acid catalysts for biodiesel production. Green Chem. 16: 2934–2957.

Su, F., S. An, D. Song, X. Zhang, B. Lu and Y. Guo. 2014b. Heteropoly acid and ZrO_2 bifunctionalized organosilica hollow nanospheres for esterification and transesterification. J. Mater. Chem. A 2: 14127–14138.

Takagaki, A., M. Toda, M. Okamura, J.N. Kondo, S. Hayashi, K. Domen et al. 2006. Esterification of higher fatty acids by a novel strong solid acid. Catal. Today 116: 157–161.

Toda, M., A. Takagaki, M. Okamura, J.N. Kondo, S. Hayashi, K. Domen et al. 2005. Green chemistry: biodiesel made with sugar catalyst. Nature 438: 178.

Trovarelli, A. and R.G. Finke. 1993. Characterization of large, polyanionic inorganic molecules: fast atom bombardment mass spectrometry of $P_2W_{15}Nb_3O_{62}^{9-}$ and of the supported organometallic catalyst precursor (1,5-COD)Ir·$P_2W_{15}Nb_3O_{62}^{8-}$. Inorg. Chem. 32: 60346039.

Wang, S.S. and G.Y. Yang. 2015. Recent advances in polyoxometalate-catalyzed reactions. Chem. Rev. 115: 4893–4962.

Watzky, M.A., E.E. Finney and R.G. Finke. 2008. Transition-metal nanocluster size vs formation time and the catalytically effective nucleus number: A mechanism-based treatment. J. Am. Chem. Soc. 130: 11959–11969.

Weiner, H., A. Trovarelli and R.G. Finke. 2003. Polyoxoanion-supported catalysis: evidence for a $P_2W_{15}Nb_3O_{62}^{9-}$-supported iridium cyclohexene oxidation catalyst starting from $[n\text{-}Bu_4N]_5Na_3[(1,5\text{-}COD)Ir{\cdot}P_2W_{15}Nb_3O_{62}]$. J. Mol. Catal. A: Chemical 191: 253–279.

Yang, Q.H., D.F. Zhou, H.C. Dai, J.F. Liu, Y. Xing, Y.H. Lin et al. 1997. Synthesis, structure and properties of undecatungstozincate containing 3A elements. Polyhedron 16: 3985–3989.

Zhao, J., H. Guan, W. Shi, M. Cheng, Z. Wang and S. Li. 2012. A Brønsted-Lewis-surfactant-combined heteropolyacid as an environmental benign catalyst for esterification reaction. Catal. Commun. 20: 103–106.

Zonnevijlle, F., C.M. Tourne and G.F. Tourne. 1982. Preparation and characterization of heteropolytungstates containing group 3a elements. Inorg. Chem. 21: 2742–2750.

Chapter 16

Design, Synthesis, and Application of Highly Photosensitive Caging Groups for Green Chemistry

Tetsuo Narumi

1. Introduction

Caged compounds are highly functional molecules whose activity is temporarily confined by chemical cages that is organically chemically attached to the original compounds including biologically active molecules and natural products and enable to exhibit the original activity by uncaging under specific conditions (Klan et al. 2013). In particular, photochemical processes are widely used as uncaging methods due to their mild conditions for the photocleavage that can proceed smoothly in the aqueous conditions. The photosensitive caged compounds facilitate the spatiotemporal dynamic analysis of numerous chemicals, including bioactive molecules (Klan et al. 2013), gases (Kitamura et al. 2016), and inorganic species (Cui et al. 2012) in living cells. Since those functions including photophysical and photochemical properties are significantly affected by the photosensitive chemical cage, numerous photolabile protecting groups have been developed so far. In this chapter, we seek to give an overview of the development of highly photosensitive protecting groups involving 8-azacoumarin-type (Narumi et al. 2014) and N-methyl-quinolinium-type (Narumi et al. 2018) that have been recently identified by our groups.

Research Institute of Green Science and Technology, Shizuoka University, 3-5-1 Johoku, Hamamatsu, Shizuoka 432-8561, Japan.
Email: narumi.tetsuo@shizuoka.ac.jp

2. 8-Azacoumarin-type Caging Group

2.1 Hydrophilicity Increasing Strategy for Coumarin-4-ylmethyls

2.1.1 Design of 8-Azacoumarin-type Caging Groups

Among the caging groups that have been developed so far, coumarin-4-ylmethyls have been widely applied to caging chemistry in recent years. Especially, the ability of two-photon photoactivation with practically useful absorption cross sections (720–900 nm) is one of the exceptional advantages of coumarin-4-ylmethyls exemplified as the 6-bromo-7-hydroxycoumarin-4-ylmethyl (Bhc) group (Furuta et al. 1999) and (7-diethylaminocoumarin-4-yl)methyl (DEAC) group (Hagen et al. 2001) (Fig. 16.1). However, one of the drawbacks of coumarins is their low aqueous solubility, which would induce the heterogeneous aggregation of hydrophobic caged compounds in the physiological conditions, resulting to the dramatical decrease of photolysis efficiency of caged compounds (Nomura et al. 2011).

To address this drawback, several structure-based strategies for increasing the hydrophilicity of coumarin-4-ylmethyls have been reported. The introduction of polar functional groups such as amino or carboxyl group(s) into the coumarin chromophore led to the identification of hydrophilic caging groups including BCMACM (Hagen et al. 2005), BBHCM (Hagen et al. 2010), and DEAC450 (Olson et al. 2013).

Our studies began with the design of novel hydrophilic chromophores. At the onset of our studies on developing hydrophilic caging groups, we postulated that the application of the alkene-to-amide isosteric switching strategy could unveil the potentials of coumarin chromophores beyond those of the reported coumarin-type caging groups. Our initial goal was the development of new strategies to increase the hydrophilicity of coumarin chromophores without diminishing the photosensitivity. Our thinking was the introduction of polar amide functionality into the coumarin chromophore (Fig. 16.2), the C6–C7 enol double bond was substituted with an amide bond, then, aromatization of the lactam moiety led to the 8-azacoumarin chromophore containing the hydroxypyridine moiety. This molecular design is based on the concept of isosteric switching of amides

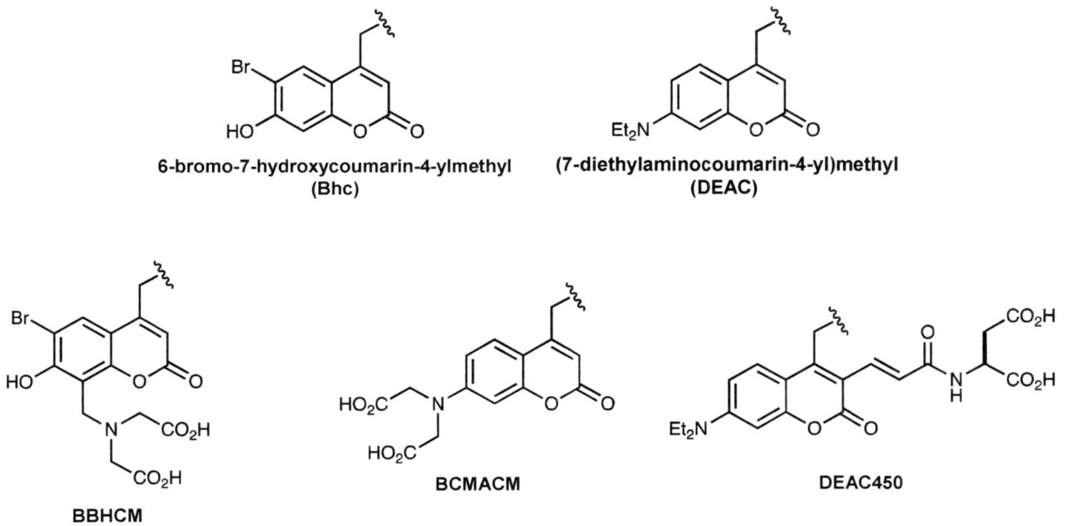

Fig. 16.1: Representative coumarin-4-ylmethyls for photolabile protecting groups.

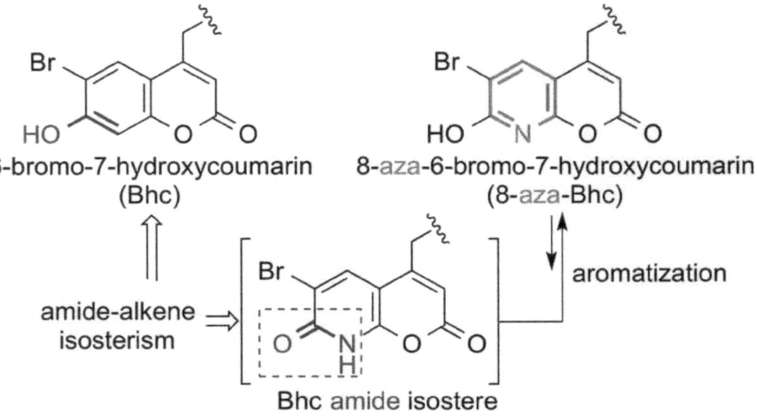

Fig. 16.2: Design of 8-azacoumarin-type chromophore based on the alkene-to-amide isosteric switching. (From Narumi et al. 2014. Org. Lett. 16: 1184–1187. With permission from the American Chemical Society.)

with alkenes (enols), familiar in the fields of chemical biology and medicinal chemistry (Meanwell 2011) and exemplified by alkene-type peptide bond isosteres (Narumi et al. 2015).

2.1.2 Synthesis of 8-Azacoumarin-type Caging Groups

For the synthesis of 8-azacoumarin-type caging groups, we have developed a reliable synthetic scheme which starts from 2,6-dichloropyridine. Successive treatments of 2,6-dichloropyridine with potassium *tert*-butoxide (*tert*-BuOK) and sodium benzyl alkoxide followed by deprotection of the benzyl moiety provided the hydroxypyridine compound. The 8-azacoumarin framework can be constructed by cyclization with dimethyl acetylenedicarboxylate (DMAD) and PPh_3, whose ester functionality at the C4 position can be reduced chemoselectively with $LiBH_4$ to provide the corresponding alcohol. Acetylation of the alcohol followed by TFA treatment gave the 4-acetoxymethyl-8-aza-7-hydroxycoumarin (aza-hc-OAc). Further synthetic effort from our group has led to the development of other substituted 8-azacoumarin-type caging groups including 8-aza-7-hydroxy-3-bromocoumarin-type (aza-3-Bhc-OAc), 8-aza-7-hydroxy-6-bromocoumarin-type (aza-Bhc-OAc) and 8-aza-7-hydroxy-6-iodocoumarin-type (aza-Ihc-OAc) (Scheme 16.1). Another group also reported an efficient synthetic approach toward 8-azacoumarins (Wang et al. 2017).

2.2 Hydrophilic Properties of 8-Aza-coumarin-4-ylmethyls

The utility of 8-aza-coumarin-4-ylmethyls as caging groups lies in their ability to be dissolved in aqueous solutions by forming hydrogen-bonding. As shown in Fig. 16.3, 8-azacoumarin derivatives showed much higher hydrophilicity than 6-bromo-7-hydroxycoumarin-4-ylmethyl (Bhc) group. For example, the saturated concentration (C_s) of 8-aza-Bhc-OAc is 10,832 mM, which is approximately 18-fold higher than that of Bhc group (C_s = 602 mM). Even 8-aza-Ihc-OAc, which has the hydrophobic iodine substituent, shows high levels of hydrophilicity (C_s = 5,014 mM). This is probably due to the higher acidity of the 7-hydroxyl group on the halogenated azacoumarin chromophore, that would

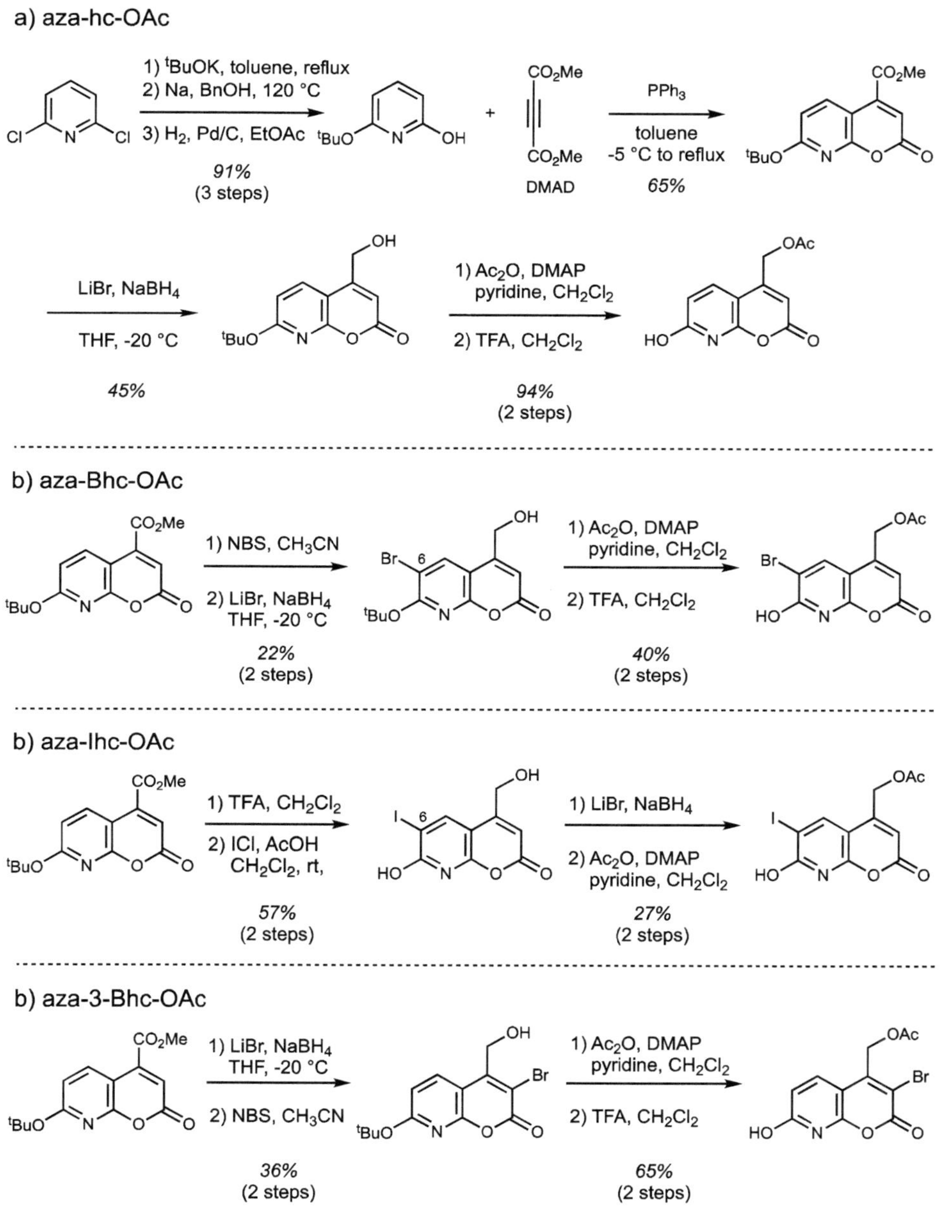

Scheme 16.1: Synthesis of 8-azacoumarin-type caging groups.

be derived from the inductive effect not only of the electron-withdrawing halogen atoms but also of a nitrogen atom. This hydrophilic potential of azacoumarin chromophores can provide new approaches for the development of hydrophilic caged compounds.

Monitoring of hydrolysis stability of synthetic compounds at room temperature by HPLC revealed that 8-aza-coumarin-4-ylmethyls are stable in the PBS under dark conditions except for aza-3-Bhc-OAc. The hydrolysis of 8-aza-coumarin-4-ylmethyls at room temperature was detected in only < 5% for 12 hours that is a sufficient time scale for

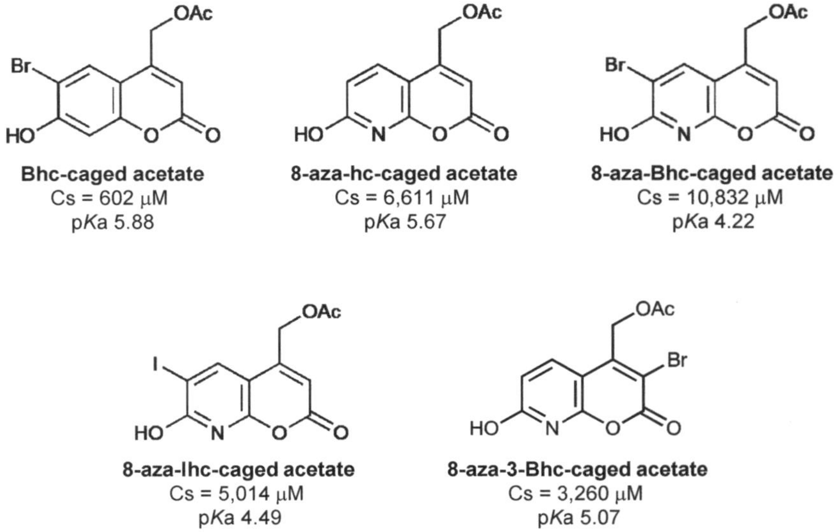

Fig. 16.3: Hydrophilic properties of Bhc-caged acetate and 8-azacoumarin-caged acetates.

the experiments with caged compounds. In particular, 8-aza-Ihc-CH_2OAc with 6-iodine substituent maintains high hydrolytic stability in the dark, whose time to reach 50% conversion ($t_{1/2}$) is 97 hours. Although aza-3-Bhc-CH_2OAc was more sensitive to hydrolysis in PBS (pH 7.4) than other 8-azacoumarines, the hydrolysis stability of aza-3-Bhc-CH_2OAc in KMOPS buffer (10 mM MOPS, 3-morpholinepropane-1-sulfonic acid, and 100 mM KCl, pH 7.1) is comparable to that of aza-Bhc-CH_2OAc with the 6-bromo substituent, indicating that dilution of the aza-3-Bhc-protected caged compounds in KMOPS buffer enables the practical use of those compounds. Furthermore, 8-aza-Bhc-OAc is more resistant than the original Bhc-OAc. Thus, the 8-azacoumarin chromophore is superior to the 6-brominated coumarin chromophore with respect not only to the hydrophilicity but also to the hydrolysis stability in the dark. This low sensitivity to spontaneous hydrolysis would be an advantage for the handling of the caged compounds, that can prevent the unexpected uncaging of those compounds.

2.3 Photophysical and Photochemical Properties of 8-Aza-coumarin-4-ylmethyls

Having recognized promising hydrophilic behaviors of 8-azacoumarin-4-ylmethyls, we evaluated the photochemical properties of azacoumarins in aqueous solutions. Photolysis of aza-hc-OAc was performed in 5 μM KMOPS buffer solution (pH 7.2) at 350 nm. As expected, photolysis of aza-hc-OAc gave the corresponding alcohol, aza-hc-OH (Fig. 16.4). HPLC analysis of the photolysis reaction of aza-hc-OAc showed that the peak corresponding to aza-hc-OAc disappeared with increasing irradiation time and the peak corresponding to the photoactivated product, aza-hc-OH, was generated simultaneously. This result suggests that the 8-azacoumarin-type cage can be utilized as a chromophore for photosensitive caging groups.

Figure 16.5 shows the time courses of photolysis reactions in terms of the consumption of the starting materials. The photolysis reaction of aza-hc-OAc follows a single-exponential decay. The time to reach 50% conversion (t_{50}) for photolysis of aza-hc-OAc is 29 seconds.

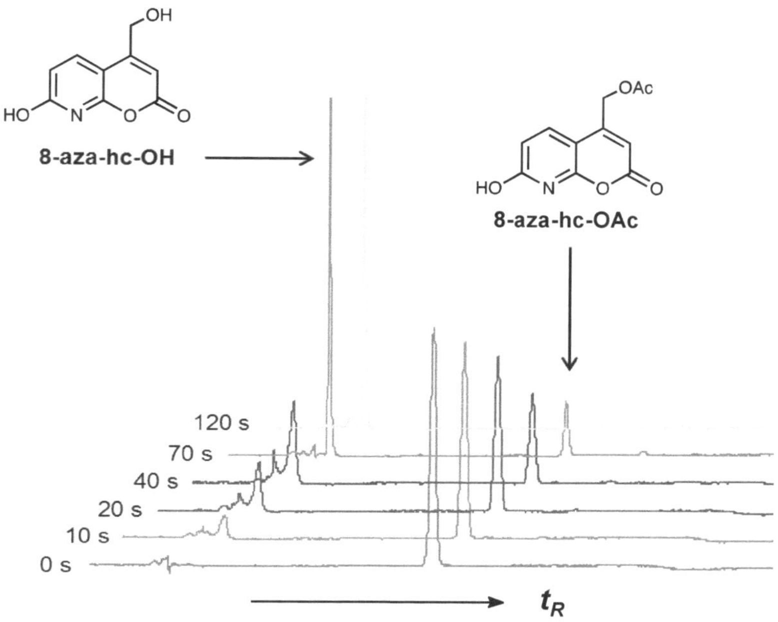

Fig. 16.4: HPLC traces for the photolysis of aza-hc-OH (measured at 340 nm). (Adapted from Narumi et al. 2014. Org. Lett. 16: 1184–1187. With permission from the American Chemical Society.)

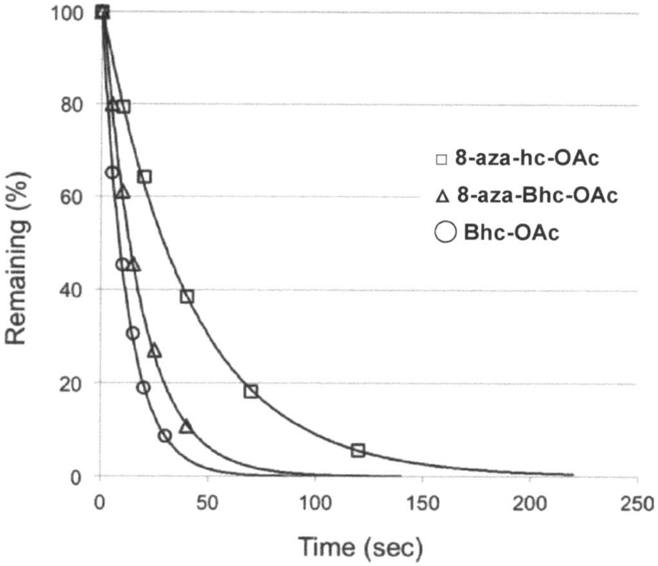

Fig. 16.5: Time courses of photolysis reactions of **aza-hc-OAc**, **aza-Bhc-OAc** and **Bhc-OAc**. (Adapted from Narumi et al. 2014. Org. Lett. 16: 1184–1187. With permission from the American Chemical Society.)

The value of t_{50} for photolysis of 8-aza-Bhc-OAc with the 6-bromo substituent is 13 seconds, which is somewhat slower than that of Bhc-OAc but is about 2.3 times shorter than that of aza-hc-OAc. In general, the photosensitivity of caged compounds cannot be discussed with the value of t_{50} if the photolysis reactions are conducted under different reaction conditions including the solvents, wavelength, temperature and instruments. In contrast, when the photolysis reactions are performed under the same conditions, it is possible to use the value of t_{50} to discuss photosensitivity. Thus, these results indicate that 6-bromo substitution delivers the increase of the photosensitivity as similar to the Bhc group. One of the reasons of bromine substitution in the Bhc group is to promote the intersystem crossing, enabling the improvement of the photosensitivity. Hence, these results also suggest that the photolytic mechanism of 8-azacoumarin-type photocage includes the intersystem crossing from the S_1 state to the triplet state.

Selected photochemical and photophysical properties of the 8-azacoumarin compounds and the original compound (Bhc-OAc) are shown in Table 16.1. Replacement of the phenyl ring of Bhc-OAc with pyridine ring led to the shift of the absorption maxima to a shorter wavelength from 370 nm for original Bhc-OAc to 356 nm for aza-hc-OAc and 362 nm for aza-Bhc-OAc. The absorption maxima of aza-Ihc-OAc with the 6-iodine substituent is 365 nm, which is comparable to that of aza-Bhc-OAc, even with the lower electron-withdrawing property of the iodine atom. On the other hand, the absorption maxima of aza-3-Bhc-OAc, which is a regiomeric isomer of aza-Bhc-OAc with the 6-bromo substituent, shifted to a longer wavelength to 378 nm, probably due to the destabilization both of HOMO and LUMO energy levels by switching the position of the electron-withdrawing Br group from the C3- to the C6-position. These results indicate that substituent(s) on 8-azacoumarin-type photocage would provide a positive effect on the physical and photochemical properties.

On the other hand, conversion to azacoumarin chromophore generally delivers the increase of molar absorptivity (ε). Although the comparison of the molar absorptivity at the absorption maxima is important for the evaluation of the superiority of the chromophore, the comparison of the molar absorptivity at the wavelength that is used for the photoactivation of the caged compounds provide the more practical evaluation. Since our initial study employed the UV light at 350 nm for the photoactivation of 8-azacoumarin-type photocages, the molar absorptivity at 350 nm (ε_{350}) was used for the discussion. The value of molar absorptivities of aza-hc-OAc and aza-Bhc-OAc are 20,175 $M^{-1}cm^{-1}$ and 20,583 $M^{-1}cm^{-1}$, respectively. Both values are higher than that of Bhc-OAc (ε_{350} = 13,774 $M^{-1}cm^{-1}$). In addition, comparison of the molar absorptivity at 365 nm, which is the wavelength also employed for the photoactivation, clearly indicates the superiority of 8-azacoumarin-type photocages including aza-Ihc-OAc. The molar absorptivity of aza-Ihc-OAc at 365 nm is 22,731 $M^{-1}cm^{-1}$, which is comparable to that of aza-Bhc-OAc (ε_{365} = 22,070 $M^{-1}cm^{-1}$) and also higher than that of Bhc-OAc (ε_{365} = 17,753 $M^{-1}cm^{-1}$). These

Table 16.1: Selected photophysical and photochemical properties of 8-azacoumarin-caged acetates and Bhc-caged acetate.

compd	λ_{max} [a] (nm)	ε_{max} [b] ($M^{-1}cm^{-1}$)	ε_{350} [c] ($M^{-1}cm^{-1}$)	Φ_{chem} [d]	$\varepsilon_{350} \cdot \Phi_{chem}$ [e]
8-aza-hc-caged acetate	356	20799	20175	0.026	526
8-aza-Bhc-caged acetate	362	23520	20583	0.059	1211
8-aza-3-Bhc-caged acetate	378	27086	20175	0.17	2667
Bhc-caged acetate	370	18071	13774	0.13	1806

[a] Long-wavelength absorption maxima. [b] Molar absorptivity at the absorption maxima. [c] Molar absorptivity at 350 nm. [d] Quantum yields for disappearance of starting materials upon 350 nm irradiation. [e] Product of the photolysis quantum yield and molar absorptivity.

high values of molar absorptivities of 8-azacoumarin-type photocages would be an advantage for the photolytic efficiency that is determined quantitatively by the product of the molar absorptivity and the photolysis quantum yield of a photolysis reaction.

The quantum yields of photolytic disappearance of starting materials were estimated by the equation $\Phi_{chem} = 1/(I \times 10^3 \varepsilon t_{90})$ (Tsien and Zuker 1986), where I is the irradiation intensity [$ein \cdot cm^{-2} \cdot s^{-1}$] and ε is the molar absorptivity of the irradiation wavelength [$cm^2 \cdot mol^{-1}$]. The quantum yield (Φ) of the disappearance of aza-hc-OAc is determined as 0.026, which is a five-fold smaller value than that of Bhc-OAc ($\Phi_{chem} = 0.13$). The quantum yield (Φ_{chem}) of the disappearance of aza-Bhc-OAc is determined as 0.059, which is larger than that of aza-hc-OAc but still two-fold smaller than that of Bhc-OAc. Theses lower quantum yields probably result from the fluorescent characters of aza-hc-OAc and aza-Bhc-OAc. An essential factor for evaluating the photosensitivity of caging groups is not quantum yield but photolysis efficiency that can be obtained from the product of the photolysis quantum yield (Φ_{chem}) and the molar absorptivity (ε) at the wavelength employed for the photoactivation. By comparing the value of the $\cdot\Phi_{chem}$ of each compound, the overall efficiency of a photolysis reaction can be quantitatively compared. In general, the value of photolysis efficiency is required to be over 100. The observed $\varepsilon_{350} \cdot \Phi_{chem}$ values of aza-hc-OAc and aza-Bhc-OAc exceeded the standard value for caged compounds. Even aza-hc-OAc with small quantum yield, the $\varepsilon_{350} \cdot \Phi_{chem}$ value was 526, whose value would be sufficient for caged compounds with practical utility in biological studies. Considering the high levels of aqueous solubility of aza-hc-OAc and aza-Bhc-OAc, 8-azacoumarin-based caging groups promise to be useful for caging chemistry.

A major challenge in this field is the development of the photosensitivity increasing strategy without any decrease of other functionalities, including hydrophilicity and hydrolytic stability. To this end for azacoumarin-type caging groups, we have developed two approaches. The first strategy is based on the substituent effect. A brief survey of the literature, along with our experience, reveals that many papers identified strong substitution effects on various caging chromophores, including quinolone-type (Petit 2012) and BODIPY-type (Slanina et al. 2017). As shown in Table 16.1, aza-3-Bhc-OAc, which is the regioisomeric 3-brominated 8-azacoumarin chromophore, has a robust set of photophysical, photochemical and hydrophilic properties for caging chemistry. The maximum wavelength of aza-3-Bhc-OAc is longer than those of other azacoumarins and Bhc-OAc, and the molar absorptivity of aza-3-Bhc-OAc at the maximum wavelength is 27,086 $M^{-1}cm^{-1}$, which is higher than those of aza-Bhc-OAc ($\varepsilon = 21107\ M^{-1}cm^{-1}$) and Bhc-OAc ($\varepsilon = 16584\ M^{-1}cm^{-1}$). In addition, the quantum yield of the disappearance of aza-3-Bhc-OAc ($\Phi_{chem} = 0.17$) is higher than those of 6-brominated coumarin chromophores, resulting in the much higher value of the photolytic efficiency of aza-3-Bhc-OAc than those of 6-brominated coumarin chromophores. These results indicated that the substitution effects including substitution pattern could positively affect photophysical and photochemical properties of the 8-azacoumarin-type caging group. Although the reason for the notable increase in those properties of aza-3-Bhc-OAc in not fully understood, one of the key features is the acidity of the 8-azacoumarin chromophore. In the case of Bhc-OAc, it is necessary to introduce the bromine atom on the C6 position of the coumarin chromophore for lowering the pKa of the adjacent hydroxyl group promoting the formation of the strongly absorbing anion. In contrast, the 7-hydroxyl groups of 8-azacoumarin chromophores are more acidic than 7-hydroxycoumarin chromophore. Even aza-hc-OAc without bromine substituents has a lower pK_a value than Bhc-OAc, whose distinctive feature can offer a new strategy to improve the photosensitivity by chemical modifications that are distinct from those in previous approaches.

The second strategy for increasing the photosensitivity is based on the heavy atom effect that can be induced by the introduction of heavy atoms such as bromine and iodine atoms promoting the intersystem crossing from the S_1 state to the triplet state. As described earlier, bromination of the coumarin chromophore provides several advantages for photolysis reactions not only to improve the photolysis quantum yield but also to lower the pKa value of the 7-hydroxyl group by the inductive effect because the deprotonated form is favorable for the photophysical and photochemical properties. This is particularly true for the development of Bhc group reported by Furuta and Thien 1999 Although 6-bromination of the 8-azacoumarin chromophore is also favorable to increase the photolytic efficiency and to lower the pKa value of the chromophore, the photolytic efficiency of the 8-aza-Bhc group was slightly lower than that of the original Bhc group. The fact that the aza-hc group without bromine substituents has a lower pK_a value than the Bhc group prompted us to invoke that the introduction of a heavier iodine substituent in the 8-azacoumarin chromophore, which would enable the improvement of photochemical properties by more efficiently promoting intersystem crossing. As shown in Table 16.2, the introduction of an iodine atom to the C6-position of the 8-azacoumarin chromophore resulted in the notable increase of the photolysis quantum yield of 8-azacoumarin-type caging groups. The value of the photolysis quantum yield of aza-Ihc-OAc is 0.52, which is remarkably higher than those of the corresponding 6-brominated coumarin- and azacoumarin-type caging groups. Furthermore, comparison of the photolysis quantum yield of aza-Ihc-OAc with aza-3-Bhc-OAc and two typical caged acetates (o-nitrobenzyl (oNB)- and p-hydroxyphenacyl (pHP)-caged acetates) revealed the superiority of aza-Ihc group. The value of the photolytic efficiency of aza-Ihc-OAc is 11,820, which is exceedingly large over the practical standard value for caged molecules and is also the highest value in the coumarin platform and > 100-fold higher than those of the other typical caged acetates.

To gain some insight into the increased photolytic efficiency of aza-Ihc-OAc, the fluorescence quantum yields competing with photolysis reaction were determined. The fluorescence quantum yields of Bhc-caged acetate and aza-3-Bhc-caged acetate are 0.22 and 0.25, which is comparable to those photolysis quantum yields, indicating that the excitation energy is used with similar efficiency for fluorescence emission and photolysis reaction. The fluorescence quantum yield of aza-Bhc-caged acetate is higher than the photolysis quantum yield, indicating that the energy for causing photolysis reaction shifts for emitting fluorescence. In contrast, the fluorescence quantum yield of aza-Ihc-caged

Table 16.2: Selected photophysical and photochemical properties of the 2nd generation of 8-azacoumarin-caged acetates.

compd	λ_{max} [a] (nm)	ε_{365} [b] ($M^{-1}cm^{-1}$)	Φ_{chem} [c]	$\varepsilon_{365} \bullet \Phi_{chem}$ [d]
8-aza-Ihc-caged acetate	356	22731	0.52	11820
8-aza-3-Bhc-caged acetate	378	23828	0.20	4766
Bhc-caged acetate	370	17753	0.17	2942
*o*NB[e]-caged acetate	265	228	0.19	43
*p*HP[f]-caged acetate	329	1408	0.096	135

[a] Long-wavelength absorption maxima. [b] Molar absorptivity at 350 nm. [c] Quantum yields for disappearance of starting materials upon 350 nm irradiation. [d] Product of the photolysis quantum yield and molar absorptivity. [e] o-Nitrobenzyl. [f] p-hydroxyphenacyl.

acetate is found to be 0.15, which is the smallest value in the compared coumarin- and aza-coumarin compounds, suggesting that the energy for emitting fluorescence shifts for causing photolysis. Therefore, the observed significant increase of the photolytic efficiency of aza-Ihc-caged acetate can be explained by the iodine-derived heavy atom effect that promotes the rate of intersystem crossing. To our knowledge, our report is the first case that an iodine atom is used as a substituent for improving the photolytic efficiency of caging groups. Following our report, Winter's group reported iodine-substituted BODIPY-type caging groups enabling the development of green light photosensitive caged compounds (Slanina et al. 2017). Taking into account the excellent aqueous solubility and hydrolytic stability of aza-Ihc-caged acetate, aza-Ihc group promise to be useful for caging chemistry. This excellent level of the photosensitivity of the aza-Ihc group would be an advantage for decreasing the photodamage of cells or tissues by shortening the photoirradiation time on uncaging.

3. *N*-Alkyl-quinolinium-type Caging Group

Despite the encouraging potentials of 8-azacoumarin-type caging groups, one of the disadvantages is the utilization of ultraviolet light (250–400 nm) for one photon activation, since ultraviolet light is phototoxic to living cells and has low tissue penetration power. This is a long-standing problem for most of the reported photosensitive caging groups including our 8-azacoumarin-types, leading to the limited application of most caged compounds *in vitro* and not *in vivo*. As part of our thinking about the development of more useful caged compounds, we recognized the potentials for the photoactivation by longer-wavelength light, providing several advantages, including lower phototoxicity and high tissue penetration power.

3.1 Design of N-alkyl-quinolinium-type Caging Groups

For the design of novel visible light-sensitive chromophores, we set several criteria. First, the chromophore should be sensitive in the visible light region (> 450 nm) for one-photon activation. At the outset of our research in 2015, only a limited number of caging groups with photoactivation capability by visible light had been developed. The caged compounds that can be photolytically activated by wavelengths > 450 nm with high photolytic efficiency include ruthenium-bipyridine (Rubi) complexes (Fino et al. 2009, Salierno et al. 2010), coumarin derivatives (Fournier et al. 2013, Lin et al. 2018, Olson et al. 2013), fluorescein analogues (Antony et al. 2013, Sebej et al. 2013), and BODIPY derivatives (Goswami et al. 2015, Slanina et al. 2017, Umeda et al. 2014) (Fig. 16.6). In particular, Rubi-caged compounds such as Rubi-Glu and Rubi-GABA have proven to be useful for the spatiotemporal dynamic analysis in the neuroscience, albeit with the practical problems including the decrease of photosensitivity by the storage at high concentrations. Second, the chromophore should possess the ability for late-stage modification of the molecular structure by elaboration such as functionalization of visible light sensitivity after the conjugation with chemicals. For the chemical synthesis of visible light-sensitive caged compounds, starting from the compounds that are sensitive to the visible light would decrease the operability significantly because it is necessary to perform all the following experimental operations under light-shielded conditions.

A brief survey of the literature implied the potential of 7-hydroxyquinoline (HQ)-type photocages such as 8-bromo-7-hydroxyquinoline-2-ylmethyl (BHQ) (Fedoryak

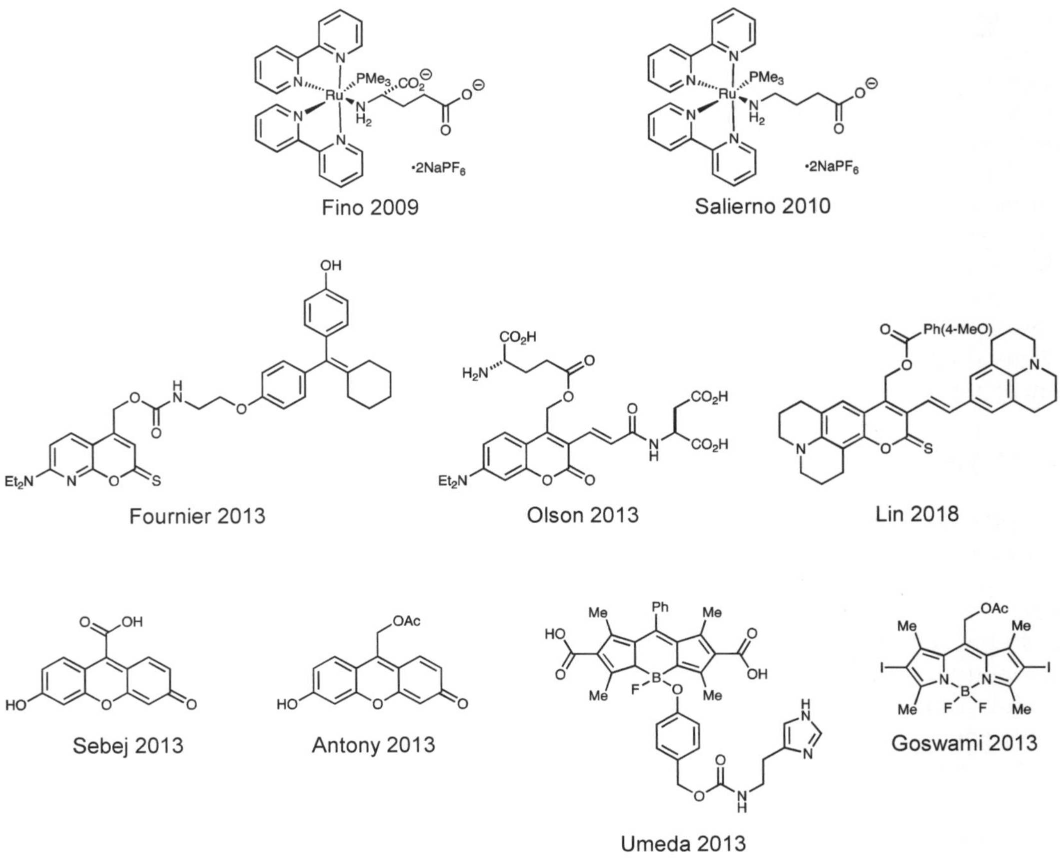

Fig. 16.6: Representative visible light sensitive caging groups.

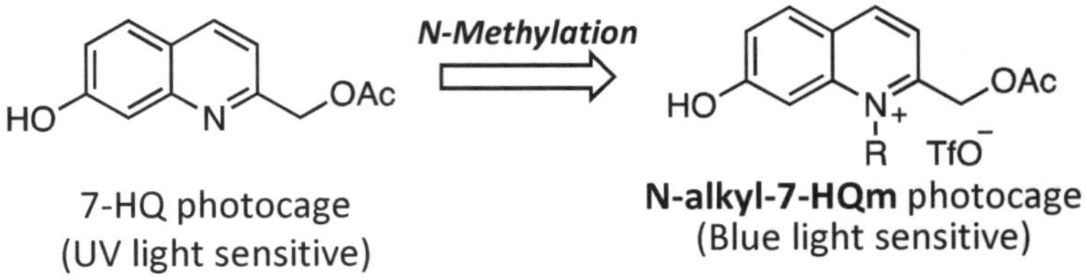

Fig. 16.7: Design of N-alkylquinolinium-type chromophore by N-alkylation.

and Dore 2002) and 8-cyano-7-hydroxyquinoline-2-ylmethyl (CyHQ) (Davis et al. 2009), which facilitate one-photon activation with high levels of water solubility and quantum efficiency. In addition, N-alkyl-7-hydroxyquinolinium derivatives, which can be prepared from 7-hydroxyquinolines by one-step N-alkylation, have proven to be useful for optical pH sensors (Badugu et al. 2008) and green fluorescent dyes (Senda et al. 2010) with large molar absorptivities. These successful examples guided us to envision that N-alkylation of the HQ cage could deliver a red shift of the absorption maximum into the visible light region and also increase the molar absorptivity with high levels of quantum efficiency of the photolysis reaction and hydrophilicity (Fig. 16.7).

3.2 Synthesis of N-methyl-quinolinium-type Caging Groups

Preliminary study on N-Alykl effect on photophysical properties of 7-hydroxyquinolinium revealed that the N-methylhydroxyquinolinium (*N*-Me-7-HQm) was a suitable chromophore with the absorption maxima at 425 nm and the high molar absorptivity (ε_{max} = 12,300 $M^{-1}cm^{-1}$) (Fig. 16.8).

N-Me-7-HQm-caged acetate was synthesized from the acetate that is the intermediate of the synthesis of BHQ group (Dore). Treatment of the acetate with methyl triflate for N-methylation followed by removal of the TBDPS group by H2O at 80 degrees gave *N*-Me-7-HQm-caged acetate (Scheme 16.2). By use of a similar procedure, the 7-O-methyl derivative (*N*-Me-7-MQm-caged acetate) and 7-dimethylamino derivatives (*N*-Me-7-DMAQm and *N*-Me-4-Et-7-DMAQm) were synthesized from the corresponding alcohols.

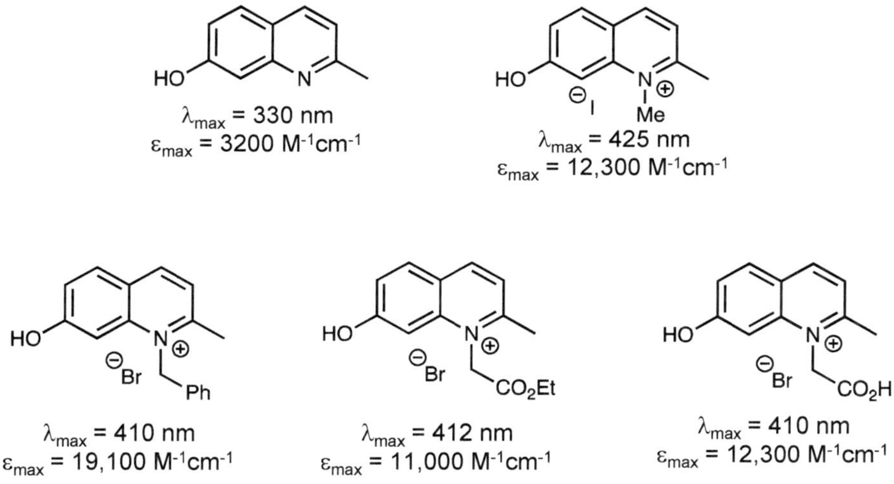

Fig. 16.8: N-Alkyl effect on photophysical properties.

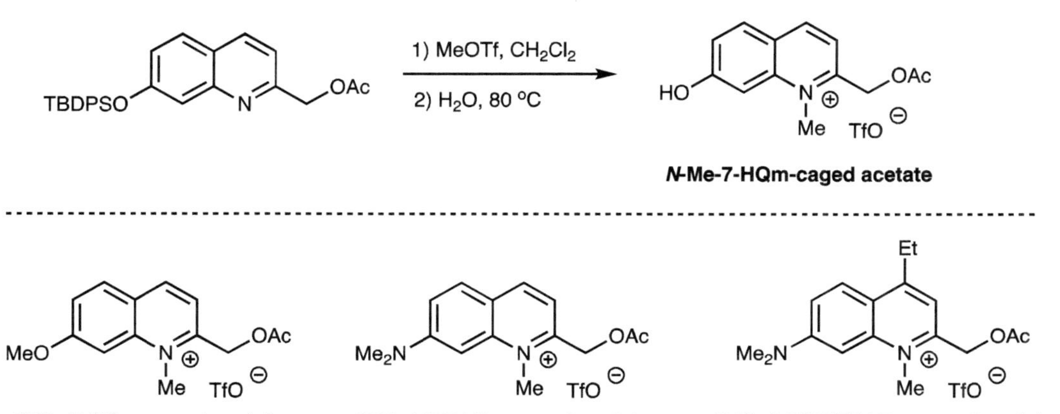

Scheme 16.2: Synthesis of *N*-methylquinolinium derivatives. (Adapted from Narumi et al. 2018. Org. Lett. 20: 4178–4182. With permission from the American Chemical Society.)

3.3 Photosensitivity of N-Methyl-quinolinium-type Caging Groups

The blue light (458 nm) photosensitivity of N-Me-7-HQm-caged acetate was evaluated under photolysis in 10 mM PBS at pH 7.2 at 458 nm. As expected, photolysis of N-Me-7-HQm-caged acetate gave the corresponding alcohol (N-Me-7-HQm-OH) via a C-O bond cleavage to release the acetate followed by the trap with H_2O (Fig. 16.9). HPLC analysis showed that the peak corresponding to N-Me-7-HQm-caged acetate disappeared with increasing time for the photoirradiation and the peak of the alcohol was produced at the same time, indicating that the *N*-Me-7-HQm cage can be photolytically activated with a blue light (458 nm). Figure 16.10 shows the time course of photolysis reactions of the synthesized N-Me-quinolinium-caged acetates in terms of the consumption of the starting materials. The photolysis reaction of N-Me-7-HQm-caged acetate follows a single-exponential decay with the time to reach 90% conversion (t_{90}) at 42 seconds. A comparison of the time courses of the photolysis reactions revealed that the substituents on N-Me-HQm chromophore significantly affect the photosensitivity. Under the tested conditions, the O-methylated derivative (*N*-Me-7-MQm-caged acetate) revealed unphotoreactive. Although the introduction of the 7-dimethylamino group led to the notable decrease of the photosensitivity, the introduction of the 4-ethyl group delivered the recovery of the photosensitivity, but which is still lower than that of *N*-Me-7-HQm-caged acetate.

The products of the photolysis quantum yield and molar absorptivity (ε 458) were determined as the photolytic efficiencies, and these data are summarized in Table 16.3 with selected absorption data. The photolytic efficiency of N-Me-7-HQm-caged acetate exceeds the standard value (100) for the useful caged compounds, whose value is 234. On the other hand, DMAQm derivatives (*N*-Me-7-DMAQm and *N*-Me-4-Et-7-DMAQm) show lower photolytic efficiencies than *N*-Me-7-HQm-caged acetate, albeit with the longer absorption maximum and the larger molar absorptivity. This is the result from the significantly lower photolysis quantum yields of DMAQm derivatives (0.0012 for *N*-Me-7-DMAQm and 0.0028 for *N*-Me-4-Et-7-DMAQm). These results indicated that the 7-hydroxy group of *N*-Me-7-HQm-caged acetate contributes significantly to the photosensitivity of the quinolinium skeleton. Although the pKa value of the 7-hydroxy group of N-Me-7-HQm-caged acetate is about 5 to 6, the 7-hydroxy group of the quinolinium structure can work as a superacid

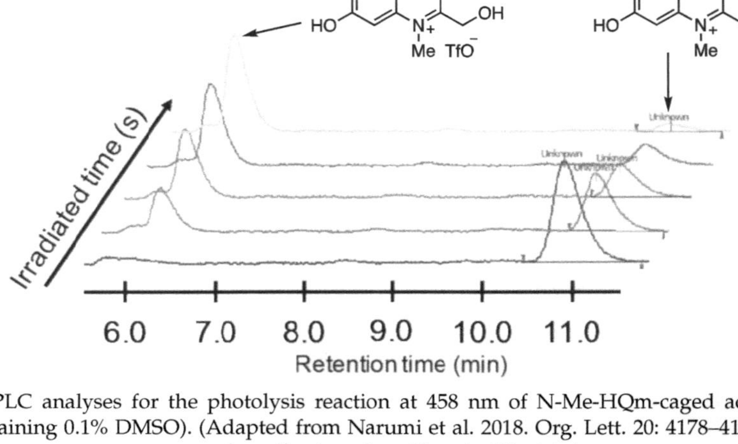

Fig. 16.9: HPLC analyses for the photolysis reaction at 458 nm of N-Me-HQm-caged acetate in PBS buffer (pH 7.4, containing 0.1% DMSO). (Adapted from Narumi et al. 2018. Org. Lett. 20: 4178–4182. With permission from the American Chemical Society.)

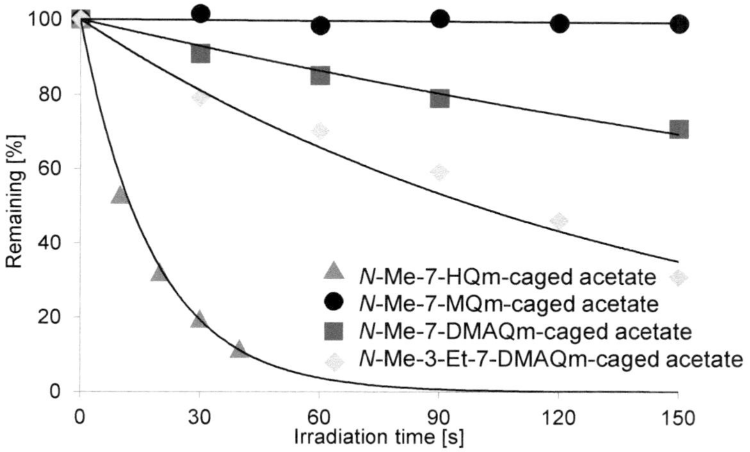

Fig. 16.10: Time courses of the photolysis reaction of *N*-methylquinolinium derivatives. (Adapted from Narumi et al. 2018. Org. Lett. 20: 4178–4182. With permission from the American Chemical Society.)

Table 16.3: Selected photophysical and photochemical properties of *N*-methylquinolinium derivatives.

compd	λ_{max} [a] (nm)	ε_{458} [b] ($M^{-1}cm^{-1}$)	Φ_{chem} [c]	$\varepsilon_{458} \bullet \Phi_{chem}$ [d]
N-Me-7-HQm-caged acetate	418	5,200	0.045	234
N-Me-7-MQm-caged acetate	357	–	–	–
N-Me-7-DMAQm-caged acetate	449	7,400	0.0012	8.9
N-Me-7-Et-DMAQm-caged acetate	441	10,400	0.0028	29

[a] Wavelength of absorption maxima in PBS (0.1% DMSO). [b] molar absorptivity at 458 nm. [c] quantum yield for photoreaction irradiated at 458 nm. [d] photolytic efficiency.

in the excited state, thereby enabling the formation of the strongly absorbing anion-type, resulting in the promotion of the intramolecular charge transfer based on their electron donating property.

3.4 Application of N-Me-7-HQm Caging Group to α-Amino Acids

Encouraged by these findings, we tried to synthesize *N*-Me-HQm-caged amino acids by a three-step procedure including coupling and N-methylation followed by cleavage of other protecting groups. First, we thought that N-Me-HQm cage could be introduced by esterification of the a-carboxyl group of the amino acid; however, our attempt revealed that the N-Me-HQm esters of amino acids were difficult to isolate in a pure form because of the immediate decomposition of N-Me-HQm ester by undesired hydrolysis. In order to overcome this problem, we synthesized the *N*-Me-HQm carbamates of amino acids, and identified that N-Me-HQm carbamates are suitable for their purification by preparative HPLC. For example, the synthetic scheme of *N*-Me-HQm carbamate of glutamic acid (Glu) is shown as Scheme 16.3. Coupling of H-Glu(tBu)-OtBu·HCl with the alcohol afforded the corresponding carbamate, which was alkylated with MeOTf leading to the corresponding glutamate with an N-Me-HQm photocage. Removal of the TBDPS group and t-butyl esters

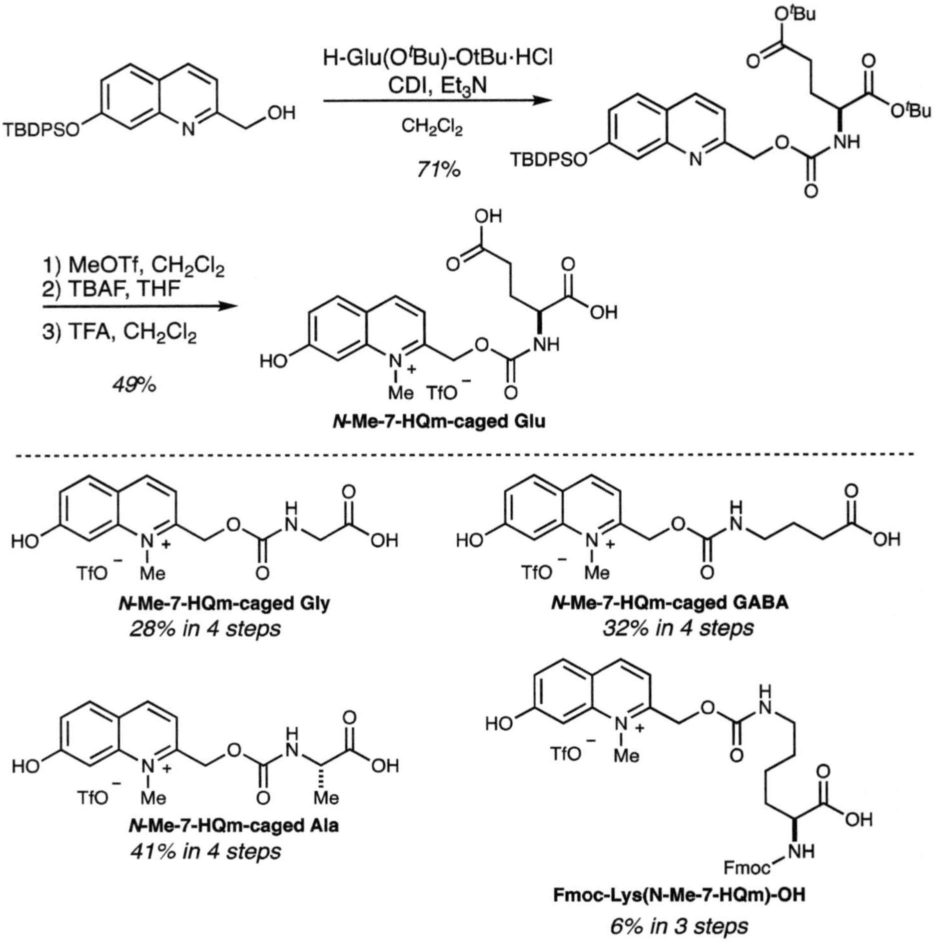

Scheme 16.3: Synthesis of *N*-Me-7-HQm-caged amino acids. (Adapted from Narumi et al. 2018. Org. Lett. 20: 4178–4182. With permission from the American Chemical Society.)

followed by preparative HPLC gave the desired N-Me-HQm carbamate. This synthetic approach can be applied to the synthesis of *N*-Me-HQm-caged amino acids from the corresponding esters of Gly, GABA, Ala, and Lys, respectively.

The synthesized *N*-Me-HQm-caged amino acids can be activated by blue light irradiation. Change of the ester conjugation to the carbamate conjugation did not lead to significant effects on optical properties (Table 16.4): the absorption maxima of *N*-Me-HQm-caged amino acids are around 415 nm, and the molar absorptivities are 7,000 to 14,000 $M^{-1}cm^{-1}$. The *N*-Me-HQm-caged Glu has the highest photolytic efficiency ($\Phi_{chem} \cdot \varepsilon_{458}$ = 173), which is a sufficiently large value to support practical use in caged compounds. *N*-Me-HQm-caged carbamates of Gly, GABA, and Ala have similar levels of photolytic efficiencies. Amongst the N-Me-HQm photocage of the a-amino group, Fmoc-Lys(N-Me-HQm)-OH, which has the photocage on the side chain amino group, is also photoreactive without any damages of the Fmoc group. We believe that the Fmoc-protected amino acids with the *N*-Me-HQm cage on the side chain will be a valuable building block

Table 16.4: Selected photophysical and photochemical properties of *N*-Me-HQm-caged amino acids.

compd	l_{max}^{a} (nm)	e_{458}^{b} ($M^{-1}cm^{-1}$)	Φ_{chem}^{d}	$e_{458} \bullet \Phi_{chem}^{e}$
N-Me-HQm-caged Glu	413	4,800	0.036	173
N-Me-HQm-caged Gly	415	2,500	0.044	110
N-Me-HQm-caged GABA	416	2,500	0.046	115
N-Me-HQm-caged Ala	416	5,400	0.025	135
Fmoc-Lys(N-Me-HQm)-OH	416	2,000	0.048	96
Rubi-Glu	450	4,000	0.068	272

[a] Long-wavelength of absorption maxima. [b] molar absorptivity at 458 nm. [c] time to reach 90% conversion, [c] quantum yield for photoreaction irradiated at 458 nm. [d] photolytic efficiency.

for the synthesis of caged peptides and related molecules that can be controlled by visible light irradiation.

4. Conclusion and Outlook

Caged compounds are today widely recognized as valuable photochemical tools for the dynamic analysis of various chemicals, including gases, bioloactive molecules and inorganic species in living cells. Much of the interest of their studies focus on the application of existing caged compounds in the fields of life science. Although the several caged compounds that can be activated by UV light or visible light have proven to be useful in the neuroscience, the development of novel caged compounds with high levels of photolytic efficiency and hydrophilic properties is one of the most important studies that can be accelerated by synthetic organic chemistry.

As described in this chapter, visible light-sensitive caged compounds that can be activated by one-photon excitation reaction with readily available LED light sources, which can be expected to reduce phototoxicity, do not require expensive lasers or special experimental environments, providing a method to control the various chemicals in living cells. Thus, the development of visible light-sensitive caged compounds will be one direction in this field. However, there is no established guideline for the design of those photosensitive molecules. Even with the great help that can be obtained by computational science, many of them depend on the researcher's experience and trial and error. From the viewpoint of the feasibility of development, as seen in the development of *N*-Me-7-HQm-type, DEAQ 450-type, and BODIPY-type, the molecular design strategy based on the structural modification of existing chromophores is one of the robust approaches. Since the organic molecules inherently have unlimited structural and functional diversity, there is the possibility that caged compounds can be further evolved by synthetic organic chemistry depending on their respective purposes. Most importantly, we anticipate that the optical control technology based on the caged compounds will make possible the development of modality for entirely controlling the function of bioactive molecules.

Acknowledgments

The major part of this work was conducted at the Institute of Biomaterials and Bioengineering, Tokyo Medical and Dental University and Research Institute of Green

Science and Technology, Shizuoka University. This research was supported by a Kurata Grant from The Kurata Memorial Hitachi Science and Technology Foundation, by a Grant-in-Aid for Young Scientists (B) and by a Grant-in-Aid for Young Scientists (A) from The Ministry of Education, Culture, Sports, Science and Technology, Japan.

References

Antony, L.A., T. Slanina, P. Sebej, T. Solomek and P. Klán. 2013. Fluorescein analogue xanthene-9-carboxylic acid: A transition-metal-free CO releasing molecule activated by green light. Org. Lett. 15: 4552–4555.

Badugu, R., Y. Kostov, G. Rao and L. Tolosa. 2008. Development and application of an excitation ratiometric optical pH sensor for bioprocess monitoring. Biotechnol. Prog. 24: 1393–1401.

Cui, J.X., R.A. Gropeanu, D.R. Stevens, J. Rettig and A. Campo. 2012. New photolabile BAPTA-based Ca2+ cages with improved photorelease. J. Am. Chem. Soc. 134: 7733–7740.

Davis, M.J., C.H. Kragor, K.G. Reddie, H.C. Wilson, Y. Zhu and T.M. Dore. 2009. Substituent effects on the sensitivity of a quinoline photoremovable protecting group to one- and two-photon excitation. J. Org. Chem. 74: 1721–1729.

Ellis-Davies, G.C.R. 2007. Caged compounds: photorelease technology for control of cellular chemistry and physiology. Nat. Methods 4: 619–628.

Fedoryak, O.D. and T.M. Dore. 2002. Brominated hydroxyquinoline as a photolabile protecting group with sensitivity to multiphoton excitation. Org. Lett. 4: 3419–3422.

Fino, E., R. Araya, D.S. Peterka, M. Salierno, R. Etch-enique and R. Yuste. 2009. Rubii-Glutamate: two-photon and visible-light photoactivation of neurons and dendritic spines. Frontiers in Neural Circuits 3: 1–9.

Fournier, L., I. Aujard, T.L. Saux, S. Maurin, S. Beaupierre, J.-B. Baudin et al. 2013. A blue-absorbing photolabile protecting group for *in vivo* chromatically orthogonal photoactivation. Chem. Eur. J. 19: 17494–17507.

Furuta, T., S.S.H. Wang, J.L. Dantzker, T.M. Dore, W.J. Bybee, E.M. Callaway et al. 1999. Brominated 7-hydroxycoumarin-4-ylmethyls: Photolabile protecting groups with biologically useful cross-sections for two photon photolysis. Proc. Natl. Acad. Sci. U.S.A. 96: 1193–1200.

Goswami, P.P., A. Syed, C.L. Beck, T.R. Albright, K.M. Mahoney, R. Unash et al. 2015. BODIPY-derived photoremovable protecting groups unmasked with green light. J. Am. Chem. Soc. 137: 3783–3786.

Hagen, V., J. Bendig, S. Fring, T. Eckardt, S. Helm, D. Reuter et al. 2001. Highly efficient and ultrafast phototriggers for cAMP and cGMP by using long-wavelength UV/Vis-activation. Angew. Chem. Int. Ed. 40: 1045–1048.

Hagen, V., B. Dekowski, V. Nache, R. Schmidt, D. Geissler, D. Lorenz et al. 2005. Coumarinylmethyl esters for ultrafast release of high concentrations of cyclic nucleotides upon one- and two-photon photolysis. Angew. Chem. Int. Ed. 44: 7887–7891.

Hagen, V., F. Kilic, J. Schaal, B. Dekowski, V.R. Schmidt and N. Kotzur. 2010. [8-[Bis(carboxymethyl) aminomethyl]-6-bromo-7-hydroxycoumarin-4-yl]methyl moieties as photoremovable protecting groups for compounds with COOH, NH2, OH, and C=O Functions. J. Org. Chem. 75: 2790–2797.

Kitamura, K., M. Kawaguchi, N. Ieda, N. Miyata and H. Nakagawa. 2016. Visible light-controlled nitric oxide release from hindered nitrobenzene derivatives for specific modulation of mitochondrial dynamics. Eur. ACS Chem. Biol. 11: 1271–1278.

Klan, P., T.C.G. Solomek, A. Bochet, R. Blanc, M. Givens, V. Rubina et al. 2013. Photoremovable protecting groups in chemistry and biology: Reaction mechanisms and efficacy. Chem. Rev. 113: 119–191.

Kobayakawa, T., T. Narumi and H. Tamamura. 2015. Remote stereoinduction in the organocuprate-mediated allylic alkylation of Allylic gem-dichlorides: Highly diastereoselective synthesis of (Z)-chloroalkene dipeptide isosteres. Org. Lett. 17: 2302–2305.

Lin, Q., L. Yang, Z. Wang, Y. Hua, D. Zhang, B. Bao et al. 2018. Coumarin photocaging groups modified with an electron-rich styryl moiety at the 3-position: Long-wavelength excitation, rapid photolysis, and photobleaching. Angew. Chem. Int. Ed. 57: 3722–3726.

Meanwell, N.A. 2011. Synopsis of some recent tactical application of bioisosteres in drug design. J. Med. Chem. 54: 2529–2591.

Narumi, T., H. Takano, N. Ohashi, A. Suzuki, T. Furuta and H. Tamamura. 2014. Isostere-based design of 8-azacoumarin-type photolabile protecting groups: A hydrophilicity-increasing strategy for coumarin-4-ylmethyls. Org. Lett. 16: 1184–1187.

Narumi, T., K. Miyata, A. Nii, K. Sato, N. Mase and T. Furuta. 2018. 7-Hydroxy-N-methylquinolinium chromophore: A photolabile protecting group for blue-light uncaging. Org. Lett. 20: 4178–4182.

Nomura, W., T. Narumi, N. Ohashi, Y. Serizawa, N.E. Lewin, P.M. Blumberg et at. 2011. Synthetic caged DAG-lactones for photochemically controlled activation of protein kinase C. ChemBioChem. 12: 535–539.

Olson, J.P., H.-B. Kwon, K.T. Takasaki, C.Q. Chiu, M.J. Higley, B.L. Sabatini et al. 2013. Optically selective two-photon uncaging of glutamate at 900 nm. J. Am. Chem. Soc. 135: 5954–5957.

Petit, M., C. Tran, T. Roger, T. Gallavardin, H. Dhimane, F. Palma-Cerda et al. 2012. Substitution effect on the one- and two-photon sensitivity of DMAQ "Caging" groups. Org. Lett. 14: 6366–6369.

Salierno, M., E. Marceca, D.S. Peterka, R. Yuste and R. Etch-enique. 2010. A fast ruthenium polypyridine cage complex photoreleases glutamate with visible or IR light in one and two photon regimes. J. Inord. Biochem. 104: 418–422.

Sebej, P., J. Winterr, P. Müller, T. Slanina, J. Al Anshori, P.A. Lovely Angel et al. 2013. Fluorescein analogues as photoremovable protecting groups absorbing at ~ 520 nm. J. Org. Chem. 78: 1833–1843.

Senda, N., Y. Miwa, J. Tanaka, A. Momotake and T. Arai. 2010. Tsukuba-green: A fluorescent dye that emits green fluorescence useful for live-cell imaging. Chem. Lett. 39: 308–310.

Shrestha, P., E. Palalo, D. Kand, J.A. Peterson, A.S. Dutton, N. Rubinstein et al. 2017. In search of the perfect photocage: structure-reactivity relationships in meso-methyl BODIPY photoremovable protecting groups. J. Am. Chem. Soc. 139: 15168–15175.

Slanina, T., P. Shrestha, E. Palao, D. Kand, J.A. Peterson, A.S. Dutton, N. Rubinstein, R. Weinstain, A.H. Winter and P. Klán. 2017. In Search of the Perfect Photocage: Structure–Reactivity Relationships in meso-Methyl BODIPY Photoremovable Protecting Groups. J. Ame. Chem. Soc. 139: 15168–15175 .

Takano, H., T. Narumi, N. Ohashi, A. Suzuki, T. Furuta and H. Tamamura. 2014. Development of the 8-aza-3-bromo-7-hydroxycoumarin-4-ylmethyl group as a new entry of photolabile protecting groups. Tetrahedron 70: 4400–4404.

Tsien, R.Y. and R.S. Zuker. 1986. Control of cytoplasmic calcium with photolabile tetracarboxylate 2-nitrobenzhydrol chelators. Biophys. J. 50: 843–853.

Umeda, N., H. Takahashi, T. Ueno, T. Komatsu, T. Terai, K. Hanaoka et al. 2014. Boron dipyrromethene as a fluorescent caging group for single-photon uncaging with long-wavelength visible light. ACS Chem. Biol. 9: 2242–2246.

Wang, D., Y. Wang, J. Zhao, M. Shen, J. Hu, Z. Liu et al. 2017. Strategic approach to 8-azacoumarins. Org. Lett. 19: 984–987.

Chapter 17

Development and Applications of Supramolecular Hydrogel

Masamichi Yamanaka

1. Introduction

Hydrogels, composed mainly of water, are highly flexible and can be applied in various fields as useful materials. Water absorbency is an important property of hydrogels enabling application in disposable diapers. Moisture retentivity of hydrogels is a fundamental function in cosmetic applications. Hydrogels are used in the preparation of artificial skin because of their high biocompatibility. Adsorption in hydrogels enables sustained release of drugs. Hydrogels can be divided into two types namely, chemical hydrogels and physical hydrogels. These hydrogels are categorized according to their cross-linking points. Hydrogels in which cross-linking points are formed by covalent bonds are referred to as chemical hydrogels and those in which cross-linking points are formed by noncovalent bonds, such as hydrogen bonds, are called physical hydrogels. Generally, covalently cross-linked chemical hydrogels have higher strength than physical hydrogels. Although many hydrogels are made of polymers, some physical hydrogels are formed by self-assembly of low molecular weight organic molecules through noncovalent interactions. These hydrogels are referred to as supramolecular hydrogels and their component molecules are called low molecular weight hydrogelators.

Supramolecular hydrogels are highly flexible because all their formation driving forces are reversible noncovalent bonds (Estroff and Hamilton 2004, Datta and Bhattacharya 2014). The general formation mechanism of a supramolecular hydrogel is as follows (Fig. 17.1): (1) Low molecular weight hydrogelators form supramolecular polymer by dimensionally controlled self-assembly. (2) The resulting supramolecular polymers are bundled to form nano-sized fibrous aggregates. (3) A supramolecular hydrogel is formed by a three-dimensional network structure of fibrous aggregates that incorporate water molecules into

Department of Chemistry, Faculty of Science, and Research Institute of Green Science and Technology, Shizuoka University, 836 Ohya, Suruga-ku, Shizuoka 422-8529, Japan.
Email: yamanaka.masamichi@shizuoka.ac.jp

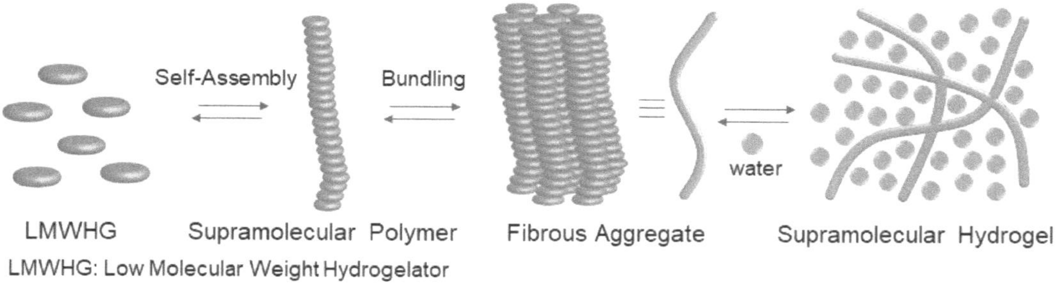

Fig. 17.1: General formation mechanism of supramolecular hydrogel.

the inner space. Supramolecular hydrogels exhibit stimuli-responsive nature because all formation steps are reversible noncovalent bonds. Supramolecular hydrogels responding to various stimuli such as light, chemicals and pH responsiveness have been reported in recent decades. Interestingly, the structural change at molecular level, in response to the external stimulus, is amplified into a macroscopic alteration of a supramolecular hydrogel's formation and collapse. It is possible to generate a supramolecular hydrogel's response to a desired stimulus, by designing low molecular weight hydrogelators rationally. It was found that a C_3-symmetric tris-urea molecule worked as a low molecular weight gelator for organic solvents (Yamanaka et al. 2007). Based on this observation, various supramolecular gels, including supramolecular hydrogels composed of C_3-symmetric tris-urea frameworks have been developed (Yamanaka 2016). In this chapter, development of stimuli-responsive supramolecular hydrogels, based on rational molecular design, has been outlined. The development of electrophoretic method of biopolymer using supramolecular hydrogel as a matrix, has also been described.

2. Development of Stimuli-responsive Supramolecular Hydrogels

2.1 Development of Low Molecular Weight Hydrogelator having Sugar Hydrophilic Group

Development of a low molecular weight hydrogelator was more difficult than the development of a low molecular weight organogelator. In the development of low molecular weight organogelators, the gelation of a wide variety of solvents needed to be investigated. In marked contrast, water was the only solvent in the development of a low molecular weight hydrogelator. Furthermore, controlling self-assembly through hydrogen bonding interactions was more difficult in an aqueous environment than in organic solvents. Various C_3-symmetric tris-urea molecules were synthesized and their gelation abilities were evaluated (Yamanaka et al. 2008). It was revealed that the C_3-symmetric tris-urea was an effective framework for low molecular weight gelators and its gelation ability was retained even after the introduction of various functional groups to the outer aromatic ring. The introduced functional groups accumulated on the surface of the fibrous aggregate, when the tris-urea molecule formed a supramolecular gel. When a hydrophilic group was introduced into the tris-urea molecule, the resulting amphiphilic molecule was expected to self-assemble in a one-dimensional manner in water, to reduce the contact area with the core hydrophobic region. As a result, this amphiphilic tris-urea molecule would work as a low molecular weight hydrogelator.

An amphiphilic tris-urea molecule **1** to which six glucosides (serving as hydrophilic groups) were introduced into the aromatic rings of the outer shell, was designed as a low molecular weight hydrogelator (Fig. 17.2) (Yamanaka et al. 2011). This amphiphilic tris-urea molecule **1** was synthesized using pentaacetyl-D-glucose and phloroglucinol as starting materials. By mixing the synthesized amphiphilic tris-urea molecule **1** with water, dissolving it under heat and then allowing it to stand at room temperature, a transparent supramolecular hydrogel was formed as expected, and its minimum gelation concentration was estimated to be 1.5 wt% (Fig. 17.2). From the Scanning Electron Microscope (SEM) observation of the xerogel, homogeneous fibrous aggregates were detected (Fig. 17.2). The amphiphilic tris-urea molecule **1** produced a transparent supramolecular hydrogel only on mixing it with water and letting it stand at room temperature. The minimum gelation concentration in this method was estimated to be 2.0 wt%. The supramolecular hydrogel prepared at room temperature, without treating it to heat, retained its transparent state for a while, but turned turbid after several weeks to months. SEM observation of the xerogel of the supramolecular hydrogel prepared at room temperature showed homogeneous fibrous aggregates, thinner than that found in supramolecular hydrogels prepared by the typical heating method. SEM observation of the xerogel prepared from a turbid supramolecular hydrogel described above, showed heterogeneous fibrous aggregates containing thick fibrous aggregates that were not observed at the initial stage. In supramolecular hydrogels prepared under heating conditions, appropriate bundling had progressed and formed stable homogeneous fibrous aggregates. Supramolecular hydrogels prepared without heating had sufficient bundling for gelation, but their fibrous aggregates were not stable enough. As a result, bundling progressed gradually, even after forming the supramolecular hydrogel, and resulted in heterogeneous fibrous aggregates that allowed formation of turbid supramolecular hydrogel.

Analysis of the FT-IR spectrum of the xerogel suggested that intermolecular hydrogen bonds were formed between the ureido groups. Therefore, the supramolecular hydrogel expected to show anion-responsive properties. When sodium salts of various anions were added to the top surface of the supramolecular hydrogel, the gel-to-sol phase transitions progressed from the contacted interface. It required two equivalents of chloride anion for the amphiphilic tris-urea molecule **1** to achieve complete gel-to-sol phase transition. Two equivalents of other typical monovalent anions were also required to complete

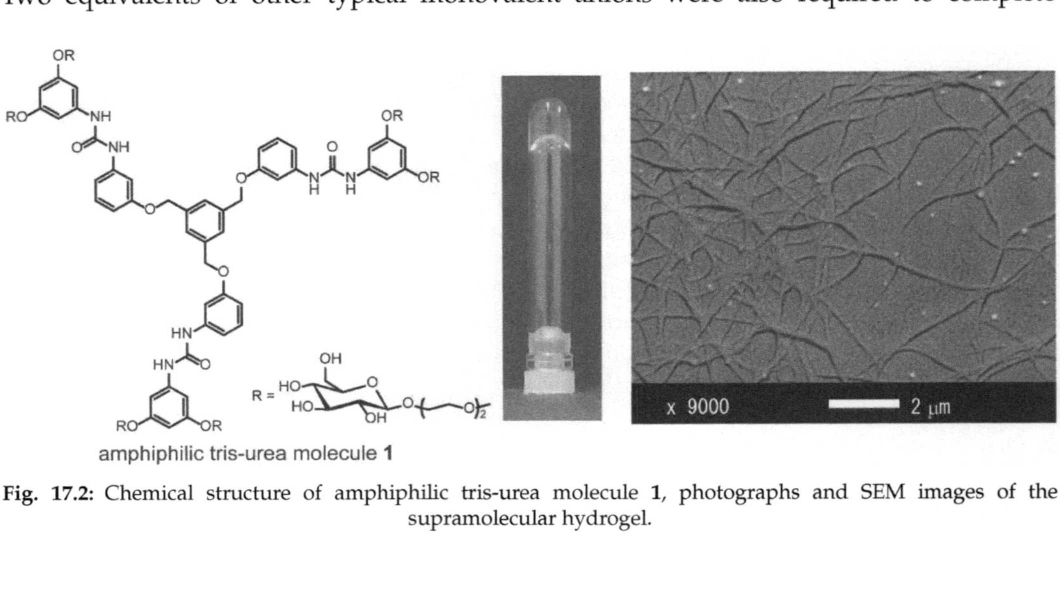

amphiphilic tris-urea molecule **1**

Fig. 17.2: Chemical structure of amphiphilic tris-urea molecule **1**, photographs and SEM images of the supramolecular hydrogel.

gel-to-sol phase transitions. As an exception, weakly coordinated anions such as iodine and perchlorate ions were required in excessive amounts, for the progress of phase transitions. In case of divalent anions, one equivalent of the anions was required for the amphiphilic tris-urea molecule **1** to complete gel-to-sol phase transition. Since the anions were hydrated in water, their association constants to the amphiphilic tris-urea would almost be identical. Therefore, the anion-responsive gel-to-sol phase transition of the supramolecular hydrogel was achieved in a manner that was nonselective to the anion species. By utilizing this nonselective anion responsiveness, visual evaluation of hardness of mineral water was achieved (Fig. 17.3). A mixture of Crystal Geyser, with a total hardness of 38 mg/L (soft water), and 2.0 wt% of **1** gave a transparent gel. Paradiso, which was classified as moderately hard water (total hardness = 290 mg/L), formed an opaque gel on mixing it with 2.0 wt% of **1**. Hazy sols were obtained from mixtures of **1** and hard water samples, such as Wattwiller (total hardness = 627 mg/L) or Contrex (total hardness = 1551 mg/L). The hardness of water was identified by the three different states, i.e., a transparent gel, an opaque gel and a hazy sol.

Glucosides introduced as hydrophilic groups to amphiphilic tris-urea were accumulated at a high density, on the surface of the fibrous aggregates. Therefore, interaction between glucoside moieties and sugar-binding proteins (lectin) caused macroscopic alteration of the supramolecular hydrogel (Fig. 17.4). Well-studied concanavalin A (ConA), which associates with glucoside at moderate strength (binding constant of Me-α-Glc to ConA is $3.0 \times 10^3\ M^{-1}$), was selected as a lectin for the gelation experiments, upon mixing with the amphiphilic tris-urea (Dam and Brewer 2002). When 0.009 equivalent of ConA was added to the supramolecular hydrogel prepared from the amphiphilic tris urea and water, the transparent hydrogel changed to an opaque hydrogel. In addition, the gel-to-sol phase transition temperature rose from 38°C to 85°C, with the addition of ConA. It was considered that the fibrous aggregate cross-linked with ConA, resulting in cloudiness of the gel and elevation of gel-to-sol phase transition temperature. Addition of 0.016 equivalents or more of ConA relative to the amphiphilic tris-urea molecule **1** caused the phase transition of the gel to suspension. Excess ConA would lead to an excessive cohesion of the fibrous aggregates, thereby obstructing any gelation. It was assumed that the supramolecular hydrogel regenerated through the formation of a complex of ConA

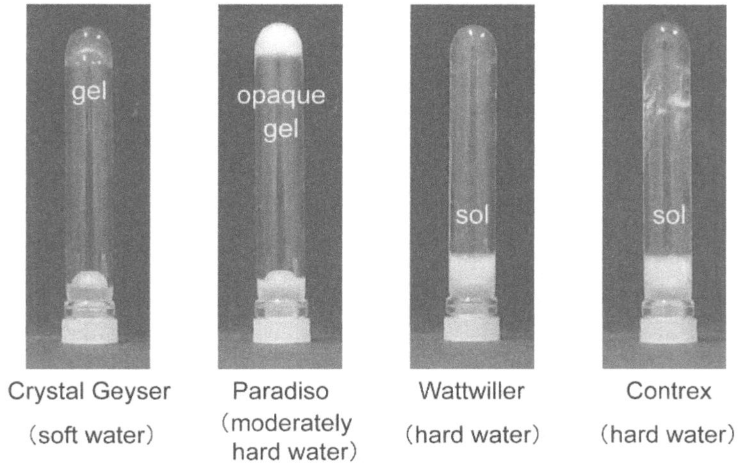

Fig. 17.3: Gelation of mineral water with different hardness using amphiphilic tris-urea molecule **1**.

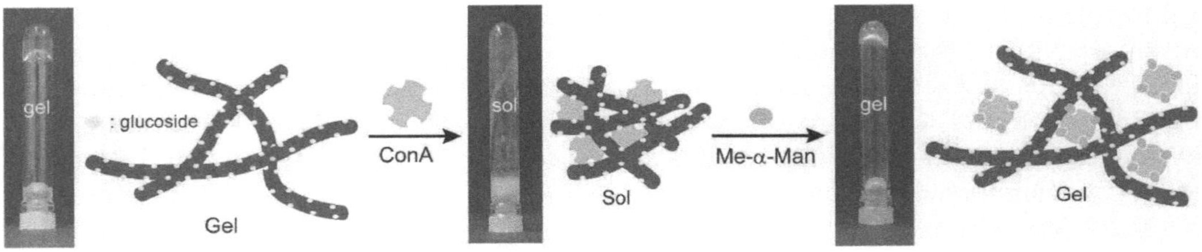

Fig. 17.4: Photographs and schematic representation of the gel-to-sol phase transition triggered by ConA and saccharide.

and another saccharide. Addition of strongly associated α-methyl-D-mannoside (Me-α-Man) (K_a = 1.1 × 10^4 M^{-1}) caused regeneration of the supramolecular hydrogel. On the other hand, excess amount of D-galactose (Gal), which did not associate with ConA, did not regenerate the supramolecular hydrogel.

2.2 Development of Low Molecular Weight Hydrogelator with Carboxy Group as Hydrophilic Group

Many low molecular weight hydrogelators have an amphiphilic structure composed of a hydrophilic part and a hydrophobic part. Generally, it requires multiple steps for synthesis of those amphiphilic low molecular weight hydrogelators. The amphiphilic tris-urea molecule **1** that was developed as part of this study also required more than 10 steps for the synthesis. Therefore, it was not easy to prepare the amphiphilic tris-urea molecule **1** on the gram-scale. This was a serious problem when considering application of supramolecular hydrogel to various fields. Due on this, a new C_3-symmetric amphiphilic tris-urea molecule was developed, which could be synthesized in a short step, as a low molecular weight hydrogelator. An amphiphilic tris-urea molecule **2**, which had carboxy groups as hydrophilic groups, was designed as a candidate for a low molecular weight hydrogelator (Fig. 17.5) (Yamanaka et al. 2015). Since the hydrophilicity of the amphiphilic tris-urea molecule **2** could be controlled by deprotonation and protonation of carboxy groups, **2** would form a supramolecular hydrogel by choosing appropriate conditions. Synthesis of the amphiphilic tris-urea molecule **2** could be achieved in a short step, since commercially available isophthalic acid derivatives could be used as a starting material, and synthesis using several gram-scales could easily be achieved.

Gelation of the amphiphilic tris-urea molecule **2** was investigated. A mixture of water (or acidic aqueous solution) and **2** yielded only a precipitate containing suspension. A slightly viscous solution was obtained by mixing a basic aqueous solution, such as an aqueous sodium hydroxide solution, and the amphiphilic tris-urea molecule **2** (Fig. 17.5). SEM observation of this solution revealed highly homogeneous fibrous aggregates having thickness of about 30 to 80 nm (Fig. 17.5). Regardless of the formation of fibrous aggregates, the reason for this solution not forming supramolecular hydrogels was assumed to be due to electrostatic repulsion between negatively charged carboxylates. Therefore, when hydrochloric acid was added to this viscous solution, a supramolecular hydrogel was formed by resolving electrostatic repulsions between fibrous aggregates (Fig. 17.5). The minimum gelation concentration of the amphiphilic tris-urea molecule **2** was 0.3 wt% (3.0 mM), and gelation proceeded with the addition of protonic acids, excluding hydrochloric acid. SEM image of a xerogel of **2** showed uniform and intertwining nanofibers of 30 to 90 nm (Fig. 17.5). Similar fibrous aggregates were found in the viscous solution and in the supramolecular hydrogel. The pH value of the viscous solution prepared from sodium hydroxide aqueous solution and **2** was 5.9. Based on the pK_a values of isophthalic acid (pK_{a1} = 3.50, pK_{a2} = 4.50), almost all of the carboxy groups would be deprotonated at this pH. When a protonic acid was added to the viscous solution, a partial gel was formed when the pH value became less than 5.0, and complete gelation was achieved when the pH value became less than 4.0.

Controlling the pH of the mixture allowed reversible gel-to-sol phase transition (Fig. 17.6). Addition of an alkaline solution to the supramolecular hydrogel gave a viscous solution. Regelation proceeded when an acidic solution was added to the viscous solution. This phase change could be repeated more than 10 times. Supramolecular hydrogel formation in mixtures of **2** and alkaline solutions, was also triggered by a calcium salt. A fluid mixture of **2** and NaOH solution was spontaneously converted into a supramolecular hydrogel, with the addition of $CaCl_2$ (Fig. 17.6). This gelation would be based on chelate cross-linking between carboxylates, by calcium ions. Addition of [2,2,2]cryptand, which formed a host–guest complex with the Ca^{2+} and prevented chelation within the fibrous aggregates of **2**, induced the gel-to-sol phase transition (Fig. 17.6). The supramolecular hydrogel spontaneously reconstructed with the addition of excess $CaCl_2$ to the solution (Fig. 17.6). Lanthanide ions also induced hydrogel formation in a mixture of **2** and alkaline solutions. Supramolecular hydrogel prepared from **2** and $Tb(OTf)_3$ exhibited

amphiphilic tris-urea molecule 2

Fig. 17.5: Chemical structure of amphiphilic tris-urea molecule **2**, photographs and SEM images of the viscous solution and the supramolecular hydrogel.

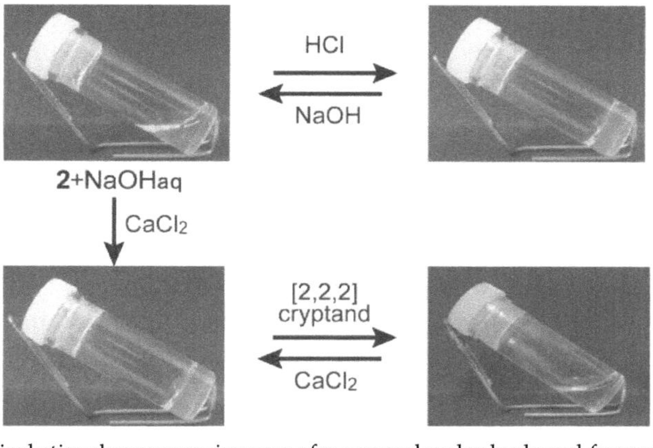

Fig. 17.6: pH and chemical stimulus responsiveness of supramolecular hydrogel formed from amphiphilic tris-urea molecule **2**.

green luminescence due to UV excitation. Furthermore, these supramolecular hydrogels were stable in water without causing swelling or shrinkage and adsorbed organic dyes efficiently (Takeshita et al. 2017).

3. Electrophoresis of Biopolymers using Supramolecular Hydrogel

3.1 History of Electrophoresis

Electrophoresis is a common experimental method in biological science for separating and analyzing biopolymers such as proteins and DNA (Kurien and Scofield 2012, Makovets 2013). Polymer gels such as polyacrylamide and agarose have been widely used as carriers for electrophoreses of these biopolymers. Here the electrophoretic method developed using self-assembled supramolecular hydrogels as carriers, has been outlined.

Protein electrophoresis is an old and most frequently used technique in life sciences research. The technique has dramatically developed, since the first electrophoresis of serum albumin reported by Tiselius in 1937, although several studies regarding protein electrophoresis have been reported in the 19th century (Tiselius 1937). Protein electrophoresis was initially performed in free boundary conditions. Subsequent electrophoretic techniques were developed using paper or gel as a carrier, i.e., zone electrophoresis during the 1950s. Polyacrylamide gel, which could be prepared by copolymerization of acrylamide and N,N′-methylenebisacrylamide (bis), was first applied as a carrier for protein electrophoresis in 1959 (Raymond and Weintraub 1959). Since then, protein electrophoresis using a polyacrylamide carrier has grown rapidly, and has become an indispensable technique for protein analysis for more than half a century. In particular, sodium dodecyl sulfate-polyacrylamide gel electrophoresis (SDS-PAGE), the procedure which was first established in the early 1970s, was the most frequently used method of protein electrophoresis (Laemmli 1970). In SDS-PAGE, negatively charged denatured proteins were electrophoresed in the polyacrylamide gel, and separated mainly according to their molecular weights. Several useful derivative electrophoretic techniques, such as affinity electrophoresis and two-dimensional (2D) electrophoresis, have also been developed. For instance, phosphorylated protein receptor (Phos-tag) copolymerized with polyacrylamide gel, allowed affinity electrophoresis of phosphorylated proteins to be performed (Kinoshita et al. 2009). 2D electrophoresis is a principal technique in proteomics research. Regardless of all the technical developments in protein electrophoresis, the development of novel carriers for protein electrophoresis was rarely reported. But the polyacrylamide was not a perfect carrier for electrophoresis and there were several drawbacks. For example, recovery of protein samples after electrophoresis often plagued researchers. Although there were several recovery methods, those methods sometimes required skilled techniques to achieve high recovery rates. Development of custom-made affinity electrophoresis, which could detect necessary arbitrary proteins, was also a difficult task. To fundamentally solve these problems, development of novel carriers for protein electrophoresis seemed crucial. Therefore, methods for protein electrophoresis using supramolecular gels, i.e., *Supramolecular Gel Electrophoresis* (SUGE) were designed (Yamanaka 2018). Utilizing the characteristics of supramolecular hydrogel flexibility and stimulus responsiveness made it possible to efficiently recover protein samples after electrophoresis. In addition, it was understood that it would be possible to develop tailor-made affinity electrophoresis, utilizing the highly designable property of low molecular weight hydrogelators.

3.2 Electrophoresis of Denatured Proteins with Supramolecular Hydrogel as Carriers (SDS-SUGE)

During the development of a variety of amphiphilic tris-urea molecules as candidates for low molecular weight hydrogelators, amphiphilic tris-urea molecule **3** was synthesized. The amphiphilic tris-urea molecule **3** formed a supramolecular hydrogel with an aqueous solution containing ionic surfactants such as SDS. The supramolecular hydrogel was formed by mixing **3** and Tris[tris(hydroxymethyl)aminoethane]-glycine-SDS (TGS) solution (Tris: 25 mM; glycine: 192 mM; SDS: 3.5 mM), which was a typical buffer used in SDS-PAGE (Fig. 17.7) (Yamamichi et al. 2011). Therefore, experiments of electrophoresis (SDS-SUGE) of denatured proteins using this supramolecular hydrogel as a carrier, were commenced.

After a range of investigations, the following experimental procedure was established for SDS-SUGE (Fig. 17.8). (I) Supramolecular hydrogel was filled in a glass capillary (inner diameter: 2 mm, length: 8 cm). (Agarose (2.0 wt%) was added to the supramolecular hydrogel of amphiphilic tris-urea molecule **3** (2.0 wt%) to improve physical strength of the hydrogel after electrophoresis.) (II) Protein sample was filled at one end of the supramolecular hydrogel-filled glass capillary. (III) Electrophoresis was carried out using a submarine type electrophoresis apparatus at 100 V for around 150 minutes. (IV) The supramolecular hydrogel was removed from the glass capillary and divided into eight equal pieces. (V) The protein sample was extracted from the divided supramolecular hydrogel. (VI) The separation pattern was analyzed by standard SDS-PAGE of the supernatants. Regardless of being a very simple operation of adding buffer solution to a gel after electrophoresis and performing centrifugal separation in the extraction outlined in procedure (V), as a characteristic feature of using supramolecular hydrogel as a carrier, 50% of the protein could be extracted. The proteins were barely extracted from the polyacrylamide gel using the same operation. It was deliberated that the aggregate structure of supramolecular hydrogel could efficiently extract protein, as a result of disintegration by physical stimulation of centrifugation. SDS-SUGEs were performed by using six proteins,

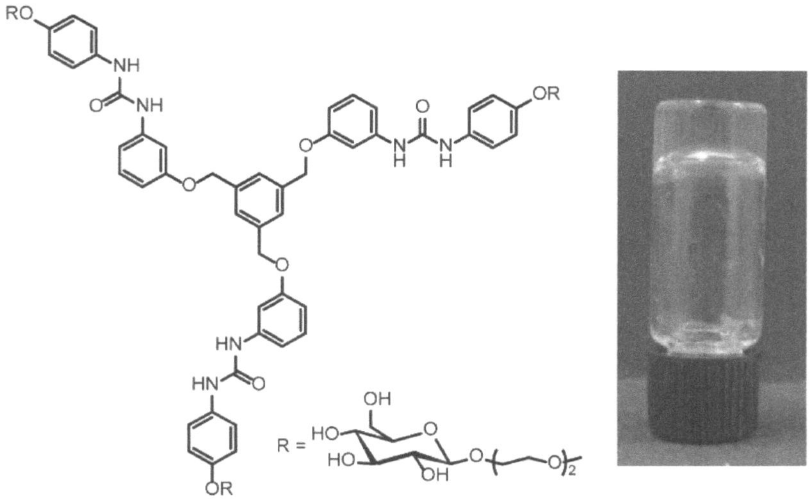

amphiphilic tris-urea molecule **3**

Fig. 17.7: Chemical structure of amphiphilic tris-urea molecule **3**, photograph of the supramolecular hydrogel of TGS solution.

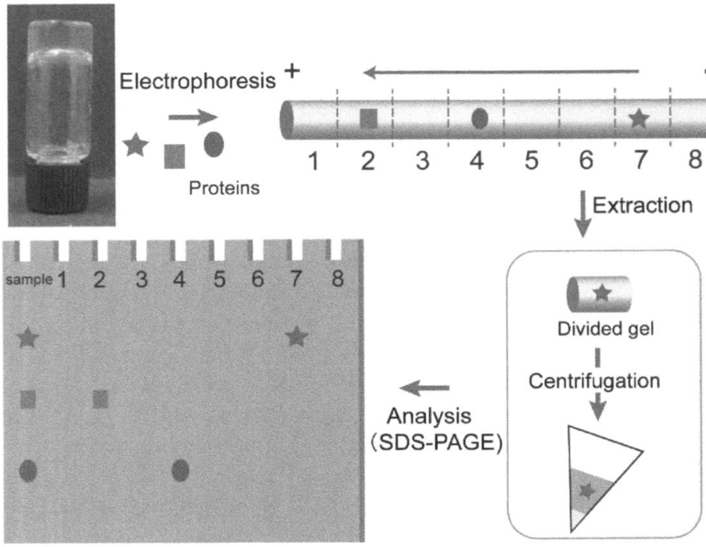

Fig. 17.8: Typical procedure of SDS-SUGE.

namely β-galactosidase (β-Gal, 116 kDa), bovine serum albumin (BSA, 66 kDa), ovalbumin (OVA, 45 kDa), carbonic anhydrase (CA, 29 kDa), lysozyme (LZM, 14 kDa), and aprotinin (APR, 6.5 kDa). A mixture of two proteins with different molecular weights was used for each experiment.

Several SDS-SUGE experiments revealed a unique separation manner of denatured proteins (Fig. 17.9). A mixture of β-Gal and OVA was employed for SDS-SUGE at 100 V for 120 minutes (Fig. 17.9a). Smaller OVA (45 kDa) was observed in lanes 4 to 7, and distinct band was detected in lane 5. Larger β-Gal (116 kDa) was found in lanes 6 to 8, with distinct band being observed in lane 7. OVA was electrophoresed to a large extent on the anode side than β-Gal, as it was in standard SDS-PAGE; however, the separation was unclear. SDS-SUGE of BSA and OVA exhibited similar results (Fig. 17.9b). Smaller OVA (45 kDa) was observed in lanes 3 and 4, and distinct band was detected in lane 3. Larger BSA (66 kDa) was found in lanes 4 and 5, with distinct band observed in lane 4. Next, a mixture of OVA and the much smaller LZM was employed for SDS-SUGE at 100 V for 170 minutes. OVA (45 kDa) was detected in lanes 3 and 4, whereas the smaller LZM (14.4 kDa) was detected in lane 5. Smaller LZM was retained closer to the cathode side compared to larger OVA, i.e., SDS-SUGE using the supramolecular hydrogel of **3** showed a different separation manner in comparison with the typical SDS-PAGE. A mixture of OVA and APR was employed for SDS-SUGE at 100 V for 150 minutes (Fig. 17.9c). OVA was observed in lanes 3 to 5, and the smaller APR (6.5 kDa) was found in lane 7. These results indicated that SDS-SUGE exhibited remarkably poor mobility for smaller proteins. Two different molecular sieve effects would work in this electrophoresis using supramolecular hydrogel. One was analogous to the typical SDS-PAGE separation mechanism. Denatured proteins can pass through the three-dimensionally intertwined fibrous network of the supramolecular hydrogel, and smaller proteins show large mobility. The other separation mechanism resembles that which is found in gel filtration. Isolated spaces constructed from self-assembled aggregates of **3** are suitable for retaining smaller proteins against an electric current. Larger proteins were mainly influenced by the former SDS-PAGE-like separation

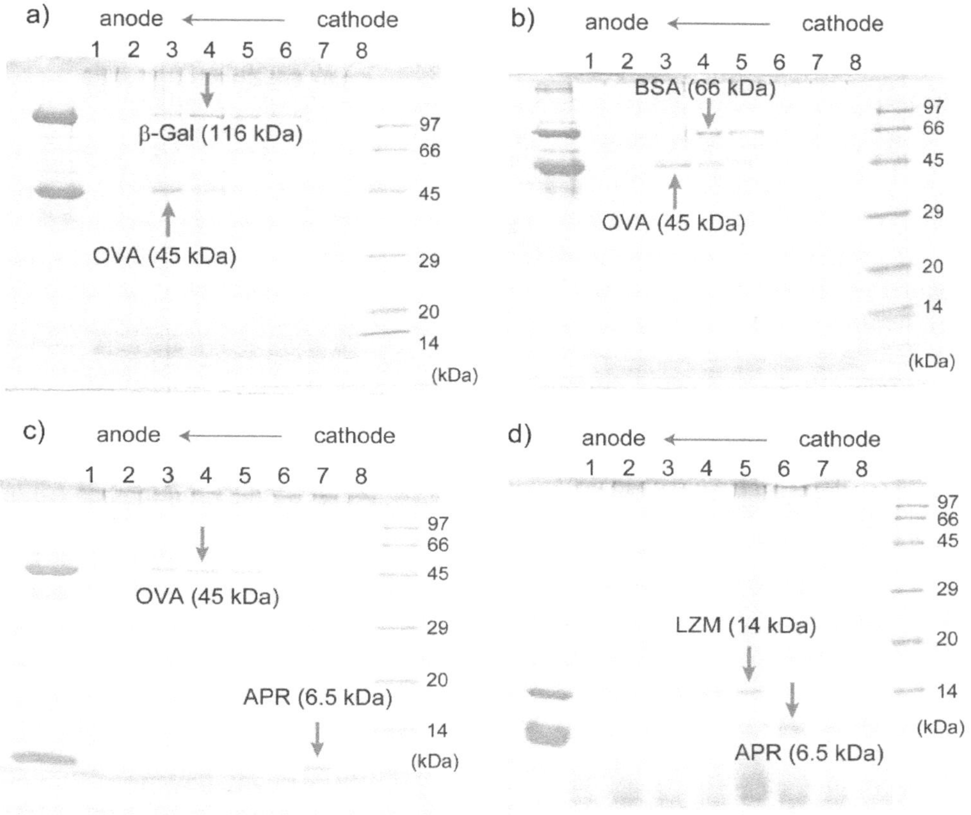

Fig. 17.9: SDS-PAGE analyses of SDS-SUGE separation using TGS solution (Tris: 25 mM; glycine: 192 mM; SDS: 3.5 mM) of (a) β-Gal (116 kDa) and OVA (45 kDa); (b) BAS (66 kDa) and OVA (45 kDa); (c) OVA (45 kDa) and APR (6.5 kDa); (d) LZM (14 kDa) and APR (6.5 kDa).

mechanism. A mixture of OVA and CA (29 kDa) was employed for SDS-SUGE at 100 V for 170 minutes. OVA was observed mainly in lanes 4 and 5, and the smaller CA was found in lane 6. The separation of small-size proteins was allowed by SDS-SUGE. Electrophoresis of LZM and APR was performed at 100 V for 150 minutes (Fig. 17.9d). SDS-PAGE analysis showed that LZM was observed in lane 5, and smaller APR was found at the more cathodic lanes 6 and 7.

The presence of an ionic surfactant such as SDS was essential for gelation of **3**, and a mixture of **3** and pure water formed an insoluble suspension. Various ionic surfactants, including anionic sodium sulfates as well as cationic surfactants such as dodecyltrimethylammonium bromide, effectively induce supramolecular hydrogel formation of **3** (Jinno and Yamanaka 2012). However, non-ionic surfactants such as *n*-octyl-β-D-glucopyranoside and tetraethyleneglycol monododecyl ether could not induce gelation. The concentration of SDS was important for the formation of supramolecular hydrogel of **3**. The mixture of **3** and a small amount of SDS formed insoluble suspensions, and the mixture of **3** and a large amount of SDS formed clear solutions. SDS solutions within the range of 0.5–4 mM yielded supramolecular hydrogels for 9 mM of **3**. The effective concentration of SDS was lower than the critical micelle concentration of SDS

(8.2 mM). It was found that hydrogels that were more transparent were obtained using higher concentrations of SDS. The rate of bundling would reflect the difference in the transparency of the hydrogels of **3**. SEM observations of a mixture of **3** and SDS indicated fibrous aggregates of varying thickness, depending on the concentration of SDS. The SEM image of a sample prepared from a suspension of **3** exhibited a shapeless morphology. The xerogel prepared from a hydrogel (9.0 mM of **3** and 0.50 mM of SDS) demonstrated intertwined fibrous aggregates whose diameters of 100–600 nm. The SEM image of the xerogel prepared from a hydrogel (9.0 mM of **3** and 4.0 mM of SDS) exhibited many homogeneous fibrous aggregates with diameters ranging from 100–250 nm. Notably, a dried sample of a solution (9.0 mM of **3** and 10 mM of SDS) displayed fibrous aggregates with a large aspect ratio, and ranging in thickness from 50–80 nm. Mixtures of **3** and the cationic surfactant (dodecyltrimethylammonium bromide) showed similar results. Amphiphilic tris-urea molecule **3** self-assembled and randomly bundled in aqueous media, but ionic surfactants appeared to restrict the rate of bundling.

The separation manner of SDS-SUGE using TGS solution (Tris: 25 mM; glycine: 192 mM; SDS: 3.5 mM) was altered by the molecular weights of proteins. Smaller proteins displayed better mobility than larger proteins in case of proteins with molecular weight > 45 kDa, as in SDS-PAGE. In contrast, for proteins with a molecular weight of < 45 kDa, smaller proteins showed lower mobility than larger proteins. Furthermore, the thickness of fibrous aggregates in the supramolecular hydrogel of **3** varied depending on the concentration of SDS. These observations led to our hypothesis that the mechanism of protein separation in SDS-SUGE is dependent on the concentration of SDS.

A new set of SDS-SUGE experiments were performed using a supramolecular hydrogel prepared from **3** (2.0 wt%), agarose (2.0 wt%), and TGS solution (Tris: 25 mM; glycine: 192 mM; SDS: 1.0 mM). A mixture of β-Gal and BSA was employed for SDS-SUGE in the TGS solution (25 mM Tris, 192 mM glycine, 1.0 mM SDS) at 100 V for 150 minutes (Fig. 17.10a) (Tazawa et al. 2016). Smaller BSA (66 kDa) was detected in lanes 5 and 6, and distinct band was found in lane 5. Larger β-Gal (116 kDa) was observed in lanes 7 and 8. A mixture of BSA and OVA was separated by SDS-SUGE in the TGS solution (25 mM Tris, 192 mM glycine, 1.0 mM SDS) at 100 V for 180 minutes. Smaller OVA (45 kDa) was found in lanes 6 to 8, with a distinct band that was observed in lane 7. Larger BSA (66 kDa) was detected in lanes 4 and 5, with a distinct band that was found in lane 5. OVA remained closer to the cathode than BSA. This result was quite different from the result of SDS-SUGE using a TGS solution containing 3.5 mM SDS. Next, a mixture of BSA and CA was separated by SDS-SUGE in the TGS solution (25 mM Tris, 192 mM glycine, 1.0 mM SDS) at 100 V for 180 minutes (Fig. 17.10b). Smaller CA (29 kDa) was found in lanes 6 to 8, with distinct a band that was detected in lane 6, and larger BSA (66 kDa) was observed in lanes 3 to 5, with a distinct band that was found in lane 4. The smaller CA remained closer to the cathode than the larger BSA. A mixture of OVA and CA also showed similar separation, and the smaller CA was found closer to the cathodic region than the larger OVA. Thus, the threshold for separation with TGS solution containing 1.0 mM SDS was estimated to be about 66 kDa. Compared to SDS-SUGE using TGS solution with 3.5 mM of SDS, the threshold for different separation modes shifted towards a larger protein molecular weight.

A supramolecular hydrogel was prepared from amphiphilic tris-urea molecule **3** (2.0 wt%), agarose (2.0 wt%), and TGS solution (Tris: 25 mM; glycine: 192 mM; SDS: 7.0 mM), and was utilized for the SDS-SUGE. A mixture of BSA and OVA was employed for SDS-SUGE in TGS solution (Tris: 25 mM; glycine: 192 mM; SDS: 7.0 mM) at 100 V for 180 minutes. Smaller OVA (45 kDa) showed a slightly larger mobility than larger BSA

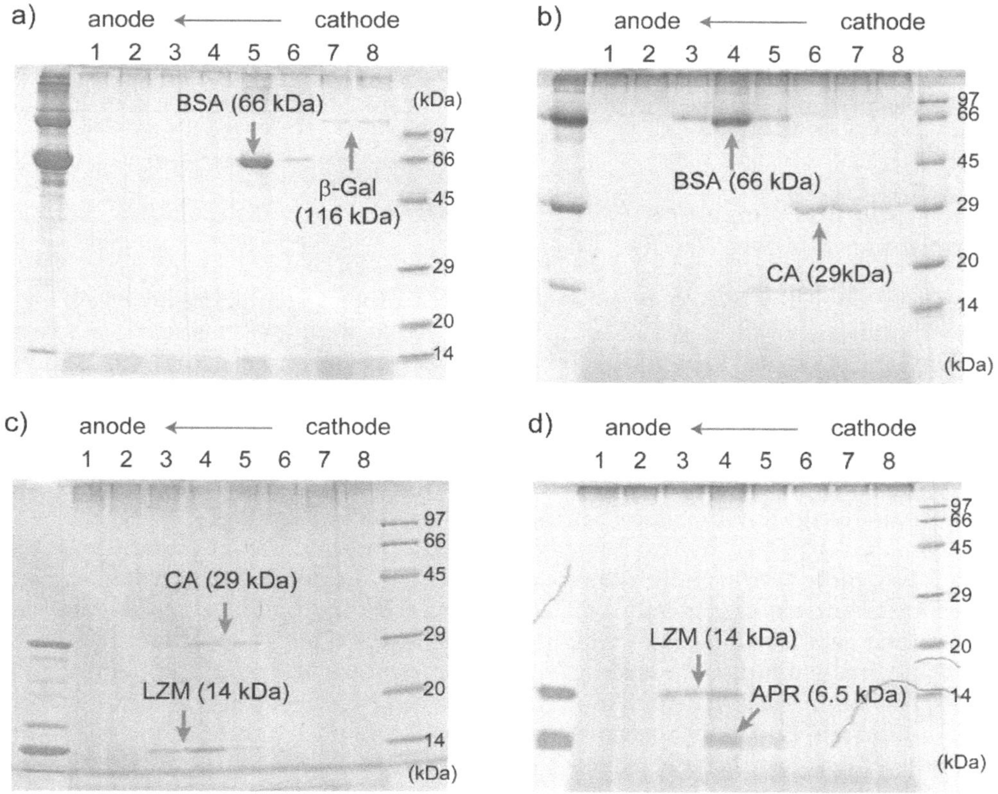

Fig. 17.10: SDS-PAGE analyses of SDS-SUGE separation using TGS solution (Tris: 25 mM; glycine: 192 mM; SDS: x mM) of (a) β-Gal (116 kDa) and BSA (66 kDa) (x = 1.0); (b) BSA (66 kDa) and CA (29 kDa) (x = 1.0); (c) CA (29 kDa) and LZM (14 kDa) (x = 7.0); (d) LZM (14 kDa) and APR (6.5 kDa) (x = 7.0).

(66 kDa). A mixture of OVA and CA was separated by SDS-SUGE in TGS solution (Tris: 25 mM; glycine: 192 mM; SDS: 7.0 mM) at 100 V for 140 minutes. Smaller CA (29 kDa) was detected in lanes 3 to 5, with a distinct band that was found in lane 4. Larger OVA (45 kDa) was detected in lanes 4 to 7, and a distinct band that was observed in lane 5. The SDS-SUGE of a mixture of CA and LZM was also performed in TGS solution (Tris: 25 mM; glycine: 192 mM; SDS: 7.0 mM) at 100 V for 120 minutes (Fig. 17.10c). Smaller LZM (14.4 kDa) was observed in lanes 3 and 4, and larger CA (29 kDa) was detected in lanes 4 and 5.

A mixture of LZM and APR was employed for SDS-SUGE in TGS solution (Tris: 25 mM; glycine: 192 mM; SDS: 7.0 mM) at 100 V for 120 minutes (Fig. 17.10d). Smaller APR (6.5 kDa) was observed in lanes 4 and 5, and larger LZM (14.4 kDa) was found in lanes 3 and 4. Smaller APR showed poor mobility than larger LZM. When the TGS solution was used with 7.0 mM of SDS, the threshold for different separation modes decreased to about 14 kDa.

The threshold for different separation mechanisms shifted according to the concentration of SDS. At low concentrations of SDS, separation mechanism similar to that of gel filtration was dominant, and the threshold observed larger molecular weight range. In contrast, a separation mechanism analogous to that of SDS-PAGE was dominant at high concentrations of SDS, and its threshold appeared in the small molecular weight range. The bundling ratio of **3** varied depending on the concentration of SDS. Thinner fibers were observed upon increasing the concentration of SDS. As a result, smaller-sized pores, which retained the proteins against the electric current, would form at higher concentrations of SDS.

3.3 Electrophoresis of Native Protein with Supramolecular Hydrogel as Carrier (Native-SUGE)

In the study of SDS-SUGE, it became clear that a protein sample could be efficiently recovered after electrophoresis when a supramolecular hydrogel was used as a carrier. Therefore, in order to exploit this characteristic, electrophoresis of native protein sample using supramolecular hydrogel (native-SUGE) was studied. For the gelation of amphiphilic tris-urea molecule **3** used in SDS-SUGE, the presence of an ionic surfactant and a TG solution (Tris: 25 mM; glycine: 192 mM) used for electrophoresis of native protein samples (native-PAGE) that was not gelled, were essential. Therefore, development of a low molecular weight hydrogelator for the TG solution was proposed.

The balance between hydrophobicity and hydrophilicity was important in order to design a low molecular weight hydrogelator. Other hydrophilic structures apart from **3** seemed to be suitable as a gelator of TG solution, and amphiphilic tris-urea molecule **4** was designed as a low molecular weight hydrogelator for the TG solution. Synthesis of **4** was achieved by a multistep reaction started between trimesic acid and pentaacetyl-D-glucose. The amphiphilic tris-urea molecule **4** showed higher gelation ability. In addition to TG solution and pure water, various aqueous solutions such as acidic aqueous solution (8 M hydrochloric acid), basic aqueous solution (7 M potassium hydroxide solution) and high salt concentration aqueous solution (saturated saline) had the ability to gel (Fig. 17.11) (Higashi et al. 2013). It was also revealed that these supramolecular hydrogels of **4** displayed thixotropic properties.

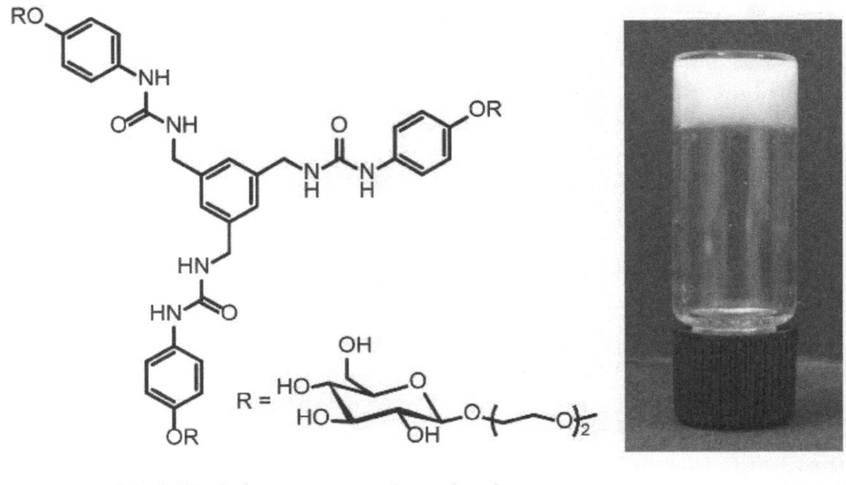

amphiphilic tris-urea molecule **4**

Fig. 17.11: Chemical structure of amphiphilic tris-urea molecule **4**, photograph of the supramolecular hydrogel of TG solution.

Electrophoresis of native proteins using supramolecular hydrogel of **4** was performed using the following procedure, which was similar to SDS-SUGE (Fig. 17.8) (Munenobu et al. 2014). (I) Supramolecular hydrogel of **4** (2.0 wt%) and TG solution (Tris: 25 mM; glycine: 192 mM; pH: 8.3) was filled in a glass capillary. (II) Sample proteins solution was adsorbed onto one end of the supramolecular hydrogel, and both ends of the capillary were filled with agarose gel (2.0 wt%). (III) The glass capillary was immersed in TG solution (Tris: 25 mM; glycine: 192 mM; pH: 8.3) in a submarine electrophoresis system and electrophoresed at 100 V for 100 minutes. (IV) The electrophoresed gel was removed from the glass capillary and divided into eight equal parts (numbered 1–8 from the anode). (V) Proteins were extracted from the supramolecular hydrogel by centrifugation. (VI) The separation pattern was analyzed by a typical SDS-PAGE.

Three kinds of acidic native proteins, D-lactate dehydrogenase (LDH, tetramer = 146 kDa, Isoelectric point (pI) = 4.0), β-galactosidase (β-Gal, tetramer = 540 kDa, pI = 4.6), and ovalbumin (OVA, monomer = 45 kDa, pI = 4.7), were used for experiments of native-SUGE with the supramolecular hydrogel of **4**. Two of the three proteins were electrophoresed by the native-SUGE system and their separations were analyzed by the typical SDS-PAGE (Fig. 17.12).

First, a mixture of LDH and β-Gal was employed for native-SUGE (Fig. 17.12a). LDH was observed in lanes 4 and 5, and a distinct band that was found in lane 4. β-Gal was detected in lanes 5 and 6, and a distinct band was found in lane 5. The smaller (146 kDa) and more acidic (pI = 4.0) LDH showed larger mobility than the larger (540 kDa) and less acidic (pI = 4.6) β-Gal in this system. It was confirmed that LDH and β-Gal retained their enzymatic activities after electrophoresis using supramolecular hydrogel of **4**. Next, native-SUGE of a mixture of LDH and OVA was performed (Fig. 17.12b). SDS-PAGE analysis showed that LDH was identified in lanes 3 and 4, and OVA was detected in lanes 5 and 6. Smaller and less acidic OVA exhibited poor mobility than the larger and more acidic LDH. Native-SUGE of β-Gal and OVA exhibited similar results (Fig. 17.12c). β-Gal was detected in lanes 4 and 5, and OVA was found in lanes 5 and 6. Smaller and less acidic OVA showed poor mobility than the larger and more acidic β-Gal. These results demonstrated that in native-SUGE, native acidic proteins were separated mainly by their isoelectric point (pI), and their molecular weights were less affected.

A mixture of green fluorescent protein (GFP, 27 kDa, pI = 5.57) and red fluorescent protein (RFP, 27 kDa, pI = 5.65) was employed for native-SUGE. UV irradiation to the supramolecular hydrogel of **4** showed green and red fluorescent bands during and after electrophoresis (Fig. 17.12d). This result indicated that GFP and RFP remained in their native form throughout the electrophoretic process. GFP was observed in lanes 5 and 6, and a distinct band was found in lane 5. RFP was detected in lanes 6 and 7, and a distinct band was found in lane 7. The less acidic RFP showed poor mobility than the more acidic GFP, regardless of having similar isoelectric points.

Glucosides, which were hydrophilic moieties of amphiphilic tris-urea molecule **4**, would cover the surfaces of the nanofibers formed through the self-assembly of **4**. Those glucosides on nanofibers could interact with appropriate lectins during electrophoresis, and the lectins would show poor mobility in comparison with the non-carbohydrate-binding proteins. ConA (tetramer = 112 kDa, pI = 4.4–5.5) was employed for affinity electrophoresis using the supramolecular hydrogel of **4**. A moderate association between **4** and ConA was expected based on the moderate association constant between α-methyl-D-glucopyranoside and ConA ($K_a = 3.0 \times 10^3\ M^{-1}$). The native ConA remained at the cathode at the starting point and was detected at lane 8 (Fig. 17.13a). In contrast, the denatured

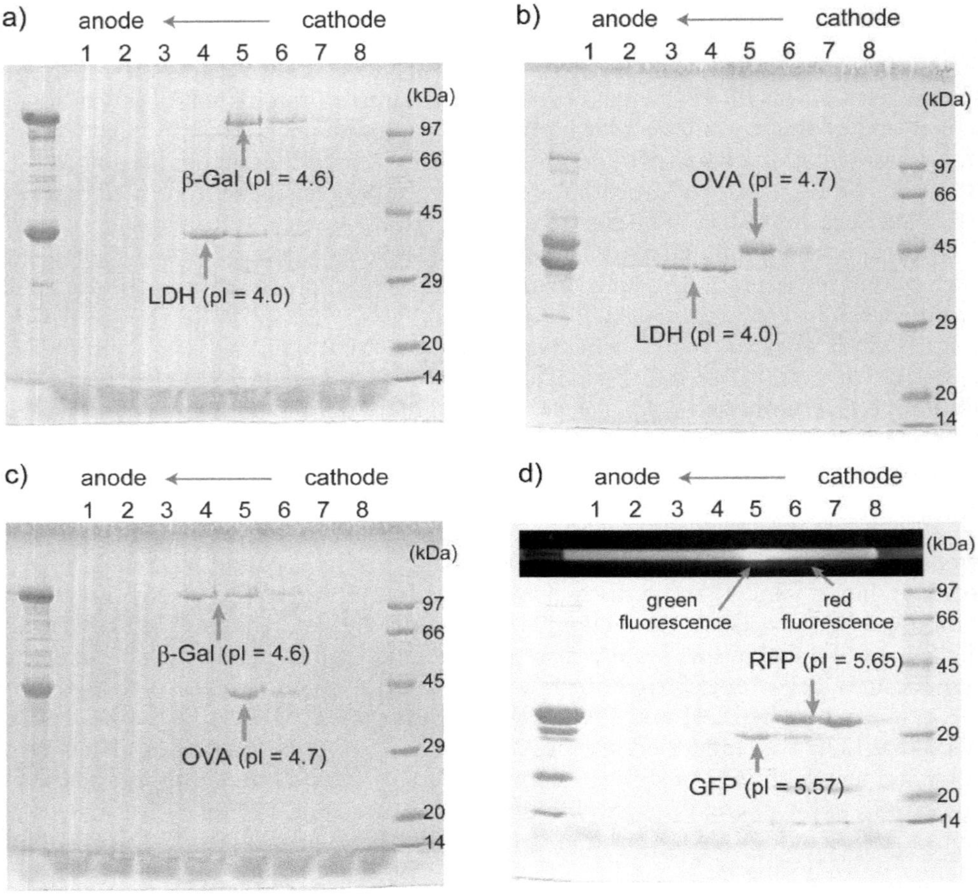

Fig. 17.12: SDS-PAGE analyses of native-SUGE of (a) LDH (146 kDa, pI = 4.0) and β-Gal (540 kDa, pI = 4.6); (b) LDH and OVA (45 kDa, pI = 4.7); (c) β-Gal and OVA; (d) GFP (27 kDa, pI = 5.57) and RFP (27 kDa, pI = 5.65) and photographs of the UV irradiated glass capillary after electrophoresis (insert).

Color version at the end of the book

ConA electrophoresed towards the anode under the same conditions, and was found in lanes 4 and 5 (Fig. 17.13b). These results indicated that glucosides on the nanofiber surfaces interacted with native ConA, and the association inhibited its electrophoretic mobility.

3.4 Electrophoresis of DNA using Supramolecular Hydrogel as a Carrier (DNA-SUGE)

Agarose and polyacrylamide gels were frequently used as matrices for electrophoresis of DNA. Small DNA fragments below 500 base pairs (bp) were separated by polyacrylamide gel, which had small pore sizes in the 5–100 nm range. In contrast, agarose gel, with pore sizes of 200–500 nm, was used for the separation of much larger DNA fragments (> 100 bp). Smaller DNA fragments generally migrated faster than larger DNA fragments in the typical DNA electrophoreses. It is still a challenging subject to separate DNA fragments larger

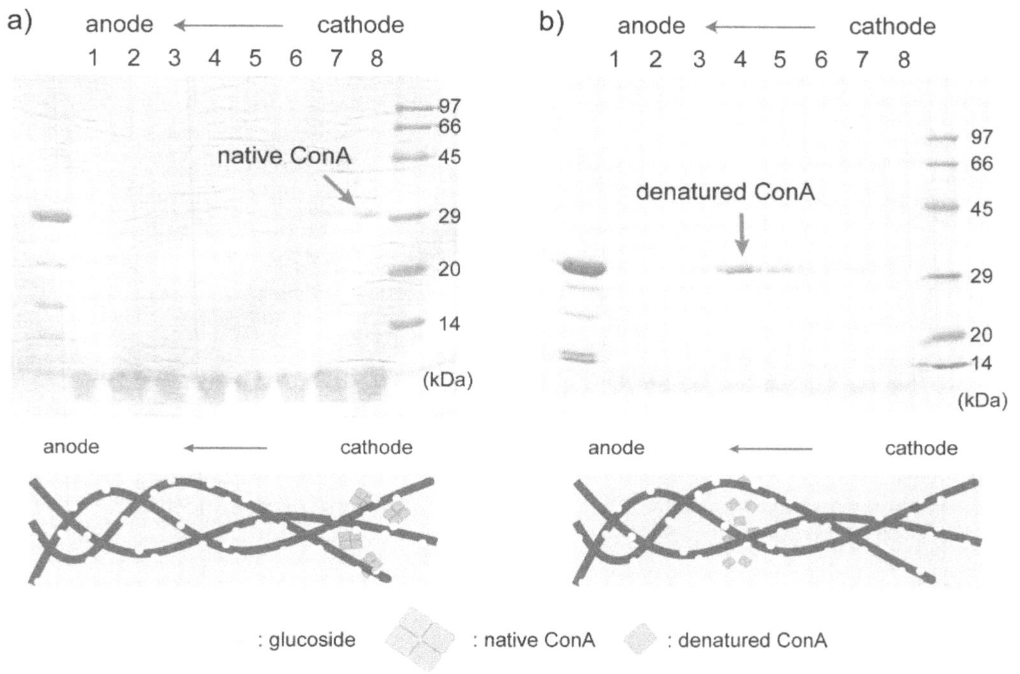

Fig. 17.13: SDS-PAGE analyses of native-SUGE of (a) ConA (112 kDa (as tetramer), pI = 4.4–5.5); (b) denatured ConA.

than 20 kbp by typical continuous field electrophoresis. Large DNA fragments migrated at comparable rates in the continuous field electrophoresis. Pulsed-field gel electrophoresis was developed to separate large DNA fragments (Schwartz and Cantor 1984). Pulsed-field gel electrophoresis was used for cloning large DNA fragments, karyotype analyses of microorganisms and epidemiological analyses of infectious disease, but it required special equipment and long analysis times. The ability of a typical continuous field electrophoresis apparatus to separate and analyze large DNA fragments proved to be a valuable alternative for pulsed-field gel electrophoresis.

Native acidic proteins were separated on the basis of their isoelectric points rather than their molecular weights in native-SUGE experiments using supramolecular hydrogel of **4**. This result indicated that the network structure of the supramolecular hydrogel formed from amphiphilic tris-urea molecule **4** was coarse and/or flexible, allowing large biopolymers to pass through. This finding encouraged the examination of supramolecular hydrogel of **4** for the electrophoretic analysis of large DNA fragments.

A supramolecular hydrogel formed from **4** and TBE solution (Tris: 45 mM; boric acid: 45 mM; EDTA: 1.0 mM) was used for the matrix of electrophoresis of large DNA fragments (DNA-SUGE) (Tazawa et al. 2017). Almost identical procedures as those mentioned in SDS-SUGE and native-SUGE were used in DNA-SUGE. Three commercial DNA markers, the λ-Hind III digest (Takara Bio Inc.; 2–23 kbp DNA fragments), Lambda DNA-Mono Cut Mix (New England Biolabs; 10–49 kbp DNA fragments), and Marker 7 GT (Nippon Gene; 10–165 kbp DNA fragments), were applied in this study. The supramolecular hydrogel was divided into 8 or 20 equal parts after the electrophoresis. Their extracts were analyzed using DNA electrophoresis using agarose H gel (Nippon Gene).

Initially, the λ-Hind III digest DNA marker, containing 2.0, 2.3, 4.4, 6.6, 9.4, and 23.1 kbp DNA fragments was employed for DNA-SUGE. The electrophoresis was performed using 2.0 wt% of the supramolecular hydrogel formed from **4** at 150 V for 90 minutes (Fig. 17.14a). The DNA fragments were separated according to their sizes, with longer DNA fragments displaying poorer mobility than shorter DNA fragments. DNA fragment of 2.0 kbp was detected in most anodic lane 2. DNA fragments of 2.3, 4.4, and 6.6 kbp were found in lanes 3, 4, and 5, respectively. Larger DNA fragments of 9.4 and 23.1 kbp were mainly observed in lane 6, and a thin band corresponding to DNA fragment of 23.1 kbp was found in lane 7. It is noteworthy that the six DNA fragments were finely separated in DNA-SUGE. DNA fragments with similar sizes, i.e., DNA fragments of 2.0 and 2.3 kbp, were fully separated.

The Lambda DNA-Mono Cut Mix, containing 10.1, 15.0, 17.1, 24.0, 24.5, 30.0, 33.5, 38.4 and 48.5 kbp DNA fragments was applied for DNA-SUGE. The suitable DNA-SUGE condition for the separation of the λ-Hind III digest was unsuitable for the separation of the Lambda DNA-Mono Cut Mix. A low concentration of supramolecular hydrogel of **4** and a low voltage were more effective for the separation of the Lambda DNA-Mono Cut Mix composed of multiple DNA fragments, although high concentrations of the

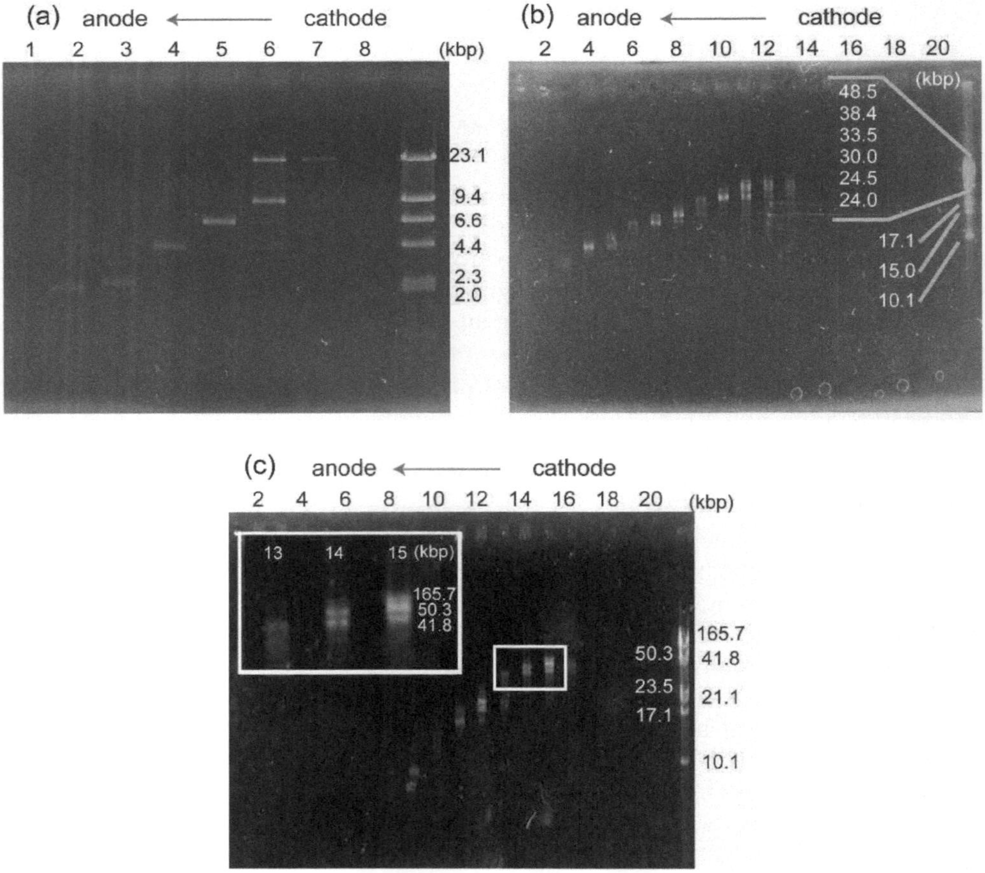

Fig. 17.14: Agarose H gel electrophoresis following DNA-SUGE separated (a) λ-Hind III digest; (b) Lambda DNA-Mono Cut Mix; (c) Marker 7 GT.

supramolecular hydrogels of **4** resulted in fine resolution. Consequently, DNA-SUGE of the Lambda DNA-Mono Cut Mix performed using 1.0 wt% of the supramolecular hydrogel of **4** at 50 V for 6 hours, and the supramolecular hydrogel was divided into 20 equal parts after the electrophoresis (Fig. 17.14b). DNA fragment of 10.1 kbp was mainly detected in lane 4, while DNA fragment of 15.0 kbp was found in lane 7, and DNA fragment of 17.1 kbp was observed in lane 8, confirming clear separation. DNA fragments of 24.0 to 48.5 kbp were detected in lanes 10 to 13; although these DNA fragments were loosely separated, their identification was problematic owing to the low resolving efficiency of agarose H gel electrophoresis.

Marker 7 GT, containing 10.1, 17.7, 21.1, 23.5, 41.8, 50.3, and 165.7 kbp DNA fragments was employed for DNA-SUGE. Suitable separation was obtained from DNA-SUGE using 2.0 wt% of supramolecular hydrogel formed from **2** at 40 V for 9 hours (Fig. 17.14c). DNA fragments of 10.1 and 17.7 kbp were detected only in lanes 9 and 11, respectively, while DNA fragments of 21.1 and 23.5 kbp were found in lane 12. DNA fragment of 41.8 kbp was observed in lanes 13 to 15, while DNA fragment of 50.3 kbp was detected in lanes 14 and 15. The largest DNA fragment of 165.7 kbp was found only in lane 15. It is worth mentioning that 50.3 and 165.7 kbp DNA fragments displayed altered mobility in DNA-SUGE.

4. Conclusion

It has been established that a C_3-symmetric tris-urea structure is an excellent framework as a low molecular weight gelator. Amphiphilic tris-urea molecules modified with hydrophilic groups realized dimensionally controlled self-assembly in an aqueous environment to form a supramolecular hydrogel. The molecular recognition ability of the functional group introduced into the low molecular weight hydrogelator is reflected as the stimulus responsiveness of the resulting supramolecular hydrogel. The input stimulus is generally amplified during the macroscopic phase transition of supramolecular hydrogels.

Electrophoreses of biopolymers using supramolecular hydrogels (SUGE) instead of traditional polymer hydrogels, has been studied. Several SUGE methods for the separation of biopolymers, namely denatured proteins (SDS-SUGE), native proteins (Native-SUGE), and DNAs (DNA-SUGE), have been developed. SUGEs are a new and promising method, but there some problems that remain to be solved. Preparing a stiff supramolecular hydrogel is an important subject because of the fragility of supramolecular hydrogels makes the SUGE operation difficult. Development of the low-molecular-weight hydrogelator that can be synthesized in large quantities can make supramolecular hydrogel an inexpensive matrix for electrophoresis. Supramolecular hydrogels have enormous potential as novel matrices for the electrophoresis of biopolymers. When these problems are overcome, SUGE can be an innovative experimental technique for life science.

References

Dam, T.K. and C.F. Brewer. 2002. Thermodynamic studies of lectin–carbohydrate interactions by isothermal titration calorimetry. Chem. Rev. 102: 387–429.

Datta, S. and S.M. Bhattacharya. 2014. Multifarious facets of sugar-derived molecular gels: molecular features, mechanisms of self-assembly and emerging applications. Chem. Soc. Rev. 44: 5596–5637.

Estroff, L.A. and A.D. Hamilton. 2004. Water gelation by small organic molecules. Chem. Rev. 104: 1201–1218.

Higashi, D., M. Yoshida and M. Yamanaka. 2013. Thixotropic hydrogel formation in various aqueous solutions through self-assembly of an amphiphilic tris-urea. Chem. Asian J. 8: 2584–2587.

Jinno, Y. and M. Yamanaka. 2012. Ionic surfactants induce amphiphilic tris(urea) hydrogel formation. Chem. Asian J. 7: 1768–1771.

Kinoshita, E., E. Kinoshita-Kikuta and T. Koike. 2009. Separation and detection of large phosphoproteins using Phos-tag SDS-PAGE. Nature Protocol 4: 1513–1521.

Kurien, B.T. and R.H. Scofield. 2012. Protein Electrophoresis Methods and Protocols. Springer Science.

Laemmli, U.K. 1970. Cleavage of structural proteins during the assembly of the head of bacteriophage T4. Nature 227: 680–685.

Makovets, S. 2013. DNA Electrophoresis Methods and Protocols. Springer Science.

Munenobu, K., T. Hase, T. Oyoshi and M. Yamanaka. 2014. Supramolecular gel electrophoresis of acidic native proteins. Anal. Chem. 86: 9924–9929.

Raymond, S. and L. Weintraub. 1959. Acrylamide gel as a supporting medium for zone electrophoresis. Science 130: 711.

Schwartz, D.C. and C.R. Cantor. 1984. Separation of yeast chromosome-sized DNAs by pulsed field gradient gel electrophoresis. Cell 37: 67–75.

Takeshita, J., Y. Hasegawa, K. Yanai, A. Yamamoto, A. Ishii, M. Hasegawa et al. 2017. Organic dye adsorption by amphiphilic tris-urea supramolecular hydrogel. Chem. Asian J. 12: 2029–2032.

Tazawa, S., K. Kobayashi and M. Yamanaka. 2016. Effect of sodium dodecyl sulfate concentration on supramolecular gel electrophoresis. ChemNanoMat. 2: 423–425.

Tazawa, S., K. Kobayashi, T. Oyoshi and M. Yamanaka. 2017. Supramolecular gel electrophoresis of large DNA fragments. Electrophoresis 38: 2662–2665.

Tiselius, A. 1937. Electrohoresis of serium globulin. I. Biochem. J. 31: 313–317.

Yamamichi, S., Y. Jinno, N. Haraya, T. Oyoshi, H. Tomitori, K. Kashiwagi et al. 2011. Separation of proteins using supramolecular gel electrophoresis. Chem. Commun. 47: 10344–10346.

Yamanaka, M., T. Nakamura, T. Nakagawa and H. Itagaki. 2007. Reversible sol-gel transition of a tris-urea gelator that responds to chemical stimuli. Tetrahedron Lett. 48: 8990–8993.

Yamanaka, M., T. Nakagawa, R. Aoyama and T. Nakamura. 2008. Synthesis and estimation of gelation ability of C_3-symmetry tris-urea compounds. Tetrahedron 64: 11558–11567.

Yamanaka, M., N. Haraya and S. Yamamichi. 2011. Chemical stimuli-responsive supramolecular hydrogel from amphiphilic tris-urea. Chem. Asian J. 6: 1022–1025.

Yamanaka, M., K. Yanai, Y. Zama, J. Tsuchiyagaito, M. Yoshida, A. Ishii et al. 2015. Cation-tuned stimuli-responsive and optical properties of supramolecular hydrogels. Chem. Asian J. 10: 1299–1303.

Yamanaka, M. 2016. Development of C_3-symmetric tris-urea low-molecular-weight gelators. Chem. Rec. 16: 768–782.

Yamanaka, M. 2018. Supramolecular gel electrophoresis. Polymer J. 50: 627–635.

Chapter 18

Chemistry of Fairy Rings

Jae-Hoon Choi and *Hirokazu Kawagishi**

1. Introduction

The circles or arcs of prompted or suppressed plant growth or of the fruiting bodies of soil-inhabiting fungi which often occur in floors of woodlands, agricultural or grassland in most parts of the world are generally called "fairy rings" (Fig. 18.1A). The term "fairy rings" has its origin in the myths and superstitions associated with their occurrence in the Middle Ages. The circles were thought by some to be the dancing sites of fairies. The stimulated grass growth which may be symptomatic of the activities of some fairy ring fungi results from the breakdown of organic nitrogen compounds as fungus mycelium or other organic material is decomposed and the nitrogen made available is then taken up by the green plants (Couch 1995, Evershed 1884, Shantz and Piemeisel 1917, Smith 1959). While fairy rings have been widely reported (Shantz and Piemeisel 1917), high resolution aerial imagery was made available in the last decade (Fig. 18.1B and C).

2. Fairy Ring

2.1 Symptoms

The symptoms of fairy ring are initially approximately 10 of centimeters or less in diameter, and expand in size year after year, reaching up to several meters in diameter in old turf stands. The mycelia of the fungus in the soil of a fairy rings grow in a concentric manner as a fungal isolate would do in pure culture on a petri dish of agar media, because of the apical growth habit of the fungi. Shantz el al. divided these into three types. Three types of fairy ring symptoms are observed in turfgrasses: Type I, Type II, and Type III (Shantz and Piemeisel 1917).

Research Institute of Green Science and Technology, Shizuoka University, 836 Ohya, Shizuoka, Shizuoka 422-8529, Japan.
Email: choi.jaehoon@shizuoka.ac.jp
* Corresponding author: kawagishi.hirokazu@shizuoka.ac.jp

Fig. 18.1: Fairy rings that appeared on turfgrass in park Hamamatsu City (A) in Japan. Fairy rings in park near campus of University of Shizuoka (B, C). Initial image capture from Google Earth (B, C). (C) is a close-up image of (B). The interaction between fungus and plant stimulates growth of the grass.

Color version at the end of the book

Type I Those where the grass is killed or badly damaged.

Type II Those where the grass growth is stimulated only.

Type III Those where there is no damage to the turf, but the fruiting bodies of the fungus are found in rings.

2.2 Fairy-ring Causing Fungi

Many fungi species, mostly Basidiomycetes, have been recorded associated with fairy rings (Shantz and Piemeisel 1917, Smith 1959). At least, 56 species of basidiomycetous fungi have been reported to be capable of developing fairy rings (Table 18.1). The size of the fruiting bodies among the species range from 1 cm to 30 cm in diameter. Tahama has recorded some types of symptoms in *Zoysia* and *Agrostis* turf in which fairy ring fungus were growing (1978). Turfgrasses used for golf courses in Japan consist mainly of two types: warm-season Zoysiagrass, such as *Zoysia matrella*, and cool-season bentgrass, such as *Agrostis palustris*. *Z. matrella* is primarily used for fairways, however *A. palustris* is usually used for putting green for resistance to low mowing height. The main species of fairy-ring fungi on golf courses are *Lycoperdon perlatum*, *M. oreades*, and *Lepista sordida* (Tani 1991).

3. Fairy Ring-forming Fungus *Marasmius oreades*

3.1 Symptoms by M. oreades

M. oreades is the most common cause of the growth-suppressed type of fairy rings in all types of turfgrass (Type I). The rings are very common on golf course fairways over a

Table 18.1: Species of basidiomyceous fungi that form fairy rings.

Fairy ring species	Reference	Fairy ring species	Reference
Agaricus arvensis	(Shantz and Piemeisel 1917)	*Hygrophoropsis aurantiaca*	(Couch 1995)
Agaricus lilaceps	(Couch 1995)	*Lactarius insulsus*	(Couch 1995)
Agaricus campestris	(Couch 1995)	*Lactarius piperatus*	(Couch 1995)
Bovista dermoxantha	(Terashima et al. 2004)	*Lactarius torminosus*	(Couch 1995)
Calocybe carnea	(Smith 1959)	*Lepista luscina*	(Couch 1995)
Calocybe gambosa	(Ramsbottom 1953)	*Lepista nuda*	(Couch 1995)
Calvatia cyathiformis	(Shantz and Piemeisel 1917)	*Lepista personata*	(Smith 1959)
Chlorophyllum molybdites	(Shantz and Piemeisel 1917)	*Lepista sordida*	(Tahama 1978)
Clavulinopsis corniculata	(Smith 1959)	*Leucoagaricus naucinus*	(Couch 1995)
Clitocybe gibba	(Couch 1995)	*Leucopaxillus giganteus*	(Couch 1995)
Clitocybe maxima	(Ramsbottom 1953)	*Lycoperdon curtisii*	(Shantz and Piemeisel 1917)
Clitocybe nebularis	(Couch 1995)	*Lycoperdon hiemale*	(Couch 1995)
Clitocybe rivulosa	(Smith 1959)	*Lycoperdon perlatum*	(Couch 1995)
Collybia butyracea	(Smith 1959)	*Marasmius oreades*	(Shantz and Piemeisel 1917)
Coprinopsis atramentaria	(Couch 1995)	*Marasmius urens*	(Couch 1995)
Coprinus comatus	(Couch 1995)	*Melanoleuca grammopodia*	(Couch 1995)
Disciseda borista	(Shantz and Piemeisel 1917)	*Melanoleuca melaleuca*	(Shantz and Piemeisel 1917)
Gymnopus confluens	(Couch 1995)	*Nolanea staurospora*	(Couch 1995)
Hebeloma crustuliniforme	(Couch 1995)	*Panaeolus campanulatus*	(Couch 1995)
Hydnellum suaveolens	(Couch 1995)	*Panaeolus foenisecii*	(Couch 1995)
Hydnum compactum	(Couch 1995)	*Paxillus involutus*	(Couch 1995)
Hydnum repandum	(Couch 1995)	*Psilocybe semilanceata*	(Smith 1959)
Hygrocybe coccinea	(Smith 1959)	*Scleroderma verrucosum*	(Couch 1995)
Hygrocybe mucronella	(Smith 1959)	*Suillus grevillei*	(Couch 1995)
Hygrocybe nivea	(Couch 1995)	*Tricholoma columbetta*	(Couch 1995)
Hygrocybe pratensis	(Smith 1959)	*Tricholoma panoeolum*	(Couch 1995)
Hygrocybe psittacina	(Smith 1959)	*Tricholoma terreum*	(Couch 1995)
Hygrocybe virginea	(Couch 1995)	*Vascellum curtisii*	(Terashima et al. 2004)

wide range of climate. Fruiting bodies of the fungus are usually produced in the bare zone or at the junction of the bare and the outer zone. The symptoms related to this fungus, a common pathogen causing the most damaging rings in North America, Europe, and the South Pacific, have been summarized (Smith 1980). With *M. oreades*-based fairy rings, then, the nature of pathogenesis is probably a complex combination of (1) depletion of available nitrogen for plant growth, (2) development of a hydrophobic soil, and (3) direct pathogenic activities of the fungus on the root system. Pathology of fairy rings by *M. oreades* has shown that this organism can infect and colonize the roots of Kentucky bluegrass (*Poa pratensis*), creeping red fescue (*Festuca rubra*) and Colonial bentgrass (*Agrostis tenuis*) (Filer 1965).

3.2 Cyanogensis and Plant Growth Regulator

Bayliss showed that the fungus produced an excretion toxic to grass roots and that roots were invaded by the fungus (Bayliss 1911). Cyanogensis was first discovered in 1871 when Losecke found hydrogen cyanide (HCN) while crushing *M. oreades* (Robbins et al. 1950). Lebeau and Hawn demonstrated that the toxin, HCN, is produced by the mycelium of the fungus in culture and under natural condition (Lebeau and Hawn 1963). Filer showed that *M. oreades* could parasitize grass seedlings and inhibit root growth and that the amount of HCN produced by the mycelium was sufficient to damage higher plants (Filer 1965). HCN concentrations were sufficiently high to inhibit grass root growth and greatly inhibit the growth of fungi (Blenis et al. 2004). Both the fruiting bodies and the mycelia of *M. oreades* produce large quantities of HCN (Stijve and De Meijer 1999). Identification of the cyanohydrin of glyoxylic acid as the precursor of hydrocyanic acid in *M. oreades.* As cyanohydrin is released into the culture medium, it should be present in the soil near *M. oreades* mycelia (Fig. 18.2). Consequently, cyanohydrin of glyoxylic acid might serve

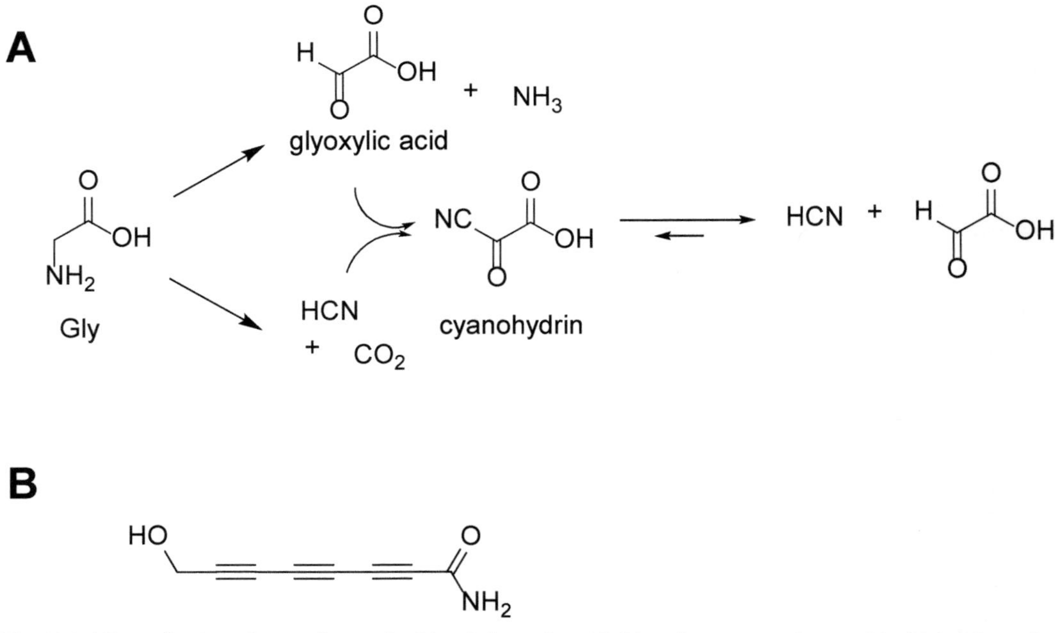

Fig. 18.2: Biosynthetic pathway of cyanohydrin of glyoxylic acid (A) and structure of agrocybin (B) in *M. oreades.*

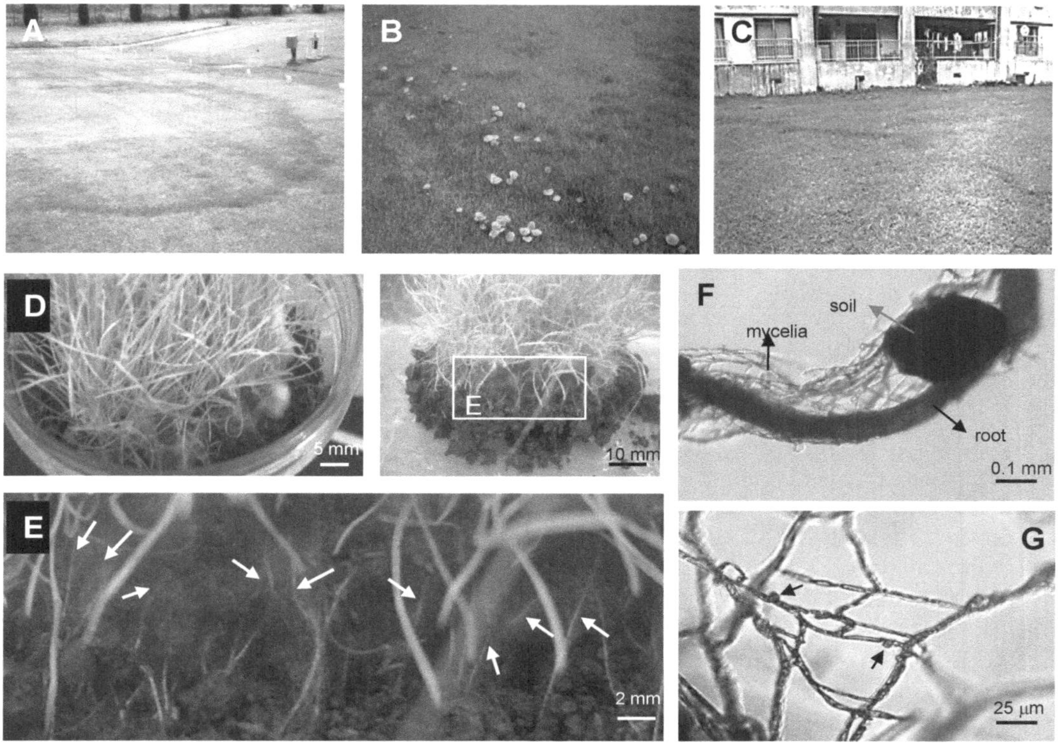

Fig. 18.3: Fairy rings on the Turfgrass Study Field, Chiba Prefectural Agriculture and Forestry Research Center (A, B) and on the campus of Shizuoka University (C), in Japan by *L. sordida.* (D) Pictures of the fungus-infected soil cultured for 3 weeks. Arrows in (E) indicate the roots surrounded by the mycelia of the fungus. (F) Microscope photograph of root-surrounding parts using the Olympus CKX41. Arrows in (G) indicate the clamp connection.

not only as the inactive precursor of hydrocyanic acid in the defense of *M. oreades* but also as an active compound in plant pathogenesis (Caspar and Spiteller 2015). Agrocybin as the phytotoxic substance, previously isolated from the fungus *Agrocybe dura*, has also been isolated from the cultures of *M. oreades* (Fig. 18.3) (Traquair and McKeen 1986). It is suggested that agrocybin related to the growth-suppression by *M. oreades*.

4. Fairy Ring-forming Fungus *Lepista sordida*

4.1 Symptoms by L. sordida

Fairy rings by *L. sordida* have been reported only in Japan (Fig. 18.3A and B). The occurrence of symptoms associated with this fungus, such as the formation of a dark green zone caused by stimulated turfgrass and a dead zone, have been reported only in turfgrass of *Z. matrella* and *Z. japonica* (Tahama 1978, Tani 1991, Terashima 2005). Terashima previously observed fairy rings of this fungus with a radius of six meters and reported that the outward movement of the ring was 125 cm annually on *Z. matrella* (Terashima 2005). The visible mycelia were mainly present in the range of 0–2 cm of the upper soil and mycelia did not observe at a depth of 3 cm or more (Terashima 2005).

4.2 Fairy Chemicals—a Candidate for a New Family of Plant Hormones

This research was begun with very personal experience of one of authors (Kawagishi) more than 10 years ago during his stay in the university's housing complex on the campus of Shizuoka University. One day, the author noticed that the turfgrass in the lawn became greener than the adjacent grass and that the green formed into an arc. After about one year, the arc became bigger and the turfgrass on the arc grew more thick (Fig. 18.3C), and then fruiting bodies of a fungus appeared on the arc. This mushroom was identified as *L. sordida*. Since the first paper on "fairy rings" was published in 1675 and the subsequent reports were introduced in Nature in 1875, the "fairy" (the cause of the turf growing up) had been a mystery, although there was an accepted theory such as a resulting accumulation of nitrogen (Couch 1995). However, the author questioned this accepted notion and started investigating the possibility of specific plant growth-regulating substance(s) being produced by the fungi.

4.2.1 Disclosure of the "Fairy"

Microorganisms often produce different secondary metabolites under artificially controlled conditions. Therefore, it was confirmed that the *L. sordida* strain we used exhibited growth-promoting activity. The cultivated mycelia were placed under bentgrass seedlings in a deep petri dish and incubated for three weeks. As a result, the infected grass grew higher than the intact plant.

The isolation of the "fairy" from the fungus extracts was guided by its growth promoting activity against bentgrass. The liquid-cultured fungus was divided into the culture broth and mycelia. The culture broth was fractionated by repeated chromatography, resulting in the isolation of the "fairy" 2-azahypoxanthine (AHX, **1**) (Fig. 18.4A) (Choi et al. 2010b). Although **1** had been chemically synthesized, this was the first isolation of **1** from nature (Vaughan et al. 2002).

Sporadically, fairy rings appear as growth-suppressed rings. We also isolated the suppressing compound, imidazole-4-carboxamide (ICA, **2**) (Fig. 18.4A) (Choi et al. 2010a). Although **2** also had been synthesized chemically, this is the first purification from a natural source like **1** (Leone-Bay and Glaser 1987).

Later, we found that **1** was metabolized to other compounds after it was taken into rice. The metabolite, 2-aza-8-oxohypoxanthine (AOH, **3**), was a novel compound, and the conversion from **1** into **3** was also observed in other plants such as Arabidopsis, tomato, and turfgrass (Fig. 18.4A). The reaction from **1** into **3** was very similar to the reaction catalyzed by xanthine oxidase (XOD) to produce uric acid from xanthine. Therefore, **1** was treated with commercially available XOD, and **3** was obtained almost quantitatively (Fig. 18.4A) (Choi et al. 2014). **3** stimulated the seedlings of bentgrass and rice as **1**. This study was covered in Nature and we named the three compounds (**1** to **3**) as fairy chemicals (FCs) after the title of the article (Mitchinson 2014).

4.2.2 Molecular Mechanism of Plant Growth Regulating Activity of FCs

FCs affected on growth of all the plants tested such as Arabidopsis, asparagus, corn, komatsuna (*Brassica rapa* L. var perviridis), lettuce, potato, rice, tea, tobacco, tomato and wheat. The molecular mechanism of the activity of FCs, especially **1**, has been investigated by using rice that belongs to the same family as the turfgrass. The effect of **1** on rice transcriptome was investigated using oligo DNA microarrays and Reverse Transcriptase-Polymerase Chain Reaction (RT-PCR).

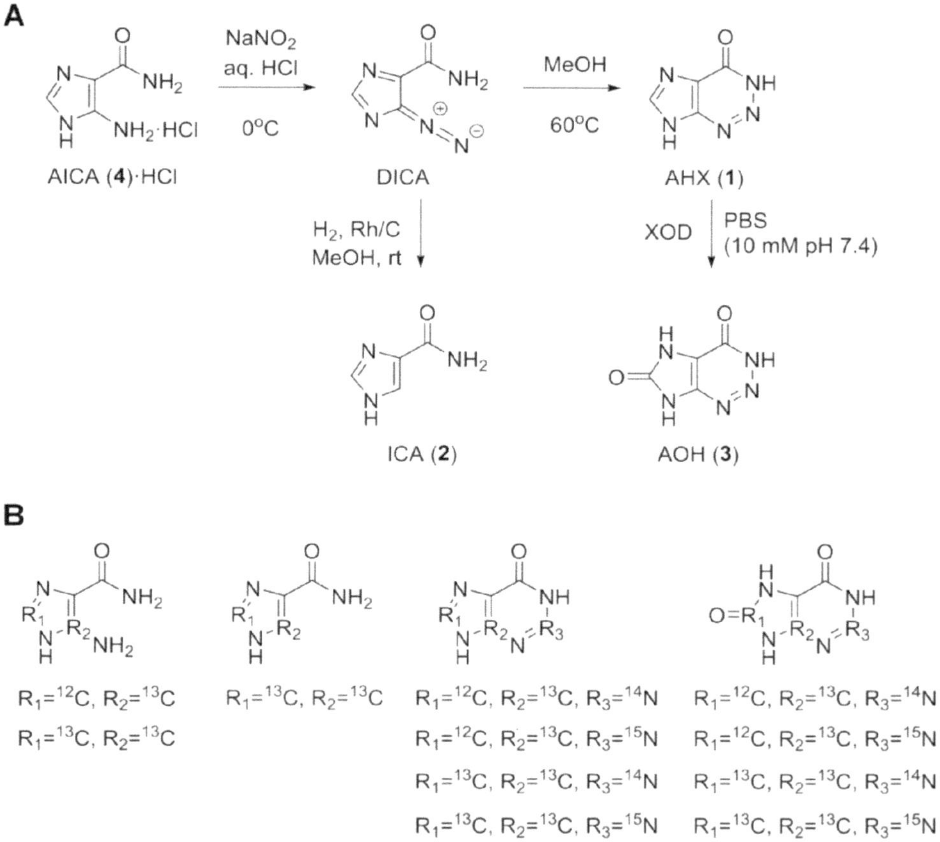

Fig. 18.4: Synthesis of FCs (A) and isotope-labeled AICA and FCs (B).

The most noticeable result was the up-regulation of an aquaporin, OsTIP2;1, Bowman-Birk type proteinase inhibitor (BBI), and glutathione S-transferases (GST) in AHX-treated seedlings (Choi et al. 2010b). Tonoplast Intrinsic Proteins (TIPs) are the major components of the vacuolar membranes and the most abundant aquaporins in plants (Tyerman et al. 2002). Aside from the water channel function, AtTIP2;1 and AtTIP2;3 of Arabidopsis and TaTIP2 of wheat have been known to transport ammonia/ammonium ion (Jahn et al. 2004, Loqué et al. 2005). To elucidate effect of **1** on absorption of ammonia/ammonium ion into rice, rice seedlings were treated with **1** and $^{15}NH_4NO_3$ as the sole nitrogen source. As a result, **1** promoted ^{15}N absorption into the plant, especially into the root where OsTIP2;1 was up-regulated by the treatment with **1**. On the other hand, when $NH_4{}^{15}NO_3$ was used as the sole nitrogen source, no increase of ^{15}N in the plant was observed (Choi et al. 2010b).

GSTs and BBI confer resistance to plants against various biotic and abiotic stresses such as herbivory, high or low temperature and salt. Previous studies have been reported that GST genes were transferred into rice and other plants to increase tolerance for various stresses such as low temperature or salt (Takesawa et al. 2002, Zhao and Zhang 2006). Treatment of rice with **1** promoted germination of the treated seeds under low temperature stress (15°C) and shoot growth of the treated seedlings under salt stress (0.1 M NaCl), similar to the previous results about the GST-transgenic rice (Takesawa et al. 2002, Zhao and Zhang 2006). Furthermore, rice seedlings gained tolerance to high temperature stress (35°C) by the treatment with **1**. These results indicated that the plant acquired tolerance for various stresses from the environment and increased the absorption of ammonia/ ammonium ion by the treatment of **1**, resulting in the growth promotion.

The up- and down-regulated genes by treatment with **1** or **3** were very similar to each other, however, treatment with **2** gave almost the opposite results to **1** and **3**. Nevertheless, these three compounds equally gave tolerance for various stresses to the plant and increased crop yields.

It has been reported that various imidazole derivatives had growth-promoting activity against some plants and that the derivatives showed the activity as the substitute for cytokinin (Cavender 1986). Tea cells were incubated with a cytokinin, 6-benzylaminopurine (BAP), an auxin, naphthaleneacetic acid (NAA), and/or **1** in order to know the relationship between **1** or **2** and plant hormones. Both hormones are requisite for normal cell proliferation. In the presence of both hormones, **1** and **2** did not affect cell proliferation. When **1** was used instead of BAP or NAA, **1** or **2** did not act as a substitute for either hormone. On the other hand, in the absence of both the hormones, **1** promoted cell proliferation dose-dependently and **2** showed no effect. These results indicated that the effect of **1** is independent of the two hormones and acted like any another hormone (Choi et al. 2010b).

4.2.3 The Structure-activity Relationship

To elucidate the Structure-Activity Relationship (SAR), growth regulating activity of **1-3**, 5-aminoimidazole-4-carboxamide (AICA, **4**), xanthine and hypoxanthine toward bentgrass was tested. As a result, **1**, **3** and **4** promoted the growth and only **2** inhibited the growth (Choi et al. 2010a, 2010b, 2014).

In order to investigate SAR of **1**-analogues and derivatives in terms of growth-promoting activity using rice, a series of **1**-derivatives by Pd-catalyzed C-H arylation were synthesized (Fig. 18.5) (Kitano et al. 2018). All the arylated analogues of **1** had a root-growth-promoting activity stronger than that of **1**. Future identification of target(s) of **1** and its arylated derivatives may provide the molecular basis of their plant-growth activity.

4.2.4 Endogenous FCs

1 is chemically synthesized from **4** via 4-diazo-4*H*-imidazole-4-carboxamide (DICA) (Fig. 18.4A) (Choi et al. 2014, Ikeuchi et al. 2014). **2** is also synthesized from **4**, and **3** is converted from **1** enzymatically (Choi et al. 2014, Ouchi et al. 2018). Notably points are that the precursor of FCs, **4**, is on the purine metabolic pathway in animals, plants and microorganisms, and further metabolism of **4** had not known (Fig. 18.6). Based on our findings and the facts mentioned above, we hypothesized that plants themselves produced FCs via a route similar to the chemical synthesis (Fig. 18.6).

To prove the endogenety of **1** and **3** in rice, we tried to detect and quantify the compound in rice that was cultivated in an aseptic condition by LC-MS/MS using [4-^{13}C, 2-^{15}N]-**1** and [4-^{13}C, 2-^{15}N]-**3** as internal standards (Fig. 18.4B). As a result, **1** and **3** were detected from the rice. The amounts of endogenous **1** (457 ± 75 ng/kg F.W. in shoot and 273 ± 52 ng/kg F.W. in root) and **3** (not detected in shoot and 1289 ± 406 ng/kg F.W. in root) in rice were similar as those of plant hormones, strigolactones and brassinosteroids (Gomez-Roldan et al. 2008, Ohnishi et al. 2012, Umehara et al. 2008). **1** and/or **3** were also detected in various living organisms (Table 18.2) (Choi et al. 2014, Kawagishi 2018,

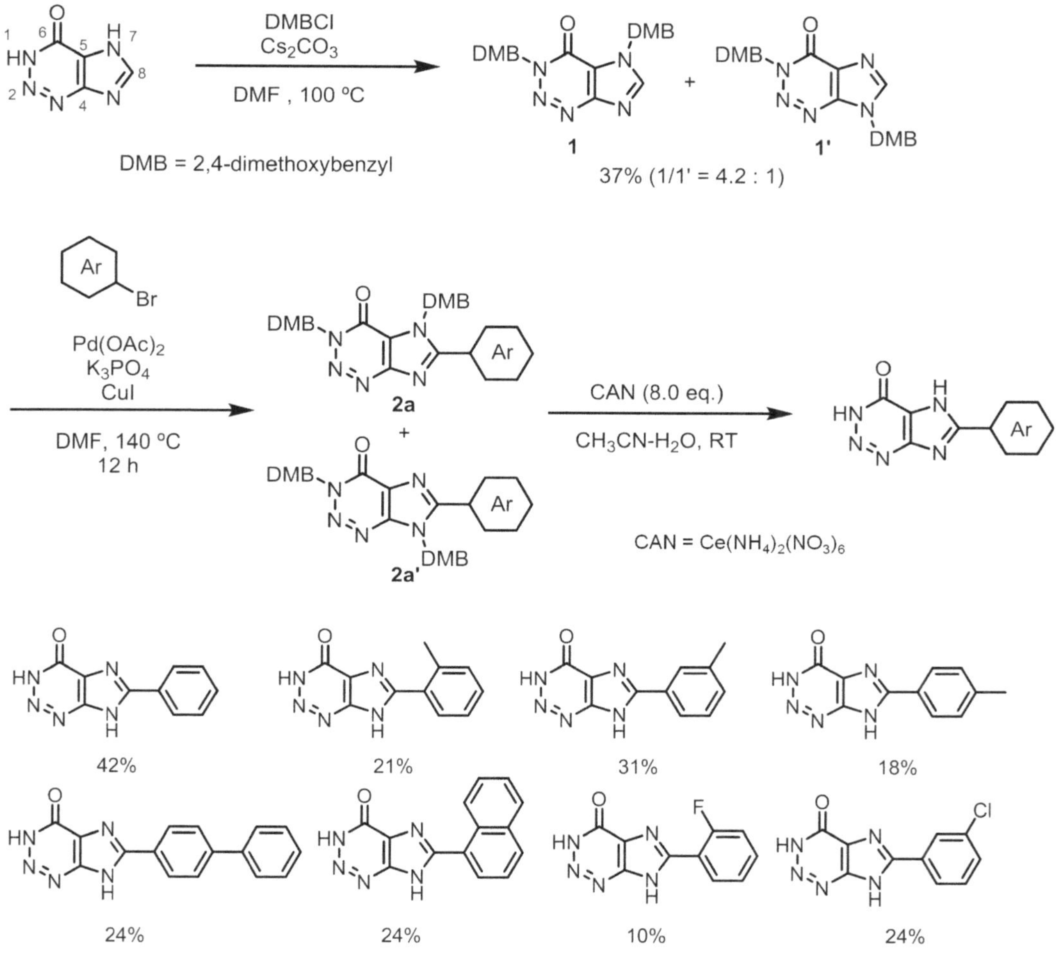

Fig. 18.5: Synthesis of arylated AHX.

2019). Since **1** was originally isolated from a fungus, it is possible that microorganisms around plants produced **1** and the compound was taken into plants. Therefore, rice and Arabidopsis were cultivated aseptically, and the compounds were also detected from the extracts of the aseptic plants (Choi et al. 2014). Recently, we developed a method for detection of FCs by LC-MS/MS and the endogenous existence of **2** in various plants was also proved (paper submitted).

4.2.5 FCs Biosynthetic and Metabolic Pathways

To verify the existence of our hypothesis of the biosynthetic route from **4** to **1** and then **3** in plants (Fig. 18.6), rice seedlings were cultivated in liquid medium with [5-^{13}C]-**4** and incorporation rate of the compound into the seedlings and the amount of [4-^{13}C]-**1** and [4-^{13}C]-**3** in the seedlings were analyzed (Choi et al. 2014). The results showed that **4** in the culture medium was incorporated into the seedlings and the absorbed **4** was converted to **3** via **1** in the plant.

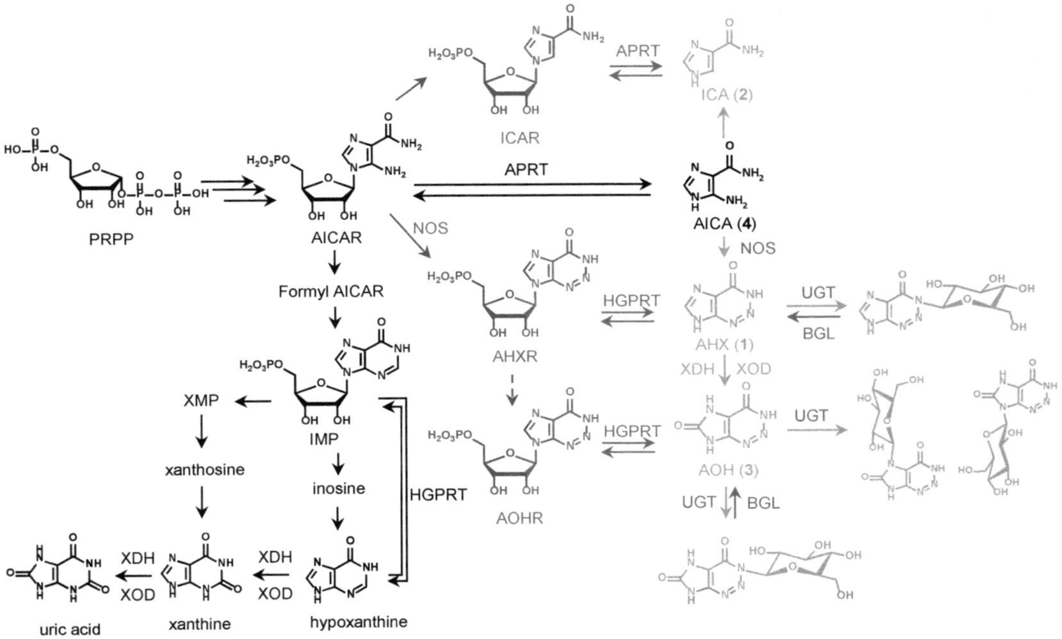

Fig. 18.6: Purine metabolic pathway in animals, plants, and microorganisms, including the novel route. The route in black was cited from KEGG (Kyoto Encyclopedia of Genes and Genomes). New metabolites and routes are indicated in red (published results). Hypothetic or unpublished routes, metabolites, and enzymes are described in blue.

Color version at the end of the book

All the results allowed us to conclude that **1** and **3** are novel metabolites in a previous unknown purine metabolic pathway in plants, at least in rice. As mentioned above, many plants, algae and mushrooms contain **1** and/or **3**, suggesting that the pathway is conserved in various living organisms (Table 18.2). In the mushroom-forming fungus *Lepista sordida* in which **1** was originally discovered, the biosynthesis of **1** was promoted by adding AICA ribonucleotide (AICAR) to the culture media. The expression level of adenine/AICA phosphoribosyltransferase (APRT) gene was also stimulated by the addition. APRT catalyzes the conversion from AICAR into **4** and this enzyme activity was detected in the fungus. These results suggested that there is probably the pathway from AICAR to **1** via **4** in other fungi and plants (Fig. 18.6) (Suzuki et al. 2016).

The enzyme, hypoxanthine-guanine phosphoribosyltransferase (HGPRT) catalyzes a reaction similar to APRT and the reversible interconversion between IMP and hypoxanthine, and GMP and guanine through addition-elimination of the phosphoribose group. Genomic information indicates that rice has an HGPRT gene, and we considered the possibility that this enzyme catalyzes the reaction from **1** and **3** to hypothetic ribotides, **1**-ribotide (AHXR) and **3**-ribotide (AOHR). Therefore, heterologous expression of the enzyme was performed using *Escherichia coli*, and treatment of **1** or **3** with the expressed HGPRT (OsHGPRT) afforded the ribotides of **1** and **3** (unpublished data) (Fig. 18.6).

Preliminary results of similar experiments concerning **2** showed that **2** also was biosynthesized from **4** and **2**-ribotide might exist endogenously as in the case of **1** and **3** (data not shown).

Table 18.2: Detection of AHX and/or AOH by LC-MS/MS.

Organism species		AHX	AOH
Poaceae	rice (Nipponbare; above-ground part)	d	nd
	rice (Nipponbare; root)	d	d
	rice grain (Nipponbare, Koshihikari)	d	d
	wheat grain (Norin No. 61)	d	d
	corn (mille-feuille, Kankanmusume)	d	nd
	bentgrass (above-ground part, root)	d	d
	zoysiagrass (above-ground part)	d	nd
	zoysiagrass (root)	d	d
Liliaceae	asparagus (above-ground part)	d	nd
Eucalyptus	*Eucalyptus pellita* (above-ground part, root)	d	nd
	Eucalyptus camaldulensis (above-ground part, root)	d	nd
Solanaceae	tomato (House Momotaro, Reika, Endeavour; above-ground part)	d	nd
	tomato (House Momotaro, Reika, Endeavour; root)	d	d
	potato (Danshaku, May Queen)	d	d
Cucurbitaceae	cucumber (above-ground part, root)	d	nd
Theaceae	tea (*Camellia sinensis*; above-ground part)	d	d
	tea (*Camellia sinensis*; root)	d	nd
Cruciferae	komatsuna (*Brassica rapa* L. var *perviridis*) (above-ground part)	d	nd
	komatsuna (*Brassica rapa* L. var *perviridis*) (root)	d	d
	Japanese radish (*Raphanus sativus* var. *longipinnatus*)	d	nd
	Arabidopsis thaliana (above-ground part)	d	nd
	Arabidopsis thaliana (root)	d	d
Araceae	eddo (*Colocasia antiquorum*, Yamasatoimo)	d	d
Algae	*Chlorella vulgaris; Parachlorella beyerinckii*	d	d
	Synechocystis sp. PCC6803	d	nd
Mushroom	*Lepista nuda*	d	d
	Lepista sordida	d	nd
	Lyophyllum connatum	d	d
	Tricholoma flavovirens	d	nd
	Lyophyllum decastes	d	nd
	Flammulina velutipes	d	nd
	Pholiota adiposa	d	nd
	Hypholoma sublateritium	d	nd
	Lyophyllum shimeji	d	nd
	Cortinarius caperatus	d	nd
	Grifola frondosa	d	nd

d: detected, nd: not detected.

In the biosynthesis of **1** from **4** and AHXR from AICAR, it is highly possible that the origin of nitrogen is Nitric Oxide (NO); incorporation of **4** and a NO generating agent, 1-hydroxy-2-oxo-3-(3-aminopropyl)-3-isopropyl-1-triazene stimulated the biosynthesis of **1** in the fungus *L. sordida*. Meanwhile, a Nitric Oxide Synthase (NOS) inhibitor, N^G-monomethyl-L-arginine, acetate, suppressed the production of **1**. In addition, when the fungus was cultivated in the presence of ^{14}N-labeled L-arginine that releases NO by NOS, ^{14}N-labeled **1** was produced (paper in preparation). Similar experiments are in progress against rice. Various effects of NO on plants have been reported, and some part of the effects might be due to **1** or **3** incorporating NO in the molecule.

Rice seedlings were treated with **1** solution in order to investigate further metabolism of **1** and **3**, and the treated plant gave metabolites (Fig. 18.6). Three metabolites were isolated and the structures were determined. They were glucosides of **1** or **3** (Choi et al. 2018). The results of LC-MS/MS indicated that all the glucosides also exist endogenously in rice. One kind of the plant hormones, cytokinins, possess a purine skeleton as **1** and **3** and are also glucosylated in plants. Those glucosides are more stable to cytokinin degradation enzymes than the corresponding free bases. In terms of their physiological activity, the glucosides show little or no activity against plants, suggesting that they are inactive forms of the hormone. The obtained *N*-glucosides also exhibited no activity against rice. Plants might regulate the activity of **1** and **3** by glucosidation as cytokinin. However, various data suggested that biological activity of FCs are completely different from that of cytokinin and FCs are recognized as completely different compounds from cytokinin by plants. Recently, it was clarified that **2** is also biosynthesized from **4** via DICA like **1** in (Fig. 18.6) (paper submitted). Other than the disclosed pathways mentioned above, we hypothesize that FCs are also biosynthesized from their ribotides as other compounds on the purine pathway. In any event, all the data mentioned above led to the conclusion that we discovered new metabolites biosynthesized on a novel purine metabolic pathway in plants (Fig. 18.6).

4.2.6 The Possibility of Practical Use of FCs

Since FCs regulated all the plants tested regardless of their families including rice, we thought that FCs can increase the yield of rice grains. Large amounts of FCs are needed for cultivation experiments. **1** and **2** are chemically synthesized readily (Leone-Bay and Glaser 1987, Vaughan et al. 2002). On the other hand, only method for synthesis of **3** was enzymatic conversion from **1** by xanthine oxidase (XOD) (Ikeuchi et al. 2014). However, commercially available XOD is not appropriate for large-scale preparation of the compound because the enzyme is expensive and the reaction with the enzymatic proceeds only with low concentration of **1**. In the course of searching for better methods for synthesis of **3**, we happened to notice that aqueous solution of **1** was accidentally contaminated by microorganisms and **1** in the contaminated solution was efficiently converted to **3**. Therefore, we tried to isolate the airborne microorganism that produced **1**. As a consequence, we succeeded in the isolation of the bacterium that converted high concentrations of **1** to **3** with a very high yield and identified the bacterium as *Burkholderia contaminans* CH-1 (Choi et al. 2016). The high-quality draft genome sequence of *B. contaminans* CH-1 was determined and the genome contained 8,065 protein-coding sequences (Choi et al. 2017).

When rice was continuously treated with **1** at 5 μM or **2** at 2 μM in greenhouse experiments, the grain yield per plant increased by 25.5 or 26.0% (Choi et al. 2010a, 2010b). The results encouraged us to do further experiments in the field. **1** or **2** was used in the following periods, (1) during the time the seedlings rose in nursery boxes (seedling treatment), (2) at the time of transplanting, (3) at the time of panicle formation. As a result, brown rice yields were increased by all the three treatments. Surprisingly, seedling treatment (for only two weeks in nursery boxes) with **1** or **2** increased the yields. In addition, the treatments did not change the eating quality and contents of free fatty acid, amylose, protein and water of the brown rice (Asai et al. 2015).

The field experiments to examine effects of **1** and **2** on wheat grain yield were also done (Tobina et al. 2014). In 2011, two-week treatment of wheat seedlings with **1** or **2** resulted in the increase of grain by 10.2 and 5.6%, respectively. In 2012, wheat seeds were soaked in **1** or **2** solution for 36 hours and germinated. The germinated seeds were sown in the field. After sowing, the plant was cultivated in the absence of the compounds. This only 36-hour treatment with **1** or **2** raised the grain yields by 20.2 and 9.8%, respectively. **1** also increased yields of potato (pot soil cultivation) by 19%, lettuce (pot cultivation) by 21%, and asparagus (hydroponic cultivation) by 100% (Choi et al. 2010b). **3** also showed the similar effects on rice and wheat to those of **1** and **2** (unpublished data).

We have several years' data for the effects of FCs on the yields of several other crops. The data indicated that FCs give higher yields without being affected by various stresses from the environment such as low temperature, high temperature, drought, salinity and so on.

The endogenous existence of FCs in the three major food crops in the world, rice, wheat and maize, and other crops provide information on the safety of these compounds for the practical use of these compounds in agriculture (Table 18.2).

5. Conclusion

"Fairy rings" is a phenomenon in which fruiting bodies of higher fungi occur after turfgrass grows and/or dies in the form of rings. We discovered three compounds (fairy chemicals, FCs) that are principles causing the phenomenon.

Plant hormones are signal molecules produced within the plants, and occur in extremely low concentrations and control all aspects of growth and development, such as embryogenesis, the regulation of organ size, defense against pathogens, stress tolerance, reproductive development, and so on. Unlike animals, each plant cell is capable of producing hormones (https://en.wikipedia.org/wiki/Plant_hormone). Based on our finding mentioned above, we think that FCs are a novel family of plant hormones and are tracing the history of a plant hormone, gibberellins. Gibberellins were first isolated as toxins produced by *Gibberella fujikuroi* that causes Bakanae disease in rice. One of the symptoms is the abnormal growth of rice internodes. Later, it was verified that gibberellins exist endogenously in all plants, and they were acknowledged as a family of plant hormones. Now, a company is conducting research to put FCs into practical use in agriculture. In the future, FCs might be contributing to resolve the world food problem.

Acknowledgments

We thank all the coworkers of this study. We are especially grateful to Prof. Toshiyuki Kan of University of Shizuoka who synthesized all the labeled compounds used in this study. We also thank Dr. Kazuo Okamoto (Ushio ChemiX Co. Ltd.) for providing large amounts of FCs. This work was partially supported by JSPS KAKENHI (Grant Nos. JP16H06192 (J.-H.C.), JP17H06402 (H.K.)).

References

Asai, T., J.-H. Choi, T. Ikka, K. Fushimi, N. Abe, H. Tanaka et al. 2015. Effect of 2-azahypoxanthine (AHX) produced by the fairy-ring-forming fungus on the growth and the grain yield of rice. Jpn. Agric. Res. Quart. 49: 45-49. Doi: DOI 10.6090/jarq.49.45.

Bayliss, J.S. 1911. Observations on *Marasmius oreades* and *Clitocybe gigantea*, as parasitic fungi causing" fairy sings". J. Econ. Biol. 6: 111–132.

Blenis, P.V., P.S. Chow, I. Duncan and N.R. Knowles. 2004. Cyanide levels near fairy rings affect the growth of grasses and fungi. Can. J. Bot. 82: 1324–1329.

Caspar, J. and P. Spiteller. 2015. A free cyanohydrin as arms and armour of *Marasmius oreades*. ChemBioChem. 16: 570–573.

Cavender, P.L. 1986. Imidazole plant growth regulators. In imidazole plant growth regulators: Patents.

Choi, J.-H., N. Abe, H. Tanaka, K. Fushimi, Y. Nishina, A. Morita et al. 2010a. Plant-growth regulator, imidazole-4-carboxamide, produced by the fairy ring forming fungus *Lepista sordida*. J. Agric. Food Chem. 58: 9956–9959.

Choi, J.-H., K. Fushimi, N. Abe, H. Tanaka, S. Maeda, A. Morita et al. 2010b. Disclosure of the "fairy" of fairy-ring-forming fungus *Lepista sordida*. ChemBioChem. 11: 1373–1377.

Choi, J.-H., T. Ohnishi, Y. Yamakawa, S. Takeda, S. Sekiguchi, W. Maruyama et al. 2014. The source of "fairy rings": 2-azahypoxanthine and its metabolite found in a novel purine metabolic pathway in plants. Angew. Chem. Int. Ed. 53: 1552–1555.

Choi, J.-H., A. Kikuchi, P. Pumkaeo, H. Hirai, S. Tokuyama and H. Kawagishi. 2016. Bioconversion of AHX to AOH by resting cells of *Burkholderia contaminans* CH-1. Biosci. Biotechnol. Biochem. 80: 2045–2050.

Choi, J.-H., H. Sugiura, R. Moriuchi, H. Kawagishi and H. Dohra. 2017. High-quality draft genome sequence of *Burkholderia contaminans* CH-1, a gram-negative bacterium that metabolizes 2-azahypoxanthine, a plant growth-regulating compound. Genome. Announc. 5: e01148–17.

Choi, J.-H., J. Wu, A. Sawada, H. Takemura, K. Yogosawa, H. Hirai et al. 2018. N-Glucosides of fairy chemicals, 2-azahypoxanthine and 2-aza-8-oxohypoxanthine, in rice. Org. Lett. 20: 312–314.

Couch, H.B. 1995. Diseases of Turfgrasses, 3rd edn. Krieger, Melbourne, FL, pp. 181–186.

Evershed, H. 1884. Fairy rings. Nature 29: 384–385.

Filer, H.J. 1965. Damage to turf grasses caused by cyanogenetic compounds produced by *Marasmius oreades*, a fairy ring fungus. Plant Dis. Rep. 49: 571–574.

Gomez-Roldan, V., S. Fermas, P.B. Brewer, V. Puech-Pagès, E.A. Dun, J-. Pillot et al. 2008. Strigolactone inhibition of shoot branching. Nature 455: 189–194.

Ikeuchi, K., R. Fujii, S. Sugiyama, T. Asakawa, M. Inai, Y. Hamashima et al. 2014. Practical synthesis of natural plant-growth regulator 2-azahypoxanthine, its derivatives, and biotin-labeled probes. Org. Biomol. Chem. 12: 3813–3815.

Jahn, T.P., A.L. Møller, T. Zeuthen, L.M. Holm, D.A. Klærke, B. Mohsin et al. 2004. Aquaporin homologues in plants and mammals transport ammonia. FEBS Lett. 574: 31–36.

Kawagishi, H. 2018. Fairy chemicals–a candidate for a new family of plant hormones and possibility of practical use in agriculture. Biosci. Biotechnol. Biochem. 82: 752–758.

Kawagishi, H. 2019. Are fairy chemicals a new family of plant hormones? Proc. Jpn. Acad. Ser. B 95: 29–38.

Kitano, H., J.-H. Choi, A. Ueda, H. Ito, S. Hagihara, T. Kan et al. 2018. Discovery of plant growth stimulants by C–H arylation of 2-azahypoxanthine. Org. Lett. 20: 5684–5687.

Lebeau, J.B. and E.J. Hawn. 1963. A simple method for the control of fairy ring caused by *Marasmius oreades*. J. Sports Turf Res. Inst. 11: 23–25.

Leone-Bay, A. and L. Glaser. 1987. An efficient method for the preparation of 4 (5)-cyanoimidazoles. Synth. Commun. 17: 1409–1412.

Loqué, D., U. Ludewig, L. Yuan and N. von Wirén. 2005. Tonoplast intrinsic proteins AtTIP2; 1 and AtTIP2; 3 facilitate NH3 transport into the vacuole. Plant Physiol. 137: 671–680.

Mitchinson, A. 2014. Plant science: fairy chemicals. Nature 505: 298–298.

Ohnishi, T., B. Godza, B. Watanabe, S. Fujioka, L. Hategan, K. Ide et al. 2012. CYP90A1/CPD, a brassinosteroid biosynthetic cytochrome P450 of Arabidopsis, catalyzes C-3 oxidation. J. Biol. Chem. 287: 31551–31560.

Ouchi, H., T. Asakawa, K. Ikeuchi, M. Inai, J.-H. Choi, H. Kawagishi et al. 2018. Synthesis of double-^{13}C-labeled imidazole derivatives. Tetrahedron Lett. 59: 3516–3518.

Ramsbottom, J. 1953. Mushrooms and toadstools. Collins, New Naturalist Series, pp. 306.

Robbins, W.J., A. Rolnick and F. Kavanagh. 1950. Production of hydrocyanic acid by cultures of a basidiomycete. Mycologia 42: 161–166.

Shantz, H.L. and R.L. Piemeisel. 1917. Fungus fairy rings in eastern Colorado and their effects on vegetation. J. Agric. Res. 11: 191–245.

Smith, J.D. 1959. Fungal diseases of turf grasses. Sports Turf Research Institute, Bingley, pp. 90.

Smith, J.D. 1980. Is biologic control of *Marasmius oreades* fairy rings possible? Plant Dis. 64: 348–354.

Stijve, T. and A.A.R. de Meijer. 1999. Hydrocyanic acid in mushrooms, with special reference to wild-growing and cultivated edible species. Dtsch. Lebensm. Rundsch. 95: 366–373.

Suzuki, T., N. Yamamoto, J.-H. Choi, T. Takano, Y. Sasaki, Y. Terashima et al. 2016. The biosynthetic pathway of 2-azahypoxanthine in fairy-ring forming fungus. Sci. Rep. 6: 39087.

Tahama, Y. 1978. On the fairy rings of turfgrass, *Zoysia matrella* L. J. Jpn. Soc. Turfgrass Sci. 7: 111–114.

Takesawa, T., M. Ito, H. Kanzaki, N. Kameya and I. Nakamura. 2002. Over-expression of ζ glutathione S-transferase in transgenic rice enhances germination and growth at low temperature. Mol. Breed. 9: 93–101.

Tani, T. 1991. Color atlas of diseases of turfgrasses (in Japanese): Soft Science, Tokyo, pp. 84–91.

Terashima, Y., T. Fukihara and A. Fujiie. 2004. Morphology and comparative ecology of the fairy ring fungi, *Vascellum curtisii* and *Bovista dermoxantha*, on turf of bentgrass, bluegrass, and Zoysiagrass. Mycoscience 45: 251–260.

Terashima, Y. 2005. Development of a fairy ring caused by *Lepista sordida* on Zoysiagrass over an eight-year period. Int. Turfgrass Soc. Res. J. 10: 251–257.

Tobina, H., J.-H. Choi, T. Asai, Y. Kiriiwa, T. Asakawa, T. Kan et al. 2014. 2-Azahypoxanthine and imidazole-4-carboxamide produced by the fairy-ring-forming fungus increase wheat yield. Field Crops Res. 162: 6–11.

Traquair, J.A. and W.E. McKeen. 1986. Fine structure of root tip cells of winter wheat exposed to toxic culture filtrates of *Coprinus psychromorbidus* and *Marasmius oreades*. Can. J. Plant Pathol. 8: 59–64.

Tyerman, S.D., C.M. Niemietz and H. Bramley. 2002. Plant aquaporins: multifunctional water and solute channels with expanding roles. Plant Cell Environ. 25: 173–194.

Umehara, M., A. Hanada, S. Yoshida, K. Akiyama, T. Arite, N. Takeda-Kamiya et al. 2008. Inhibition of shoot branching by new terpenoid plant hormones. Nature 455: 195–200.

Vaughan, K., D.E.V. Wilman, R.T. Wheelhouse and M.F.G. Stevens. 2002. A ^{15}N NMR investigation of a series of benzotriazinones and related antitumour heterocycles. Magn. Reson. Chem. 40: 300–302.

Zhao, F. and H. Zhang. 2006. Salt and paraquat stress tolerance results from co-expression of the *Suaeda salsa* glutathione S-transferase and catalase in transgenic rice. Plant Cell Tiss. Org. 86: 349–358.

Chapter 19

Production of Viral Nanoparticles in Silkworms

*Tatsuya Kato** and *Enoch Y. Park*

1. Introduction

Baculovirus is a double-stranded DNA virus which infects invertebrates, especially insects. Among baculoviruses, *Autographa californica* multiple nucleopolyhedrovirus (AcMNPV) is the most famous for researches. This AcMNPV has been widely used for protein expression in insect cells, Sf-9, Sf-21 and High Five cells, and several commercial kits for the construction of recombinant AcMNPV are now available. Until now, commercial products based on this baculovirus expression technology has been approved and marketed. Cervarix (GlaxoSmithKline) and Flublok (Protein Sciences) are available as a human vaccine and Porcilis Pesti (MSD Animal Health), Bayovac CSF E2 (Bayer Biologicals) and CircoFLEX (Ingelvac) are also available as a pig vaccine (van Oers et al. 2015). Many kinds of recombinant proteins have been expressed using this baculovirus expression system and very useful in not only academia but also industry. Recently, AcMNPV has been developed as a carrier of foreign genes into mammalian cells for therapeutic application based on gene delivery, utilizing the property of the capability of its entering mammalian cells without any amplification (Mansouri and Berger 2018, Ono et al. 2018). On the other hand, *Bombyx mori* nucleopolyhedrovirus (BmNPV), which infects silkworms, belongs to the same family as AcMNPV and has been also used for recombinant protein expression in silkworms. This system is used especially in eastern Asia where is historically the center of silk production (Kato et al. 2010, Usami et al. 2010). Especially, silkworm larvae and pupae, which have the high capacity to produce recombinant proteins compared to cultured cells, can be used as a host for recombinant protein production. In addition, raising silkworm larvae and pupae is easier than cultivating cultured insect cells and its scale-up for the mass production of

Laboratory of Biotechnology, Research Institute of Green Science and Technology, Shizuoka University, Shizuoka, 422-8529, Japan.
Email: park.enoch@shizuoka.ac.jp;
* Corresponding author: kato.tatsuya@shizuoka.ac.jp

recombinant proteins can be achieved simply. This chapter describes the how the silkworm expression system is carried out for recombinant protein production, especially Virus-Like Particles (VLPs), and recombinant BmNPV particles to be used as a vaccine candidate.

2. Production of Virus-like Particle (VLP) in Silkworms

2.1 VLP

VLPs are composed of one or several structural proteins of viruses and have almost the same morphology as viruses, but VLPs do not have any viral genetic materials. Therefore, VLPs are not infectious and have been developed as a vaccine to viruses and a carrier of drugs for Drug Delivery System (DDS) (Ding et al. 2018, Zelitins 2013). Especially, as a vaccine, VLPs can provoke humoral and cellular responses in hosts because of its virus-like morphology. In addition, VLPs are safer as a vaccine than the conventional vaccines such as inactivated and live-attenuated vaccines. Until now, many kinds of VLPs has been produced in various expression system such as bacteria, yeast, plants, insect cells and mammalian cells (Fuenmayor et al. 2017). Commercial vaccines to human papillomavirus (HPV), Gardasil (Merck) and Cervarix (GlaxoSmithKline), are produced in yeasts and insect cells, respectively. This chapter introduces the production of VLPs in silkworms, dividing them into two types of VLPs, enveloped and non-enveloped VLPs.

2.2 Enveloped-VLPs Production in Silkworms

Viruses are divided into two types based on its morphology, enveloped and non-enveloped viruses. Enveloped viruses have the envelope composed of lipid membrane. Therefore, enveloped VLPs also have the envelope derived from lipid membrane of host cells.

In the case of retrovirus VLPs, Gag protein forms enveloped VLPs spontaneously when the gene is solely expressed in mammalian cells and insect cells (Jalaguier et al. 2011, Lynch et al. 2010). Moreover, viral envelope proteins and heterologous transmembrane proteins can be displayed on the retrovirus VLPs when Gag protein and these proteins are co-expressed in host cells (Yang et al. 2012, Zhang et al. 2010). Silkworm larvae can also produce the retrovirus VLP by expressing Gag protein in the same manner as insect and mammalian cells. Rous Sarcoma Virus (RSV)-like particles (RSV-LPs) were produced in silkworm larvae when its Gag protein was solely expressed (Deo et al. 2011). In silkworm larvae, RSV-LPs were secreted into hemolymph of silkworm larvae. In RSV-LPs, Gag proteins were processed into several fragments because Gag protein is composed of MA, p2, p10, CA and NC domains as a polypeptide. As well as in insect and mammalian cells, silkworm larvae produce heterologous protein-displaying RSV-LPs in the hemolymph by co-expressing RSV Gag protein and heterologous proteins. For example, human (pro)renin receptor (hPRR) was displayed on the surface of RSV-LPs when these proteins were co-expressed in silkworm larvae (Tsuji et al. 2010). Display of hPRR was confirmed by immuno-Transmission Electron Microscopy (immunoTEM) and ELISA. These hPRR-displaying RSV-LP was applied to the biding assay to its ligand, human (pro)renin by ELISA. SRS2 from *Neospora caninum* (NcSRS2) was displayed on the surface of RSV-LP through a glycosylphosphatidylinositol (GPI) anchor when these proteins were co-expressed (Deo et al. 2013). NcSRS2 is modified with a GPI anchor at its C-terminus of *N. caninum* in insect cells (Nishikawa et al. 2002). Deo et al. reported that silkworm recognized the GPI-anchor motif in NcSRS2 (Deo et al. 2013). Using the GPI-anchor motif in NcSRS2, a single chain variable fragment (scFv) was displayed on the surface of RSV-LPs produced in silkworm larvae by the co-expression method (Deo et al. 2014). Instead of the GPI-anchor, the transmembrane and cytoplasmic domains of some viral envelope proteins facilitate the display of heterologous proteins on the surface of RSV-LPs. C-terminal region of Glycoprotein 64 from BmNPV (BmGP64) was used for the display of the scFv, which recognizes the Tumor-Associated Glycoprotein-72 (TAG-72) on the surface of RSV-LPs produced in silkworm larvae by the co-expression method (Kato et al. 2015). These VLPs specifically carried FITC and doxorubicin (DOX) to human colon adenocarcinoma cells, LS174T cells and DOX-loaded scFv-displaying VLPs specifically killed these cells compared to HEK293 cells. In addition, to display human interleukin 2 (hIL2) on the surface of RSV-LPs, the C-terminal region of hemagglutinin (HA) from influenza A virus was used (Deo et al. 2016). In this case, together with hIL2, scFv was also displayed through a GPI anchor on the surface of RSV-LPs using the C-terminal region of NcSRS2 when RSV gag protein, scFv and IL2 were co-expressed in silkworm larvae. These RSV-LPs specifically targeted to LS174T cells and simultaneously attracted macrophages to these cells. These results indicate that the C-terminal region of envelope proteins from enveloped viruses containing the transmembrane and cytoplasmic domains can be used in the silkworm expression system for the display of heterologous proteins on the surface of RSV-LPs as well as the use of the GPI anchor motif.

Apart from RSV-LPs, full-length HA from influenza A viruses can form virus-like particles even though M1, which is a structural protein in influenza A viruses, was not co-expressed in plants and mammalian cells (Chen et al. 2007, D'Aoust et al. 2010). In silkworms, HA-based VLPs were also produced in silkworm pupae when only HA from influenza A virus was expressed (Nerome et al. 2015, Nerome et al. 2017). Needless to say, M1 protein forms influenza virus-like particles when M1 protein was expressed in silkworm larvae (Fig. 19.1) as well as in mammalian (Gómez-Puertas et al. 2000). Based

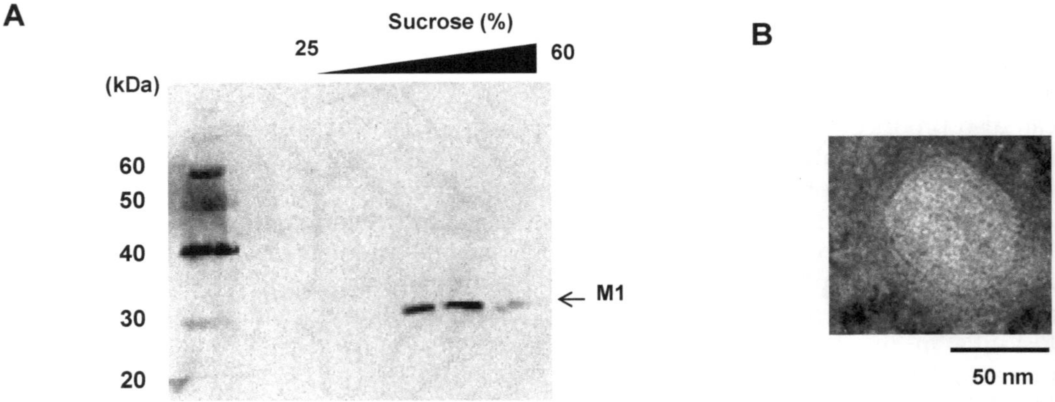

Fig. 19.1: Expression of M1 protein from influenza A H1N1 virus (A/Swine/Iowa/15/30) in silkworm larvae. (A) Sucrose density gradient centrifugation of silkworm hemolymph where recombinant M1 protein was expressed. The recombinant M1 protein was expressed in silkworm larvae using a recombinant BmNPV bacmid DNA containing the M1 gene expression cassette. M1 protein was detected using rabbit anti-influenza A virus M1 antibody (GeneTex, Irvine, CA, USA) and goat anti-rabbit IgG antibody conjugated with horse radish peroxidase (MEDICAL & BIOLOGICAL LABORATORIES, Nagoya, Japan) as primary and secondary antibodies. (B) A picture of influenza VLP composed of recombinant M1 protein in the faction of the sucrose density gradient centrifugation by a transmission electron microscope.

on these RSV-LPs, chimeric VLPs were produced in silkworm larvae by the co-expression of RSV Gag protein and M1 protein of influenza A virus (Deo et al. 2015). The diameter of these chimeric VLPs was a little small compared to that of RSV-LPs.

Japanese encephalitis virus VLPs were also produced in silkworm pupae by the expression of the gene coding prM and E proteins from this virus (Matsuda et al. 2017). In this study, VLPs composed of E protein was not secreted into the culture supernatant in cultured Bm cells, but was purified from the homogenate of silkworm pupae by sucrose density gradient centrifugation. Japanese encephalitis virus VLPs composed of prM and E proteins were secreted into the culture supernatant in High Five cells transformed with the expression vector (Yamaji et al. 2013). Few VLPs were secreted into silkworm hemolymph when the gene coding prM and E proteins from dengue virus type 2, which is also a flavivirus as well as Japanese encephalitis virus, was expressed as a polypeptide using BmNPV bac-to-bac system (paper in preparation) (Motohashi et al. 2005).

Most of enveloped VLPs have been produced in insect cells and insects using recombinant BmNPV. BmNPV is a baculoviruse which is also an enveloped virus and has its envelope outside. Therefore, it is difficult to separate enveloped VLPs from recombinant BmNPVs used for VLP production in silkworms (Minkner and Park 2018). For enveloped VLP purification from silkworms, affinity chromatography and density gradient centrifugation have been used. In the baculovirus expression system using insect cells, some methods, such as adsorptive membranes, have been developed to remove contaminated recombinant baculoviruses (Grein et al. 2014). In addition, HIV VLPs in the culture supernatant of CHO cell culture was separated from extracellular vesicles, such as exosomes or microvesicles, using an anion-exchange monolith (Steppert et al. 2016). These methods are also promising for the removal of recombinant BmNPVs from enveloped VLPs expressed in silkworms.

2.3 Non-envelope-VLP Production in Silkworms

Non-enveloped VLPs can be divided into two kinds of VLPs, such as single- and multiple-capsid VLPs. Single-capsid VLPs are composed of a single-capsid protein, but multiple-capsid VLPs are composed of several capsid proteins. Therefore, to produce multiple-capsid VLPs, co-expression system should be established.

The licensed human papillomavirus (HPV) vaccines are single-capsid VLPs. Gardasil® is composed of L1 proteins, which are structural proteins of HPVs, from HPV6, 11, 16 and 18 and Cervarix® is composed of L1 proteins from HPV16 and 18. Gardasil® and Cervarix® have been produced in yeasts and insect cells. HPV VLPs were also produced in silkworms using recombinant BmNPVs. L1 protein from HPV 6b formed VLPs in fat body when this L1 protein was expressed in silkworm larvae using the recombinant BmNPV (Palaniyandi et al. 2012, Minkner et al. 2018). L1 protein was not observed in hemolymph, indicating that L1 protein was not secreted extracellularly even though this protein forms HPV VLPs. This result corresponds to the expression of HPV L1 protein in insect cells (Baek et al. 2011, Senger et al. 2010). The size of HPV VLPs purified from hemolymph was not uniform (Fig. 19.2). After its purification from hemolymph using one step affinity chromatography, the contamination of recombinant BmNPV particles were still observed (Minkner and Park 2018). Purification of recombinant proteins including VLPs is one of bottlenecks in the silkworm expression system even though the expression level of recombinant proteins is comparatively higher than other eukaryotic expression system. Silkworm pupae is an alternative host in the silkworm expression system for the VLP preparation because porcine

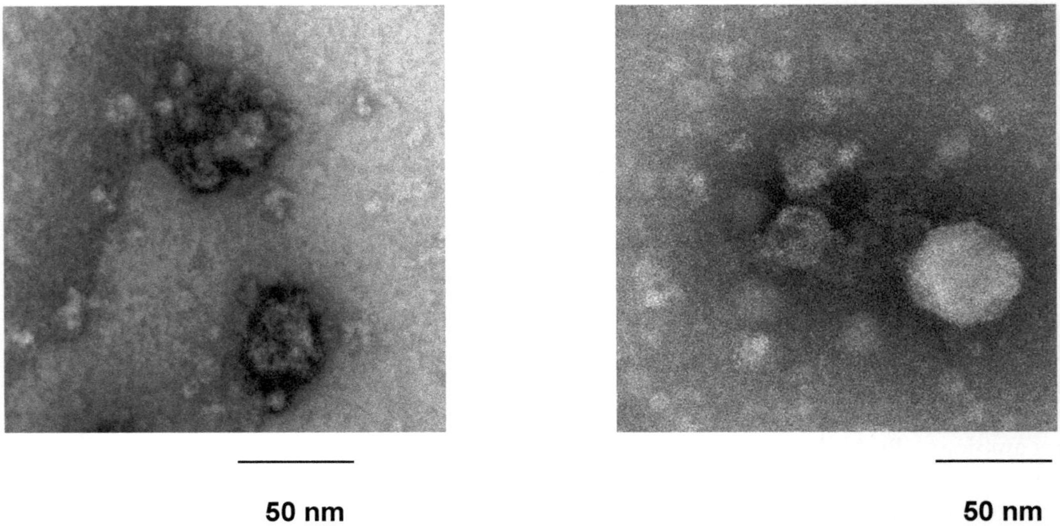

Fig. 19.2: TEM images of HPV 6b L1 protein partially purified from silkworm fat body. The L1 proteins purified from the fat body by MonoQ column chromatography formed ununiformed particles. Some particles were disrupted into small ones.

circovirus types VLPs were purified from silkworm pupae using a three step purification, ammonium sulfate precipitation, anion exchange chromatography and size-exclusion chromatography (Masuda et al. 2018). Spontaneous VLP formation of this capsid protein in silkworm pupae is advantageous because it cannot be self-assembled in *Escherichia coli* and yeasts. In silkworm pupae, L1 protein of bovine papillomavirus 6 was expressed using a recombinant hybrid NPV and purified as a VLP (Watanabe et al. 2017). In addition, L1 protein of HPV 16 was also expressed as a VLP in *Trichoplusia ni* larvae using a recombinant AcMNPV (Millán et al. 2010). Normally, the size of HPV VLPs is around 50–60 nm in diameter, but small particles were often observed. To obtain HPV VLPs with the uniform size, the disassembly and reassembly steps are required (Zhao et al. 2012).

Canine parvovirus VLPs were produced in silkworm pupae by the expression of canine parvovirus capsid protein, VP2 (Feng et al. 2014). In this study, the homogenate of silkworm pupae expressing VP2 was used in dogs as an oral vaccine to canine parvovirus. The oral administration of the homogenate into dogs induce the production of antibodies to CPV. This result indicates it is possible that canine VLPs expressed in silkworm pupae can be used as an oral vaccine in animals without any purification steps. It is an advantage for the use of VLP production in silkworms. Canine VLPs fuse with linear virus-neutralizing epitope of rabies virus were also produced in silkworm larvae and pupae (Feng et al. 2011).

Recombinant baculovirus contamination is also problematic in non-enveloped VLP preparation. However, unlike enveloped VLPs, the use of detergents facilitates the complete removal of contaminated baculoviruses because detergents can destroy the envelope of recombinant baculoviruses (Peixoto et al. 2007, Ono et al. 2018). This method can be applied to the VLP production in silkworms. In another paper, the contaminated baculoviruses worked as an adjuvant during the immunization of recombinant antigen produced in baculovirus expression system (Heinimäki et al. 2017). This suggests that residual baculoviruses in VLPs may be used as an adjuvant.

3. Co-expression of Several Proteins from Viruses for the Preparation of Complex VLPs in Silkworms

Viruses are normally composed of several structural and non-structural proteins. The expression of single structural protein from viruses sometimes forms its VLPs. However, several proteins have to be expressed simultaneously to prepare the more complex VLPs, for example, several (recombinant) protein-displaying VLPs. To achieve its production, the efficient co-expression system is needed. In the insect cell expression system, some co-expression systems have been developed. A vector (pFastbac™dual, ThermoFisher Scientific) for simultaneous two protein expression in insect cells is commercially available and this vector has two promoters, polyhedrin and p10 promoter. For the co-expression of several proteins, MultiBac system has been estabolished (Trowitzsch et al. 2010). Several gene expression cassettes can be integrated into the baculovirus genome with Tn7 transposon system and Cre-loxP recombination system. Using this system, protein complexes including complex VLPs were expressed in insect cells (Aylett et al. 2016, Bertipaglia et al. 2016, Pelosse et al. 2017). This MultiBac system can be applied to the expression of protein complexes in silkworms. Human DNA polymerase δ, which is composed of four subunits, was expressed in silkworm larvae using MultiBac system as an active form (Zhou et al. 2011). Additionally, the expression of rotavirus VLPs, which is composed of three structural proteins, were produced in silkworm larvae by the co-expression of VP2, VP6 and VP7 using MultiBac system (Yao et al. 2012). The VLP was purified from hemolymph using $CsCl_2$ density gradient centrifugation and its yield was 12.7 μg per larva. This MultiBac system facilitates the co-expression of several recombinant proteins in insect cells and insects such as silkworms and very useful for the production of not only multiple-layer VLPs but also recombinant protein-displaying enveloped VLPs. Uracil-specific excision reagent (USER) cloning method combined with the MultiBac system facilitated the co-expression of 13 genes in insect cells, leading to the purification of anaphase promoting complex or cyclosome (APC/C) of *Schizosaccharomyces pombe* (Zhang et al. 2016). Apart from the MultiBac system, biGBac system has been also established for the co-expression of multiple expression using a single recombinant baculovirus in insect cells (Weissmann et al. 2016). This system is based on restriction enzyme digestion and Gibson assembly, which is a seamless DNA assembly method by three enzymes, 5′-T5 exonuclease, Phusion® DNA polymerase and Taq DNA ligase (Gibson 2011). Using the biGBac system, the expression of several protein complexes was achieved by a single recombinant baculovirus, respectively. This biGBac system can theoretically allow 25 gene expression cassettes to be inserted into a single baculovirus genome in two steps. Recently, some ligation-free cloning systems in addition to Gibson assembly and USER system have been developed (Celie et al. 2016). As well as the MultiBac system, these technologies enable in expressing protein complexes including VLPs easily in silkworms at a large-scale for its structural analysis.

4. Production of Recombinant BmNPVs in Silkworms

AcMNPV has been developed as a carrier of foreign genes into mammalian cells for therapeutic application based on gene delivery (Mansouri and Berger 2018, Ono et al. 2018). Therefore, the production and purification of recombinant AcMNPVs have been developed. Especially, for the purification of recombinant AcMNPVs, ion exchange membrane chromatography, heparin column affinity chromatography and size exclusion column chromatography were used (Grein et al. 2012, Nasimuzzaman et al. 2016, Transfiguracion et al. 2007). In the case of BmNPV, BmNPV can infect silkworm larvae and a large amount of recombinant BmNPV is observed in the hemolymph in parallel with the expression of recombinant proteins, compared to the culture supernatant in the insect cell culture (Kato et al. 2009). Therefore, BmNPV is easier to be prepared than AcMNPV. It is very advantageous to use BmNPV as a baculovirus instead of AcMNPV. In addition, BmNPV can be purified from hemolymph by Sephacryl S-1000 or S-500 column chromatography in the presence of low concentration of Triton X-100 (Kato et al. 2011).

BmNPV has also been developed as a carrier of foreign genes into mammalian cells as well as AcMNPV. However, some modifications of BmNPV or transduction condition are needed for the gene delivery to mammalian cells using BmNPV (Imai et al. 2016, Kato et al. 2016). In the case of our group, BmGP64 was replaced with GP64 of AcMNPV (AcGP64) on the surface and AcGP64-displaying BmNPV was constructed (Fig. 19.3, Kato et al. 2016). This the addition of Phosphate-Buffered Saline (PBS) that enhances the gene transduction efficiency of not only AcMNPV but also BmNPV during its transduction, but why PBS facilitates the gene transduction by baculoviruses has not been solved yet (Hsu et al. 2004). In addition, the envelope protein has the crucial role in the entry of baculoviruses into mammalian cells to transduce the foreign genes (Barsoum et al. 1997, Tani et al. 2003). In these studies, Vesicular Stomatitis Virus G protein (VSVG) on the baculoviral envelope enhanced the gene transduction efficiency. However, it was also reported that recombinant BmNPV without any modification of its envelope protein can transduce foreign genes into mammalian cells (Kenoutis et al. 2006, Liu et al. 2017).

As well as AcMNPV, a BmNPV display system has been already established (Grabherr et al. 2001). Soluble proteins, such as GFP and antibodies, can be displayed by the fusion with BmGP64, which is a major envelop protein in BmNPV (Dojima et al. 2010, Rahman and Gopinathan 2003). In addition, the ectodomains of envelope proteins from viruses and

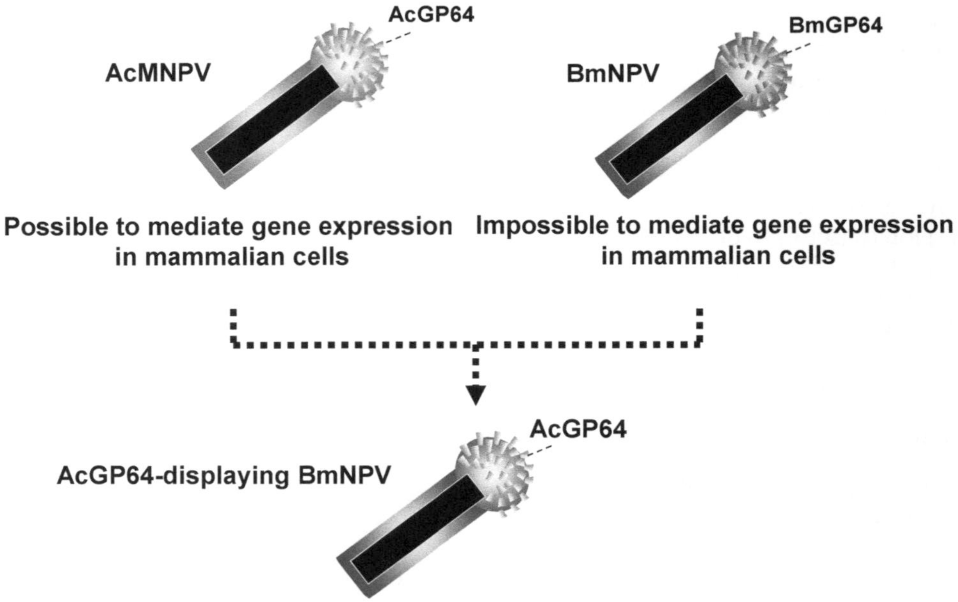

Fig. 19.3: Construction of AcGP64-displaying BmNPV.

transmembrane proteins from parasites can be also displayed on the surface of BmNPV particles (Kato et al. 2015, Rahman et al. 2003). These antigen-displaying BmNPVs induce the production of antibodies to each antigen by its immunization into mice without any adjuvant. Especially, the immunization of glycoprotein F or hemagglutinin-displaying BmNPVs into mice induced the production of the neutralizing antibodies to *Peste des petits* ruminants virus and Rinderpest virus (Raman et al. 2003). Transmembrane proteins can be also displayed on the surface of baculoviruses through its native transmembrane domains instead of the use of that from GP64 (Loisel et al. 1997). Human (pro)renin receptor, which has the transmembrane and cytoplasmic domains at its C-terminus, was displayed on the surface of BmNPV particles with its function (Kato et al. 2011, Kato et al. 2012). These BmNPV particles were prepared in Bm5 cell culture and silkworm larvae. Using this AcGP64-displaying BmNPV, dual functional AcGP64-displaying BmNPV was constructed (Fig. 19.4). NcSAG1 was displayed with the transmembrane and cytoplasmic domains of BmGP64 on the surface of the BmNPV particles together with AcGP64 and the NcSRS2 expression cassette under the control of CMV promoter was inserted into the genome of the recombinant BmNPV. The display of NcSAG1 was confirmed by western blot and ELISA (Fig. 19.5A and B). The expression of NcSRS2 under the control of CMV promoter in mammalian cells were checked by immunofluorescence microscopy (Fig. 19.5C). This recombinant BmNPV, which has the ability of both subunit vaccine and DNA vaccine, can be used for the development of vaccines to *N. caninum*.

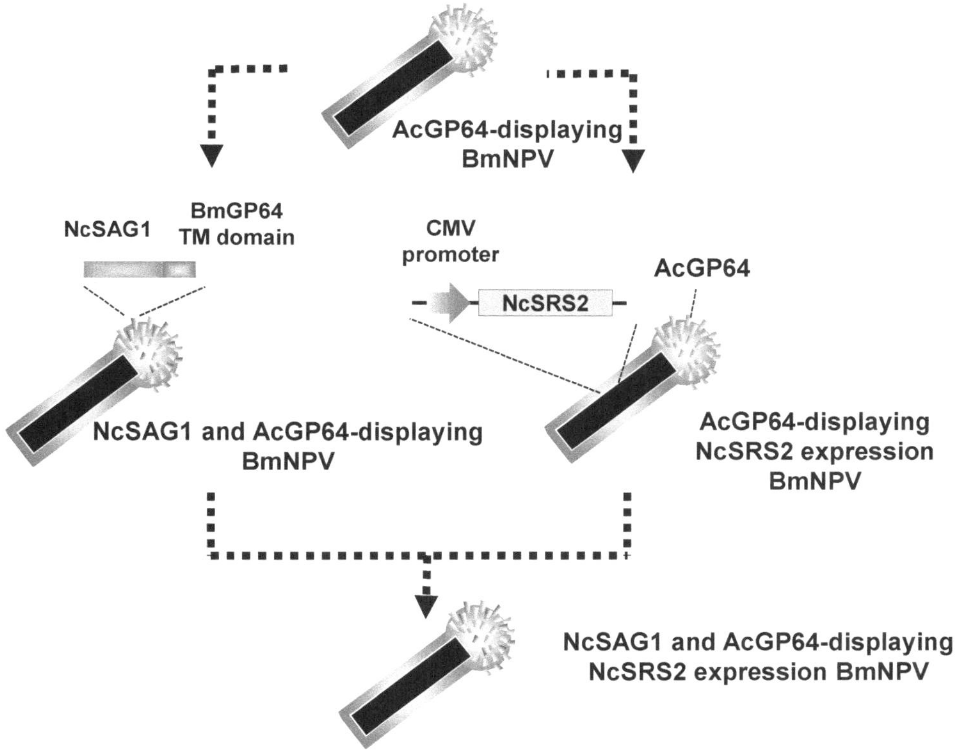

Fig. 19.4: Construction of NcSAG1and AcGP64-displaying BmNPV. This recombinant BmNPV has the NcSRS2 expression cassette under the control of CMV promoter in its genome. This recombinant BmNPV can work as both subunit vaccine and DNA vaccine.

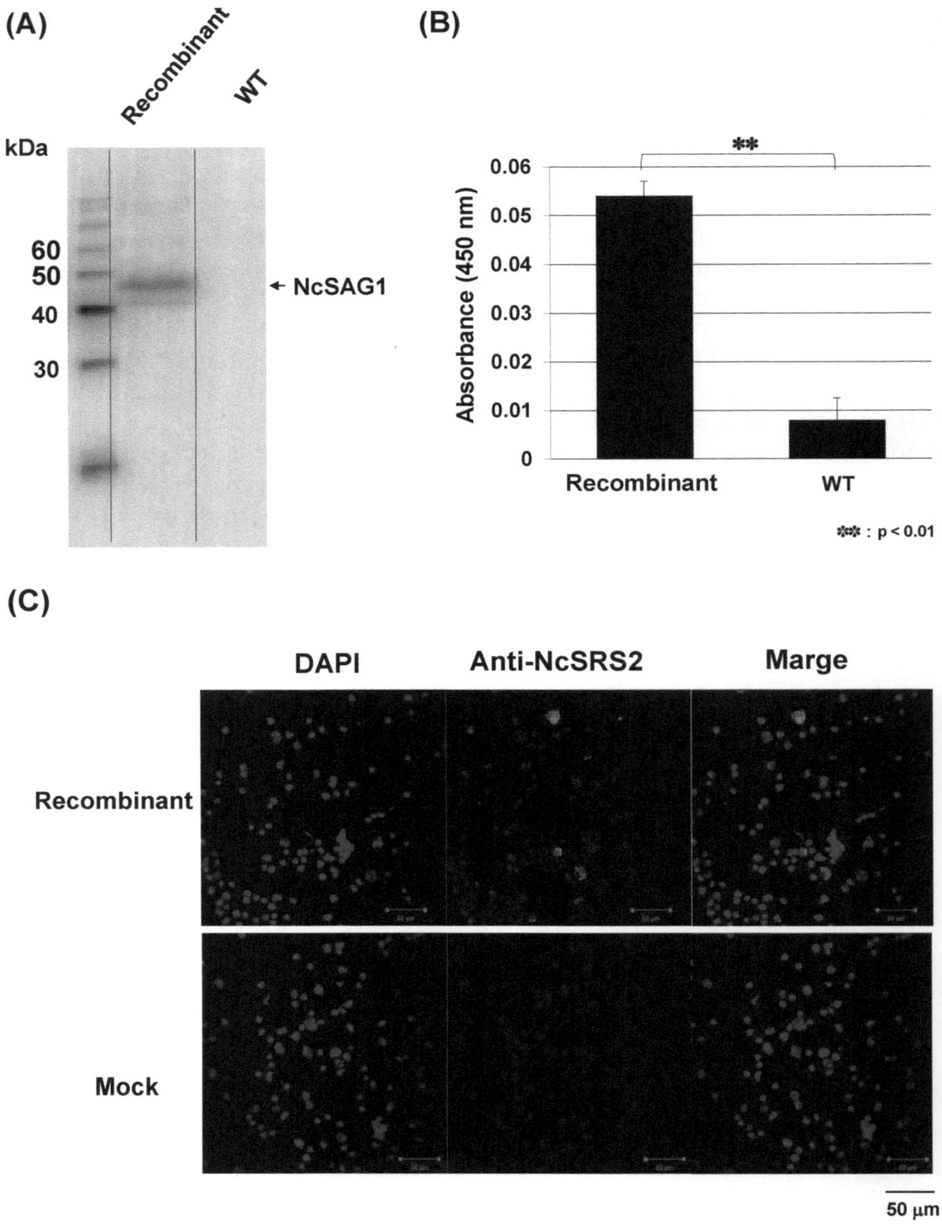

Fig. 19.5: Properties of NcSAG1and AcGP64-displaying BmNPV carrying the NcSRS2 expression cassette. (A) Western blot to detect NcSAG1 in this recombinant BmNPV sample. Mouse anti-DDDDK tag antibody and HRP-conjugated goat anti-mouse IgG were used as primary and secondary antibodies. NcSAG1 has FLAG sequence at its N-terminus. (B) ELISA to detect NcSAG1 on the surface of the recombinant BmNPV. Onto the well in a 96 well plate, 1×10^6 pfu of the recombinant BmNPV was immobilized. Mouse anti-DDDDK tag antibody and HRP-conjugated goat anti-mouse IgG were used as primary and secondary antibodies. (C) Immnofluorescent microscopy of NcSRS2 expression in HEK293T cells transfected with the recombinant BmNPV. The recombinant BmNPV was transfected into 2.5×10^5 cells of HEK293T cells at M.O.I. 150. NcSRS2 expression was detected by mouse anti-NcSRS2 antibody and Alexa Fluor 594-conjugated goat anti-mouse IgG.

5. Perspectives

Now silkworm larvae and pupae have been used for a large-scale production of recombinant proteins, especially mammalian proteins. VLPs, irrespective of the presence of the envelope, are composed of structural proteins of viruses and can be easily produced in silkworm larvae and pupae on a large scale, compared to cell cultivation (Fig. 19.6). Using recombinant BmNPVs, its contamination in purified VLPs is one of problems. However, recombinant BmNPVs can be also used as not only a vaccine but also a carrier of foreign genes into mammalian cells for the gene delivery (Fig. 19.6). In addition, the transgenic silkworm system for recombinant protein production has been established without the use of recombinant BmNPV even though it takes long time to produce recombinant proteins (Tomita 2011, Xu 2014). Silkworm is a promising host for the large-scale production of not only recombinant proteins but also viral nanoparticles.

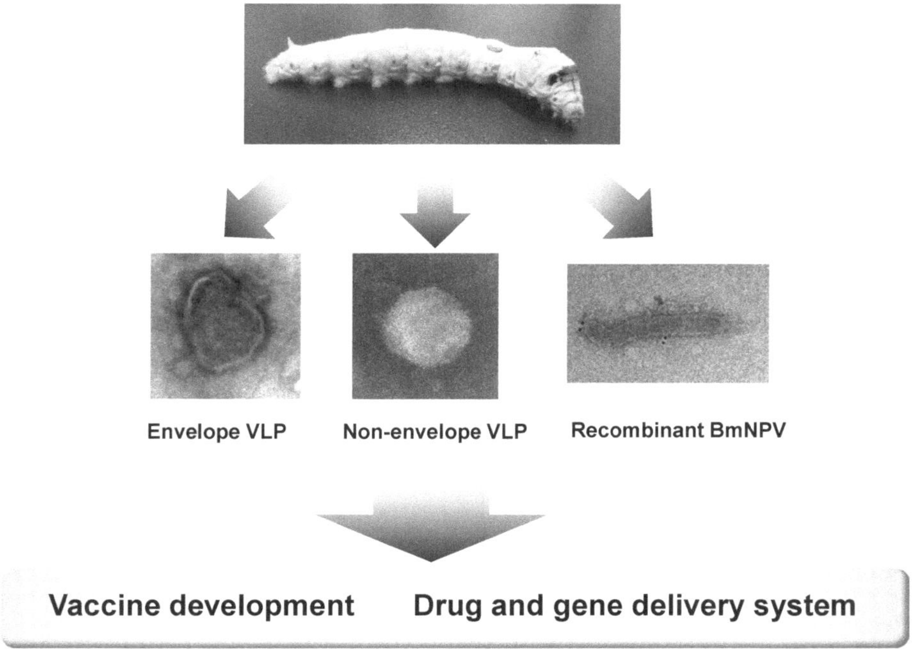

Fig. 19.6: A picture of the perspective for the production of nanoparticles in silkworms.

References

Aylett, C.H., E. Sauer, S. Imseng, D. Boehringer, M.N. Hall, N. Ban et al. 2016. Architecture of human mTOR complex 1. Science 351: 48–52.

Baek, J.O., J.W. Seo, I.H. Kim and C.H. Kim. 2011. Production and purification of human papillomavirus type 33 L1 virus-like particles from *Spodoptera frugiperda* 9 cells using two-step column chromatography. Protein Expr. Purif. 75: 211–217.

Barsoum, J., R. Brown, M. McKee and F.M. Boyce. 1997. Efficient transduction of mammalian cells by a recombinant baculovirus having the vesicular stomatitis virus G glycoprotein. Hum. Gene Ther. 8: 2011–2018.

Bertipaglia, C., S. Schneider, A.J. Jakobi, A.K. Tarafder, Y.S. Bykov, A. Picco et al. 2016. Higher-order assemblies of oligomeric cargo receptor complexes form the membrane scaffold of the Cvt vesicle. EMBO Rep. 17: 1044–1060.

Celie, P.H., A.H. Parret and A. Perrakis. 2016. Recombinant cloning strategies for protein expression. Curr. Opin. Struct. Biol. 38: 145–154.

Chen, B.J., G.P. Leser, E. Morita and R.A. Lamb. 2007 Influenza virus hemagglutinin and neuraminidase, but not the matrix protein, are required for assembly and budding of plasmid-derived virus-like particles. J. Virol. 81: 7111–7123.

D'Aoust, M.A., M.M. Couture, N. Charland, S. Trépanier, N. Landry, F. Ors et al. 2010. The production of hemagglutinin-based virus-like particles in plants: a rapid, efficient and safe response to pandemic influenza. Plant Biotechnol. J. 8: 607–619.

Deo, V.K., Y. Tsuji, T. Yasuda, T. Kato, N. Sakamoto, H. Suzuki et al. 2011. Expression of an RSV-gag virus-like particle in insect cell lines and silkworm larvae. J. Virol. Methods 177: 147–152.

Deo, V.K., K. Yoshimatsu, T. Otsuki, J. Dong, T. Kato and E.Y. Park. 2013. Display of *Neospora caninum* surface protein related sequence 2 on Rous sarcoma virus-derived gag protein virus-like particles. J. Biotechnol. 165: 69–75.

Deo, V.K., M. Yui, J. Alam, M. Yamazaki, T. Kato and E.Y. Park. 2014. A model for targeting colon carcinoma cells using single-chain variable fragments anchored on virus-like particles via glycosyl phosphatidylinositol anchor. Pharm. Res. 31: 2166–2177.

Deo, V.K., T. Kato and E.Y. Park. 2015. Chimeric virus-like particles made using GAG and M1 capsid proteins providing dual drug delivery and vaccination platform. Mol. Pharm. 12: 839–845.

Deo, V.K., T. Kato and E.Y. Park 2016. Virus-like particles displaying recombinant short-chain fragment region and interleukin 2 for targeting colon cancer tumors and attracting macrophages. J. Pharm. Sci. 105: 1614–1622.

Ding, X., D. Liu, G. Booth, W. Gao and Y. Lu. 2018. Virus-like particle engineering: from rational design to versatile applications. Biotechnol. J. 13: e1700324.

Dojima, T., T. Nishina, T. Kato, H. Ueda and E.Y. Park. 2010. Production of scFv-displaying BmNPV in silkworm larvae and its efficient purification. Biotechnol. Appl. Biochem. 57: 63–69.

Feng, H., M. Liang, H.L. Wang, T. Zhang, P.S. Zhao, X.J. Shen et al. 2011. Recombinant canine parvovirus-like particles express foreign epitopes in silkworm pupae. Vet. Microbiol. 154: 49–57.

Feng, H., G.Q. Hu, H.L. Wang, M. Liang, H. Guo, P. Zhao et al. 2014. Canine parvovirus VP2 protein expressed in silkworm pupae self-assembles into virus-like particles with high immunogenicity. PLoS One 9: e79575.

Fuenmayor, J., F. Gòdia and L. Cervera. 2017. Production of virus-like particles for vaccines. N. Biotechnol. 39: 174–180.

Gibson, D.G. 2011. Enzymatic assembly of overlapping DNA fragments. Methods Enzymol. 498: 349–361.

Gómez-Puertas, P., C. Albo, E. Pérez-Pastrana, A. Vivo and A. Portela. 2000. Influenza virus matrix protein is the major driving force in virus budding. J. Virol. 74: 11538–11547.

Grabherr, R., W. Ernst, C. Oker-Blom and I. Jones. 2001. Developments in the use of baculoviruses for the surface display of complex eukaryotic proteins. Trends Biotechnol. 19: 231–236.

Grein, T.A., R. Michalsky, M. Vega Lopez and P. Czermak. 2012. Purification of a recombinant baculovirus of Autographa californica M nucleopolyhedrovirus by ion exchange membrane chromatography. J. Virol. Methods 183: 117–124.

Grein, T.A., R. Michalsky and P. Czermak. 2014. Virus separation using membranes. Methods Mol. Biol. 1104: 456–491.

Heinimäki, S., K. Tamminen, M. Malm, T. Vesikari and V. Blazevic. 2017. Live baculovirus acts as a strong B and T cell adjuvant for monomeric and oligomeric protein antigens. Virology 511: 114–122.

Hsu, C., Y. Ho, K. Wang and Y. Hu. 2004. Investigation of optimal transduction conditions for baculovirus-mediated gene delivery into mammalian cells. Biotechnol. Bioeng. 88: 42–51.

Imai, A., T. Tadokoro, S. Kita, M. Horiuchi, H. Fukuhara and K. Maenaka. 2016. Establishment of the BacMam system using silkworm baculovirus. Biochem. Biophys. Res. Commun. 478: 580–585.

Jalaguier, P., K. Turcotte, A. Danylo, R. Cantin and M.J. Tremblay. 2011. Efficient production of HIV-1 virus-like particles from a mammalian expression vector requires the N-terminal capsid domain. PLoS One 6: e28314.

Kato, T., S.L. Manoha, S. Tanaka and E.Y. Park. 2009. High-titer preparation of *Bombyx mori* nucleopolyhedrovirus (BmNPV) displaying recombinant protein in silkworm larvae by size exclusion chromatography and its characterization. BMC Biotechnol. 9: 55.

Kato, T., M. Kajikawa, K. Maenaka and E.Y. Park. 2010. Silkworm expression system as a platform technology in life science. Appl. Microbiol. Biotechnol. 85: 459–470.

Kato, T., F. Suzuki and E.Y. Park. 2011. Purification of functional baculovirus particles from silkworm larval hemolymph and their use as nanoparticles for the detection of human prorenin receptor (PRR) binding. BMC Biotechnol. 11: 60.

Kato, T., F. Suzuki and E.Y. Park. 2012. Display of the human (pro)renin receptor on *Bombyx mori* nucleopolyhedrovirus (BmNPV) particles using Bm cells. J. Biosci. Bioeng. 114: 564–569.

Kato, T., M. Yui, V.K. Deo and E.Y. Park. 2015. Development of Rous sarcoma virus-like particles displaying hCC49 scFv for specific targeted drug delivery to human colon carcinoma cells. Pharm. Res. 32: 3699–3707.

Kato, T., T. Otsuki, M. Yoshimoto, K. Itagaki, T. Kohsaka, Y. Matsumoto et al. 2015. *Bombyx mori* nucleopolyhedrovirus displaying *Neospora caninum* antigens as a vaccine candidate against *N. caninum* infection in mice. Mol. Biotechnol. 57: 145–154.

Kato, T., S. Sugioka, K. Itagaki and E.Y. Park. 2016. Gene transduction in mammalian cells using *Bombyx mori* nucleopolyhedrovirus assisted by glycoprotein 64 of *Autographa californica* multiple nucleopolyhedrovirus. Sci. Rep. 6: 32283.

Kenoutis, C., R.C. Efrose, L. Swevers, A.A. Lavdas, M. Gaitanous, R. Matsas et al. 2006. Baculovirus-mediated gene delivery into mammalian cells does not alter their transcriptional and differentiating potential but is accompanied by early viral gene expression. J. Virol. 80: 4135–46.

Liu, X., Y. Li, X. Hu, Y. Yi and Z. Zhang. 2017. Gene delivery and gene expression in vertebrate using baculovirus *Bombyx mori* nucleopolyhedrovirus vector. Oncotarget. 8(62): 106017–106025.

Loisel, T.P., H. Ansanay, S. St-Onge, B. Gay, P. Boulanger, A.D. Strosberg et al. 1997. Recovery of homogeneous and functional beta 2-adrenergic receptors from extracellular baculovirus particles. Nat. Biotechnol. 15: 1300–1304.

Lynch, A.G., F. Tanzer, M.J. Fraser, E.G. Shephard, A.L. Williamson and E.P. Rybicki. 2010. Use of the piggyBac transposon to create HIV-1 gag transgenic insect cell lines for continuous VLP production. BMC Biotechnol. 10: 30.

Mansouri, M. and P. Berger. 2018. Baculovirus for gene delivery to mammalian cells: Past, present and future. Plasmid 98: 1–7.

Masuda, A., J.M. Lee, T. Miyata, T. Sato, S. hayashi, M. Hino et al. 2018. Purification and characterization of immunogenic recombinant virus-like particles of porcine circovirus type 2 expressed in silkworm pupae. J. Gen. Virol. in press.

Matsuda, S., R. Nerome, K. Maegawa, A. Kotaki, S. Sugita, K. Kawasaki et al. 2017. Development of a Japanese encephalitis virus-like particle vaccine in silkworms using codon-optimised prM and envelope genes. Heliyon. 3: e00286.

Millán, A.F., S. Gómez-Sebastián, M.C. Nuñez, J. Veramendi and J.M. Escribano. 2010. Human papillomavirus-like particles vaccine efficiently produced in a non-fermentative system based on insect larva. Protein Expr. Purif. 74: 1–8.

Minkner, R. and E.Y. Park. 2018. Purification of virus-like particles (VLPs) expressed in the silkworm *Bombyx mori*. Biotechnol. Lett. 40: 659–666.

Minkner, R., R. Baba, Y. Kurosawa, S. Suzuki, T. Kato, S. Kobayashi et al. 2018. Purification of human papillomavirus-like particles expressed in silkworm using a *Bombyx mori* nucleopolyhedrovirus bacmid expression system. J. Chromatogr. B Analyt. Technol. Biomed. Life Sci. 1096: 39–47.

Motohashi, T., T. Shimojima, T. Fukagawa, K. Maenaka and E.Y. Park. 2005. Efficient large-scale protein production of larvae and pupae of silkworm by *Bombyx mori* nuclear polyhedrosis virus bacmid system. Biochem. Biophys. Res. Commun. 326: 564–569.

Nasimuzzaman, M., J.C. van der Loo and P. Malik. 2016. Purification of baculovirus vectors using heparin affinity chromatography. Mol. Ther. Methods Clin. Dev. 3: 16071.

Nishikawa, Y., K. Tragoolpua, L. Makala and H. Hasegawa. 2002. *Neospora caninum* NcSRS2 is a transmembrane protein that contains a glycosylphosphatidylinositol anchor in insect cells. Vet. Parasitol. 109: 191–201.

Nerome, K., S. Sugita, K. kuroda, T. Hirose, H.S. Matsuda, K. Majima et al. 2015. The large-scale production of an artificial influenza virus-like particlevaccine in silkworm pupae. Vaccine 33: 117–125.

Nerome, K., S. Matsuda, K. Maegawa, S. Sugita, K. Kuroda, K. Kawasaki et al. 2017. Quantitative analysis of the yield of avian H7 influenza virus haemagglutinin protein produced in silkworm pupae with the use of the codon-optimized DNA: A possible oral vaccine. Vaccine 35: 738–746.

Ono, C., T. Okamoto and Y. Matsuura. 2018. Baculovirus as a tool for gene delivery and gene therapy. Viruses 10: E510.

Ono, C., J. Hirano, T. Okamoto and Y. Matsuura. 2018. Evaluation of viral contamination in a baculovirus expression system. Microbiol. Immunol. 62: 200–204.

Palaniyandi, M., T. kato and E.Y. Park. 2012. Expression of human papillomavirus 6b L1 protein in silkworm larvae and enhanced green fluorescent protein displaying on its virus-like particles. Springerplus 1: 29.

Peixoto, C., M.F. Sousa, A.C. Silva, M.J. Carrondo and P.M. Alves. 2007. Downstream processing of triple layered rotavirus like particles. J. Biotechnol. 127: 452–461.

Pelosse, M., H. Crocker, B. Gorda, P. Lemaire, J. Rauch and I. Berger. 2017. MultiBac: from protein complex structures to synthetic viral nanosystems. BMC Biol. 15: 99.

Rahman, M.M. and K.P. Gopinathan. 2003. *Bombyx mori* nucleopolyhedrovirus-based surface display system for recombinant proteins. J. Gen. Virol. 84: 2023–2031.

Rahman, M.M., M.S. Shaila and K.P. Gopinathan. 2003. Baculovirus display of fusion protein of *Peste des petits* ruminants virus and hemagglutination protein of Rinderpest virus and immunogenicity of the displayed proteins in mouse model. Virology 317: 36–49.

Senger, T., L. Schädlich, L. Gissmann and M. Müller. 2010. Enhanced papillomavirus-like particle production in insect cells. Virology 388: 344–353.

Steppert, P., D. Burgdtaller, M. Klausberger, E. Berger, P.P. Aguilar, T.A. Schneider et al. 2016. Purification of HIV-1 gag virus-like particles and separation of other extracellular particles. J. Chromatogr. A 1455: 93–101.

Tani, H., C.K. Limn, C.C. Yap, M. Onishi, M, Nozaki, Y. Nishimune et al. 2003. *In vitro* and *in vivo* gene delivery by recombinant baculoviruses. J. Virol. 77: 9799–9808.

Tomita, M. 2011. Transgenic silkworms that weave recombinant proteins into silk cocoons. Biotechnol. Lett. 33: 645–654.

Transfiguracion, J., H. Jorio, J. Meghrous, D. Jacob and A. Kamen. 2007. High yield purification of functional baculovirus vectors by size exclusion chromatography. J. Virol. Methods 142: 21–28.

Trowitzsch, S., C. Bieniossek, Y. Nie, F. Garzoni and I. Berger. 2010. New baculovirus expression tools for recombinant protein complex production. J. Struct. Biol. 172: 45–54.

Tsuji, Y., V.K. Deo, T. Kato and E.Y. Park. 2010. Production of Rous sarcoma virus-like particles displaying human transmembrane protein in silkworm larvae and its application to ligand-receptor binding assay. J. Biotechnol. 2011 155: 185–192.

Usami, A., T. Suzuki, H. Nagaya and S. Ishiyama. 2010. Silkworm as a host of baculovirus expression. Curr. Pharm. Biotechnol. 11: 246–250.

van Oers, M., G.P. Pijlman and J.M. Vlak. 2015. Thirty years of baculovirus-insect cell protein expression: from dark horse to mainstream technology. J. Gen. Virol. 96: 6–23.

Watanabe, S., T. Iizuka, S. Hatama, T. Kanno, M. Mase and T. Shibata. 2017. Production of highly immunogenic virus-like particles of bovine papillomavirus type 6 in silkworm pupae. Vaccine 35: 5878–5882.

Weissmann, F., G. Petzold, R. VanderLinden, P.J. Huis In 't Veld, N.G. Brown, F. Lampert et al. 2016. biGBac enables rapid gene assembly for the expression of large multisubunit protein complexes. Proc. Natl. Acad. Sci. USA 113: E2564–E2569.

Xu, H. 2014. The advances and perspectives of recombinant protein production in the silk gland of silkworm *Bombyx mori*. Transgenic Res. 23: 697–706.

Yamaji, H., M. Nakamura, M. Kuwahara, Y. Takahashi, T. Katsuda and E. Konoshi. 2013. Efficient production of Japanese encephalitis virus-like particles by recombinant lepidopteran insect cells. Appl. Microbiol. Biotechnol. 97: 1071–1079.

Yang, L., Y. Song, X. Li, X. Huang, J. Liu, H. Ding et al. 2012. HIV-1 virus-like particles produced by stably transfected *Drosophila* S2 cells: a desirable vaccine component. J. Virol. 86: 7662–76.

Yao, L., S. Wang, S. Su, N. Yao, J. He, L. Peng et al. 2012. Construction of a baculovirus-silkworm multigene expression system and its application on producing virus-like particles. PLoS One 7: e32510.

Zhang, R., S. Zhang, M. Li, C. Chen and Q. Yao. 2010. Incorporation of CD40 ligand into SHIV virus-like particles (VLP) enhances SHIV-VLP-induced dendritic cell activation and boosts immune responses against HIV. Vaccine 28: 5114–5127.

Zhang, Z., J. Yang and D. Barford. 2016. Recombinant expression and reconstitution of multiprotein complexes by the USER cloning method in the insect cell-baculovirus expression system. Methods 95: 13–25.

Zhao, Q., M.J. Allen, Y. Wang, B. Wang, N. Wang, L. Shi et al. 2012. Disassembly and reassembly improves morphology and thermal stability of human papillomavirus type 16 virus-like particles. Nanomedicine 8: 1182–1189.

Zeltins, A. 2013. Construction and characterization of virus-like particles: a review. Mol. Biotechnol. 53: 92–107.

Zhou, Y., H. Chen, X. Li, Y. Wang, K. Chen, S. Zhang et al. 2011. Production of recombinant human DNA polymerase delta in a *Bombyx mori* bioreactor. PLoS One 6: e22224.

Chapter 20

Glycobiology and Glycoengineering in Silkworm

*Takatsugu Miyazaki** and *Enoch Y. Park*

1. Introduction

Glycosylation is the most common posttranslational modification of proteins found in all domains of life. In eukaryotes protein glycosylation occurs on over 50% of all proteins (Apweiler et al. 1999) and plays important roles in various physiological functions including cell proliferation, differentiation, migration, cellular interactions, host–pathogen interactions, immune modulation and signal transduction (Varki 1993). Glycosylation is divided into several classes such as *N*-glycosylation (asparagine-linked) and *O*-glycosylation (e.g., serine/threonine-linked) based on linkage types between sugars and side-chains of amino acid residues. *N*-glycosylation occurs on an asparagine (Asn) residue in Asn-X-Ser/Thr sequon (where X is any amino acid except for proline) and has been reported to be involved in protein folding, stability, localization, cellular interaction and molecular half-life.

In recent years, with the increasing demand for biomedicine, numerous kinds of biopharmaceuticals and their production methods have been developed at both laboratory and industry levels. Most of biopharmaceuticals are glycoproteins that are *N*-glycosylated, and the modification is one of the most important factors for their bioactivity, transport, and half-life. Since glycosylation and other posttranslational modifications are essential for these therapeutic proteins, most of them are produced in mammalian cells such as Chinese Hamster Ovary (CHO) cells, but the expression process using CHO cells is relatively expensive (Farid 2006). Therefore, alternative production platforms using other eukaryotic expression systems such as yeasts, plants and insect cells have been developed (Gupta and Shukla 2018). All the developed eukaryotic expression system can produce glycoproteins,

Laboratory of Biotechnology, Research Institute of Green Science and Technology, Shizuoka University, 836 Ohya, Suruga-ku, Shizuoka 422-8529, Japan.
Email: park.enoch@shizuoka.ac.jp
* Corresponding author: miyazaki.takatsugu@shizuoka.ac.jp

but the architectures of *N*-glycans were quite different from mammalian species, affecting biological activity, immunogenicity and stability of biopharmaceuticals. So, conversion of *N*-glycan structures produced in non-mammalian eukaryotic expression system into human-like glycans is necessary for biopharmaceutical application.

We are focusing on domestic silkworm, *Bombyx mori*, which has been domesticated for silk production for more than five thousand years. The insect is a research model organism of Lepidoptera in the fields of genetics, physiology and biochemistry and its genome has been sequenced by whole-genome shotgun sequencing projects (Mita et al. 2004, Xia et al. 2004). *B. mori* has been developed as a bioreactor to produce recombinant proteins since Maeda et al. (1985) reported the production of human α-interferon in silkworm using recombinant baculovirus. We developed the *B. mori* nucleopolyhedrovirus bacmid expression system to facilitate protein production in *B. mori* culture cells, larvae and pupae. In this chapter, we describe glycoprotein biosynthesis pathway in insects including silkworm and studies on production and glycoengineering of glycoproteins in silkworm.

2. N-glycan Biosynthesis

2.1 N-glycan Pathway in the ER Common to all Eukaryotes

N-glycans are synthesized and processed in the ER and Golgi apparatus and the biosynthetic pathway in the ER is conserved among all eukaryotes (Aebi 2013). The pathway starts at the cytoplasmic side of the Endoplasmic Reticulum (ER). In the first step, glycosyltransferase Alg7 transfer *N*-acetylglucosamine (GlcNAc) from uridine phosphate-GlcNAc donor substrate to dolichol phosphate carrier to produce dolichol pyrophosphate-GlcNAc (GlcNAc-PP-Dol) (Fig. 20.1). Subsequently, Alg13/Alg14 complex adds a second GlcNAc residue to form *N,N'*-diacetylchitobiose core, followed by attaching five mannose

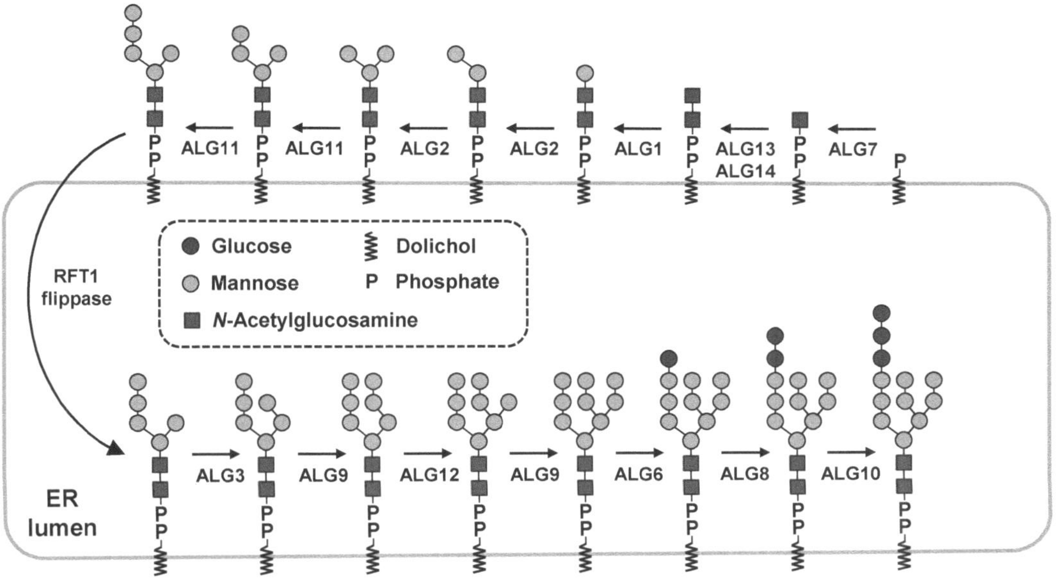

Fig. 20.1: Biosynthetic pathway of lipid-linked oligosaccharide (LLO) precursor: LLO is synthesized at the cytoplasmic side of the ER until $Man_5GlcNAc_2$-PP-Dol is produced, and then the glycan intermediate is flipped into the ER lumen, followed by further modifications catalyzed by glycosyltransferases in the ER.

(Man) residues by Alg1, Alg2 and Alg11 mannosyltransferases. The resultant Lipid Linked Oligosaccharide (LLO) $Man_5GlcNAc_2$-PP-Dol is flipped to the lumen of the ER, where it is further elongated with Man residues by Alg3, Alg12, and Alg9 mannosyltransferases to produce $Man_9GlcNAc_2$-PP-Dol. In the final steps of LLO precursor, Alg6 and Alg10 glucosyltransferases add three glucose (Glc) residues on $Man_9GlcNAc_2$-PP-Dol to form $Glc_3Man_9GlcNAc_2$-PP-Dol.

The precursor tetradecasaccharide $Glc_3Man_9GlcNAc_2$ *en bloc* transferred to an Asn residue of Asn-X-Ser/Thr sequon in a nascent polypeptide by oligosaccharyltransferase (OST) complex (Fig. 20.2). The two outermost glucose residues of the $Glc_3Man_9GlcNAc_2$ glycan are sequentially trimmed by α-glucosidase I (EC 3.2.1.106) and α-glucosidase II (EC 3.2.1.207), resulting in formation of $Glc_1Man_9GlcNAc_2$, which acts as a ligand for lectin chaperones, calnexin and calreticulin (Miyazaki et al. 2011, Moremen et al. 1994). These chaperones collaborate disulfide isomerase ERp57, and they prevent newly synthesized polypeptides from hydrophobic aggregation and non-productive disulfide formation during their folding process (Ellgaard et al. 2016). Iterative cycles of further glucose trimming by α-glucosidase II and re-glucosylation by UDP-Glc:glycoprotein glucosyltransferase (UGGT) facilitate the lectin chaperone-mediated folding of nascent polypeptides. After removal of a mannose residue by ER α-mannosidase I (EC 3.2.1.113), folded glycoproteins are transported to Golgi apparatus and further processed by Golgi-

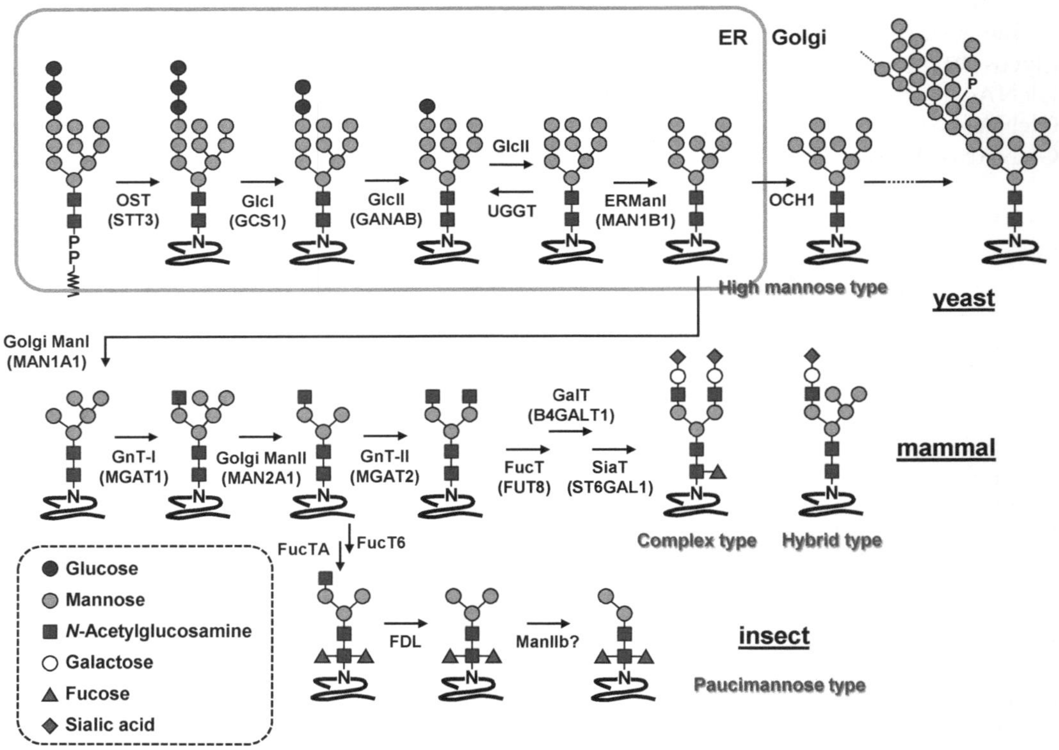

Fig. 20.2: Representative pathway of *N*-glycan processing in eukaryotes: *N*-glycan processing is highly conserved among all eukaryotes while its maturation in Golgi apparatus varies among mammals, insects, and yeasts. After OST transfers a $Glc_3Man_9GlcNAc_2$ precursor to Asn (N) residue in Asn-X-Ser/Thr sequon of nascent polypeptide, glycoside hydrolases and glycosyltransferases modify *N*-glycoproteins with various glycan architectures.

resident glycoside hydrolases and glycosyltransferases. These enzymes produce a variety of *N*-glycan structures, which are significantly different among eukaryotic lineages.

2.2 N-glycan Biosynthesis in Mammals

In case of mammalian cells, *N*-glycans are elaborated to hybrid-type or complex-type glycans in Golgi apparatus (Moremen et al. 2012). Glycoproteins that reach the Golgi apparatus are further de-mannosylated by Golgi α-mannosidase I to produce $Man_5GlcNAc_2$ glycans (Fig. 20.2). Then, β-1,2-*N*-acetylglucosaminyltransferase I (GnT-I, MGAT1, EC 2.4.1.101) catalyzes the transfer of GlcNAc residue from uridine diphosphate-GlcNAc to the α-1,3-branched mannose residue of the glycan, resulting in production of $GlcNAcMan_5GlcNAc_2$, followed by elongation on the GlcNAc residue with galactose and sialic acid residues catalyzed by galactosyltransferase (β4GalT1) and sialyltransferase, respectively, resulting in production of hybrid-type glycans. Alternatively, Golgi α-mannosidase II (EC 3.2.1.114) trims two mannose residues of $GlcNAcMan_5GlcNAc_2$ and then β-1,2-*N*-acetylglucosaminyltransferase II (GnT-II, MGAT2, EC 2.4.1.143) adds GlcNAc to the α-1,6-branched mannose residue of the glycan. β4GalT1 and sialyltransferase also extend this branch and produce biantennary complex-type glycans. β-1,4-*N*-acetylglucosaminyltransferase IV (GnT-IV, MGAT4, EC 2.4.1.145) and β-1,6-*N*-acetylglucosaminyltransferase V (GnT-V, MGAT5, EC 2.4.1.155) produce tri- and/or tetraantennary *N*-glycans by adding more GlcNAc residues to the trimannosyl core, and the highly branched glycans can influence bioactivity of glycoproteins such as erythropoietin. β-1,4-*N*-acetylglucosaminyltransferase III (GnT-III, MGAT3, EC 2.4.1.144) and fucosyltransferase FUT8 (EC 2.4.1.68) add a bisecting GlcNAc residue on the core β-mannose and a fucose residue on the inner most GlcNAc, respectively, both of which have been reported to correlate to Antibody-Dependent Cell Cytotoxicity (ADCC) activity (Hossler et al. 2009).

2.3 N-glycan Biosynthesis in Silkworm and other Insects

The ER and early Golgi processes in insect *N*-glycosylation pathway are highly similar to those of mammals; however, *B. mori* and other insect cells mostly produce high mannose type and shorter 'paucimannose' type glycans, the latter of which are often core-fucosylated (Zhu et al. 2019) (Fig. 20.2). The terminally GlcNAcylated glycans which are produced after trimming mannose residues by Golgi α-mannosidase II are then processed by FucT6 that catalyzes α-1,6-fucosylation on the innermost GlcNAc residue of the chitobiose core. Furthermore, the α-1,6-fucosylated *N*-glycans can be processed by another fucosyltransferase FucTA that catalyzes α-1,3-fucosylation on the same GlcNAc residue. The core difucosylated glycans are unique to insects, nematodes and plants, but not mammals. FucTA can also attach a fucose residue to $GlcNAcMan_3GlcNAc_2$ glycan but the resulting α-1,3-fucosylated glycans cannot be further fucosylated by FucT6 (Paschinger et al. 2005, Staudacher and Marz 1998).

An insect-specific β-*N*-acetylglucosaminidase FDL (EC 3.2.1.52) hydrolyzes β-1,2-linkage in $GlcNAcMan_3GlcNAc_2$ to produce $Man_3GlcNAc_2$, which is a key process to produce paucimannosidic *N*-glycans (Fig. 20.2). FDL removes the terminal GlcNAc residue from both fucosylated and non-fucosylated forms and results in the higher rate of paucimannose type glycans than complex type glycans. FDL enzymes from *D. melanogaster*, *Spodoptera frugiperda*, *Trichoplusia ni*, and *B. mori* have been enzymatically characterized and exhibited the same substrate specificity (Geisler et al. 2008, Geisler and

Jarvis 2010, Léonard et al. 2006, Nomura et al. 2010), suggesting that FDL is the key enzyme for production of paucimannose type *N*-glycans conserved among insects. *fdl* mutant of *D. melanogaster* was reported to produce more complex glycans and to influence its nervous system phenotypes (Léonard et al. 2006, Sarkar et al. 2006).

In addition, shorter paucimannose type *N*-glycans consisting of one or two mannose residues attached to the chitobiose core are also observed in several insects (Aoki et al. 2007, Kurz et al. 2015, Stanton et al. 2017). These glycans may be produced by another class II α-mannosidase, referred to as ManIIb or ManIII (Rosenbaum et al. 2014), but the enzymes have not been identified to generate $Man_{1-2}GlcNAc_2$ glycans to date. ManIIb (ManIII) from *S. frugiperda* and *D. melanogaster* reportedly trims mannose residues of $Man_5GlcNAc_2$ to produce $Man_{3-4}GlcNAc_2$ (Kawar et al. 2001, Nemčovičová et al. 2013), and they seem to contribute to a bypass of paucimannose type glycan production (Walski et al. 2017).

In recent decades complex *N*-glycan structures have been found in some insect *N*-glycomes (Aoki et al. 2007, Kurz et al. 2015, Stanton et al. 2017) (Fig. 20.3). The structures of the complex type *N*-glycans found in insect are quite different from those in mammals and there is a small amount (less than 20%) of them in whole *N*-glycomes of insects, but they seem to have important physiological functions in insects. GnT-II, which adds a GlcNAc residue on α-1,6-mannose branch to form biantennary glycans, is conserved in insect including *B. mori* as well as mammals but very few amounts of biantennary GlcNAcylated glycans have been found in insect cells. Altmann et al. (1993) reported that GnT-II catalytic activity in lepidopteran cells including *B. mori* Bm-N cells, is quite lower than GnT-I activity in the cells, which is one of the reasons insects produce paucimannose type glycans rather than complex type glycans. Recombinant *S. frugiperda* and *B. mori* GnT-II enzymes had the catalytic activity and the same substrate specificity that the GlcNAc residue on α-1,3-arm is necessary as mammalian orthologs (Geisler and Jarvis 2012a, Miyazaki et al. 2019). Additionally, Geisler and Jarvis demonstrated that the *S. frugiperda* GnT-II co-localizes

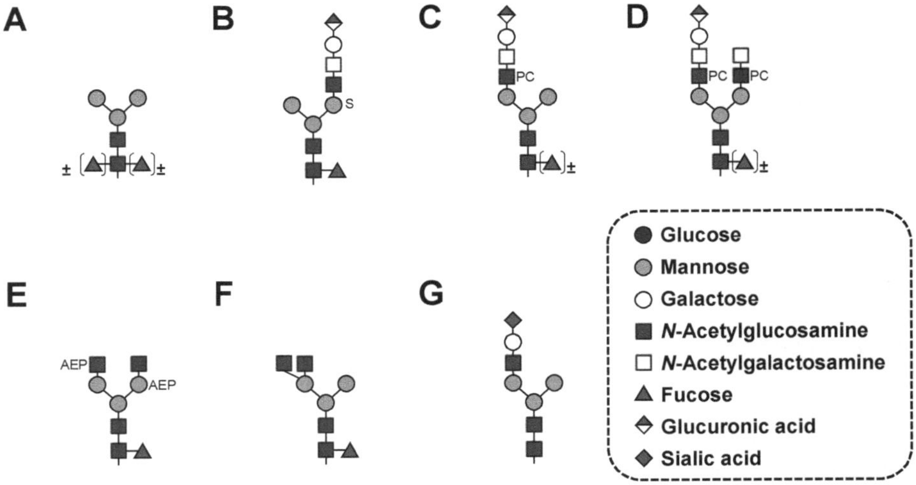

Fig. 20.3: Unique *N*-glycan structures found in insects: (A) General paucimannose *N*-glycan observed in insects. (B–E) *N*-glycan modifications with *N*-acetylgalactosamine, galactose, glucuronic acid, sulfate (S), aminoethyl phosphonate (AEP), and phosphorylcholine (PC) found in *T. ni* larva and culture cells (B and C), *L. dispar* larva (C and D), and apolipophorin III from *Lousta migratoria* (E). (F and G) Complex type *N*-glycans with an unusual branch and sialylation found in *D. melanogaster* embryo.

with its GnT-I and FDL in the Golgi apparatus (Geisler and Jarvis 2012a). On the other hand, we found that the *GnT-II* gene is ubiquitously expressed in larval and pupal stages of *B. mori* (Miyazaki et al. 2019). These findings imply that the *GnT-II* and *FDL* expression level may vary among different cells and tissues and the translation of GnT-II protein may be regulated.

Interestingly, triantennary complex type glycans have been found in dipteran larvae *N*-glycomes (Kurz et al. 2015) and honeybee royal jelly glycoproteins (Hykollari et al. 2018, Kimura et al. 2003), suggesting that these insects have glycosyltransferases that catalyze the transfer of GlcNAc residues on tri-mannosyl core of *N*-glycans like mammalian GnT-IV and GnT-V. *D. melanogaster* possesses putative glycosyltransferease genes homologous to mammalian GnT-III and GnT-IV, and *B. mori* also has a gene homologous to GnT-IV (Table 20.1). Dojima et al. (2009) reported that *N*-glycan containing a bisecting GlcNAc was found in the recombinant human β-1,3-*N*-acetylglucosaminyltransferase 3 expressed in silkworm larvae hemolymph. The gene products may be involved in *D. melanogaster* development (Yamamoto-Hino et al. 2015) but it is still unclear whether they catalyze the formation of tri- or more branched *N*-glycans.

Furthermore, insect GlcNAcylated glycans can be elongated with galactose, *N*-acetylgalactosamine (GalNAc), and glucuronic acid (GlcA) (Fig. 20.3B–D). β-1,4-*N*-acetylgalactosaminyltransferase (β4GalNAcT) that transfers a GalNAc residue to the non-reducing end GlcNAc residues with a β-1,4 linkage has been biochemically studied in *T. ni* and *D. melanogaster* (Haines and Irvine 2005, Vadaie and Jarvis 2004, van Die et al. 1996). Although insect β4GalNAcT is homologous to mammalian β4GalT1 that belongs to glycosyltransferase family 7 (GT7) in the CAZy database (http://www.cazy.org), β4GalNAcT prefers UDP-GalNAc to UDP-galactose as a donor substrate. Galactose- or GalNAc-containing *N*-glycans were found in *D. melanogaster* embryo (Aoki and Tiemeyer 2010), mosquito larvae (Kurz et al. 2015), and lepidopteran culture cells and larvae (Stanton et al. 2017). It is still unclear whether insect β4GalNAcT catalyzes galactosylation of terminal GlcNAc residues of *N*-glycans *in vivo* because the enzymes exhibited low activity toward UDP-Gal *in vitro* (Haines and Irvine 2005, Vadaie and Jarvis 2004). We characterized an ortholog from *B. mori* and identified it as β4GalNAcT with the similar substrate specificity and its details will be published elsewhere (Miyazaki et al. unpublished data). Only limited amounts of GalNAc-containing glycans exist in the whole insect *N*-glycomes but mutations in the *D. melanogaster β4GalNAcT* genes affected its phenotypes such as the decrease in neuromuscular junction bouton number, abnormal movement, and egg defects (Chen et al. 2007, Haines and Irvine 2005, Haines and Stewart 2007). GlcA-terminated glycans were found in culture cells or larvae of dipterans *Anopheles gambiae*, *Aedes aegypti*, and *D. melanogaster*, and lepidopterans *T. ni* and *Lymantria dispar* (Aoki and Tiemeyer 2010, Kurz et al. 2015, Stanton et al. 2017). *Drosophila* glycosyltransferases GlcAT-P and GlcAT-S were identified to catalyze the β-1,3-glucuronylation of terminal galactose residues of *N*-glycoproteins (Kim et al. 2003). Although there is no report on GalNAc- and GlcA-capped *N*-glycans and glycosyltransferases involved in these modifications in silkworm to date, silkworm has genes orthologous to β4GalNAcT, GlcAT-P, and GlcAT-S (Table 20.1) and may have similar pathways as the other insects.

Among insects, the sialylated *N*-glycans have been only reported to be found in *D. melanogaster* adult heads and embryos (Aoki et al. 2007, Aoki and Tiemeyer 2010, Koles et al. 2007) (Fig. 20.3G). In those cases, the sialylated *N*-glycans accounted for less than 0.1% of the total *N*-glycome. *D. melanogaster* possesses an active sialyltransferase (ST6Gal)

Table 20.1: Enzymes involved in N-glycosylation pathway of human and insects.

Enzyme name	EC number	CAZy[a]	GenBank ID		
			Homo sapiens	*D. melanogaster* (locus tag)	*B. mori*
Alg7	2.7.8.15	–	NP_001373.2	NP_609608.1 (CG5287)	XP_004925546.1
Alg13	2.4.1.141	GT1	NP_060936.1	NP_651673.1 (CG14512)	XP_021203208.1
Alg14	2.4.1.141	GT1	NP_659425.1	NP_573031.1 (CG6308)	XP_004932471.1
Alg1	2.4.1.142	GT33	NP_061982.3	NP_650662.1 (CG18012)	XP_004924175.1
Alg2	2.4.1.257 2.4.1.132	GT4	NP_149078.1	NP_647772.1 (CG1291)	XP_004931280.1
Alg11	2.4.1.131	GT4	NP_001004127.2	NP_649340.1 (CG11306)	XP_004933869.3
Alg3	2.4.1.258	GT58	NP_005778.1	NP_523829.2 (CG4084)	XP_012544505.2
Alg9	2.4.1.259 2.4.1.261	GT22	NP_079016.2	NP_651353.1 (CG11851)	XP_012552866.1
Alg12	2.4.1.260	GT22	NP_077010.1	NP_649939.1 (CG8412)	XP_004921841.1
Alg6	2.4.1.267	GT57	NP_037471.2	NP_609393.1 (CG5091)	XP_004928988.1
Alg8	2.4.1.265	GT57	NP_076984.2	NP_572355.1 (CG4542)	XP_004927866.1
Alg10	2.4.1.256	GT59	NP_116223.3	NP_729680.1 (CG32076)	XP_004926344.1
STT3A	2.4.99.18	GT66	NP_689926.1	NP_608425.1 (CG1518)	XP_004923047.1
GCS1	3.2.1.106	GH63	NP_006293.1	NP_572707.4 (CG1597)	XP_004925158.1
GANAB	3.2.1.207	GH31	NP_938149.2	NP_652145.1 (CG14476)	NP_001268822.1
MAN1B1/ ERManI	3.2.1.113	GH47	NP_057303.2	NP_651667.1 (CG11874)	XP_004929829.1
MAN1A1/Golgi ManI	3.2.1.113	GH47	NP_005898.2	NP_727407.1 (CG42275)	XP_004931141.1
MGAT1/GnT-I	2.4.1.101	GT13	NP_002397.2	NP_525117.2 (CG13431)	NP_001233218.1
MAN2A1/GMII	3.2.1.114	GH38	NP_002363.2	NP_524291.2 (CG18474)	XP_004921902.1
MGAT2/GnT-II	2.4.1.143	GT16	NP_002399.1	NP_001014684.1 (CG7921)	BBD84862.1
FUT8/FucT6	2.4.1.68	GT23	NP_835370.1	NP_572740.1 (CG2448)	NP_001280620.1
FucTA	2.4.1.214	GT10	–[b]	NP_648754.2 (CG6869)	XP_012550361.1
β4GalT1	2.4.1.38	GT7	NP_001488.2	–[b]	–[b]
β4GalNAcT	2.4.1.–	GT7	NP_848632.2 NP_775864.3	NP_610946.1 (CG8536) NP_651657.1 (CG14517)	XP_012545893.1 XP_021205072.1
ST6Gal/SiaT	2.4.99.1	GT29	NP_003023.1	NP_726474.1 (CG4871)	XP_004928745.2
MGAT3/GnT-III	2.4.1.144	GT17	NP_002400.3	NP_723787.1 (CG31849)	–
MGAT4/GnT-IV	2.4.1.145	GT54	NP_036346.1 NP_055090.1	NP_648720.2 (CG9384) NP_648721.1 (CG17173)	XP_021208029.1
MGAT5/GnT-V	2.4.1.155	GT18	NP_002401.1	–[b]	–[b]
GlcAT-P	2.4.1.135	GT43	–[b]	NP_648448.1 (CG6207)	NP_001124376.2
GlcAT-S	2.4.1.135	GT43	–[b]	NP_609303.1 (CG3881)	XP_012551722.1

[a] Classification based on the CAZy database (http://www.cazy.org; Lombard et al. 2014). GHxx, glycoside hydrolase family xx; GTyy, glycosyltransferase family yy.

[b] Enzymes expected to possess the same activities are not found.

that catalyzes the α-2,6-sialylation of Gal or GalNAc in several oligosaccharides and prefers a GalNAc residue to a Gal residue as an acceptor (Koles et al. 2004). In contrast a sialyltransferase (BmST) is also conserved in *B. mori* and is active with the similar substrate specificity as *Drosophila* ST6Gal but no sialylated glycans could be detected in silkworm (Kajiura et al. 2015). The functional CMP-sialic acid synthetases, which produce a donor substrate for sialyltransferases, have been found in *D. melanogaster* (Mertsalov et al. 2016, Viswanathan et al. 2006), *A. aegypti*, and beetle *Tribolium castaneum* (Di et al. 2017), supporting the sialoglycan biosynthesis in insect species. *B. mori* also possesses a gene homologous to the reported insect CMP-sialic acid synthetases but there is no report on its functionality to date.

2.4 N-glycan in other Eukaryotes

N-glycosylation pathway in yeast has been very well studied among all eukaryotes. *Saccharomyces cerevisiae* and *Pichia pastoris* (currently re-classified as *Komagataella pastoris* or *Komagataella phaffii*) are documented yeasts, and *P. pastoris* is widely used for high-level heterologous expression of various glycoproteins. These yeasts produce high mannose type *N*-glycans with a higher degree of mannose chains compared with other eukaryotes. Och1p mannosyltransferase catalyzes the transfer of an α-1,6 mannose unit to high mannose type glycan on glycoproteins which are transported into Golgi apparatus, and then the glycans are elongated by mannan polymerase complex to form the α-1,6 outer chain backbone containing up to 50 additional mannose residues (Striebeck et al. 2013).

In plants, complex type *N*-glycans are usually observed and their *N*-glycan processing is identical to mammals up to the formation of the $GlcNAc_2Man_3GlcNAc_2$ intermediate, but the resultant *N*-glycan structure of plants is different from that of mammals. The innermost GlcNAc residue is α-1,3-fucosylated and the core β-mannose is β-1,2-xylosylated (Jin et al. 2006, 2008). In higher plants a typical *N*-glycan usually contains a Lewis a (Le^a) structure, which is formed by attachment of β-1,3 galactose and α-1,4 fucose to the non-reducing end GlcNAc (Gomord et al. 2010).

3. Glycoengineering

3.1 Glycoprotein Production in Silkworm

Many recombinant protein expression systems have been developed. *E. coli* is the most frequently used expression host owing to easiness to handle for gene manipulation, culture and protein production. However, many mammalian proteins, especially glycoproteins, are difficult to express as active forms in *E. coli* because the bacterium does not have the ability of *N*-glycosylation like eukaryotes. The methylotrophic yeast *P. pastoris* is often used for heterologous expression of *N*-glycoproteins, but its *N*-glycan structures are different from mammals as described above.

Insect cell expression systems are now widely used to produce proteins from higher eukaryotes because insect cells have the capacity of posttranslational modifications as mammalian cells, including glycosylation and phosphorylation. Insect cell can be used for both transient and stable protein expression. Baculovirus expression system can transiently produce a large amount of proteins after infection of baculovirus harboring a desired gene into insect cells. The system also can be used not only in cultured cell lines but also insect larvae and pupae.

In recent decades the silkworm *B. mori* has also been used as a host for protein production using baculovirus vectors. Since Maeda et al. (1985) reported the production of human α-interferon in silkworm larvae for the first time, many proteins have been successfully produced using recombinant *B. mori* nucleopolyhedrovirus (BmNPV) (Kato et al. 2010). We have developed a BmNPV bacmid, which is a shuttle vector for *E. coli* and *B. mori*, to facilitate heterologous protein production in silkworm cultured cells, larvae and pupae without baculovirus construction or amplification (Motohashi et al. 2005, Park et al. 2008). The bacmid system enables us to prepare recombinant proteins in silkworm larvae and pupae within two or three weeks while the conventional silkworm–baculovirus system requires more than a month (Kato et al. 2010). Using the silkworm–BmNPV bacmid expression system, we succeeded in producing many mammalian proteins including glycoproteins, peptide hormones and membrane proteins (Du et al. 2008, 2009, Hwang et al. 2014, 2015, Ishikiriyama et al. 2009, Kajikawa et al. 2009, Miyazaki et al. 2017, 2018, Nakajima et al. 2009, Ogata et al. 2009, Park et al. 2009, Sasaki et al. 2009). This system provided a high yield of the glycoproteins from larval hemolymph with the range of 36–220 μg per larva (Table 20.2). Therefore, we believe that the silkworm-based BmNPV bacmid system is a promising tool for large-scale production of glycoproteins such as immunoglobulin and erythropoietin, which are used as biopharmaceuticals.

As described earlier paucimannose type and high mannose type *N*-glycans account for a large fraction of *N*-glycomes in insects, and also, glycoproteins expressed in insect cells are mainly modified with paucimannose type *N*-glycans. In the cases of silkworm expression system, the recombinant glycoproteins are almost paucimannose or high mannose types (Iizuka et al. 2009, Misaki et al. 2003, Nomura et al. 2015). We expressed human glycoproteins using the silkworm–BmNPV bacmid expression system and analyzed their *N*-glycan structures using high performance liquid chromatography and Matrix-Assisted Laser Desorption/Ionization-Yime Of Flight Mass Spectrometry (MALDI-TOF-MS) (Takahashi and Kato 2003, Yagi et al. 2005). The *N*-glycan structures which were identified in the recombinant glycoproteins expressed in silkworm larvae are listed in Table 20.3. In the case of recombinant human anti-BSA IgG1 expressed using the BmNPV bacmid, most of its *N*-glycans were identified as core-α1,6-fucosylated paucimannose type (Park et al. 2009). Also, human killer cell Ig-like receptor 2DL1 (KIR2DL1) prepared using the BmNPV bacmid system had paucimannose type *N*-glycans with a similar ratio as in the IgG1 (Sasaki et al. 2009). However, green fluorescent protein-fused β-1,3-*N*-acetylglucosaminyltransferase 2 (GFP-β3GnT2) that were expressed and purified from silkworm larval hemolymph contained various types of *N*-glycans, not only paucimannose type but also terminally GlcNAcylated and galactosylated type glycans. Additionally, bisecting GlcNAc, which is attached to core β-mannose by GnT-III, was

Table 20.2: Recombinant glycoproteins expressed using the silkworm–BmNPV bacmid expression system.

Proteins	Yields	Reference
GFP-β3GnT2	91 μg per larva	Park et al. 2007
Human anti-BSA IgG1	36 μg per larva	Park et al. 2009
Anti-BSA scFv	188 μg per larva	Ishikiriyama et al. 2009
KIR2DL1	200 μg per larva	Sasaki et al. 2009
Human ST6Gal1	220 μg per larva	Ogata et al. 2009

Table 20.3: N-glycan structures of human glycoproteins expressed in silkworm larvae.

Glycan structure	Relative quantity (%)		
	IgG1 (Park et al. 2008)	KIR2DL1 (Sasaki et al. 2009)	GFP-β3GnT2 (Dojima et al. 2009)
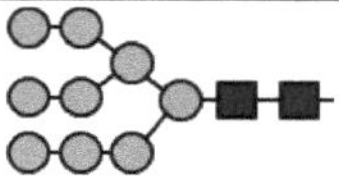	–	–	4.5
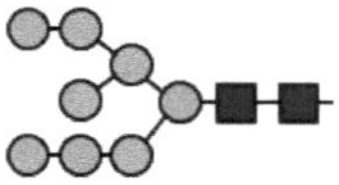	–	–	3.9
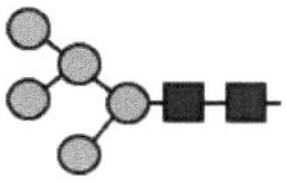	–	–	0.9
	–	–	14.0
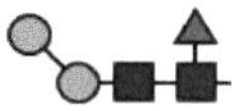	77.5	83–90	29.4
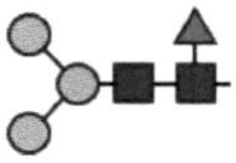	12.7	10–17	–
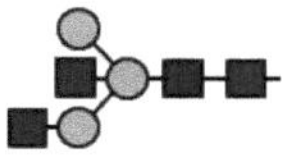	–	–	16.2
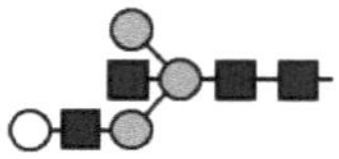	–	–	21.3
Others	9.8	–	12.5

found in recombinant GFP-β3GnT2 (Dojima et al. 2009). These observations suggested that enzymes which catalyzes these modifications might exist in silkworm larvae, though genes for the enzymes have not been identified yet.

3.2 Glycoengineering in Silkworm

As described earlier, glycoproteins expressed in insect cells and silkworm possess mainly paucimannose or high mannose type *N*-glycans. For pharmaceutical applications, the difference in *N*-glycan structure between insects and mammals causes recombinant glycoproteins to be biologically ineffective. To change the *N*-glycan architectures of glycoproteins expressed in insect cells into human type *N*-glycans, largely two strategies have been employed: knock-in of glycosyltransferases and other enzymes that are involved in biosynthesis of complex type *N*-glycans, and suppression of endogenous β-*N*-acetylglucosaminidase FDL, which is a key enzyme that leads to paucimannose type glycan biosynthesis.

The former strategy has been carried out in insect culture cells, silkworm larvae and pupae using stable expression systems or baculovirus vectors. Jarvis and Finn (1996) firstly constructed the baculovirus vector harboring *β4GalT1* gene under the control of immediate early *ie1* promoter and showed that glycoproteins expressed in lepidopteran cell could be elongated with galactose residue using the vector. Jarvis and Finn established various baculovirus vectors and lepidopteran cell lines to 'humanize' glycoproteins by co-expression of glycosyltransferases including GnT-I, GnT-II, β4GalT1, and sialyltransferases (Breitbach and Jarvis 2001, Hollister et al. 1998, 2002, Hollister and Jarvis 2001, Jarvis et al. 2001, Tomiya et al. 2003). To improve the rate of sialylation in insect cells, insect cell lines that produce cytidine-5′-monophospho (CMP)-sialic acid and efficiently transport it to the Golgi apparatus have also been developed because they have little intracellular sialic acid and no detectable CMP-sialic acid that is a substrate of sialyltranferases (Aumiller et al. 2003, 2012, Geisler and Jarvis 2012b, Mabashi-Asazuma et al. 2013).

A transgenic silkworm coexpressing human GnT-II and bovine β4GalT1 under the control of Posterior Silk Gland (PSG)-specific promoter was established and up to 17% of *N*-glycans were found to be terminally galactosylated in the PSG (Mabashi-Asazuma et al. 2015). In this case, however, recombinant proteins must be expressed in the PSG together with glycosyltransferases using a transgenic method, which is time consuming and labor intensive to obtain recombinant glycoproteins. Therefore, we attempted to 'humanize' glycoproteins expressed in silkworm pupae via coexpression of human GnT-II and β4GalT1 using the BmNPV bacmid expression system (Kato et al. 2017a). In this study, the polyhedron promoter from *Autographa californica* multiple nucleopolyhedrovirus and the actin A3 promoter from *B. mori* were used to coexpress GnT-II and β4GalT1 in silkworm pupae. When human IgG1 was coexpressed with GnT-II, approximately 15% of all *N*-glycans in the recombinant IgG1 were biantennary GlcNAcylated. Moreover, when IgG1 was coexpressed with both GnT-II and β4GalT1 under the control of the actin A3 promoter or polyhedron promoter, 16–52% of all *N*-glycans were biantennary and GlcNAcylated but not galactosyalted, and 28–35% of total *N*-glycans carried one or two galactose residues (Table 20.4). Suganuma et al. (2018) reported to achieve the production of sialylated glycoprotein in silkworm larvae using baculovirus vectors harboring glycosyltransferase genes including sialyltransferases although excessive CMP-sialic acid was supplemented to larvae by injection or oral administration. Because milligram order of recombinant protein can be produced from a few pupae (Hwang et al. 2014, 2015), our study suggested that silkworm pupae have the potential ability to produce a large quantity of glycoproteins with 'humanized' *N*-glycans.

The other approach by suppression of FDL activity has been also examined using insect cell lines and silkworm. Suppression of FDL activity was carried out in insect cells by adding a specific inhibitor, 2-acetamido-1,2-dideoxynojirimycin (ADN) (Watanabe et al. 2002) and using RNA interference (RNAi) technique (Kim et al. 2009, 2011). Both of ADN-treatment and RNAi toward *fdl* gene could partially decrease *N*-acetylglucosaminidase activity and increase the rate of GlcNAcylated *N*-glycans. In the case of silkworm, our group and the other researchers individually have performed conversion of *N*-glycan structures in glycoproteins expressed in silkworm by using ADN-inhibition and/or RNAi suppression of *fdl* expression (Kato et al. 2017b, Nagata et al. 2013, Nomura et al. 2015) successfully converted the paucimannose type glycans of the recombinant puffer fish saxitoxin- and tetrodotoxin-binding protein 3 and human erythropoietin expressed in silkworm larvae by RNAi while Nomura et al. (2015) improved the rate of the GlcNAc-terminated glycans in the recombinant calf intestine alkaline phosphatase. Additionally, we found that the

Table 20.4: *N*-glycan structures of human IgG1 coexpressed with human GnT-II and β4GalT1 in silkworm pupae.

Glycan structure	Relative quantity (%)			
	Without coexpression	Coexpression with		
		P_{Act}-GnT-II[a] P_{Pol}-β4GalT1	P_{Pol}-GnT-II P_{Act}-β4GalT1	P_{Pol}-GnT-II P_{Pol}-β4GalT1
High mannose $Man_{5-9}GlcNAc_2$	9.1	15.3	11.4	14.3
Paucimannose Man_{2-3}(±Fuc)$GlcNAc_2$	70.9	39.7	27.5	34.1
Monoantennary GlcNAcylated (non-galactosylated)	19.1	10.1	9.9	7.6
Biantennary GlcNAcylated (non-galactosylated)	–	–	52.2	15.5
Galactosylated	–	34.7	–	27.8
Others	0.9	0.7	0.4	0.7

[a] P_{Act} and P_{Pol} mean actin A3 promoter and polyhedron promoter used, respectively.

ratios of Man_3GlcNAc(±Fuc)GlcNAc and Man_2GlcNAc(±Fuc)GlcNAc in the recombinant human IgG1 was reversed by suppression of *fdl* using RNAi although the relationship between *fdl* expression and trimming of core-mannose is still unclear (Kato et al. 2017b). In *Drosophila* S2 cells, paucimannosidic structures remained despite the complete deletion of the *fdl* gene, indicating that other hexosaminidase could active on the terminal GlcNAc of *N*-glycans (Léonard et al. 2006). Three *B. mori* β-*N*-acetylglucosaminidase paralogs BmFDL, BmGlcNAcase1, and BmGlcNA case2 were reported to be localized in mainly fat body, silk gland, and hemolymph, respectively (Nomura et al. 2015). Considering these facts, combination of the knock-in of glycosyltransferase genes and suppression of β-*N*-acetylglucosaminidase activity could further improve the rate of complex type *N*-glycans in silkworm.

3.3 Glycoengineering in other Expression System

Yeasts such as *S. cerevisiae* and *P. pastoris* are widely used for glycoprotein production due to ease of handling, rapid growth, availability of various expression vectors and glycosylation machinery, etc. However, recombinant proteins expressed in yeasts contain hypermannosylated *N*-glycans, which reduces serum half-life and compromises the efficacy of therapeutic glycoproteins (Wildt and Gerngross 2005). To reduce hypermannosylation the knockout of α-mannosyltransferases OCH1 and MNN1 was introduced in *S. cerevisiae* and *P. pastoris* (Krainer et al. 2013, Nagasu et al. 1992). In addition, introduction of 14 heterologous genes of mammalian glycosyltransferases and sialic acid biosynthetic enzymes enabled *P. pastoris* to produce the recombinant erythropoietin with sialylated *N*-glycans (Hamilton et al. 2006).

Plants also have been also utilized for producing mammalian proteins because of their ability of posttranslational modifications and low cost. To 'humanize' *N*-glycans in plants, the plant-specific modifications of α-1,3-fucose and β-1,2-xylose, which are potentially immunogenic in human, have been eliminated by gene knockout strategies: site-directed mutagenesis, homologous recombination and RNAi (Cox et al. 2006, Koprivova et al. 2004, Shin et al. 2011). Since plants lack sialylation pathway of N-glycans, the related genes have been introduced in *Nicotiana benthamiana*, succeeding in the production of monoclonal antibody with sialylated *N*-glycans (Castilho et al. 2010). Coexpression of GnT-III, GnT-IV, and GnT-V resulted in the production of bisected, tri- and tetraantennary complex *N*-glycans in *N. benthamiana* (Castilho et al. 2011).

4. Conclusion

Silkworm is one of the industrially useful insects that have the ability to produce glycoproteins such as immunoglobulin and erythropoietin for pharmaceutical applications because the insects have the capability of cotranslational and posttranslational modifications including glycosylation like mammals. The structural difference of *N*-glycans between silkworm and human causes immunogenicity and biological inactivity when recombinant glycoproteins expressed in silkworm are used as pharmaceuticals. To solve this problem, using the BmNPV bacmids, we constructed the coexpression system for glycoprotein IgG1 and human glycosyltransferases and succeeded to convert its *N*-glycans to the complex type observed in humans. Recent studies revealed that insects including silkworm synthesize insect-specific complex type *N*-glycans, some of which are similar to mammalian complex type *N*-glycans. In addition, the silkworm possesses putative genes for glycosyltransferases involved in complex type *N*-glycan biosynthesis. Further studies on silkworm glycobiology will lead to more efficient production of 'humanized' glycoproteins and contribute to pharmaceutical industry.

Acknowledgement

We thank Dr. Koichi Kato and Dr. Hirokazu Yagi for structural analysis of *N*-glycans produced in silkworm.

References

Aebi, M. 2013. N-linked protein glycosylation in the ER. Biochim. Biophys. Acta 1833: 2430–2437.

Altmann, F., G. Kornfeld, T. Dalik, E. Staudacher and J. Glössl. 1993. Processing of asparagine-linked oligosaccharides in insect cells. *N*-Acetylglucosaminyltransferase I and II activities in cultured lepidopteran cells. Glycobiology 3: 619–625.

Aoki, K., M. Perlman, J.M. Lim, R. Cantu, L. Wells and M. Tiemeyer. 2007. Dynamic developmental elaboration of *N*-linked glycan complexity in the *Drosophila melanogaster* embryo. J. Biol. Chem. 282: 9127–9142.

Aoki, K. and M. Tiemeyer. 2010. The glycomics of glycan glucuronylation in *Drosophila melanogaster*. Methods Enzymol. 480: 297–321.

Apweiler, R., H. Hermjakob and N. Sharon. 1999. On the frequency of protein glycosylation, as deduced from analysis of the SWISS-PROT database. Biochim. Biophys. Acta 1473: 4–8.

Aumiller, J.J., J.R. Hollister and D.L. Jarvis. 2003. A transgenic insect cell line engineered to produce CMP-sialic acid and sialylated glycoproteins. Glycobiology 13: 497–507.

Aumiller, J.J., H. Mabashi-Asazuma, A. Hillar, X. Shi and D.L. Jarvis. 2012. A new glycoengineered insect cell line with an inducibly mammalianized protein *N*-glycosylation pathway. Glycobiology 22: 417–428.

Breitbach, K. and D.L. Jarvis. 2001. Improved glycosylation of a foreign protein by Tn-5B1-4 cells engineered to express mammalian glycosyltransferases. Biotechnol. Bioeng. 74: 230–239.

Castilho, A., R. Strasser, J. Stadlmann, J. Grass, J. Jez, P. Gattinger et al. 2010. In planta protein sialylation through overexpression of the respective mammalian pathway. J. Biol. Chem. 285: 15923–15930.

Castilho, A., P. Gattinger, J. Grass, J. Jez, M. Pabst, F. Altmann et al. 2011. *N*-glycosylation engineering of plants for the biosynthesis of glycoproteins with bisected and branched complex *N*-glycans. Glycobiology 21: 813–823.

Chen, Y.W., J.W. Pedersen, H.H. Wandall, S.B. Levery, S. Pizette, H. Clausen et al. 2007. Glycosphingolipids with extended sugar chain have specialized functions in development and behavior of *Drosophila*. Dev. Biol. 306: 736–749.

Cox, K.M., J.D. Sterling, J.T. Regan, J.R. Gasdaska, K.K. Frantz, C.G. Peele et al. 2006. Glycan optimization of a human monoclonal antibody in the aquatic plant *Lemna minor*. Nat. Biotechnol. 24: 1591–1597.

Di, W., A. Fujita, K. Hamaguchi, P. Delannoy, C. Sato and K. Kitajima. 2017. Diverse subcellular localizations of the insect CMP-sialic acid synthetases. Glycobiology 27: 329–341.

Dojima, T., T. Nishina, T. Kato, T. Uno, H. Yagi, K. Kato et al. 2009. Comparison of the *N*-linked glycosylation of human β1,3-*N*-acetylglucosaminyltransferase 2 expressed in insect cells and silkworm larvae. J. Biotechnol. 143: 27–33.

Du, D., T. Kato, A.H. Nabi, F. Suzuki and E.Y. Park. 2008. Expression of functional human (pro)renin receptor in silkworm (*Bombyx mori*) larvae using BmMNPV bacmid. Biotechnol. Appl. Biochem. 49: 195–202.

Du, D., T. Kato, F. Suzuki and E.Y. Park. 2009. Expression of protein complex comprising the human prorenin and (pro)renin receptor in silkworm larvae using *Bombyx mori* nucleopolyhedrovirus (BmNPV) bacmids for improving biological function. Mol. Biotechnol. 43: 154–161.

Ellgaard, L., N. McCaul, A. Chatsisvili and I. Braakman. 2016. Co- and post-translational protein folding in the ER. Traffic 17: 615–638.

Farid, S.S. 2006. Established bioprocesses for producing antibodies as a basis for future planning. Adv. Biochem. Eng. Biotechnol. 101: 1–42.

Geisler, C., J.J. Aumiller and D.L. Jarvis. 2008. A fused lobes gene encodes the processing β-*N*-acetylglucosaminidase in Sf9 cells. J. Biol. Chem. 283: 11330–11339.

Geisler, C. and D.L. Jarvis. 2010. Identification of genes encoding *N*-glycan processing β-*N*-acetylglucosaminidases in *Trichoplusia ni* and *Bombyx mori*: Implications for glycoengineering of baculovirus expression systems. Biotechnol. Prog. 26: 34–44.

Geisler, C. and D.L. Jarvis. 2012a. Substrate specificities and intracellular distributions of three *N*-glycan processing enzymes functioning at a key branch point in the insect *N*-glycosylation pathway. J. Biol. Chem. 287: 7084–7097.

Geisler, C. and D.L. Jarvis. 2012b. Innovative use of a bacterial enzyme involved in sialic acid degradation to initiate sialic acid biosynthesis in glycoengineered insect cells. Metab. Eng. 14: 642–652.

Gomord, V., A.C. Fitchette, L. Menu-Bouaouiche, C. Saint-Jore-Dupas, C. Plasson, D. Michaud et al. 2010. Plant-specific glycosylation patterns in the context of therapeutic protein production. Plant Biotechnol. J. 8: 564–587.

Gupta, S.K. and P. Shukla. 2018. Glycosylation control technologies for recombinant therapeutic proteins. Appl. Microbiol. Biotechnol. 102: 10457–10468.

Haines, N. and K.D. Irvine. 2005. Functional analysis of *Drosophila* β1,4-*N*-acetlygalactosaminyltransferases. Glycobiology 15: 335–346.

Haines, N. and B.A. Stewart. 2007. Functional roles for β1,4-*N*-acetlygalactosaminyltransferase-A in Drosophila larval neurons and muscles. Genetics 175: 671–679.

Hamilton, S.R., R.C. Davidson, N. Sethuraman, J.H. Nett, Y. Jiang, S. Rios et al. 2006. Humanization of yeast to produce complex terminally sialylated glycoproteins. Science 313: 1441–1443.

Hollister, J.R., J.H. Shaper and D.L. Jarvis. 1998. Stable expression of mammalian β 1,4-galactosyltransferase extends the *N*-glycosylation pathway in insect cells. Glycobiology 8: 473–480.

Hollister, J.R. and D.L. Jarvis. 2001. Engineering lepidopteran insect cells for sialoglycoprotein production by genetic transformation with mammalian β1,4-galactosyltransferase and α2,6-sialyltransferase genes. Glycobiology 11: 1–9.

Hollister, J., E. Grabenhorst, M. Nimtz, H. Conradt and D.L. Jarvis. 2002. Engineering the protein *N*-glycosylation pathway in insect cells for production of biantennary, complex *N*-glycans. Biochemistry 41: 15093–15104.

Hossler, P., S.F. Khattak and Z.J. Li. 2009. Optimal and consistent protein glycosylation in mammalian cell culture. Glycobiology 19: 936–949.

Hwang, I.W., Y. Makishima, T. Kato, S. Park, A. Terzic and E.Y. Park. 2014. Human acetyl-CoA carboxylase 2 expressed in silkworm *Bombyx mori* exhibits posttranslational biotinylation and phosphorylation. Appl. Microbiol. Biotechnol. 98: 8201–8209.

Hwang, I.W., Y. Makishima, T. Suzuki, T. Kato, S. Park, A. Terzic et al. 2015. Phosphorylation of Ser-204 and Tyr-405 in human malonyl-CoA decarboxylase expressed in silkworm *Bombyx mori* regulates catalytic decarboxylase activity. Appl. Microbiol. Biotechnol. 99: 8977–8986.

Hykollari, A., D. Malzl, B. Eckmair, J. Vanbeselaere, P. Scheidl, C. Jin et al. 2018. Isomeric separation and recognition of anionic and zwitterionic N-glycans from royal jelly glycoproteins. Mol. Cell. Proteomics 17: 2177–2196.

Iizuka, M., S. Ogawa, A. Takeuchi, S. Nakakita, Y. Kubo, Y. Miyawaki et al. 2009. Production of a recombinant mouse monoclonal antibody in transgenic silkworm cocoons. FEBS J. 276: 5806–5820.

Ishikiriyama, M., T. Nishina, T. Kato, H. Ueda and E.Y. Park. 2009. Human single-chain antibody expression in the hemolymph and fat body of silkworm larvae and pupae using BmNPV bacmids. J. Biosci. Bioeng. 107: 67–72.

Jarvis, D.L. and E.E Finn. 1996. Modifying the insect cell N-glycosylation pathway with immediate early baculovirus expression vectors. Nat. Biotechnol. 14: 1288–1292.

Jarvis, D.L., D. Howe and J.J. Aumiller. 2001. Novel baculovirus expression vectors that provide sialylation of recombinant glycoproteins in lepidopteran insect cells. J. Virol. 75: 6223–6227.

Jin, C., M. Bencúrová, N. Borth, B. Ferko, E. Jensen-Jarolim, F. Altmann et al. 2006. Immunoglobulin G specifically binding plant *N*-glycans with high affinity could be generated in rabbits but not in mice. Glycobiology 16: 349–357.

Jin, C., F. Altmann, R. Strasser, L. Mach, M. Schähs, R. Kunert et al. 2008. A plant-derived human monoclonal antibody induces an anti-carbohydrate immune response in rabbits. Glycobiology 18: 235–241.

Kajikawa, M., K. Sasaki, Y. Wakimoto, M. Toyooka, T. Motohashi, T. Shimojima et al. 2009. Efficient silkworm expression of human GPCR (nociceptin receptor) by a *Bombyx mori* bacmid DNA system. Biochem. Biophys. Res. Commun. 385: 75–79.

Kajiura, H., Y. Hamaguchi, H. Mizushima, R. Misaki and K. Fujiyama. 2015. Sialylation potentials of the silkworm, *Bombyx mori*; *B. mori* possesses an active α2,6-sialyltransferase. Glycobiology 25: 1441–1453.

Kato, T., M. Kajikawa, K. Maenaka and E.Y. Park. 2010. Silkworm expression system as a platform technology in life science. Appl. Microbiol. Biotechnol. 85: 459–470.

Kato, T., N. Kako, K. Kikuta, T. Miyazaki, S. Kondo, H. Yagi et al. 2017a. *N*-Glycan modification of a recombinant protein via coexpression of human glycosyltransferases in silkworm pupae. Sci. Rep. 7: 1409.

Kato, T., K. Kikuta, A. Kanematsu, S. Kondo, H. Yagi, K. Kato et al. 2017b. Alteration of a recombinant protein *N*-glycan structure in silkworms by partial suppression of *N*-acetylglucosaminidase gene expression. Biotechnol. Lett. 39: 1299–1308.

Kawar, Z., K. Karaveg, K.W. Moremen and D.L. Jarvis. 2001. Insect cells encode a class II α-mannosidase with unique properties. J. Biol. Chem. 276: 16335–16340.

Kim, B.T., K. Tsuchida, J. Lincecum, H. Kitagawa, M. Bernfield and K. Sugahara. 2003. Identification and characterization of three *Drosophila melanogaster* glucuronyltransferases responsible for the synthesis of the conserved glycosaminoglycan-protein linkage region of proteoglycans. Two novel homologs exhibit broad specificity toward oligosaccharides from proteoglycans, glycoproteins, and glycosphingolipids. J. Biol. Chem. 278: 9116–9124.

Kim, Y.K., K.R. Kim, D.G. Kang, S.Y. Jang, Y.H. Kim and H.J. Cha. 2009. Suppression of β-*N*-acetylglucosaminidase in the *N*-glycosylation pathway for complex glycoprotein formation in *Drosophila* S2 cells. Glycobiology 19: 301–308.

Kim, Y.K., K.R. Kim, D.G. Kang, S.Y. Jang, Y.H. Kim and H.J. Cha. 2011. Expression of β-1,4-galactosyltransferase and suppression of β-*N*-acetylglucosaminidase to aid synthesis of complex *N*-glycans in insect *Drosophila* S2 cells. J. Biotechnol. 153: 145–152.

Kimura, M., Y. Kimura, K. Tsumura, K. Okihara, H. Sugimoto, H. Yamada et al. 2003. 350-kDa royal jelly glycoprotein (apisin), which stimulates proliferation of human monocytes, bears the β1-3galactosylated *N*-glycan: analysis of the *N*-glycosylation site. Biosci. Biotechnol. Biochem. 67: 2055–2058.

Koles, K., K.D. Irvine and V.M. Panin. 2004. Functional characterization of *Drosophila* sialyltransferase. J. Biol. Chem. 279: 4346–4357.

Koles, K., J.M. Lim, K. Aoki, M. Porterfield, M. Tiemeyer, L. Wells et al. 2007. Identification of *N*-glycosylated proteins from the central nervous system of *Drosophila melanogaster*. Glycobiology 17: 1388–1403.

Koprivova, A., C. Stemmer, F. Altmann, A. Hoffmann, S. Kopriva, G. Gorr et al. 2004. Targeted knockouts of *Physcomitrella* lacking plant-specific immunogenic *N-glycans*. Plant Biotechnol. J. 2: 517–523.

Krainer, F.W., C. Gmeiner, L. Neutsch, M. Windwarder, R. Pletzenauer, C. Herwig et al. 2013. Knockout of an endogenous mannosyltransferase increases the homogeneity of glycoproteins produced in *Pichia pastoris*. Sci. Rep. 3: 3279.

Kurz, S., K. Aoki, C. Jin, N.G. Karlsson, M. Tiemeyer, I.B. Wilson et al. 2015. Targeted release and fractionation reveal glucuronylated and sulphated N- and O-glycans in larvae of dipteran insects. J. Proteomics 126: 172–188.

Léonard, R., D. Rendic, C. Rabouille, I.B. Wilson, T. Préat and F. Altmann. 2006. The *Drosophila* fused lobes gene encodes an *N*-acetylglucosaminidase involved in *N*-glycan processing. J. Biol. Chem. 281: 4867–4875.

Lombard, V., H. Golaconda Ramulu, E. Drula, P.M. Coutinho and B. Henrissat. 2014. The carbohydrate-active enzymes database (CAZy) in 2013. Nucleic Acids Res. 42: D490–95.

Mabashi-Asazuma, H., X. Shi, C. Geisler, C.W. Kuo, K.H. Khoo and D.L. Jarvis. 2013. Impact of a human CMP-sialic acid transporter on recombinant glycoprotein sialylation in glycoengineered insect cells. Glycobiology 23: 199–210.

Mabashi-Asazuma, H., B.H. Sohn, Y.S. Kim, C.W. Kuo, K.H. Khoo, C.A. Kucharski et al. 2015. Targeted glycoengineering extends the protein *N*-glycosylation pathway in the silkworm silk gland. Insect Biochem. Mol. Biol. 65: 20–27.

Maeda, S., T. Kawai, M. Obinata, H. Fujiwara, T. Horiuchi, Y. Saeki et al. 1985. Production of human alpha-interferon in silkworm using a baculovirus vector. Nature 315: 592–594.

Mertsalov, I.B., B.N. Novikov, H. Scott, L. Dangott and V.M. Panin. 2016. Characterization of *Drosophila* CMP-sialic acid synthetase activity reveals unusual enzymatic properties. Biochem. J. 473: 1905–1916.

Misaki, R., H. Nagaya, K. Fujiyama, I. Yanagihara, T. Honda and T. Seki. 2003. N-linked glycan structures of mouse interferon-β produced by *Bombyx mori* larvae. Biochem. Biophys. Res. Commun. 311: 979–986.

Mita, K., M. Kasahara, S. Sasaki, Y. Nagayasu, T. Yamada, H. Kanamori et al. 2004. The genome sequence of silkworm, *Bombyx mori*. DNA Res. 11: 27–35.

Miyazaki, T., Y. Matsumoto, K. Matsuda, Y. Kurakata, I. Matsuo, Y. Ito et al. 2011. Heterologous expression and characterization of processing α-glucosidase I from *Aspergillus brasiliensis* ATCC 9642. Glycoconj. J. 28: 563–571.

Miyazaki, T., M. Ishizaki, H. Dohra, S. Park, A. Terzic, T. Kato et al. 2017. Insulin-like peptide 3 expressed in the silkworm possesses intrinsic disulfide bonds and full biological activity. Sci. Rep. 7: 17339.

Miyazaki, T., T. Kato and E.Y. Park. 2018. Heterologous expression, purification and characterization of human β-1,2-*N*-acetylglucosaminyltransferase II using a silkworm-based *Bombyx mori* nucleopolyhedrovirus bacmid expression system. J. Biosci. Bioeng. 126: 15–22.

Miyazaki, T., R. Miyashita, S. Mori, T. Kato and E.Y. Park. 2019. Expression and characterization of silkworm *Bombyx mori* β-1,2-*N*-acetylglucosaminyltransferase II, a key enzyme for complex-type *N*-glycan biosynthesis. J. Biosci. Bioeng. 127: 273–280.

Moremen, K.W., R.B. Trimble and A. Herscovics. 1994. Glycosidases of the asparagine-linked oligosaccharide processing pathway. Glycobiology 4: 113–125.

Moremen, K.W., M. Tiemeyer and A.V. Nairn. 2012. Vertebrate protein glycosylation: diversity, synthesis and function. Nat. Rev. Mol. Cell Biol. 13: 448–462.

Motohashi, T., T. Shimojima, T. Fukagawa, K. Maenaka and E.Y. Park. 2005. Efficient large-scale protein production of larvae and pupae of silkworm by *Bombyx mori* nuclear polyhedrosis virus bacmid system. Biochem. Biophys. Res. Commun. 326: 564–569.

Nagasu, T., Y. Shimma, Y. Nakanishi, J. Kuromitsu, K. Iwama, K. Nakayama et al. 1992. Isolation of new temperature-sensitive mutants of *Saccharomyces cerevisiae* deficient in mannose outer chain elongation. Yeast 8: 535–547.

Nagata, Y., J.M. Lee, H. Mon, S. Imanishi, S.M. Hong, S. Komatsu et al. 2013. RNAi suppression of β-*N*-acetylglucosaminidase (BmFDL) for complex-type *N*-linked glycan synthesis in cultured silkworm cells. Biotechnol. Lett. 35: 1009–1016.

Nakajima, M., T. Kato, S. Kanamasa and E.Y. Park. 2009. Molecular chaperone-assisted production of human α-1,4-*N*-acetylglucosaminyltransferase in silkworm larvae using recombinant BmNPV bacmids. Mol. Biotechnol. 43: 67–75.

Nemčovičová, I., S. Šesták, D. Rendić, M. Plšková, J. Mucha and I.B. Wilson. 2013. Characterisation of class I and II α-mannosidases from *Drosophila melanogaster*. Glycoconj. J. 30: 899–909.

Nomura, T., M. Ikeda, S. Ishiyama, K. Mita, T. Tamura, T. Okada et al. 2010. Cloning and characterization of a β-*N*-acetylglucosaminidase (BmFDL) from silkworm *Bombyx mori*. J. Biosci. Bioeng. 110: 386–391.

Nomura, T., M. Suganuma, Y. Higa, Y. Kataoka, S. Funaguma, H. Okazaki et al. 2015. Improvement of glycosylation structure by suppression of β-*N*-acetylglucosaminidases in silkworm. J. Biosci. Bioeng. 119: 131–136.

Ogata, M., M. Nakajima, T. Kato, T. Obara, H. Yagi, K. Kato et al. 2009. Synthesis of sialoglycopolypeptide for potentially blocking influenza virus infection using a rat α2,6-sialyltransferase expressed in BmNPV bacmid-injected silkworm larvae. BMC Biotechnol. 9: 54.

Park, E.Y., A. Kageshima, M.S. Kwon and T. Kato. 2007. Enhanced production of secretory β1,3-*N*-acetylglucosaminyltransferase 2 fusion protein into hemolymph of *Bombyx mori* larvae using recombinant BmNPV bacmid integrated signal sequence. J. Biotechnol. 129: 681–688.

Park, E.Y., T. Abe and T. Kato. 2008. Improved expression of fusion protein using a cysteine-protease- and chitinase-deficient *Bombyx mori* (silkworm) multiple nucleopolyhedrovirus bacmid in silkworm larvae. Biotechnol. Appl. Biochem. 49: 135–140.

Park, E.Y., M. Ishikiriyama, T. Nishina, T. Kato, H. Yagi, K. Kato et al. 2009. Human IgG1 expression in silkworm larval hemolymph using BmNPV bacmids and its *N*-linked glycan structure. J. Biotechnol. 139: 108–114.

Paschinger, K., E. Staudacher, U. Stemmer, G. Fabini and I.B. Wilson. 2005. Fucosyltransferase substrate specificity and the order of fucosylation in invertebrates. Glycobiology 15: 463–474.

Rosenbaum, E.E., E. Vasiljevic, K.S. Brehm and N.J. Colley. 2014. Mutations in four glycosyl hydrolases reveal a highly coordinated pathway for rhodopsin biosynthesis and N-glycan trimming in *Drosophila melanogaster*. PLoS Genet. 10: e1004349.

Sarkar, M., P.A. Leventis, C.I. Silvescu, V.N. Reinhold, H. Schachter and G.L. Boulianne. 2006. Null mutations in *Drosophila N*-acetylglucosaminyltransferase I produce defects in locomotion and a reduced life span. J. Biol. Chem. 281: 12776–12785.

Sasaki, K., M. Kajikawa, K. Kuroki, T. Motohashi, T. Shimojima, E.Y. Park et al. 2009. Silkworm expression and sugar profiling of human immune cell surface receptor, KIR2DL1. Biochem. Biophys. Res. Commun. 387: 575–580.

Stanton, R., A. Hykollari, B. Eckmair, D. Malzl, M. Dragosits, D. Palmberger et al. 2017. The underestimated N-glycomes of lepidopteran species. Biochim. Biophys. Acta Gen. Subj. 1861: 699–714.

Staudacher, E. and L. März. 1998. Strict order of (Fuc to Asn-linked GlcNAc) fucosyltransferases forming core-difucosylated structures. Glycoconj. J. 15: 355–360.

Striebeck, A., D.A. Robinson, A.W. Schüttelkopf and D.M. van Aalten. 2013. Yeast Mnn9 is both a priming glycosyltransferase and an allosteric activator of mannan biosynthesis. Open Biol. 3: 130022.

Suganuma, M., T. Nomura, Y. Higa, Y. Kataoka, S. Funaguma, H. Okazaki et al. 2018. *N*-glycan sialylation in a silkworm-baculovirus expression system. J. Biosci. Bioeng. 126: 9–14.

Takahashi, N. and K. Kato. 2003. GALAXY (Glycoanalysis by the three axes of MS and chromatography): a web application that assists structural analyses of *N*-glycans. Trends Glycosci. Glycotechnol. 15: 235–251.

Tomiya, N., D. Howe, J.J. Aumiller, M. Pathak, J. Park, K.B. Palter et al. 2003. Complex-type biantennary *N*-glycans of recombinant human transferrin from *Trichoplusia ni* insect cells expressing mammalian β-1,4-galactosyltransferase and β-1,2-*N*-acetylglucosaminyltransferase II. Glycobiology 13: 23–34.

Vadaie, N. and D.L. Jarvis. 2004. Molecular cloning and functional characterization of a Lepidopteran insect β4-*N*-acetylgalactosaminyltransferase with broad substrate specificity, a functional role in glycoprotein biosynthesis, and a potential functional role in glycolipid biosynthesis. J. Biol. Chem. 279: 33501–33518.

van Die, I., A. van Tetering, H. Bakker, D.H. van den Eijnden and D.H. Joziasse. 1996. Glycosylation in lepidopteran insect cells: identification of a β1→4-*N*-acetylgalactosaminyltransferase involved in the synthesis of complex-type oligosaccharide chains. Glycobiology 6: 157–164.

Varki, A. 1993. Biological roles of oligosaccharides: all of the theories are correct. Glycobiology 3: 97–130.

Viswanathan, K., N. Tomiya, J. Park, S. Singh, Y.C. Lee, K. Palter et al. 2006. Expression of a functional *Drosophila melanogaster* CMP-sialic acid synthetase. Differential localization of the *Drosophila* and human enzymes. J. Biol. Chem. 281: 15929–15940.

Walski, T., K. De Schutter, E.J.M. Van Damme and G. Smagghe. 2017. Diversity and functions of protein glycosylation in insects. Insect Biochem. Mol. Biol. 83: 21–34.

Watanabe, S., T. Kokuho, H. Takahashi, M. Takahashi, T. Kubota and S. Inumaru. 2002. Sialylation of *N*-glycans on the recombinant proteins expressed by a baculovirus-insect cell system under β-*N*-acetylglucosaminidase inhibition. J. Biol. Chem. 277: 5090–5093.

Wildt, S. and T.U. Gerngross. 2005. The humanization of N-glycosylation pathways in yeast. Nat. Rev. Microbiol. 3: 119–128.

Xia, Q., Z. Zhou, C. Lu, D. Cheng, F. Dai, B. Li et al. 2004. A draft sequence for the genome of the domesticated silkworm (*Bombyx mori*). Science 306: 1937–1940.

Yagi, H., N. Takahashi, Y. Yamaguchi, N. Kimura, K. Uchimura, R. Kannagi et al. 2005. Development of structural analysis of sulfated *N*-glycans by multidimensional high performance liquid chromatography mapping methods. Glycobiology 15: 1051–1060.

Yamamoto-Hino, M., H. Yoshida, T. Ichimiya, S. Sakamura, M. Maeda, Y. Kimura et al. 2015. Phenotype-based clustering of glycosylation-related genes by RNAi-mediated gene silencing. Genes Cells 20: 521–542.

Zhu, F., D. Li and K. Chen. 2019. Structures and functions invertebrate glycosylation. Open Biol. 9: 180232.

Chapter 21

Detection of Infectious Viruses using Advanced Nanobiotechnology for Green Society

Ankan Dutta Chowdhury and *Enoch Y. Park**

1. Introduction

Infectious viruses have been recognized as a deadly threat for mankind since the first civilization established on earth. After advancement of several technologies and prevention methods, there have still been several examples of serious virus outbreaks every year around the world, especially from high populated countries like India, China and North Africa's region (Calder et al. 2003, Carroll et al. 2015, Cotterelle et al. 2005, Hjertqvist et al. 2006, Rahman et al. 2012, Yadav et al. 2012). There are considerable fears that such kinds of outbreaks can rapidly spread worldwide to become epidemics with devastating effects on populations and their social and economic impact. To achieve the goal, UNESCO's latest vision encourages us to counter the above challenges by providing a 'new way forward' through the development of an inclusive green society (Didham and Paul 2015). This report covers the main priority area of health safety by the detection of virus methodologies and its recent progress for the transition to a green society. Several significant ways to the techniques for analyzing different virus samples are discussed here.

To prevent the viral outbreak, the rapid identification and quantification of contagious virus in all possible samples as well as in food and water is of great significance in prompt treatment and effective management for illness. Conventional virus detection methods which are the most reliable method of diagnosis yet, such as Polymerase Chain Reaction (PCR) and branched-chain DNA test are not well-suited for point-of-care diagnosis as they are time-consuming, expensive and also require intensive sample preparation with high

Laboratory of Biotechnology, Research Institute of Green Science and Technology, Shizuoka University, Shizuoka, 422-8529, Japan.
Email: dc_ankan@yahoo.co.in
* Corresponding author: park.enoch@shizuoka.ac.jp

skilled technicians (Lau et al. 2004, Saiki et al. 1988). This has been the key inspiration behind the increased demand for the development of an alternative method of virus detection, recently reflected in scientific literatures. To achieve such a goal, scientists have been trying to use the new horizons of nanobiotechnologies in the virus detection approach since the last two decades. After emerging the nanotechnology where the application is mainly concerned with extremely small things in nano-dimension with at least one dimension must have less than 100 nm in size, can be used across all the other scientific fields, such as chemistry (Chowdhury et al. 2012), biology (Chowdhury et al. 2014b, Sarikaya et al. 2003), physics (Hyeon et al. 2015), materials science and engineering (Hu and Niemeyer 2019, Khademhosseini et al. 2017). The properties of many bulk materials change drastically when formed as a nanoparticle which is typical because nanoparticles have a greater surface area per weight ratio than bulk particles which causes them to be more interactive toward other molecules (Gottschalk et al. 2013). Applying this phenomenon of nanoparticles, several approaches for ultrasensitive detection of biomarkers or pathogens have been reported based on fluorescence (Dutta Chowdhury et al. 2017, Ganganboina et al. 2017), electrochemistry (Chowdhury et al. 2014a, Chowdhury et al. 2018a, Chowdhury et al. 2018b), electrochemiluminescence (Larson et al. 2016), Surface Enhanced Raman Scattering (SERS) (Fan et al. 2014), etc. Proper blend of nanotechnologies with the conventional techniques for pathogen detection can make some extremely useful biosensors which can be served for the practical application, we need.

Biosensors are portable devices which are used to detect the presence or concentration of any biological analytes such as a biomolecule, a biological structure, different pathogens even virus particles in a rapid process. It consists of three distinct parts: a component that specifically recognizes the analyte and produces signal, a transducer which can convert the signal to a detectable one and a reader device (Cao et al. 2014, Scheller et al. 2001). It can enable sensors to detect very small amounts of analytes using various types of detecting elements, such as carbon matrixes (Ahmed et al. 2017, Wang et al. 2013b), noble metal nanoparticles (Ahmed et al. 2016, Draz et al. 2018), metal oxide nanowires (Pang et al. 2015, Xiao et al. 2012) or organic nano-framework (Lei et al. 2014) which can be used in nanotechnology-based biosensors. Due to the small dimension of nanotubes or nanoparticles a few target molecules are sufficient to change the chemical properties of the sensing elements and transduce to the detecting signals. Though there are plenty of researches already published in last two decades on biosensing for virus detection however very few are commercialized as yet. Keeping in the mind the necessity and viral evolution, the journey still needs to continue to reach the desired goal for green society.

Generally, the virus detection targets can be easily categorized into two sections. Firstly, the direct method; where the whole virus particle or its extracted genome or structural protein can be targeted as analyte. As in some cases, few virus particles of less than 100, can also be sufficient for fatality, it is extremely difficult to build such a kind of technology to detect direct analytes. In that case, indirect methods are relatively easier to develop. As the target analytes are collected from the generated antibodies of infected body, the target concentration always is sufficient for analysis. However, in this case for quantification of the disease, selectivity becomes a crucial problem to overcome due to the obvious cross-reactivity with other infectious viruses of similar family. Therefore, false positive or negative results are quite consistent for those rapid kits. In brief, the scenario can be represented in the Fig. 21.1 where the reliability and the difficulty both are proportionally increasing from the indirect sensing to direct virus particle method.

Over these past two decades, scientists are trying to overcome the limitations in both ways and published several methods of virus detection. On summarizing the whole

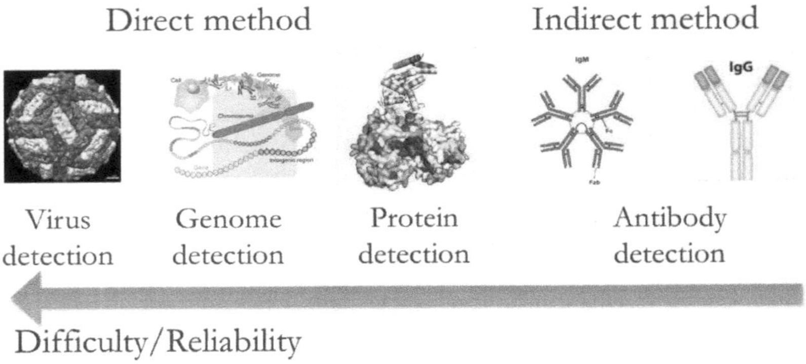

Fig. 21.1: Target for virus detection for sensing: direct targets from virus molecules and indirect targets from the infected bodies.

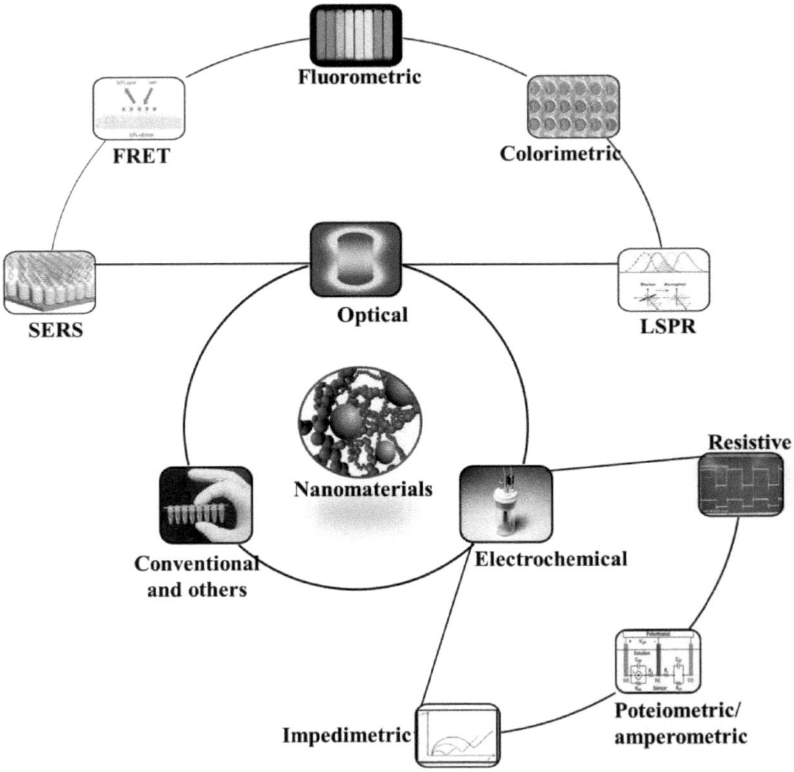

Fig. 21.2: Application of nanotechnology on virus detection: Beside the conventional virus detection protocols, the nanotechnology has been assisted two main detection protocols of optical and electrochemical.

Color version at the end of the book

scenario, all detection principles can be designated into three different ways as depicted in Fig. 21.2; (1) the conventional methods of PCR (Payungporn et al. 2006), Nucleic Acid Sequence Based Amplification (NASBA) (Collins et al. 2003), ELISA (Boonham et al.

2014), etc., based where the protocols are tuned and upgraded through some innovative and incremental studies. However, the limitations of time consuming and high cost still remain a challenging task for their real application. (2) The most extensively investigated method of optical studies which can be sub-sectioned in five main parts of absorption based colorimetric (Khoris et al. 2019), fluorometric (Lee et al. 2017), Foster Resonance Electron Transfer (FRET)-based fluorometric (Stanisavljevic et al. 2015), Localized Surface Plasmon Resonance (LSPR)-based fluorometric (Nasrin et al. 2018) and Surface Enhanced Raman Spectroscopic (SERS) (Durmanov et al. 2018). (3) The electrochemical sensing mainly comprised by the resistive (Nuzaihan et al. 2016), amperometric or potentiometric (Nuzaihan et al. 2016) and impedimetric methods (Jarocka et al. 2014).

This chapter discusses the recent advancement of nanobiotechnology for the application of virus detection. Among several methods of detection, we have emphasized mostly well studied parts including colorimetric, fluorometric, electrochemical and SERS-based techniques. The synthesis of the nanomaterials and their design for the biosensor has been discussed in brief. In addition, the advancement of these methods towards the point of care diagnosis, also been discussed at the end of this chapter.

2. Detection of Virus

2.1 Optical

Optical biosensors are the most common type of biosensors due to their simplicity. Here we provide a brief classification and description of their mechanism and their applications for virus detection. The main focus is placed on the mostly used fluorometric and colorimetric biosensors including LSPR, FRET and SERS. Colorimetric assay shows obvious superiority among all other sensing applications for its simplicity, convenience, cost effectiveness and the possibility to be read out by naked eyes. However, the relatively high detection limit of these conventional colorimetric biosensors restricts their application in low level virus concentration. Noble metal nanoparticles such as gold nanoparticle (AuNP), silver nanoparticle (AgNP) have attracted great interest due to their excellent physical and chemical properties, particularly in the visible range (Holme et al. 2018, Khoris et al. 2019). Of all the kinds of noble metal nanomaterials, AuNPs with different nano-dimensional structures like nanorod, nanosphere, core-shell, etc., show great prospective for the assembly of different kinds of biosensors due to their highly controllable shape and tunable LSPR peak from the visible to near-infrared range (Zhao et al. 2008). AgNP also has been used in various applications with or without AuNPs, such as SERS detection (Xu et al. 2004), mass spectrometry (Liu et al. 2011) and TMB-H_2O_2 enzymatic reaction (Khoris et al. 2019). In a recent study, highly enhanced peroxidase-like activity for the detection of Norovirus was obtained using silver-coated AuNP core/shell nanohybrid structure with hydrogen peroxide and 3,3′,5,5′-tetramethylbenzidine (TMB) as the substrate and chromogenic agent, respectively (Khoris et al. 2019). The H_2O_2/TMB solution was used to develop an intense blue color due to the synergistic effect of the peroxidase-like activity with Au/Ag nano-hybrid, as shown in Fig. 21.3a. The detection limit was found 13.2 copies mL^{-1}. In a similar type of study, Ahmed et al. (2017) described an AuNP based nanocomposites mixed with graphene sheets (Grp-AuNP) which was used as a robust nanoprobe for the detection of Norovirus particles through the superior peroxidase-like activity. Though these methods are quite satisfactory in the UV-visible measuring tool, however, they regretfully fail to show distinct color changes to the naked eye. To improve the visual detection, it is

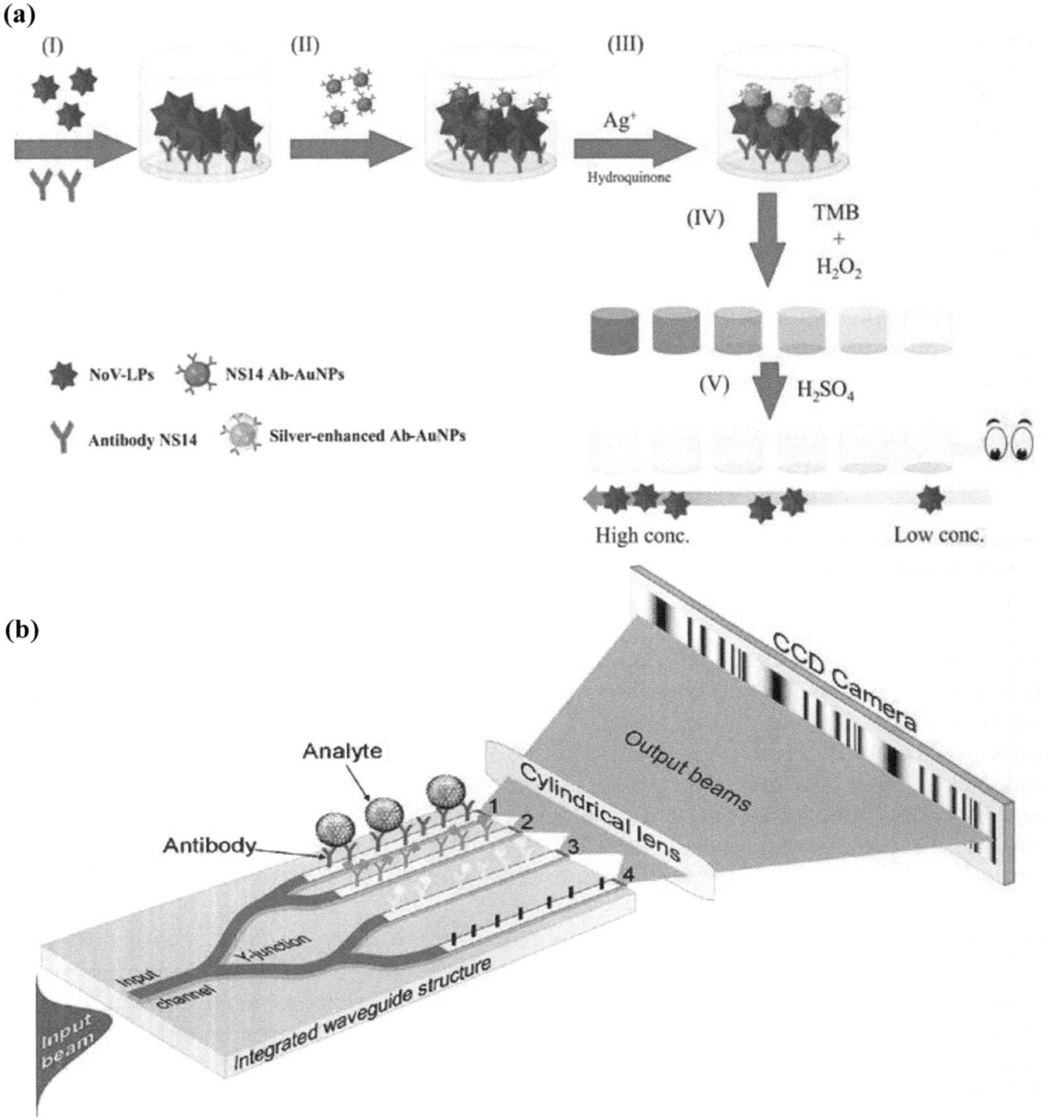

Fig. 21.3: (a) Schematic illustration of silver-enhanced nanozyme-based AuNP immunoassay for the detection of Norovirus, (b) Schematic of the four-channel integrated optical sensor: 1, 2, and 3 indicate the measuring channels, and 4 is the reference channel.

suggested to develop an immunoassay approach that shows multicolor changes. In this regard, De La Rica and Stevens (2013) used hydrogen peroxide as the reductant for the growth of AuNP, and found a detection limit of 10–18 g mL^{-1} of a target protein. Zhang et al. (2016) fabricated a gold nanoparticle-aggregation based biosensor for the detection of influenza virus A/H7N9. The detection limit was found as low as 25 pg mL^{-1}. All these approaches exhibit good sensitivity in buffer or ultrapure water samples however not in the complex matrix like serum. In addition, the accuracy for the determination of targets with the naked eyes still is a big challenge. Therefore, in spite of their excellent portability, the relatively lower stability in harsh medium and the high affinity towards sulfurated molecules of interfering substances of noble metal nanoparticles may be limiting the application of these colorimetric sensing.

Recently, Ymeti et al. (2007) described a new way of optical sensing using four different channels that can be used to monitor different analytes in a single reaction by coating the channels with their corresponding antibodies. One channel has been used for the reference and other three for different analytes (Fig. 21.3b). Specific binding of analyte to the antibody-coated sensing surface which is probed by the guided modes causes a corresponding phase change. The change is measured as the interference pattern. Though this type of sensing process needs highly sophisticated instruments, however proper improvement of this system can make a potential platform for precise virus detection in the near future.

2.2 Fluorometric

After incorporation of nanotechnology in biosensing during last two decades, an exponential growth has been observed for the direct, real-time and label-free optical biosensors of many chemical and biological substances. However, there are very few reports for the practical applications. There are a huge number of variations in the construction of optical biosensors and this chapter will focus on a few that have been selected on the basis of their application on virus with relatively new techniques. Fluorometric assays are the most applied methods in optical sensing with a lot of variety of proposed techniques due to its easy methodologies and reliable outcomes. Generally, analyzed parameters are being used as fluorescence intensity and its quenching or enhancement efficiency and turn off or regeneration of fluorescence energy transfer (Mehrotra 2016). Several nanomaterials have already been introduced and applied, such as metal nanoparticle (Pang et al. 2015), metallic or graphitic quantum dot (Adegoke et al. 2016a, Adegoke and Park 2016), magnetic nanoparticle (Oh et al. 2018), and carbon-based nanomaterial like carbon nanotubes (Ahmed et al. 2016), graphene (Dutta Chowdhury et al. 2018), etc. To enhance the fluorescence behavior or sensitivity of the sensor successful hybrid nanomaterials have been also reported including core-shell or alloyed nanoparticle and modified carbon nanomaterials. Proper blending of nanocomposite has been created often which possess the merits of their organic and inorganic components, and hence exhibit new or multiple functionalities that one component does not have. For the biosensor, these properties of nanocomposites plays the key role as the signal transduces the interactions between the specific nanomaterial and its corresponding target virus which induces the changes in the properties of the nanomaterial, resulting in the detection of the specific target virus. Adegoke et al. (2016b) demonstrated a SiO_2-coated highly fluorescent CdZnSeS inorganic QD, conjugated molecular beacon as a probe for the detection of extremely low concentrations of Norovirus RNA of 8.2 copies mL^{-1}. Similarly, Liang et al. (2017) described an 'off-on' biosensor using mitoxantrone as a quencher of the CdTe QD's fluorescence via an electron transfer mechanism as depicted in Fig. 21.4a, switching the red fluorescence into an "off" state. When the mitoxantrone intercalated into the double helix structure of DNA, it was peeled off the CdTe QD surface, resulting in the recovery of QD's fluorescence. They also reported high sensitivity with the Limit Of Detection (LOD) of 3.038 nM. In another study, Tian et al. (2012) aimed to construct a FRET-based sensing platform using QDs-ssDNA/CNTs inherent properties of CdTe QDs and CNTs for the detection of H5N1 influenza virus (Fig. 21.4b). Here, the CdTe QDs-labeled ssDNA was used as the donor molecule whereas the CNT was used

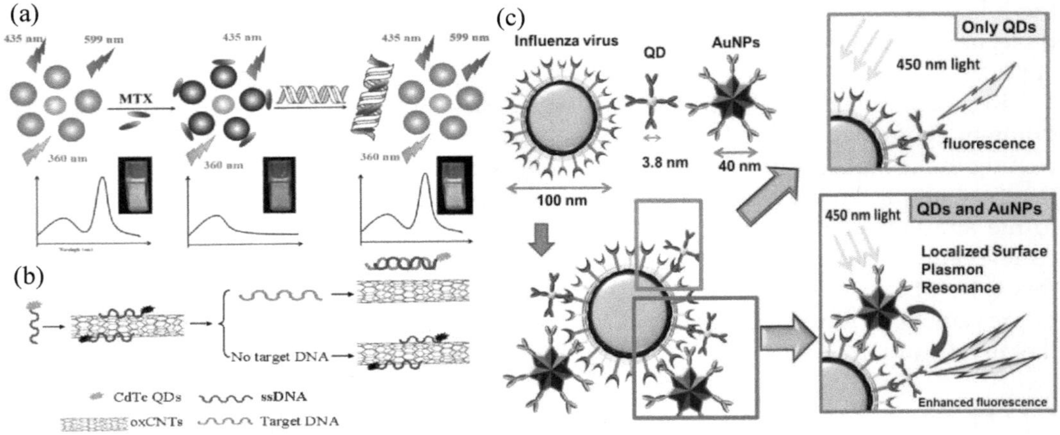

Fig. 21.4: (a) Schematic diagram of the CdTe–CDs–MTX system, (b) FRET-based platform of CNT and CdTe QDs for the influenza H5N1 detection, (c) Detection principle of the influenza virus using the LSPR-induced fluorescence biosensor.

as the acceptor in FRET couple. They confirmed that the fluorescence of the QDs-ssDNA probe is quenched by CNTs initially which is recovered proportionally with respect to the concentration of target DNA by forming hybridized dsDNA and free QDs. Takemura et al. (2017) reported a rapid and ultrasensitive detection of H1N1 influenza virus using LSPR induced optical transduction from AuNP-labeled antibody to adjacent alloyed CdSeTeS QD. The mechanism of the biosensor has been schematically presented in Fig. 21.4c where an antigen-antibody interaction has shown in which LSPR induced from AuNPs is used to enhance the fluorescence of adjacent CdSe QDs for the detection of the influenza virus. To get the maximum specificity, they used two different antibodies to capture the virus molecule from both sides as a sandwich assay, resulting in the LSPR induced fluorescent enhancement.

In recent years, carbon-based graphene quantum dots have emerged as a potential substitute of these environmentally toxic inorganic quantum dots. Dengue virus has four serotypes, and researches have made a few attempts to identify them with high sensitivity. Based on nitrogen, sulfur doped graphene quantum dots with gold nanoparticle nanoassembly (N,S-GQDs@AuNP) where the N,S-GQDs acted as the fluorescent molecule, a new class of identification method of dengue virus serotype was developed by Dutta Chowdhury et al. (2018). Based on the similar principle, Pang et al. (2015) described a new method using anti-rHA aptamers, immobilized onto the surface of Ag@SiO_2 nanoparticles. In this case, Triazole Orange (TO) was used as a fluorescent reporter. In the absence of target recombinant hemagglutinin (rHA) protein, the aptamer kept the single stranded state so that TO was free in the solution and showed almost no fluorescence emission. After addition of the target rHA protein, the aptamer strand formed the guanine-quadruplex with rHA, so that TO could insert into the G-quadruplex and changed to the excited state as for intramolecular twisting. In the intervening time, the intercalated TO have got closer to the surface of the Ag@SiO_2 by the G-quadruplex complex, immobilized on the nanoparticles. As a result, the fluorophore experienced a surface plasmon resonance enhancement, creating more efficient fluorescence emission. The plausible mechanism is represented in Fig. 21.5a. However, in these conventional fluorometric sensors, the applications are

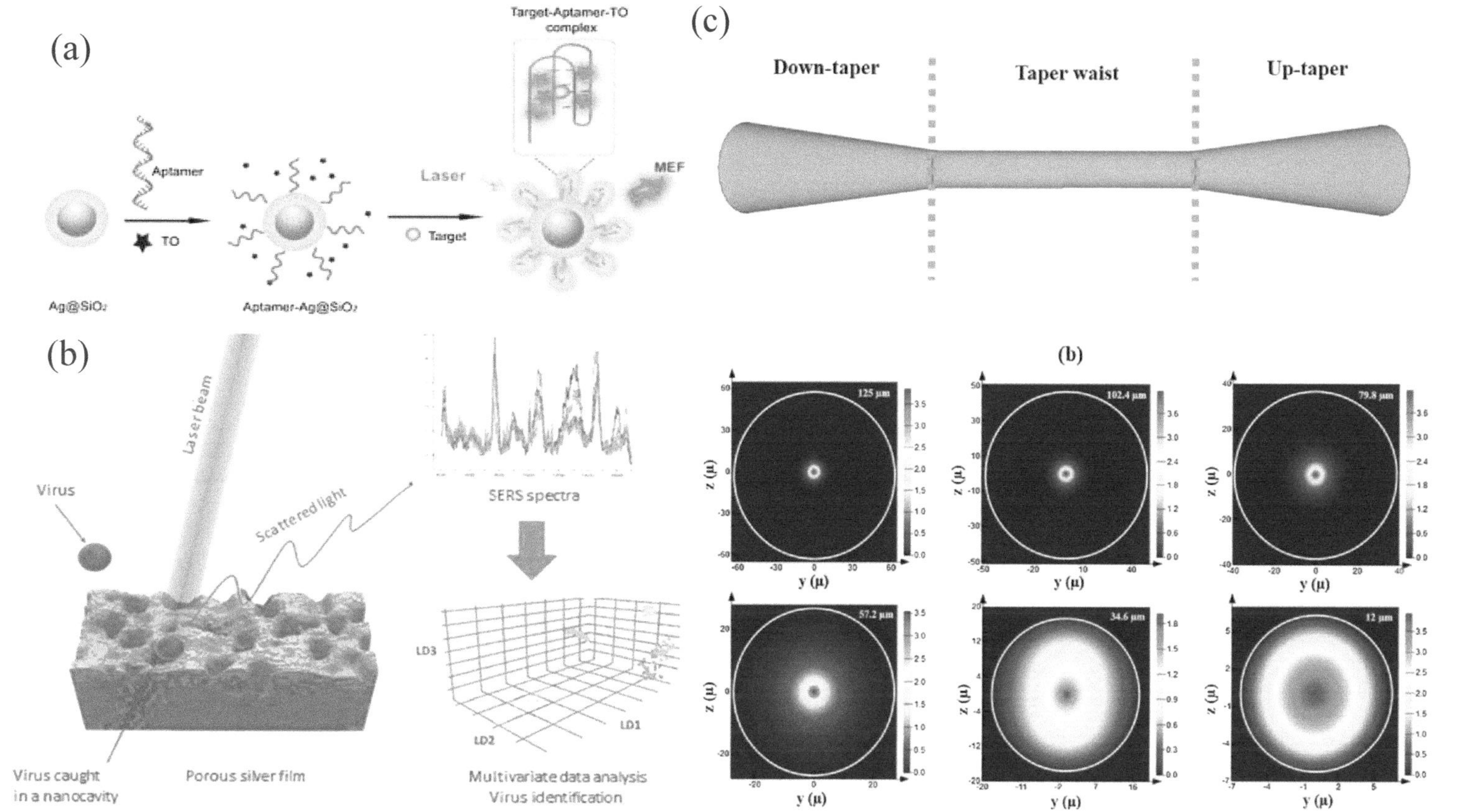

Fig. 21.5: (a) preparation of aptamer-based Ag@SiO_2 biosensor and the detection principle of rHA protein of influenza H5N1 virus, (b) silver based SERS substrate fabrication and its spectral analysis for virus detection, (c) a schematic diagram of a tapered fiber and its modal evolution during transition from non-tapered region to taper waist.

Color version at the end of the book

still restricted in ultrasensitive detection due to its inability to measure extremely small changes of target. In spite of complex methodologies, SERS based technique can be a good alternative using Raman active molecule as probe. Durmanov et al. (2018) described a novel SERS substrate which was tested on its capabilities in acquiring spectra of four viral particles consecutively, as shown in Fig. 21.5b. Two enveloped viruses of Myxoma and Canine Distemper virus and two non-enveloped viruses of Tobacco Mosaic and Potato virus X, were separately tested where the target analytes were entirely different in sizes, morphologies, structural composition and physicochemical properties. The silver substrate contains nanoscale pore-like structures and cavities, separated by rough surface which was proposed for detection of four viruses. Apart from these fluorophore-quencher principles, Kamil et al. (2018) proposed an optical fiber based sensor for the quantitative detection of Dengue E proteins utilizing a functionalized tapered fiber as transducer. According to the detection, as presented in Fig. 21.5c, the tapered fiber with Dengue E protein specific antibodies immobilized onto its silane-coated surface and after the introduction of Dengue E proteins that are specifically bound to the antibodies.

In brief, it can be concluded from the above-mentioned optical biosensors that they can provide new analytical tools for sensitive screening of a wide range of virus samples. Various sensitivity-enhancement techniques for optical biosensors have been already developed with improved sensitivity and optical biosensors have been successfully tested in several fields including the commercialized test strips for home-testing, point-of-care testing or laboratory use, such as home pregnancy kits or glucose monitoring strips. However, to achieve the goal in virus monitoring, these are still under development and limited mostly to academic and pharmaceutical environments.

2.3 Electrochemical

The growing interest in electrochemical methods for virus detection arises from their ability of measurement in high sensitivity with low cost and their excellent applicability with microfabrication. Among the different electrochemical process, mostly used approaches are generally based on the enzyme or antibody labeling in order to generate amplified electrical signals. Depending on the detecting signals, the electrochemical biosensors can be divided in to three clear categories of potentiometric/amperometric where the current or potential are measured, resistive and impedimetric. Though large signal amplification can be attained from the labeling of an oligonucleotide by an enzyme, however the process is not an easy one. Moreover, after preparation, the activity of the conjugated complex must be controlled owing to the poor stability of enzymes. Various electrochemical biosensors have been developed for the detection of viral DNA sequences (Chowdhury et al. 2014a, Chowdhury et al. 2014b). The easy operational instrumental techniques and the simplicity of the fabrication method, DNA sensors always have some advantages over the other methods. In addition, excellent stability in the inhibitory components from environmental and clinical samples makes the electrochemical DNA sensors as an excellent choice for development of an inexpensive point-of-care platform. Manzano et al. (2018) developed an electrochemical DNA biosensor for specific nucleic acid sequence of hepatitis A virus with a LOD of 0.15 fg μL^{-1}. In this case, DNA hybridization was identified by the simple Cyclic Voltammetric (CV) technique, measuring the changes in the difference of peak potential (ΔEp) of the used redox indicator of tripropylamine (TPA), as depicted in Fig. 21.6a. Wang et al. (2013a) demonstrated that a horizontally aligned single walled carbon nanotube (SWCNT) array coated with AuNPs which was a very promising platform for

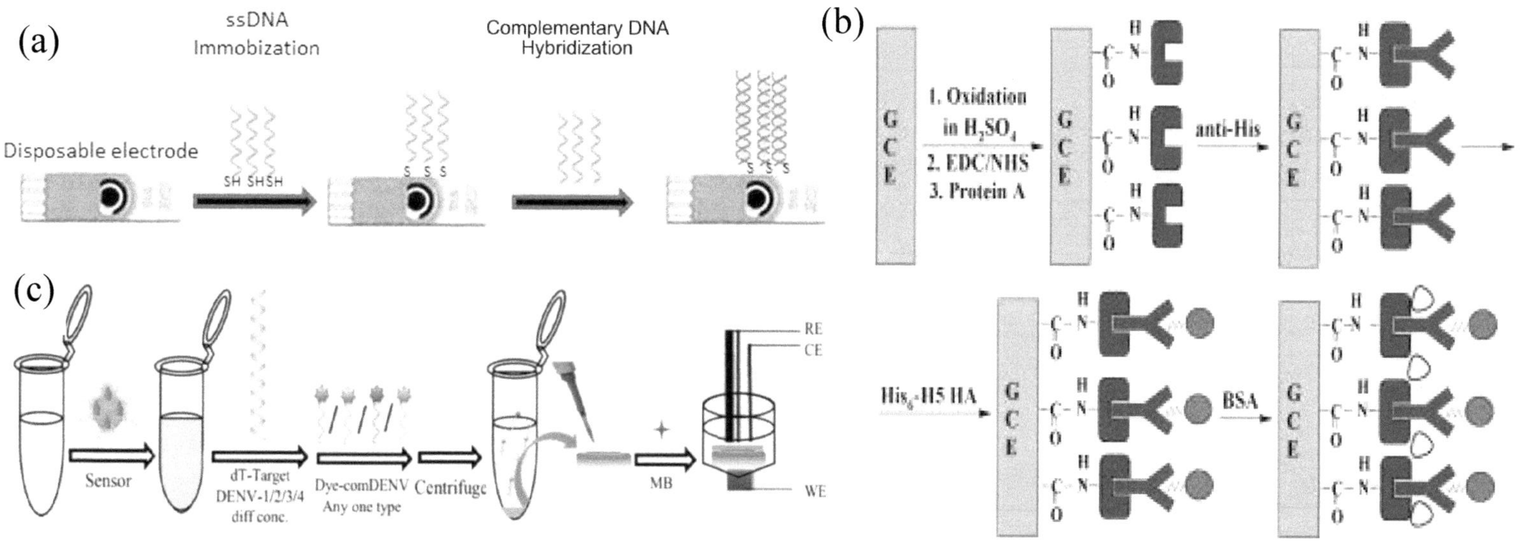

Fig. 21.6: (a) Quantitative detection of hepatitis A virus using specific DNA probe, (b) Steps of immunosensor formation of recombinant His-tagged hemagglutinin antigen for impedimetric immunosensor against avian influenza, (c) electrochemical quantification method of Dengue DNA detection using Au||AuNP@N,S-GQDs-dA nanocomposites.

Color version at the end of the book

DNA detection for Hepatitis B and papilloma virus. They designed a novel sandwich-type electrochemical sensor using Fe_3O_4@GQDs nanohybrid and apoferritin-encapsulated Cu (Cu-apoferritin) as the electrochemical transducer for the detection of avian leukosis virus (Wang et al. 2013b). Jarocka et al. (2014) presented a similar type of immunosensor for the detection of antibodies against hemagglutinin (HA) from highly pathogenic avian influenza virus A/H5N1 with a LOD of 2.1 pg mL^{-1}. Electrochemical impedance spectroscopy has been applied to identify the specific interaction between the recombinant his-tagged hemagglutinin (His-H5 HA) and the appropriate monoclonal antibody where the $[Fe(CN)_6]^{3-/4-}$ has been used as the electroactive signal indicator, presented in Fig. 21.6b. In another report of Dutta Chowdhury et al. (2018), detected multiple serotypes of Dengue RNA by employing Differential Pulse Voltammetric (DPV) method with a linear range of 10^{-14} to 10^{-6} M of DNA, demonstrating the applicability of developed convenient system for clinical diagnosis and biomolecular interaction studies. The intercalation property of methyelene blue redox indicator has been used here as the desired probe for DNA quantification employing the different binding ability of methyelene blue with single and double stranded DNA, as shown in Fig. 21.6c. In some cases, DPV or impedimetric electrochemical sensors have already been applied in a pathological laboratory, however, to extend its application in point of care diagnosis, more attention is still required.

3. Application of Virus Detection in Point of Care Diagnosis

In addition, with less technical expertise, simple, inexpensive and rapid diagnostic kits with simple laboratory equipment are the ultimate goal to prevent epidemic infectious diseases. Though there are thousands of reports published every year in the field of biosensing application, but their extension to point of care diagnosis are rarely reported. In an article of Li et al. (2015), combined combination of hollow channels in the SlipPAD methodology and electrochemical detection has been provided to obtain a single step, full sandwich, quantitative DNA assay for Hepatitis B Virus (HBV) detection. The limit of detection was obtained as low as 85 pM, which is quite satisfactory for early detection of HBV without applying any further DNA amplification process. Nevertheless, this is a very good LOD using a simple and user-friendly device costing just US$0.36 including reagents. On the other hand, to make the portable point-of-care, user friendly diagnostic tools, recently smartphones have appeared as a promising device for the attachment with the optical as well as electrochemical sensors. The Complementary Metal Oxide Semiconductor (CMOS) camera of phone can be used for detecting optical signals, including fluorescence for applications such as isothermal nucleic acid amplification tests (Fig. 21.7a). Priye et al. (2018) presented a smartphone-based technique using CIExyY (chromaticity-luminance) color space to measure the luminance of fluorescent signals arising from nucleic acid amplification targets. The sensitivity was achieved more than an order of magnitude of the traditional RGB intensity-based analysis. Moreover, they applied this chromaticity-luminance method for the detection of Zika and Chikungunya virus RNA, separately by the end point reverse transcription loop-mediated isothermal amplification (RT-LAMP). In a similar study, Wei et al. (2013) described a cell-phone-based field-portable imaging platform to detect a single virus or deeply sub-wavelength objects. Rong et al. (2018) developed a smartphone-based fluorescent Lateral flow immunoassay (LFIA) platform for detection of Zika virus NS1 protein (Fig. 21.7b). Apart from these, recently, motion-based biosensing technologies have been paying significant attention due

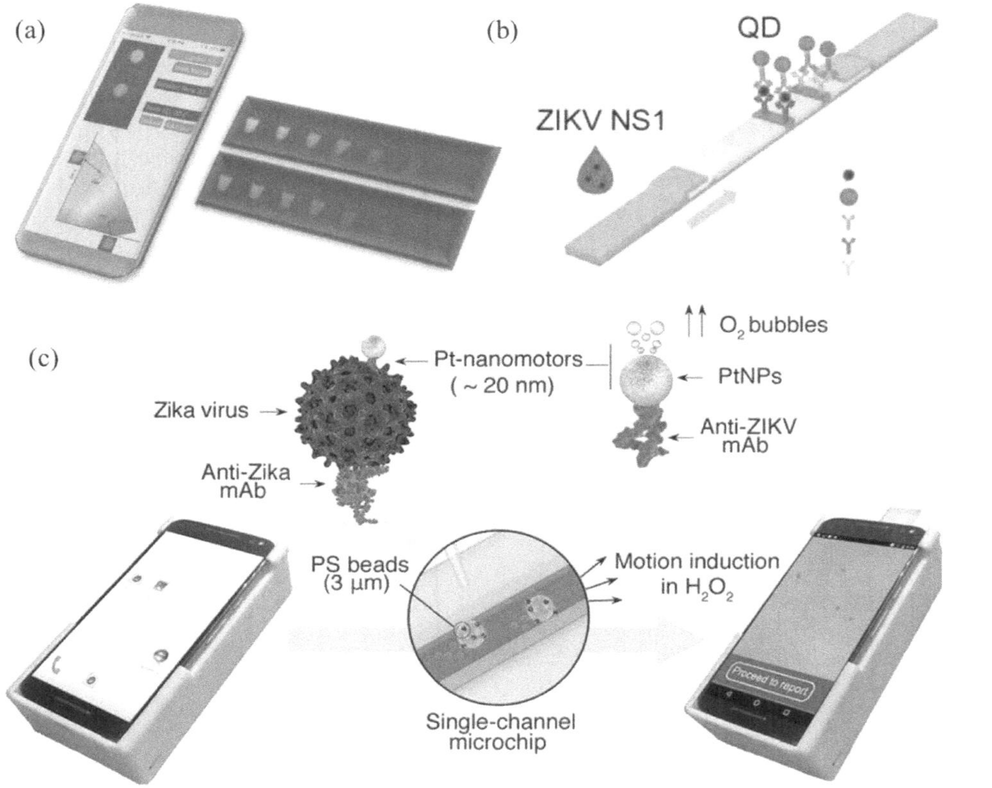

Fig. 21.7: (a) General display of a smart phone based biosensor, (b) detection principle of Zika protein, inside the biosensor strip, (c) motion-based immunological detection of Zika virus using Pt-nanomotors and a smartphone where the presence of the target virus in the testing sample results in the accumulation of platinum nanomotors on the surface of beads.

to the fact that the difference of motion as signal readout is a simple and can offer high resolution and real-time detections. Several reports already have been described using the motion-based signals for the detection of different DNAs and protein as targets. Though the biosensors have demonstrated sufficiently good performances for target detection, most of them depend on the highly sophisticated optical microscopic techniques for tracking the changes in the motion and speed of motor structures in the presence of target analytes. Draz et al. (2018) reported a newly developed method for virus detection through monitoring the catalytic-based motion of nanostructures, as shown in Fig. 21.7c. The virus was captured on the surface of a 3 µm modified bead with anti-Zika virus monoclonal antibody. A Pt-nanomotors was also conjugated with the assembly to form the Pt-Zika-bead nanocomplex. The presence of Pt-nanomotors on the surface of the bead induces the required motion in 15% H_2O_2 solution. The change in the beads velocity can be measured by a Nanomotor-based Bead-motion Cellphone (NBC) system through a motion tracking cellphone application. After proper development in terms of simplification, these types of systems have great potential for usage in the development of point-of-care diagnostics for future analysis.

4. Summary

Nanotechnology has gained a huge importance in biosensing due to the improved detection sensitivity of biosensors through various analytical techniques. In this chapter, nanomaterials with their superior properties in combination with optical and electrochemical methods have discussed with designing of the new and improved sensors for low level virus detection. Despite the instability in complex medium, noble metals like Au, Ag can be used for the colorimetric detection by the naked eye, though plenty of research would be needed for their up gradation for point of care diagnosis. In addition, significant color generation for the naked eye is also a big challenge for their practical application. Whereas the electrochemical and fluorometric assays have already made some advancement to reach their goal. For the fluorometric detection, QDs including carbon-based quantum dots-based sensors are already giving a great contribution to the miniaturization of the biosensors. Paper based electrochemical sensors which are already commercialized have been successful. The fluorometric sensors using smart phones as the detectors have a great potential for the development of cost effective and user-friendly detections, which can lead to affordable tools and perhaps applied in everyday life in the future's green society.

Acknowledgments

A.D.C. sincerely thanks the Japan Society for the Promotion of Science (JSPS) for a postdoctoral fellowship (Grant Nos. P17359).

References

Adegoke, O. and E.Y. Park. 2016. Gold nanoparticle-quantum dot fluorescent nanohybrid: Application for localized surface plasmon resonance-induced molecular beacon ultrasensitive DNA detection. Nanoscale Res. Lett. 11: 523.

Adegoke, O., T. Kato and E.Y. Park. 2016a. An ultrasensitive alloyed near-infrared quinternary quantum dot-molecular beacon nanodiagnostic bioprobe for influenza virus RNA. Biosens. Bioelectron. 80: 483–490.

Adegoke, O., M.-W. Seo, T. Kato, S. Kawahito and E.Y. Park. 2016b. An ultrasensitive SiO_2-encapsulated alloyed CdZnSeS quantum dot-molecular beacon nanobiosensor for norovirus. Biosens. Bioelectron. 86: 135–142.

Ahmed, S.R., J. Kim, T. Suzuki, J. Lee and E.Y. Park. 2016. Enhanced catalytic activity of gold nanoparticle-carbon nanotube hybrids for influenza virus detection. Biosens. Bioelectron. 85: 503–508.

Ahmed, S.R., K. Takemeura, T.C. Li, N. Kitamoto, T. Tanaka, T. Suzuki et al. 2017. Size-controlled preparation of peroxidase-like graphene-gold nanoparticle hybrids for the visible detection of norovirus-like particles. Biosens. Bioelectron. 87: 558–565.

Boonham, N., J. Kreuze, S. Winter, R. van der Vlugt, J. Bergervoet, J. Tomlinson et al. 2014. Methods in virus diagnostics: from ELISA to next generation sequencing. Virus Res. 186: 20–31.

Calder, L., G. Simmons, C. Thornley, P. Taylor, K. Pritchard, G. Greening et al. 2003. An outbreak of hepatitis A associated with consumption of raw blueberries. Epidemiol Infect. 131: 745–751.

Cao, J., T. Sun and K.T. Grattan. 2014. Gold nanorod-based localized surface plasmon resonance biosensors: A review. Sensors Actuators B: Chem. 195: 332–351.

Carroll, M.W., D.A. Matthews, J.A. Hiscox, M.J. Elmore, G. Pollakis, A. Rambaut et al. 2015. Temporal and spatial analysis of the 2014–2015 Ebola virus outbreak in West Africa. Nature 524: 97.

Chowdhury, A.D., A. De, C.R. Chaudhuri, K. Bandyopadhyay and P. Sen. 2012. Label free polyaniline based impedimetric biosensor for detection of *E. coli* O157: H7 Bacteria. Sensors Actuators B: Chem. 171: 916–923.

Chowdhury, A.D., N. Agnihotri, A. De and M. Sarkar. 2014a. Detection of positional mismatch in oligonucleotide by electrochemical method. Sensors Actuators B: Chem. 202: 917–923.

Chowdhury, A.D., R. Gangopadhyay and A. De. 2014b. Highly sensitive electrochemical biosensor for glucose, DNA and protein using gold-polyaniline nanocomposites as a common matrix. Sensors Actuators B: Chem. 190: 348–356.

Chowdhury, A.D., A.B. Ganganboina, E.Y. Park and R.A. Doong. 2018a. Impedimetric biosensor for detection of cancer cells employing carbohydrate targeting ability of Concanavalin A. Biosens. Bioelectron. 122: 95–103.

Chowdhury, A.D., A.B. Ganganboina, Y.C. Tsai, H.C. Chiu and R.A. Doong. 2018b. Multifunctional GQDs-ConcanavalinA@Fe_3O_4 nanocomposites for cancer cells detection and targeted drug delivery. Anal. Chim. Acta 1027: 109–120.

Collins, R.A., L.S. Ko, K.Y. Fung, K.Y. Chan, J. Xing, L.T. Lau et al. 2003. Rapid and sensitive detection of avian influenza virus subtype H7 using NASBA. Biochem. Biophys. Res. Commun. 300: 507–515.

Cotterelle, B., C. Drougard, J. Rolland, M. Becamel, M. Boudon, S. Pinede et al. 2005. Outbreak of norovirus infection associated with the consumption of frozen raspberries, France, March 2005. Weekly releases (1997–2007) 10: 2690.

De La Rica, R. and M.M. Stevens. 2013. Plasmonic ELISA for the detection of analytes at ultralow concentrations with the naked eye. Nat. Protoc. 8: 1759.

Didham, R.J. and O.M. Paul. 2015. The role of education in the sustainable development agenda: Empowering a learning society for sustainability through quality education; Achieving the Sustainable Development Goals: From agenda to action. IGES, 95–134.

Draz, M.S., N.K. Lakshminaraasimulu, S. Krishnakumar, D. Battalapalli, A. Vasan, M.K. Kanakasabapathy et al. 2018. Motion-based immunological detection of Zika virus using Pt-nanomotors and a cellphone. ACS Nano. 12: 5709–5718.

Durmanov, N.N., R.R. Guliev, A.V. Eremenko, I.A. Boginskaya, I.A. Ryzhikov, E.A. Trifonova et al. 2018. Non-labeled selective virus detection with novel SERS-active porous silver nanofilms fabricated by electron beam physical vapor deposition. Sensors Actuators B: Chem. 257: 37–47.

Dutta Chowdhury, A., N. Agnihotri, R.A. Doong and A. De. 2017. Label-free and nondestructive separation technique for isolation of targeted DNA from DNA–protein mixture using magnetic Au–Fe_3O_4 nanoprobes. Anal. Chem. 89: 12244–12251.

Dutta Chowdhury, A., A.B. Ganganboina, F. Nasrin, K. Takemura, R.A. Doong, D.I.S. Utomo et al. 2018. Femtomolar detection of dengue virus DNA with serotype identification ability. Anal. Chem. 90: 12464–12474.

Fan, W., Y.H. Lee, S. Pedireddy, Q. Zhang, T. Liu and X.Y. Ling. 2014. Graphene oxide and shape-controlled silver nanoparticle hybrids for ultrasensitive single-particle surface-enhanced Raman scattering (SERS) sensing. Nanoscale 6: 4843–4851.

Ganganboina, A.B., A. Dutta Chowdhury and R.-A. Doong. 2017. N-doped graphene quantum dots-decorated V_2O_5 nanosheet for fluorescence turn off-on detection of cysteine. ACS Appl. Mater. Interfaces 10: 614–624.

Gottschalk, F., T. Sun and B. Nowack. 2013. Environmental concentrations of engineered nanomaterials: review of modeling and analytical studies. Environ. Pollut. 181: 287–300.

Hjertqvist, M., A. Johansson, N. Svensson, P. Abom, C. Magnusson, M. Olsson et al. 2006. Four outbreaks of norovirus gastroenteritis after consuming raspberries, Sweden, June–August 2006. Weekly Releases (1997–2007) 11: 3038.

Holme, M.N., S. Rana, H.M. Barriga, U. Kauscher, N.J. Brooks and M.M. Stevens. 2018. A robust liposomal platform for direct colorimetric detection of sphingomyelinase enzyme and inhibitors. ACS Nano. 12: 8197–8207.

Hu, Y. and C.M. Niemeyer. 2019. From DNA nanotechnology to material systems engineering. Adv. Mater. 1806294.

Hyeon, T., L. Manna and S.S. Wong. 2015. Sustainable nanotechnology. Chem. Soc. Rev. 44: 5755–5757.

Jarocka, U., R. Sawicka, A. Góra-Sochacka, A. Sirko, W. Zagórski-Ostoja, J. Radecki et al. 2014. Electrochemical immunosensor for detection of antibodies against influenza A virus H5N1 in hen serum. Biosens. Bioelectron. 55: 301–306.

Kamil, Y.M., M.A. Bakar, M. Mustapa, M. Yaacob, N. Abidin, A. Syahir et al. 2018. Label-free Dengue E protein detection using a functionalized tapered optical fiber sensor. Sensors Actuators B: Chem. 257: 820–828.

Khademhosseini, A., W.W. Chan, M. Chhowalla, S.C. Glotzer, Y. Gogotsi, J.H. Hafner et al. 2017. Nanoscience and Nanotechnology Cross Borders. ACS Publications.

Khoris, I.M., K. Takemura, J. Lee, T. Hara, F. Abe, T. Suzuki et al. 2019. Enhanced colorimetric detection of norovirus using *in-situ* growth of Ag shell on Au NPs. Biosens. Bioelectron. 126: 425–432.

Larson, C., B. Peele, S. Li, S. Robinson, M. Totaro, L. Beccai et al. 2016. Highly stretchable electroluminescent skin for optical signaling and tactile sensing. Science 351: 1071–1074.

Lau, L.T., J. Banks, R. Aherne, I.H. Brown, N. Dillon, R.A. Collins et al. 2004. Nucleic acid sequence-based amplification methods to detect avian influenza virus. Biochem. Biophys. Res. Commun. 313: 336–342.

Lee, J., K. Takemura and E. Park. 2017. Plasmonic nanomaterial-based optical biosensing platforms for virus detection. Sensors 17: 2332.

Lei, J., R. Qian, P. Ling, L. Cui and H. Ju. 2014. Design and sensing applications of metal–organic framework composites. Trends Anal. Chem. 58: 71–78.

Li, X., K. Scida and R.M. Crooks. 2015. Detection of hepatitis B virus DNA with a paper electrochemical sensor. Anal. Chem. 87: 9009–9015.

Liang, S.S., L. Qi, R.L. Zhang, M. Jin and Z.Q. Zhang. 2017. Ratiometric fluorescence biosensor based on CdTe quantum and carbon dots for double strand DNA detection. Sensors Actuators B: Chem. 244: 585–590.

Liu, R., X. Liu, Y. Tang, L. Wu, X. Hou and Y. Lv. 2011. Highly sensitive immunoassay based on immunogold–silver amplification and inductively coupled plasma mass spectrometric detection. Anal. Chem. 83: 2330–2336.

Manzano, M., S. Viezzi, S. Mazerat, R.S. Marks and J. Vidic. 2018. Rapid and label-free electrochemical DNA biosensor for detecting hepatitis A virus. Biosens. Bioelectron. 100: 89–95.

Mehrotra, P. 2016. Biosensors and their applications—A review. J. Oral. Biol. Craniofac. Res. 6: 153–159.

Nasrin, F., A.D. Chowdhury, K. Takemura, J. Lee, O. Adegoke, V.K. Deo et al. 2018. Single-step detection of norovirus tuning localized surface plasmon resonance-induced optical signal between gold nanoparticles and quantum dots. Biosens. Bioelectron. 122: 16–24.

Nuzaihan, M., U. Hashim, M.M. Arshad, S. Kasjoo, S. Rahman, A. Ruslinda et al. 2016. Electrical detection of dengue virus (DENV) DNA oligomer using silicon nanowire biosensor with novel molecular gate control. Biosens. Bioelectron. 83: 106–114.

Oh, S., J. Kim, V.T. Tran, D.K. Lee, S.R. Ahmed, J.C. Hong et al. 2018. Magnetic nanozyme-linked immunosorbent assay for ultrasensitive influenza A virus detection. ACS Appl. Mater. Interfaces 10: 12534–12543.

Pang, Y., Z. Rong, J. Wang, R. Xiao and S. Wang. 2015. A fluorescent aptasensor for H5N1 influenza virus detection based-on the core–shell nanoparticles metal-enhanced fluorescence (MEF). Biosens. Bioelectron. 66: 527–532.

Payungporn, S., S. Chutinimitkul, A. Chaisingh, S. Damrongwantanapokin, C. Buranathai, A. Amonsin et al. 2006. Single step multiplex real-time RT-PCR for H5N1 influenza A virus detection. J. Virol. Methods 131: 143–147.

Priye, A., C.S. Ball and R.J. Meagher. 2018. Colorimetric-luminance readout for quantitative analysis of fluorescence signals with a smartphone CMOS sensor. Anal. Chem. 90: 12385–12389.

Rahman, M.A., M.J. Hossain, S. Sultana, N. Homaira, S.U. Khan, M. Rahman et al. 2012. Date palm sap linked to Nipah virus outbreak in Bangladesh, 2008. Vector Borne Zoonotic Dis. 12: 65–72.

Rong, Z., Q. Wang, N. Sun, X. Jia, K. Wang, R. Xiao et al. 2018. Smartphone-based fluorescent lateral flow immunoassay platform for highly sensitive point-of-care detection of Zika virus nonstructural protein 1. Anal. Chim. Acta, 140–147.

Saiki, R.K., D.H. Gelfand, S. Stoffel, S.J. Scharf, R. Higuchi, G.T. Horn et al. 1988. Primer-directed enzymatic amplification of DNA with a thermostable DNA polymerase. Science 239: 487–491.

Sarikaya, M., C. Tamerler, A.K.Y. Jen, K. Schulten and F. Baneyx. 2003. Molecular biomimetics: Nanotechnology through biology. Nat. Mater. 2: 577.

Scheller, F.W., U. Wollenberger, A. Warsinke and F. Lisdat. 2001. Research and development in biosensors. Curr. Opin. Biotechnol. 12: 35–40.

Stanisavljevic, M., S. Krizkova, M. Vaculovicova, R. Kizek and V. Adam. 2015. Quantum dots-fluorescence resonance energy transfer-based nanosensors and their application. Biosens. Bioelectron. 74: 562–574.

Takemura, K., O. Adegoke, N. Takahashi, T. Kato, T.C. Li, N. Kitamoto et al. 2017. Versatility of a localized surface plasmon resonance-based gold nanoparticle-alloyed quantum dot nanobiosensor for immunofluorescence detection of viruses. Biosens. Bioelectron. 89: 998–1005.

Tian, J., H. Zhao, M. Liu, Y. Chen and X. Quan. 2012. Detection of influenza A virus based on fluorescence resonance energy transfer from quantum dots to carbon nanotubes. Anal. Chim. Acta 723: 83–87.

Wang, S., L. Li, H. Jin, T. Yang, W. Bao, S. Huang et al. 2013a. Electrochemical detection of hepatitis B and papilloma virus DNAs using SWCNT array coated with gold nanoparticles. Biosens. Bioelectron. 41: 205–210.

Wang, X., L. Chen, X. Su and S. Ai. 2013b. Electrochemical immunosensor with graphene quantum dots and apoferritin-encapsulated Cu nanoparticles double-assisted signal amplification for detection of avian leukosis virus subgroup J. Biosens. Bioelectron. 47: 171–177.

Wei, Q., H. Qi, W. Luo, D. Tseng, S.J. Ki, Z. Wan et al. 2013. Fluorescent imaging of single nanoparticles and viruses on a smart phone. ACS Nano. 7: 9147–9155.

Xiao, F., Y. Li, X. Zan, K. Liao, R. Xu and H. Duan. 2012. Growth of metal–metal oxide nanostructures on freestanding graphene paper for flexible biosensors. Adv. Funct. Mater. 22: 2487–2494.

Xu, S., X. Ji, W. Xu, X. Li, L. Wang, Y. Bai et al. 2004. Immunoassay using probe-labelling immunogold nanoparticles with silver staining enhancement via surface-enhanced Raman scattering. Analyst 129: 63–68.

Yadav, P.D., C.G. Raut, A.M. Shete, A.C. Mishra, J.S. Towner, S.T. Nichol et al. 2012. Detection of Nipah virus RNA in fruit bat (Pteropus giganteus) from India. Am. J. Trop. Med. Hyg. 87: 576–578.

Ymeti, A., J. Greve, P.V. Lambeck, T. Wink, S.W. van Hövell, T.A. Beumer et al. 2007. Fast, ultrasensitive virus detection using a young interferometer sensor. Nano. Lett. 7: 394–397.

Zhang, H., X. Ma, S. Hu, Y. Lin, L. Guo, B. Qiu et al. 2016. Highly sensitive visual detection of Avian Influenza A (H7N9) virus based on the enzyme-induced metallization. Biosens. Bioelectron. 79: 874–880.

Zhao, W., M.A. Brook and Y. Li. 2008. Design of gold nanoparticle based colorimetric biosensing assays. Chem. Bio. Chem. 9: 2363–2371.

Index

A

Abscisic acid 175
anion removal 200, 206–209
antagonists 176–180, 182, 183
Aroma 145–154
Artificial intelligence (AI) 186
2-aza-8-oxohypoxanthine 274
2-Azahypoxanthine 274

B

Biomass 30, 21, 34, 36–44, 104
Biopharmaceutical 299
Bombyx mori 284
Bombyx mori nucleopolyhedrovirus bacmid 299
Bubble 89–92, 95, 97

C

Cage compounds 194, 196, 200, 204, 210, 212, 232, 233, 235–239, 241, 244, 246, 247
Camellia sinensis 147, 151
Carbon dioxide 47, 100, 101
Chaperone 164–171
chemical defense 145, 146
computation 96
coumarin 233–236, 239–241
Cyanobacteria 104, 116–121, 123
Cyanobacteriochrome 119

D

deep ocean sequestration 85
diffusion coefficient 46–48, 50–55, 59
diglycoside 147, 153

E

Ecosystem 101, 107–109, 112
electrochemical 318, 319, 324–326, 328
electrophoresis 256–259, 262–267
encapsulation 196, 199
energy consumption 83, 84, 87
environmental water 194, 195
Esterification 220–223, 225
EWT 134, 138, 139
extracellular electron transfer 1

F

fairy chemicals 274, 281
fairy rings 269–274, 281
Fermentation 14, 15, 19–23, 25, 26
Fine bubbles 65, 66, 71, 80
flow reactor 78
fluorescence 317, 321, 322, 326
food quality 146

G

gas-lift pump 85, 88, 89, 94–96
geldanamycin 161, 164, 166–169, 171
gels 251, 256, 264
glycoprotein 298–301, 303, 305–310
glycosidically bound volatile 146, 147
glycosyltransferase 299–301, 303, 304, 307–310
green organic chemistry 66
groundwater 101, 103, 104, 112

H

health care 316
heat shock protein 159, 165
heat shock response 161, 163–169, 171
heat stress 241, 242
heat tolerance 158–165, 168–171
heat tolerance enhancers 158, 160, 169
heat wave 158–160, 171
heterogeneous catalyst 215, 221, 223–225
hydroxy quinoline 241, 242
hyperspectral 127–134, 136, 137, 139

I

imidazole-4-carboxamide 274
infectious virus 316, 317
isothiocyanates 161–171

K

Kakita River 102–106, 109, 112
Kano River 103–107, 109, 112

L

Lepista sordid 270, 271, 278, 279
lignin degradation 16–19

M

Marasmius oreades 270, 271
Meerwein-Ponndorf-Verley reduction 224
microbial cosystems 1–3, 8
microwave 65, 78–80
molecular structure 215, 216, 225
monitoring 128, 130, 131
monoglycoside 148, 149, 151, 153
monoterpenes 161, 162, 167, 168, 171
Mt. Fuji 100, 102–104, 112
multi-modal deep learning 189, 191, 192

N

Nanoparticles 284, 293
nanotechnology 317, 318, 321, 328
N-glycan 299–303, 305–310
nucleopolyhedrovirus 284
nutrient 100, 101, 104–110, 112

P

perchlorate ion 194
photoactivation 233, 238, 239, 241
photolabile protecting group 232, 233
photoreceptor 116, 118–122
phytochrome 116, 119, 120
plant growth regulator 272
plant hormone signaling 175
plant hormones 175, 176
plant wilt detection 189
PLSR 129, 130, 133–139
polyoxometalate 215–217, 219, 221–223, 225
primeveroside 147–153
protein-protein interaction 175, 176
purine metabolism 276, 278, 280

R

reflectance 128–139

S

sanguinarine 160–165, 168–171
seawater 101, 102, 104, 108, 109
self-assembly 250, 251, 263, 267
Silkworm 284–291, 293, 298, 299, 301, 303, 305–310
soft sensor 186–192
spring 103–106, 112
subcritical fluids 30
supercritical fluid 30, 46, 51, 53, 65, 68, 80
Suruga Bay 100, 103–112

T

tea 146–154
terpene 149
trace metal 101, 105
Transpiration 127–132, 134–139

U

ultrasensitive detection 317, 322, 324
urea 251–255, 257, 260, 262, 263, 265, 267

V

vegetation indices 129, 133, 138, 139
volatile organic compound 145

W

water 65–68, 71–73, 75, 80
white-rot fungi 14–26

Z

zirconocene 215, 216, 225

About the Editors

Prof. Enoch Y. Park is a professor of Research Institute of Green Science and Technology (RIGST), Shizuoka, Japan. He received his Ph.D. degree in 1990 at the University of Tokyo. He was assistant professor at Nagoya University, Japan in 1990 and associate professor at Shizuoka University in 1994 and professor in 1999. He was the first person in the world to develop a bacmid expression system in silkworm. He has published 271 original articles, 36 reviews and 11 book chapters.

Prof. Takayuki Saito is a professor of Research Institute of Green Science and Technology (RIGST), Shizuoka, Japan. He received his Ph.D. degree in 1995 at Tohoku University. He researched environmental engineering at Research Institute for Resources and Environment belonging to METI from 1979 to 1999. He transferred to Shizuoka University as a professor in 2000. He has been investigating processes of CO_2 capture and separation, and sequestration, on the basis of the multiphase flows.

Prof. Hirokazu Kawagishi is a professor of Research Institute of Green Science and Technology (RIGST), Shizuoka, Japan. He received his Ph.D. degree in 1985 at Hokkaido University. He was appointed as assistant professor in Shizuoka University (1985), and was promoted to associate professor (1989) and professor (1999). He has been studying natural products chemistry especially about mushrooms. He is author/co-author of 225 of original articles, 61 of reviews, and 26 of book chapters.

Prof. Masakazu Hara is a professor of Research Institute of Green Science and Technology (RIGST), Shizuoka, Japan. He received his Ph.D. degree in 1995 at Kyoto University. He was hired as assistant professor (1997) and then as associate professor (2003) at Shizuoka University. He conducted research and development of heat tolerance enhancers (HTLEs) to improve plant growth under climate change. He also has expertise on LEA proteins which are common stress-related proteins in plants.

Color Plate Section

Chapter 13

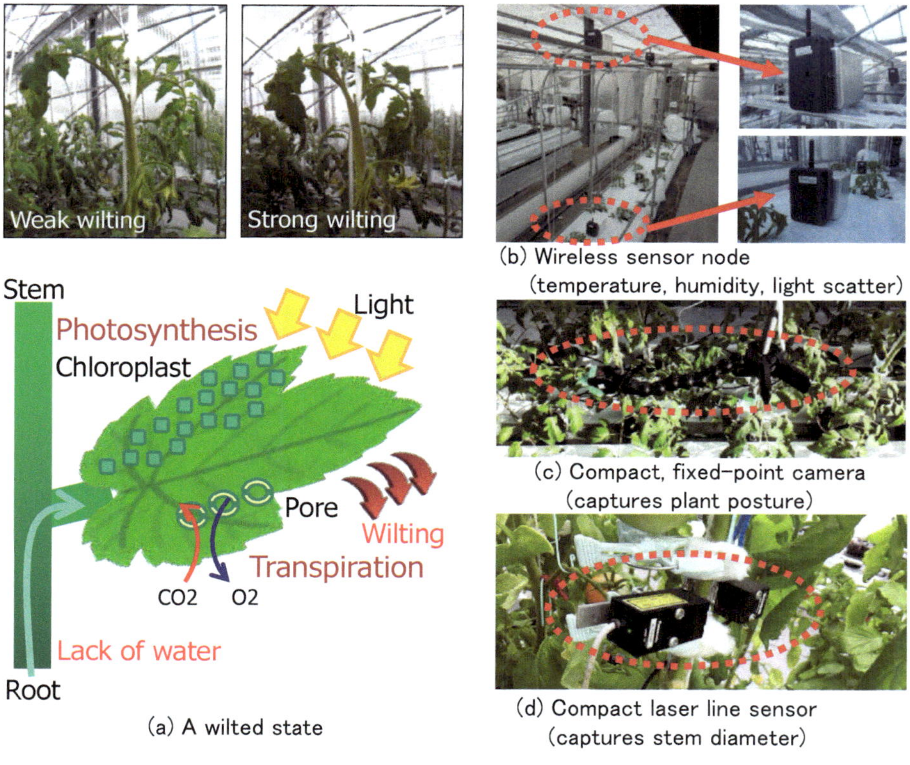

Fig. 13.1: Sensors for modelling plant wilting.

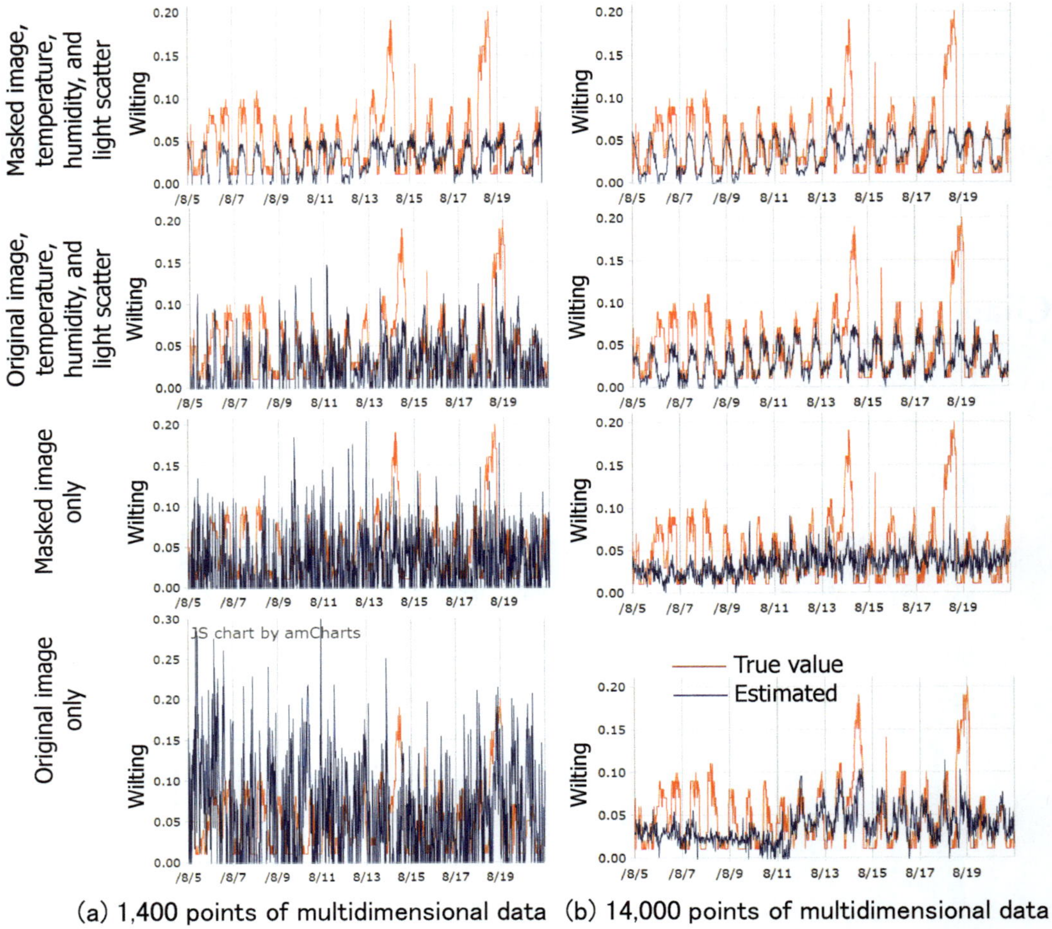

Fig. 13.4: Results from analyzing the quantity and quality of training data.

Chapter 17

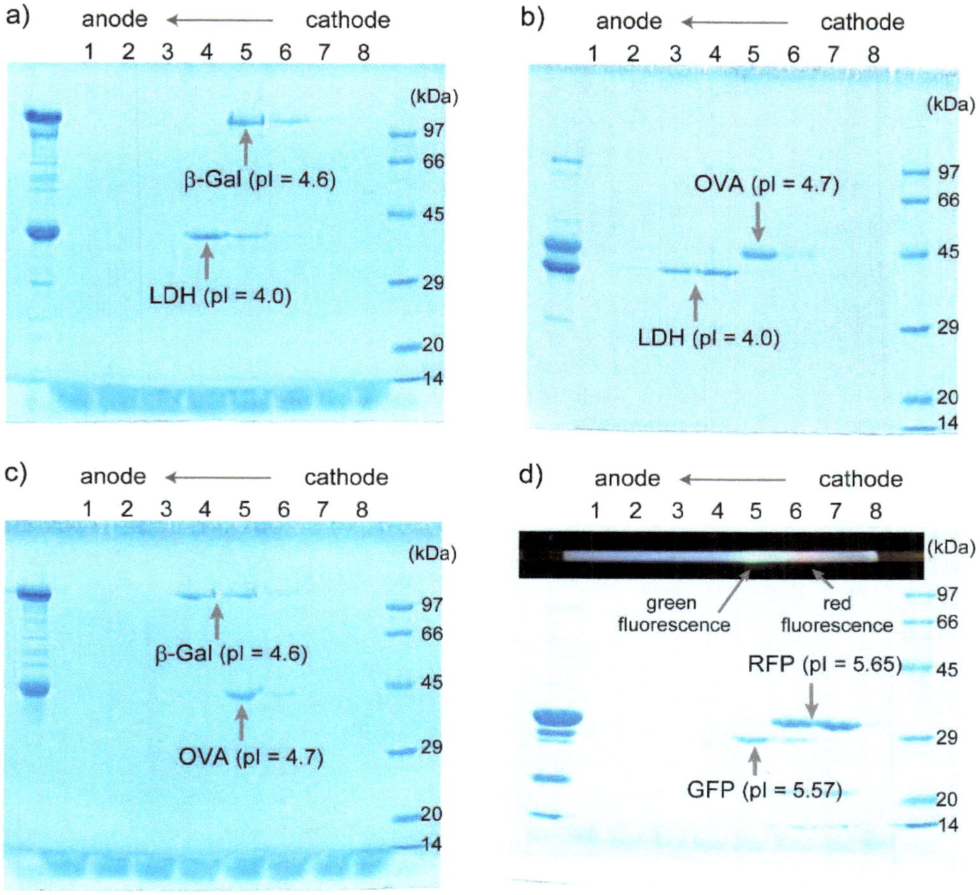

Fig. 17.12: SDS-PAGE analyses of native-SUGE of (a) LDH (146 kDa, pI = 4.0) and β-Gal (540 kDa, pI = 4.6); (b) LDH and OVA (45 kDa, pI = 4.7); (c) β-Gal and OVA; (d) GFP (27 kDa, pI = 5.57) and RFP (27 kDa, pI = 5.65) and photographs of the UV irradiated glass capillary after electrophoresis (insert).

Chapter 18

Fig. 18.1: Fairy rings that appeared on turfgrass in park Hamamatsu City (A) in Japan. Fairy rings in park near campus of University of Shizuoka (B, C). Initial image capture from Google Earth (B, C). (C) is a close-up image of (B). The interaction between fungus and plant stimulates growth of the grass.

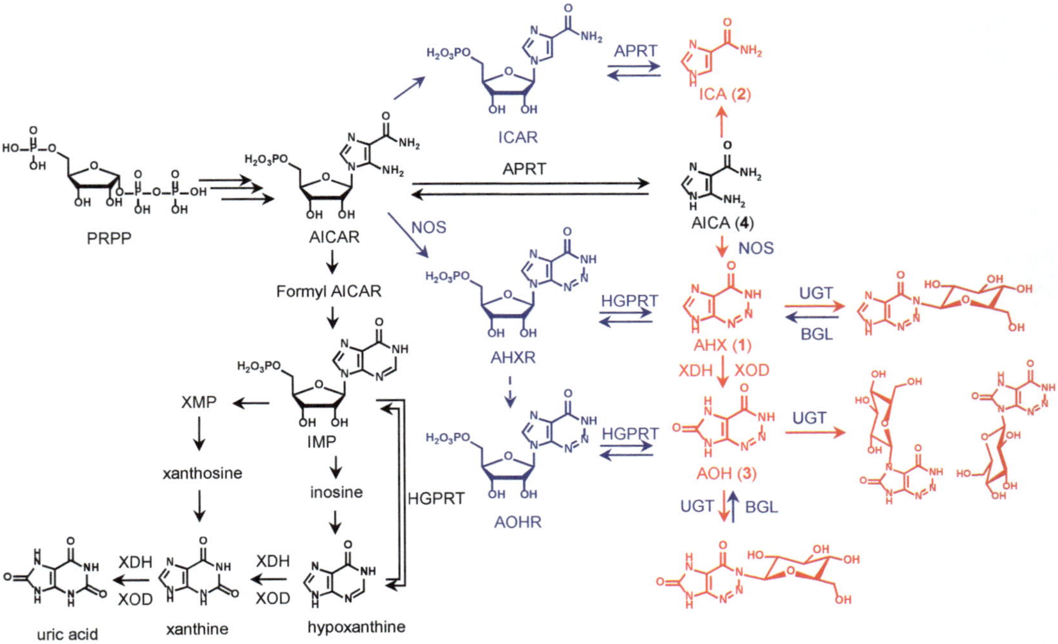

Fig. 18.6: Purine metabolic pathway in animals, plants, and microorganisms, including the novel route. The route in black was cited from KEGG (Kyoto Encyclopedia of Genes and Genomes). New metabolites and routes are indicated in red (published results). Hypothetic or unpublished routes, metabolites, and enzymes are described in blue.

Chapter 21

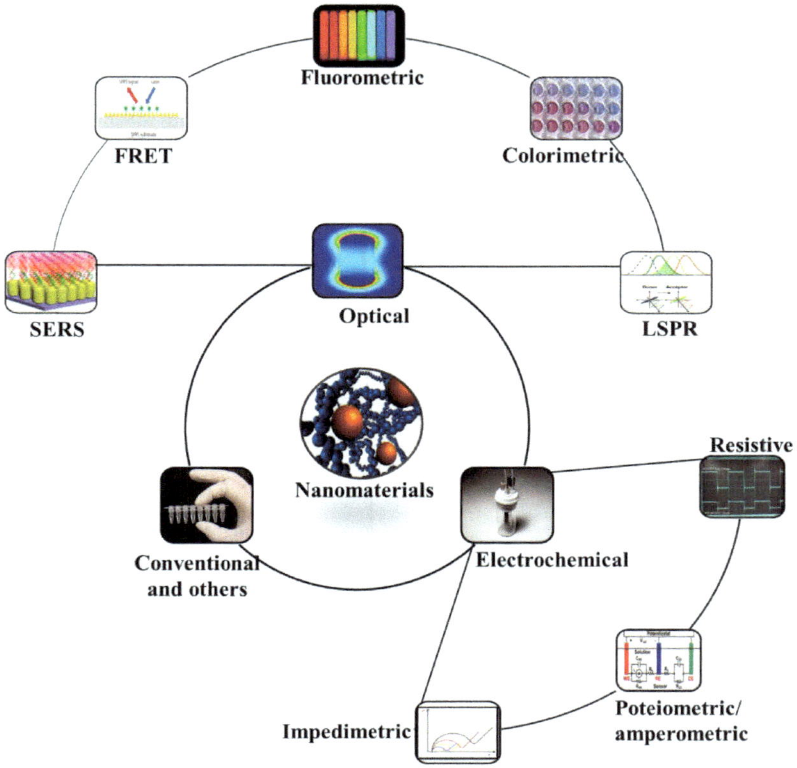

Fig. 21.2: Application of nanotechnology on virus detection: Beside the conventional virus detection protocols, the nanotechnology has been assisted two main detection protocols of optical and electrochemical.

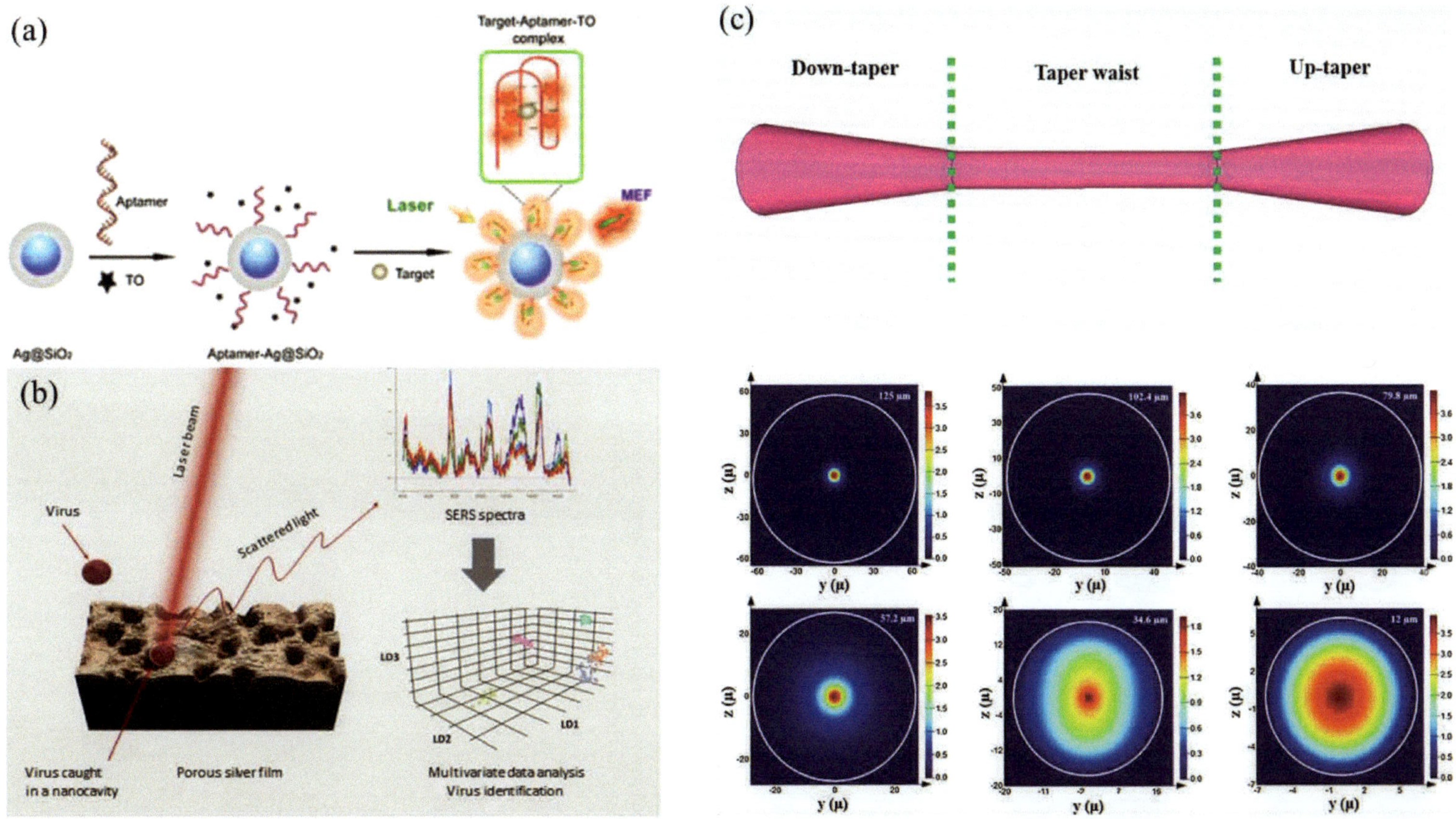

Fig. 21.5: (a) preparation of aptamer-based Ag@SiO_2 biosensor and the detection principle of rHA protein of influenza H5N1 virus, (b) silver based SERS substrate fabrication and its spectral analysis for virus detection, (c) a schematic diagram of a tapered fiber and its modal evolution during transition from non-tapered region to taper waist.

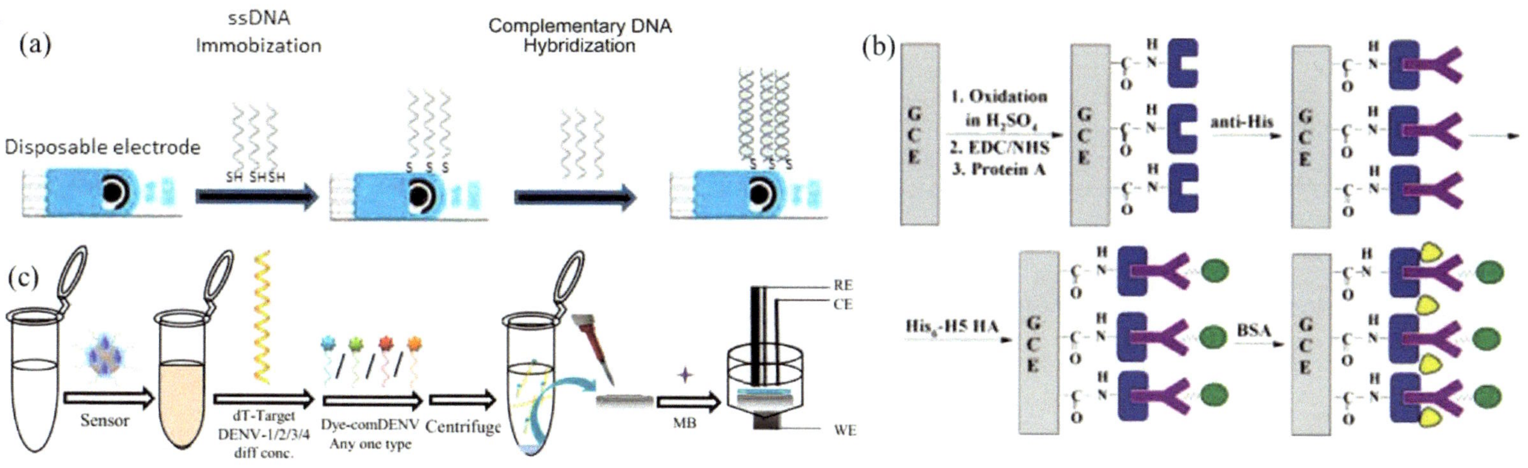

Fig. 21.6: (a) Quantitative detection of hepatitis A virus using specific DNA probe, (b) Steps of immunosensor formation of recombinant His-tagged hemagglutinin antigen for impedimetric immunosensor against avian influenza, (c) electrochemical quantification method of Dengue DNA detection using Au||AuNP@N,S-GQDs-dA nanocomposites.